TITANIUM

Physical Metallurgy Processing and Applications

F.H. Froes, editor

ASM
INTERNATIONAL

ASM International®
Materials Park, Ohio 44073-0002
asminternational.org

Comments, criticisms, and suggestions are invited, and should be forwarded to ASM International.

Prepared under the direction of the ASM International Technical Book Committee (2013–2014), Margaret Bush, Chair.

ASM International staff who worked on this project include Scott Henry, Director, Content & Knowledge-Based Solutions; Karen Marken, Senior Managing Editor; Sue Sellers, Editorial Assistant; Madrid Tramble, Manager of Production; Kate Fornadel, Senior Production Coordinator; Diane Whitelaw, Production Coordinator.

Library of Congress Control Number: 2014957773
ISBN-13: 978-1-62708-079-8
ISBN 10: 1-62708-079-1
SAN: 204-7586

ASM International®
Materials Park, OH 44073-0002
asminternational.org

Printed in the United States of America

Contents

Preface ix

About the Editor xi

Chapter 1
History and Extractive Metallurgy 1

    Historical Background 1
    The Early Titanium Industry and More Recent Developments 2
    Ores and Their Occurrences 9
    The Metal Titanium 9
    General Properties and Alloying Behavior 15
    Mechanical Properties 16
    Special Chemical and Physical Properties 17
    Economics 17
    Process Challenges 18
    Extractive Metallurgy 19
    Summary 27

Chapter 2
Introduction to Solidification and Phase Diagrams 31

    Atoms 31
    Solidification of Metals 32
    Alloying 35
    Phase Diagrams 37
    Summary 47
    Glossary 48

Chapter 3
Principles of Alloying Titanium 51

    Atomic Structure of Titanium 51
    Alloying Elements 56
    Titanium Alloys 66
    Terminal Alloy Formulation 71
    Intermetallic compounds Ti_3Al and TiAl 71
    Summary 72

Chapter 4
Principles of Beta Transformation and Heat Treatment of
Titanium Alloys **75**

    Beta Transformation 75
    Metastable Phases and Metastable Phase Diagrams 77
    Transformation Kinetics 84
    Heat Treatment 86
    Summary 93

Chapter 5
Deformation and Recrystallization of Titanium and Its Alloys **95**

    Deformation. 95
    Development of Texture in Titanium. 98
    Texture Strengthening 99
    Strain Hardening 102
    Strain Effects 104
    Superplasticity. 104
    Internal Changes 105
    Annealing 106
    Neocrystallization 108
    Gamma Titanium Aluminide 110
    Summary 110

Chapter 6
Mechanical Properties and Testing of Titanium Alloys **113**

    Effect of Alpha Morphology on Titanium Alloy Behavior 113
    Hardness 116
    Tensile Strength. 116
    Ductility. 118
    Creep and Stress Rupture 119
    Fatigue Strength. 121
    Toughness 125
    Fatigue Crack Growth Rate 129
    High-Temperature Near-Alpha Alloys 130
    Alpha-Beta Alloys 131
    Beta Alloys 133
    Titanium Aluminides 133
    Metal-Matrix Composites 136
    Shape Memory Alloys. 138
    Summary 139

Chapter 7
Metallography of Titanium and Its Alloys **141**

    Review of Physical Metallurgy—Alpha and Beta. 141
    Terminology Used to Describe Titanium Alloys Structures 143
    Metastable Phases 146
    Related Terms 149
    Ordered Intermetallic Compounds 151
    Effect of Fabrication and Thermal Treatment on Microstructure. 152

Metallographic Specimen Preparation 154
Summary 154
Glossary 155
Appendix—Metallographic Preparation 157

Chapter 8
Melting, Casting, and Powder Metallurgy 161

Melting 161
Casting 168
Titanium Powder Metallurgy 176
Safety 200
Future Developments in Titanium Powder Metallurgy 202
Summary 203
Glossary of Acronyms 203

Chapter 9
Primary Working 207

Crystal Structure 207
Forging 208
Ingot Breakdown 208
Forged Billets and Bars 210
Rolling 211
Radial Precision Forging Machines 212
Rolled Rod and Bar 212
Plate, Sheet, Coil, and Foil Rolling 214
Extrusion 216
Wire and Tube Processing 219
Summary 221

Chapter 10
Secondary Working of Bar and Billet 225

Physical Metallurgy 225
Forging 226
Classes of Forgings 228
Extrusion 231
Microstructure and Mechanical Properties 234
Surface Effects of Heating 239
Modeling 240
Summary 240
Glossary 241

Chapter 11
Forming of Titanium Plate, Sheet, Strip, and Tubing 243

Forming Considerations 243
Preparation for Forming 245
Heating Methods 246
Forming Lubricants 247
Tooling Materials 247
Forming Processes 247
Summary 262

Chapter 12
Joining Titanium and Its Alloys 265

Welding 265
Welding Procedures. 269
Brazing 283
Soldering 285
Adhesive Bonding 286
Mechanical Fastening 287
Summary 290

Chapter 13
Machining and Chemical Shaping of Titanium 293

Machinability. 293
General Machining Requirements. 295
Scrap Prevention 299
Hazards and Safety Considerations. 300
Milling Titanium 300
Turning, Facing, and Boring 305
Drilling Titanium. 313
Surface Grinding 318
Broaching 319
Tapping 319
Recent Advances in Machining. 320
Flame Cutting 322
Chemical Machining 323
Electrochemical Machining. 327
Summary 328

Chapter 14
Corrosion 331

Corrosion Behavior of Titanium 331
Forms of Corrosion 335
Alloying for Corrosion Prevention 345
Chemical and Related Applications. 347
Summary 350

Chapter 15
Applications of Titanium 353

Early Applications 353
Material Availability 353
Aerospace Applications. 354
Sheet Metal Applications. 362
Industrial Applications. 363
Engineering Properties 363
Medical Applications. 370
Consumer Applications 372
Armor Applications 373
Automotive Applications. 374

Building Applications 375
Power Utility Applications 375
Marine Applications 378
Miscellaneous Applications. 378
Summary 379

Index. **381**

Preface

THE TITANIUM INDUSTRY has been in existence for approximately 60 years, and a great amount of information on the science and technology of this "wonder" metal has been compiled in that relatively short time. This reference book is based on an education course developed by ASM International in the early 1980s, which has been revised several times as new technical information became available, the latest revision in 2014 by F.H. (Sam) Froes, an expert in titanium and titanium alloy technology.

This book is a comprehensive compilation of the science and technology of titanium and its alloys. It details the history of the titanium industry and discusses various extraction processes, including the Kroll and Hunter processes and others. The fundamentals of solidification and phase diagrams are discussed, numerous detailed descriptions of beta (β)-to-alpha (α) transformations are included, and there are extensive discussions on processing, characteristics, and performance of the different classes of titanium alloys, including alpha (α), alpha-beta (α-β), beta (β), and intermetallic compounds. There are chapters devoted to alloying, deformation and recrystallization, mechanical properties and testing, and metallography. The following are also covered: melting and casting; forming of plate, sheet, strip, and tubing; joining; and machining. Practical aspects of primary and secondary processing are given, including a comprehensive description of superplastic forming. Details of expanding powder metallurgy techniques are included. The relationship of microstructure to mechanical properties is addressed in detail. A detailed description of corrosion behavior is included, and a comprehensive section on current applications of titanium and its alloys, documenting why certain alloys are used in various applications as well as their limitations, is also addressed.

Permeating the book are examples of how lowering the cost of titanium can lead to increased use. I believe that this book will be of considerable value to persons new to the industry as well as practitioners, and that it will significantly increase your knowledge of the science and technology of titanium.

Dr. F.H. (Sam) Froes
Tacoma, Washington, August 2014

About the Editor

Dr. F.H. (Sam) Froes has been involved in the titanium field for more than 40 years. After receiving a B.S. from Liverpool University, M.S. and Ph.D. degrees from Sheffield University, he was employed by a primary titanium producer, Crucible Steel Company, where he was leader of the Titanium Group. He spent time at the United States Air Force (USAF) Materials Laboratory, where he was a branch chief and supervisor of the Light Metals Group, which included titanium. While at the USAF Laboratory, Dr. Froes co-organized the landmark TMS-sponsored Conference on Titanium Powder Metallurgy in 1980. This was followed by 17 years at the University of Idaho, where he was director and department head of the Materials Science and Engineering Department. During this tenure, Dr. Froes was Chairman of the World Titanium Conference held in San Diego in 1992. He has over 800 publications, in excess of 60 patents, and has edited almost 30 books, the majority on various aspects of titanium. Recent publications include a comprehensive review of titanium powder metallurgy and an article on titanium additive manufacturing. He has organized more than 10 symposia on various aspects of titanium science and technology, including in recent years co-sponsorship of four TMS symposia on cost-effective titanium. Since the early 1980s, Dr. Froes has taught the ASM International education course "Titanium and Its Alloys." He is an ASM Fellow, a member of the Russian Academy of Science, and was awarded the Service to Powder Metallurgy by the Metal Powder Association.

CHAPTER 1

History and Extractive Metallurgy*

THE ELEMENT TITANIUM (Ti) is a unique metal. It is the fourth-most abundant structural metal in the Earth's crust (~0.5%). It has a desirable combination of physical, chemical (corrosion resistance, low bioreactivity), and mechanical properties that make it attractive for many aerospace, medical, and industrial applications (Ref 1.1, 1.2). Early use was geared toward the aerospace industry, but later applications included industrial, automotive, sports, and medicine due to its unique characteristics. Titanium is a transitional metal, distinct from other light metals such as aluminum and magnesium. It has a high solubility for a number of other elements and high reactivity with interstitial elements (oxygen, nitrogen, hydrogen, and carbon). Titanium has a relatively short production history, with the first commercial quantities of the metal produced in 1950. By 2011, worldwide annual sponge production increased to 186,000 metric tons (excluding U.S. production) and capacity increased to 283,000 metric tons. Production of titanium ores and concentrates is approaching 10 million metric tons. Current and historic production and data on titanium are maintained by the United States Government (Ref 1.3, 1.4, 1.5).

Historical Background

The element titanium was discovered in England by the Reverend William Gregor in 1790. In 1791, Gregor presented a description and chemical composition of some black magnetic sands found on the southern Cornish coast. His analysis of the black sand corresponded roughly to that of the mineral ilmenite ($FeTiO_3$).

Little interest was shown in the discovery until 1795, when M.H. Klaproth noticed close agreement between Gregor's account and results of his own investigation of oxide extracted from rutile (impure TiO_2) from Hungary. The identity of the two materials was established; Klaproth acknowledged priority to Gregor and applied the name titanium to the new element.

Early attempts to prepare pure titanium from its compounds resulted in the formation of nitrides (TiN), carbides (TiC), or carbonitrides (TiCN), which, because of their metallic luster and appearance, were often mistaken for metal. In 1887, L.F. Nilson and O. Peterson obtained a product of 97.4% purity by reducing titanium tetrachloride with sodium in an airtight steel cylinder.

Another early worker was H. Moissan, who reduced titanium dioxide with carbon in a lime crucible at the temperature of a powerful electric arc. The product contained 5% C, but on reheating with additional TiO_2 this was reduced to 2%.

The first pure titanium metal was prepared in the United States by M.A. Hunter at the General Electric Company in 1906. Hunter followed the methods of Nilson and Peterson and excluded air from the apparatus. He obtained metallic titanium practically free of impurities.

In Holland in 1925, A.C. Van Arkel and J.H. deBoer produced titanium by thermal decomposition of titanium tetrachloride. Titanium made by this procedure was very expensive but pure.

The start of the present large-scale titanium industry can be traced to the work of W.J. Kroll. He produced ductile titanium metal by reacting titanium tetrachloride with magnesium metal in a closed pressureless system with an inert gas (argon) atmosphere. The first display of cold ductile titanium in the United States (produced by Kroll at the Bureau of Mines in Albany, Oregon) took place in October 1938.

---

*Adapted and revised from Eldon R. Poulsen and Francis H. Froes, originally from Richard A. Wood, *Titanium and Its Alloys*, ASM International.

The Degussa Company was working on titanium at approximately the same time as Kroll. They produced over 400 kg (880 lb) of titanium by sodium reduction of titanium tetrachloride. However, the material contained up to 2% Fe.

In approximately 1940, the United States Bureau of Mines became interested in the characteristics and production of titanium metal. After reviewing all the known processes, the Bureau selected the Kroll process as the one most likely to economically produce ductile titanium, and it set up a series of reactors for making titanium. A Bureau publication in 1946 described a Kroll unit capable of making 7 kg (15 lb) batches of good-quality titanium powder by magnesium reduction, followed by acid leaching to remove the excess magnesium and MgCl$_2$ (Fig. 1.1).

In 1949, the Bureau reported the successful operation of a magnesium-reduction unit for making 40 kg (90 lb) batches of titanium. This unit was similar to the one previously reported, except for the batch size. In 1952, the Bureau reported the removal of magnesium and magnesium chloride from titanium sponge by vacuum distillation.

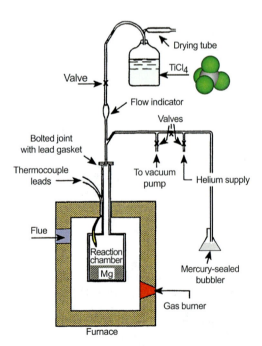

Fig. 1.1 Reaction vessel for making titanium powder using the Kroll process. The equipment was capable of producing 7 kg (15 lb) batches.

The Early Titanium Industry and More Recent Developments

In the early 1950s, a number of large companies helped to meet the challenge to produce titanium. In most cases, new organizations were formed to combine technical expertise with metal production facilities. The companies formed were normally a combination of a pigment company with the chemical expertise and a stainless steel company with the vacuum melting and mill processing capabilities.

In the United States in the late 1940s/early 1950s, a number of companies entered the titanium business, with strong government support. A pilot unit created in 1947 at DuPont expanded production to 800,000 kg (1.8 million lb) of sponge per year by 1952. Remington Arms, a 60%-owned DuPont subsidiary, used iodide titanium and DuPont sponge. Studies by Battelle Columbus and Remington Arms included melting, alloy development, physical metallurgy, and the production of mill products as early as 1948. Powder approaches were evaluated, but vacuum arc melting proved to be most successful. The first titanium for actual flight was ordered from Remington Arms in 1949 by the Douglas Company. Unalloyed (A70) sheet was rolled by Republic Steel and formed into 578 different parts for the Douglas Mach 2 X -3 Stiletto. The technical cadre assembled at Remington Arms later became the nucleus for Rem-Cru Titanium Inc., formed equally by Remington Arms and Crucible Steel in June 1950.

Allegheny Ludlum Steel Corporation started a semicommercial titanium-melting facility in 1949. They and the National Lead Company organized Titanium Metals Corporation of America (TMCA) in January 1950 on an equal basis. Later, TMCA became the first fully integrated company for producing titanium from ore to finished products. National Lead operated the Bureau of Mines sponge plant in 1951 and constructed pilot plants for sponge in 1949 and 1951. Subsequently the name TMCA was changed to TIMET.

P.R. Mallory Company entered the market in 1947 when they began work on the powder metallurgy of titanium under Navy sponsorship. Along with the Sharon Steel Corporation, they organized the Mallory-Sharon Titanium Corporation on an equal basis in 1951 to produce and market the metal. National Distillers and Chemical Corporation started sponge production in 1957. They acquired P.R. Mallory's interest in Mallory-Sharon in 1958, changing the name to

Reactive Metals, Inc. In 1964, U.S. Steel acquired 50% of the company, and the name eventually was changed to RMI Company.

Cramet, Inc. was organized in 1953 as a wholly owned subsidiary of the Crane Company to supply sponge for Republic Steel. Republic organized a titanium division in 1950 to produce mill products. Dow Chemical Company started sponge production in 1954, and the Electrometallurgical Company of Union Carbide began in 1956. All of these companies dropped out of the market during the 1958 downturn, which resulted in large part from the change in emphasis by the U.S. military from the use of manned aircraft to missiles.

However, production in the early and mid-1950s accelerated rapidly, although not as rapidly as the overly enthusiastic predictions that production would reach 180 million kg (400 million lb) by 1960, which would have exceeded the production of stainless steel and magnesium, and approached the production levels of aluminum by

1965. Some early production capacities in the United States are given in Table 1.1.

Other U.S. producers became active at later dates, including Oregon Metallurgical Corporation, Dow-Howmet, International Titanium Incorporated of Washington, Western Zirconium Company, and Albany Titanium Company. Early U.S. sponge producers are listed in Table 1.2 along with capacities.

Numerous other companies became active participants in the titanium industry, including forgers, rollers, extruders, foundries, tube manufacturers, and fabricators.

The development and growth of titanium production also occurred outside the United States. Work in the United Kingdom, for example, centered at Imperial Chemical Industries Ltd. (ICI) where sponge production began in 1948. A few hundred kilograms (pounds) of magnesium-reduced Kroll sponge were produced. Plant capacity was increased to 9000 kg (20,000 lb) per year in 1951, and a 90,000 kg (200,000 lb) per

Table 1.1 Early capacities of major U.S. producers

Company	Mill product capacity, est. 1956		Ingot melting capacity, est. 1957	
	$\times 10^6$ kg/yr	$\times 10^6$ lb/yr	$\times 10^6$ kg/yr	$\times 10^6$ lb/yr
Titanium Metals Corp. of America	2.3	5.0	10.0	22.0
Rem-Cru Titanium Inc.	1.8	4.0	6.0	13.2
Mallory-Sharon Titanium Corp.	1.4	3.0	5.4	12.0
Republic Steel Corp.	0.5	1.1	5.4	12.0

Source: TIMET records

Table 1.2 U.S. titanium sponge producers and capacity, 1947 to 1987

Organization	Process(a)	Approximate capacity, 1000 kg (2000 lb) per year				
		Capacity	Initial year	1958	1984	1987
U.S. Bureau of Mines	Mg, V	1.8 (4)	1947	...	(b)	...
E.I. du Pont de Nemours & Co., Inc.	Mg, V	2.3(5)	1947	6,500 (14,400)	(b)	...
Titanium Metals Corporation of America	Mg, L	3,200 (7,200)	1951	8,200 (18,000)	14,500 (32,000)	14,500 (32,000)
Dow Chemical Company	Mg, L	100 (216)	1954	1,600 (3,600)(b)	(b)	...
Cramet, Inc.	Mg, L	470 (1,034)	1955	5,500 (12,000)	(b)	...
Union Carbide Corp.	Na, L	6,800 (15,000)	1956	6,800 (15,000)	(b)	...
National Distillers and Chemical Corporation(c)	Na, L	4,500 (10,000)	1958	4,500 (10,000)	8,600 (19,000)	8,600 (19,000)
Oregon Metallurgical Corporation	Mg, L	...	1966	...	4,100 (9,000)	4,100 (9,000)
D-H Titanium Company(d)	E, L	90 (200)	1981	...	(b)	...
Teledyne Wah Chang Albany	Mg, V	910 (2,000)	1980	...	1,400 (3,000)	...
Western Zirconium Company	Mg	455 (1,000)	1982	...	5,900 (13,000)(e)	...
Albany Titanium, Inc.	K_2TiF_6	...	1982	...	...	...

(a) Mg = Kroll (magnesium) process; V and L = vacuum distillation or leaching; Na = sodium process; E = electrolytic process. (b) Operations discontinued. (c) Now RMI Company, owned by ND&CC (50%) and U.S. Steel (50%). (d) Pilot plant operation from 1979 to 1982. (e) Estimate. Source: TIMET Records

year plant was planned. However, following development of a modified production process, subsequent plants were changed to a sodium-reduction technique, with capacities of 45,000 kg (100,000 lb) per year in 1953 and 1.5 million kg (3 million lb) in 1955. At the same time, ICI Metals Division (subsequently IMI) started melting titanium at a capacity of 150,000 kg (300,000 lb) per year in 1954. By 1955, ICI was producing 1.5 million kg (3 million lb) and became the principal European manufacturer of titanium and titanium alloy mill products.

In continental Europe, ingot melting and fabrication started in approximately 1955 and has continued since at companies in France, Germany, and Sweden. Sponge was manufactured for a few years in France, but the process was discontinued in 1963.

The birth of the Soviet titanium industry occurred in 1950, and Kroll sponge production began in 1954. Major expansions have been made since that date.

Several Japanese firms also became early sponge producers, as shown in Table 1.3, supplying metal to other countries, including the United States.

By 1987, U.S. sponge manufacturers had been reduced to three: TIMET, RMI, and Oremet. The Japanese by then had become major sponge producers, with limited capacity in melting and processing.

Early challenges of production included development of inert double-consumable melting in cold-mold furnace, circumvention of hydrogen embrittlement due to inadequate vacuum melting, chemical cleaning and use of gas furnaces, and hot salt stress-corrosion cracking due to chlorides on stressed specimens above 3000 °C (5400 °F). During the early 1950s, the value of aluminum, manganese, and vanadium as alloy additions was established in alloys such as Ti-8Mn, Ti-4Al-4Mn (1951), and the "workhorse alloy" Ti-6Al-4V (1954), patented by Crucible Steel. The first beta alloy, B120VCA (Ti-13V-11Cr-3Al), was also developed by Crucible Steel and was used extensively on the SR-71 (1955). Silicon additions for elevated-temperature use were introduced in Britain (1956). McDonnell (later McDonnell Douglas) used just 13.6 kg (30 lb) of titanium on the F3H airframe (1951), increasing the use to 136 kg (300 lb) in 1954. The experimental X-15 high-flying supersonic aircraft was composed of 17.5% by weight titanium alloy. Engine use was also established in the mid-1950s, with first use on the PWA J57 in 1954, with an increased use on the GE J73 (6% in 1954). The Rolls-Royce Avon engine used Ti-2Al-2Mn starting in 1954. At the same time, use of titanium in corrosion applications and for orthopedic devices was occurring. In 1957, the U.S. titanium industry had an annual capacity of 20.4 million kg (45 million lb) of sponge and a capacity of greater than 9 million kg (20 million lb) of mill products. (Despite this capacity, only 4.5 million kg, or 10 million lb, were shipped in 1957.) The late-1957 decision by the U.S. military to emphasize missiles over manned aircraft resulted in a thinning out of the titanium industry.

By 1970, space exploration and the launching of a number of new civilian jets during the 1960s resulted in a tripling of mill product shipments in the United States to 13.5 million kg (30 million lb). Over 90% of these shipments went to non-military aerospace systems such as the B747, DC10, and L1011. Engine use also increased the GE4 (slated for use on the U.S. Supersonic Transport, or SST), which was composed of 32% Ti by weight. Advances in quality occurred with the development of triple melting for rotating components and the avoidance of inclusions by more careful cleaning of scrap. Nonaerospace use of titanium also developed; desalination plants, power plants, and other fresh- and saltwater applications made use of corrosion-resistant grades containing small additions from platinum group metals.

Just one year later, in 1971, with the cancellation of the U.S. SST project, the titanium market reached another low, with just 9.26 million kg

Table 1.3   Early Japanese titanium sponge production (1000 kg, or 2000 lb, per year)

Company	1952	1953	1954	1987
Osaka Titanium Manufacturing Company	8.2 (18)	54.4 (120)	307 (676)	18,000 (40,000)
Toho Titanium Industry Company, Ltd.	...	4.5 (10)	239 (526)	12,000 (26,000)
Nippon Soda Company Ltd.	...	5.4 (12)	34 (74)	4,500 (10,000)
Showa Ti	...	...	...	2,700 (6000)
Nippon Electric Metallurgical Company, Ltd.	...	...	25.4 (56)	...
Mitsui Mining and Smelting Company	...	...	6.4 (14)	...
Total	8.2 (18)	64.4 (142)	611 (1346)	37,000 (82,000)

Source: TIMET records

(20.4 million lb) of mill product shipments. In 1974, production rebounded to 15.8 million kg (34.8 million lb), and by 1978 a new peak of 18.1 million kg (40.0 million lb) was established. There were new alloys available, such as Ti-6Al-6V-2Sn (higher strength than Ti-6Al-4V), Ti-8Al-1Mo-1V (a high-modulus alloy), and Ti-6Al-2Sn-4Zr-2Mo-0.1Si (an elevated-temperature alloy). Also, in addition to flat products castings, extrusions and tubing were being fabricated. There was increased use in high-bypass turbofan engines in large transports such as the C5A and Boeing 747. The temperature at which titanium alloys were used increased to 600 °C (1115 °F). At the same time, the percentage of titanium in military airframes increased to 20 to 30% for systems such as the F14, F15, and the B1 bomber. New cost-effective processing/fabrication techniques were introduced, including superplastic forming (SPF), diffusion bonding (DB), combined SPF/DB, and hot isostatic pressing (of castings to remove porosity, and powders to achieve full density).

A rise in titanium shipments occurred in the 1980s in large part due to the U.S. military buildup during President Reagan's term and the increase of aerospace and other nonmilitary uses. In 1989, a new record of U.S. mill shipments of 25 million kg (55 million lb) was achieved. With the formation of the Titanium Development Association (renamed The International Titanium Association) and the World Titanium Conferences, held at four-year intervals (initiated in London, 1968), and the development of the ASM

International course on *Titanium and Its Alloys* in the late 1970's, designers became better informed on the characteristics and use of titanium. Alloy processes and fabrication techniques matured with the development of another beta alloy (Ti-15V-3Cr-3Al-3Sn) and a forgeable near-beta alloy (Ti-10V-2Fe-3Al). Isothermal forging, SPF castings, and extensive use of scrap were now accepted practices. The aerospace industry increased the use of titanium on Boeing and Airbus commercial systems, as shown in Fig. 1.2. The cyclic nature of the U.S. titanium market is shown in Fig. 1.3 for a similar time period as in Fig. 1.2.

In the early 1990s, with the ending of the Cold War and the collapse of the former Soviet Union, defense expenditures declined as did the military demand for titanium. The U.S. titanium mill shipments dropped precipitously from the record 25 million kg (55 million lb) in 1989 to 15.4 million kg (34 million lb) in 1991, predominantly due to the greatly reduced military procurement. This occurred despite a projected greater than 30% Ti on the airframe of the Advanced Tactical Fighter (F-22) and in excess of 10% expected on the Boeing 777. The actual percentages are up to 45% on the F-22 and 15% on the Boeing 777, including a large amount of Ti-10V-2Fe-3Al on the landing gear. Sponge capacity in the United States dropped from 34.4 to 19.5 million kg (67 to 43 million lb) with the closing of the RMI facility in Ashtabula, Ohio. In the early 1990s, new alloys such as TiB21S and the titanium aluminides (Ti$_3$Al and TiAl) received increasing attention.

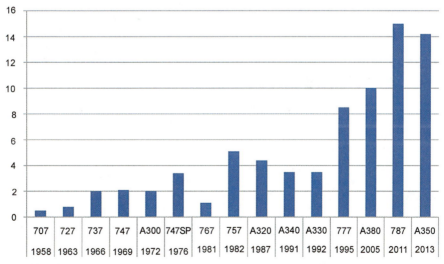

Fig. 1.2 Growth in titanium use as a percentage of total gross empty weight on Boeing and Airbus aircraft. Note the decreased use on the 767 was due to a perceived shortage in titanium when this plane was designed. Designers substituted other materials for titanium (such as steel and aluminum).

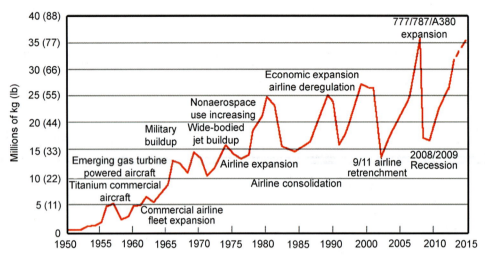

Fig. 1.3 U.S. titanium industry shipments from 1950 to present. Adapted from National Materials Advisory Board data

Sports applications such as lightweight bicycle frames and the explosive demand for titanium golf drivers occurred. Use of titanium by Boeing for commercial aircraft alone increased to an estimated 11 million kg (25 million lb) per year by 1996.

By the mid-to-late 1990s, a new record 27 million kg (60 million lb) of mill products shipped per year in the United States, with 15% Ti being used on the Boeing 777. This use occurred for weight savings, volume constraints, operating temperatures, compatibility with polymeric composites, and corrosion resistance. Titanium alloys used included Ti-10V-2Fe-3Al on the main landing gear; Ti-15V-3Cr-3Al-3Sn for ducts, fittings, and nut clips; and beta-21S (Ti-15Mo-2.7Zr-3Al-0.2Si) on the engine nacelle, cowls, and exhaust assembly (Ref 1.1). The new record was followed by a downturn after the September 2011 attack, with the U.S. market reaching 1975 levels of 13.5 million kg (30 million lb) of shipments per year. During this time period, an interesting new alloy, Ti-4Al-2.5V-1.5Fe (designated alloy 425) was developed by Allegheny Technologies Incorporated Wah Chang. This alloy exhibited many of the characteristics of the Ti-6Al-4V alloy, but it was cold workable.

Between 2003 and 2007, with the advent of the Airbus A380, the Joint Strike Fighter (JSF, F-35), and the Boeing 787 (plus military conflicts in Iraq and Afghanistan), U.S. mill products reached a new record of 35.8 million kg (79 million lb) per year. The 787 uses in excess of 20% Ti, including the high-strength, high-

toughness Ti-5Al-5V-5Mo-3Cr alloy in the landing gear, wing structure, and nacelle area (further details of this use are given in Chapter 15, "Applications of Titanium," in this book). This new record was followed by the global recession and banking crisis of 2008 to 2009, precipitating a fall in U.S. mill products to 24.5 million kg (54 million lb). TIMET also developed two new alloys: a low-cost alpha-beta alloy with iron replacing vanadium (Ti-6Al-2Fe-0.1Si) and a low-cost beta alloy (Ti-6.8Mo-4.5Fe-1.5Al).

Demand increased during the 2010 to 2014 time period, with large titanium purchases for the A380, JSF, and 787. Titanium use in engines and airframes is increasing. Major commercial airplane producers Boeing and Airbus use extensive amounts of titanium (Table 1.4), because titanium is compatible with carbon-fiber-reinforced composites. (Aluminum has a galvanic corrosion problem with composites.) Titanium alloys are increasingly used to reduce the weight of aircraft structures. The new alpha-beta alloy TIMETAL 54M (Ti-5Al-4V-0.75Mo-0.5Fe, developed by TIMET) has demonstrated 30% faster machinability over Ti-6Al-4V, with similar mechanical properties and tool wear. TIMETAL 54M has been evaluated for aerospace components manufactured by several vendors. Product forms, including forgings, forged billet, and round bar, have been processed using standard aerospace techniques. A large database of tensile mechanical properties has been successfully developed for electron beam single-melt forged products for use in industrial applications.

Ores and Their Occurrences

The most economical ore for titanium metal is the mineral rutile (Ref 1.3). Rutile deposits exist in North America, Africa, India, Brazil, and Australia. Australian deposits of principal importance are in the provinces of New South Wales, Queensland, and Western Australia. They occur in beach sands and in alluvial deposits and are mined either by dredging or by open-pit techniques. The beneficiated sands are shipped to metal-extraction centers.

Of the many minerals containing the element titanium as one of the constituents, ilmenite (or iron titanate), an iron-titanium oxide ($FeTiO_3$), is the most abundant and of great importance as a possible future ore for titanium metal. Ilmenite is found on every continent. Huge deposits are worked in the United States and in Canada to recover titanium dioxide (TiO_2) for use as a pigment and coating for welding rod. Ilmenite is used in the production of titanium tetrachloride ($TiCl_4$) by converting the ore to a synthetic rutile. In this operation, iron is leached out of the ilmenite using various methods.

Other minerals containing high enough concentrations of titanium to be of economic interest are leucoxene (weathered ilmenite, plus rutile mixture) and perovskite ($CaTiO_3$). Sphene ($CaTiSiO_5$) and pyrophanite ($MnTiO_3$) are minerals of the type that probably never will be ores for the recovery of titanium metal due to the cost of processing.

Rutile is the preferred ore for titanium metal because it contains the highest TiO_2 content of all the titanium-bearing minerals. Table 1.5 shows the typical TiO_2 content of several titanium minerals and of two slags produced from ilmenite concentrates. The latter are potential ores of titanium metal and are currently used to produce $TiCl_4$ in pigment plants.

Other materials containing less TiO_2 than rutile and ilmenite are not considered ores for titanium metal at the present time, because there is no known economical way to extract the TiO_2 content for the eventual production of metal.

The Metal Titanium

The metal titanium has a density between that of aluminum and steel. It is the ninth-most abundant element in the Earth's crust and the fourth-most abundant structural metal. Titanium has the strength of alloyed steels and the density of aluminum. Table 1.6 shows the physical and mechanical properties of pure titanium.

As a structural metal, titanium is still in its infancy, especially when compared with iron and steel. The first commercial titanium was produced in 1948. The total production that year was just over 1800 kg (4000 lb). By 1955, the production

Table 1.4 Raw titanium mill products in commercial airframes

Aircraft	Titanium, metric tons	Aircraft	Titanium, metric tons
Boeing 787	116	Airbus A380	77
Boeing 777	58	Airbus A340	24
Boeing 747	76	Airbus A330	17
Boeing 737	18	Airbus A320	12

Table 1.5 Typical TiO₂ content of several titanium minerals and slags

Material	TiO₂ content, %
Rutile	97.0–98.5
Ilmenite (from Quilon)	57.3–61.0
Ilmenite (from Florida)	58.0–63.0
Ilmenite (from Macintyre)	43.0–50.0
Ilmenite (from Baie-St. Paul)	38.5–41.5
Magnetite	8.0–11.0
Perovskite	0.0–54.0
Sorel slag (ilmenite from Quebec)	68.0–72.0
Osaka slag (ilmenite from Japan)	90.0–92.0

Table 1.6 Some physical and mechanical properties of pure titanium

Property	Value	
Density, g/cm³ (lb/in.³)	4.51	(0.163)
Melting temperature, °C (°F)	1660	(3020)
Specific heat, cal/g · °C (J/kg · K)	519	(0.124)
Thermal conductivity, Btu · ft²/ft · h · °F		~9
Thermal expansion (0–315 °C, or 32–600 °F), μm/m · °C (μin./in. · °F)	~3	(~5)
Electrical resistivity, μΩ · cm (circular mil · Ω/ft)	42	(250)
Magnetic susceptibility, emu/g		3.17
Tensile modulus, ×10³ MPa (×10⁶ psi)	~101	(~14.7)
Compression modulus, ×10³ MPa (×10⁶ psi)	~103	(~15)
Shear modulus, ×10³ MPa (×10⁶ psi)	~44	(~6.4)
Poisson's ratio		~0.4
Tensile strength, MPa (ksi)	240	(35)
Tensile yield strength (0.2%), MPa (ksi)	170	(25)
Compression yield strength (0.2%), MPa (ksi)	170	(25)
Elongation in tension, %		35
Shear strength, MPa (ksi)	~140	(~20)
Charpy impact strength, J (ft · lbf)	27–54	(20–40)
Fatigue strength, F_{tu}		0.5–0.6
Notched fatigue strength, ($K_t \geq 4$), F_{tu}		0.2–0.3
Creep strength (315 °C, or 600 °F, 10 h, 0.2%), MPa (×10³ psi)	<83	(<12)
Rupture strength (315 °C, or 600 °F, 1000 h), MPa (×10³ psi)	<69	(<10)
Bend radius, R/t		~4
Hardness (1500 kg load), HB		~65

Source: Ref 1.5

had grown to over 9000 kg (20,000 lb). In 1986 (excluding the former Communist bloc countries), the ingot capacity had increased to approximately 90,000 kg (200,000 lb). Also in 1986, it was estimated that the former USSR produced an additional 36,000 kg (80,000 lb) per year. Table 1.7 shows the U.S. titanium consumption between 2007 and 2011. Reference 1.3 includes government statistics on titanium ore and titanium sponge production and capacity.

Uses for titanium have expanded, based on its inherent properties as well as on the development of new alloys. The major use is still in airborne applications, such as engines, airframes, missiles, and spacecraft. Aerospace applications are based on the low density and high strength-to-weight ratio of alloyed titanium at elevated temperatures. The corrosion resistance of titanium makes it a natural material for use in seawater, marine, and naval applications. In addition, use of titanium is extensive in seawater-cooled power plant condensers. Over 70 million m (200 million ft) of welded titanium tubing has been used in power plant surface seawater condensers with no corrosion-related failures. This performance led one manufacturer to offer a 40 year corrosion guarantee for surface seawater condenser tubing.

Titanium is also extensively used in oil refineries, paper and pulp bleaching operations, nitric acid plants, and certain organic synthesis production.

Titanium has found use in the medical field. The largest use of titanium and titanium alloys as surgical implants has been for bone plates, screws, intramedullary rods, and hip nails. Partial and total joint replacements for the hip, knee, elbow, jaw, finger, and shoulder are commercially produced from unalloyed titanium and from the Ti-6Al-4V alloy. Unalloyed titanium heart valves and titanium mesh mandibular bone grafting trays are also available. The corrosion-resistant metal is used widely as a hermetically sealed container for pacemakers and as an encapsulating material for iodine-125 interstitial implants used to treat various tumors. Titanium is also used to produce near-net shape components for body implants.

General Properties and Alloying Behavior

Table 1.8 lists the titanium grades as defined by ASTM International and the American Society of Mechanical Engineers. Most of the grades are alloys with various additions of aluminum, vanadium, nickel, ruthenium, molybdenum, chromium, or zirconium for the purpose of improving and/or combining various mechanical characteristics, heat resistance, conductivity, microstructure, creep, ductility, and corrosion resistance.

In many applications, titanium and titanium alloys are naturally protected against corrosion due to the metals forming a stable and substantially inert protective oxide film on its surface. However, palladium (Pd), ruthenium (Ru), nickel (Ni), and molybdenum (Mo) are elements that can be added to titanium alloys to obtain a significant corrosion-resistance improvement, particularly when they are used in slightly reducing environments, where titanium may not form the necessary protective oxide film on the metal surface to prevent corrosion. Some of the physical and mechanical properties of titanium are given in Tables 1.9(a and b). Because several grades of unalloyed titanium are available and because all the properties listed have not been measured or reported for any single grade, the list contains property values for a mixture of grades. The variable quantity of oxygen is the

Table 1.7   United States titanium consumption

	2007	2008	2009	2010	2011 (estimate)
Titanium sponge					
Imports for consumption	25,900	23,900	16,600	20,500	32,000
Exports	2,000	2,370	820	293	200
Consumption, reported	33,700	W	W	34,900	49,000
Price, dollars/kg, year end	14.76	15.64	15.58	10.74	10.30
Stocks, industry, year end	7,820	14,200	15,300	10,500	8,500
Employment, number	400	350	300	300	300
Titanium dioxide					
Production	1,440,000	1,350,000	1,230,000	1,320,000	1,420,000
Imports for consumption	221,000	183,000	175,000	204,000	180,000
Exports	682,000	733,000	649,000	758,000	815,000
Consumption, apparent	979,000	800,000	757,000	767,000	785,000
Producer price index, year end	162	170	164	194	252
Stocks, producer, year end	NA	NA	NA	NA	NA
Employment, number	4,300	4,200	3,800	3,400	3,400

Note: W, numbers withheld by individual companies; NA, not available

Table 1.8   Common ASTM International/American Society of Mechanical Engineers titanium grades

Grade	Description
1	Unalloyed titanium, low oxygen (0.18 wt% max), low strength
2	Unalloyed titanium, standard oxygen (0.25 wt% max), medium strength
3	Unalloyed titanium, medium oxygen (0.35 wt% max), high strength
4	Unalloyed titanium, high oxygen (0.40 wt% max), extra-high strength
5	Titanium alloy (6% Al, 4% V)
7	Unalloyed titanium plus 0.12 to 0.25% Pd, standard oxygen, medium strength
9	Titanium alloy (3% Al, 2.5% V), high strength; mainly aerospace applications
11	Unalloyed titanium plus 0.12 to 0.25% Pd, low oxygen, low strength
12	Titanium alloy (0.3% Mo, 0.8% Ni), high strength
13	Titanium alloy (0.5% Ni, 0.05% Ru), low oxygen
14	Titanium alloy (0.5% Ni, 0.05% Ru), standard oxygen
15	Titanium alloy (0.5% Ni, 0.05% Ru), medium oxygen
16	Unalloyed titanium plus 0.04 to 0.08% Pd, standard oxygen, medium strength
17	Unalloyed titanium plus 0.04 to 0.08% Pd, low oxygen, low strength
18	Titanium alloy (3% Al, 2.5% V, plus 0.04 to 0.08% Pd)
19	Titanium alloy (3% Al, 8% V, 6% Cr, 4% Zr, 4% Mo)
20	Titanium alloy (3% Al, 8% V, 6% Cr, 4% Zr, 4% Mo) plus 0.04 to 0.08% Pd
21	Titanium alloy (15% Mo, 3% Al, 2.7% Nb, 0.25% Si)
23	Titanium alloy (6% Al, 4% V, extra-low interstitial)
24	Titanium alloy (6% Al, 4% V) plus 0.04 to 0.08% Pd
25	Titanium alloy (6% Al, 4% V) plus 0.3 to 0.8% Ni and 0.04 to 0.08% Pd
26	Unalloyed titanium plus 0.08 to 0.14% Ru, standard oxygen, medium strength
27	Unalloyed titanium plus 0.08 to 0.14% Ru, low oxygen, low strength
28	Titanium alloy (3% Al, 2.5% V) plus 0.08 to 0.14% Ru
29	Titanium alloy (6% Al, 4% V, with extra-low interstitial elements plus 0.08 to 0.14% Ru

chief difference between commercially pure grades.

Very small amounts of interstitial (see Chapter 2, Figure 6 for an explanation of interstitial) contaminants such as oxygen (O), carbon (C), nitrogen (N), and hydrogen (H) can change the mechanical properties of titanium quite markedly. Generally, they impart strength at the expense of ductility. For example, approximately 0.1% N more than doubles the strength of titanium but cuts ductility in half. Large additions of the interstitials (still less than 1%) can embrittle titanium, making it unusable. Thus, in seeking beneficial changes in properties by alloying, the interstitial additions are controlled. They are useful when limited to low levels. The several strength levels of unalloyed titanium, noted previously, are thus produced commercially by controlling the level of interstitial contaminants. Similarly, several grades of selected titanium alloys (such as Ti-6Al-4V and Ti-5Al-2.5Sn) are produced by controlling the interstitial element (e.g., oxygen, nitrogen, carbon) content.

Mechanical Properties

Drastic changes in the mechanical properties of titanium are achieved by alloying with various metallic elements, including aluminum, tin, zirconium, manganese, vanadium, molybdenum, and chromium. As discussed in other chapters in this book, some of these additions can result in titanium alloys capable of achieving even greater strength by heat treatment.

The room-temperature properties and characteristics of titanium and titanium alloys given in Tables 1.6 and 1.9(a and b) indicate the usefulness of titanium as a structural metal. The material offers even greater advantages at higher and lower temperatures.

The classic comparison between titanium and other aerospace metals is on a strength-to-weight ratio basis. Such a comparison is shown in Fig. 1.4. Here, tensile strengths for the upper-use-temperature range of titanium are compared with the tensile strengths of 7075-T6 aluminum and with the precipitation-hardening-type stainless steels. Note that strength comparisons without density considerations are not so favorable to titanium.

The low-temperature properties of titanium alloys are usually compared with those of metals frequently used in cryogenic applications. Such a comparison is given in Fig. 1.5. Here, the tensile yield strengths of the materials are compared after adjusting for density.

While exhibiting low levels of ductility, the reduced density and excellent creep behavior (Fig. 1.6) of the titanium aluminides (especially equiatomic TiAl) make them look attractive for use in elevated-temperature applications (see Chapter 15, "Applications of Titanium," in this book).

Special Chemical and Physical Properties

Corrosion Resistance (Oxidizing Environments). Titanium, being a reactive metal, depends on the formation of a very thin, tightly adherent, protective oxide film for its corrosion resistance. This TiO_2 film is corrosion resistant to a wide array of chemical environments, most notably, seawater and other chloride-brine media.

There are a few special situations where corrosive environments combined with states of stress in titanium lead to degradation known as stress-corrosion cracking (hot salt stress corrosion and accelerated crack propagation in aqueous solutions). These phenomena are described further in Chapter 14, "Corrosion," in this book.

Table 1.9A Typical physical properties of wrought titanium alloys

Nominal composition, %	Coefficient of linear thermal expansion, μm/m · K (μin./in. · °F)							Electrical resistivity, μΩ · m(a)	Thermal conductivity, W/m · K(a)	Density(a) g/cm³	Density(a) lb/in.³
	20–100 °C (70–212 °F)	20–205 °C (70–400 °F)	20–315 °C (70–600 °F)	20–425 °C (70–800 °F)	20–540 °C (70–1000 °F)	20–650 °C (70–1200 °F)	20–815 °C (70–1500 °F)				
Commercially pure titanium											
ASTM grades 1, 2, 3, 4, 7, and 11	8.6 (4.8)	...	9.2 (5.1)	...	9.7 (5.4)	10.1 (5.6)	10.1 (5.6)	0.42–0.52	16	4.51	0.163
α alloys											
5Al-2.5Sn	9.4 (5.2)	...	9.5 (5.3)	...	9.5 (5.3)	9.7 (5.4)	10.1 (5.6)	1.57	7.4–7.8	4.48	0.162
5Al-2.5Sn (low O$_2$)	9.4 (5.2)	...	9.5 (5.3)	...	9.7 (5.4)	9.9 (5.5)	10.1 (5.6)	1.80	7.4–7.8	4.48	0.162
Near α											
8Al-1Mo-1V	8.5 (4.7)	...	9.90 (5.0)	...	10.1 (5.6)	10.3 (5.7)	...	1.99	...	4.37	0.158
11Sn-1Mo-2.25Al-5.0Zr-1Mo-0.2Si	8.5 (4.7)	...	9.2 (5.1)	...	9.4 (5.3)	...	...	1.62	6.9	4.82	0.174
6Al-2Sn-4Zr-2Mo	7.7 (4.3)	...	8.1 (4.5)	...	8.1 (4.5)	...	...	1.9	7.1 at 100 °C (212 °F)	4.54	0.164
5Al-5Sn-2Zr-2Mo-0.25Si	...	...	...	...	...	...	10.3 (5.7)	...	...	4.51	0.163
6Al-2Nb-1Ta-1Mo	...	...	...	...	...	...	...	...	6.4	4.48	0.162
IMI 685	9.8 (5.4)	9.3 (5.2)	9.5 (5.3)	9.8 (5.4)	10.1 (5.6)	9.0 (5.0)	...	1.68	4.2	4.45	0.161
IMI 829	...	9.45 (5.3)	...	9.8 (5.4)	...	9.98 (5.5)	...	...	...	4.54	0.164
IMI 834	...	10.6 (5.9)	...	10.9 (6.1)	...	11 (6.1)	...	...	...	4.55	0.164
α-β alloys											
8Mn	8.6 (4.8)	9.2 (5.1)	9.7 (5.4)	10.3 (5.7)	10.8 (6.0)	11.7 (6.5)	12.6 (7.0)	0.92	10.9	4.73	0.171
3Al-2.5V	9.5 (5.3)	9.9 (5.5)	9.9 (5.5)	...	9.9 (5.5)	...	...	...	...	4.48	0.162
6Al-4V	8.6 (4.8)	9.0 (5.0)	9.2 (5.1)	9.4 (5.2)	9.5 (5.3)	9.7 (5.4)	...	1.71	6.6–6.8	4.43	0.160
6Al-4V (low O$_2$)	8.6 (4.8)	9.0 (5.0)	9.2 (5.1)	9.4 (5.2)	9.5 (5.3)	9.7 (5.4)	...	1.71	6.6–6.8	4.43	0.160
6Al-6V-2Sn	9.0 (5.0)	...	9.4 (5.2)	...	9.5 (5.3)	...	...	1.57	6.6(b)	4.54	0.164
7Al-4Mo	9.0 (5.0)	9.2 (5.1)	9.4 (5.2)	9.7 (5.4)	10.1 (5.6)	10.4 (5.8)	11.2 (6.2)	1.7	6.1	4.48	0.162
6Al-2Sn-4Zr-6Mo	9.0 (5.0)	9.2 (5.1)	9.4 (5.2)	9.5 (5.3)	9.5 (5.3)	...	...	...	7.7(c)	4.65	0.168
6Al-2Sn-2Zr-2Mo-2Cr-0.25Si	...	...	9.2 (5.1)	...	...	...	...	...	...	4.57	0.165
IMI 550	8.8 (4.9)	9.0 (5)	9.2 (5.1)	9.3 (5.2)	9.7 (5.4)	10.1 (5.6)	...	1.58	7.5	4.60	0.166
IMI 679	8.2 (4.6)	8.9 (4.9)	9.3 (5.2)	9.4 (5.2)	9.6 (5.3)	...	...	...	...	4.84	0.175
β alloys											
13V-11Cr-3Al	9.4 (5.2)	9.9 (5.5)	10 (5.55)	10.1 (5.6)	10.2 (5.7)	10.4 (5.8)	...	...	...	4.82	0.174
8Mo-8V-2Fe-3Al	...	...	...	...	...	...	...	...	...	4.84	0.175
3Al-8V-6Cr-4Mo-4Zr	8.7 (4.8)	9 (5)	9.4 (5.2)	9.6 (5.3)	...	...	...	...	...	4.82	0.174
11.5Mo-6Zr-4.5Sn	7.6 (4.2)	8.1 (4.5)	8.5 (4.7)	8.7 (4.8)	8.7 (4.8)	...	...	1.56	...	5.06	0.183
15V-3Cr-3Al-3Sn	8.5 (4.7)	8.7–9 (4.8–5)	9.2 (5.1)	9.4 (5.3)	9.7 (5.4)	...	...	1.47	8.08	4.71	0.170
5Al-2Sn-2Zr-4Cr	9 (5)	9.2 (5.1)	9.4 (5.2)	9.5 (5.3)	...	...	...	...	...	...	...

(a) Room temperature. (b) At 93 °C (200 °F) (c) In solution-treated and aged condition

Table 1.9B   Minimum and average mechanical properties of wrought titanium alloys at room temperature

Composition, %	Condition	Minimum and average tensile properties(a)						Average or typical properties			
		Ultimate tensile strength, MPa (ksi)	0.2% yield strength, MPa (ksi)	Elongation, %	Reduction in area, %	Charpy impact strength, J (ft·lbf)	Hardness	Modulus of elasticity, GPa (10⁶ psi)	Modulus of rigidity, GPa (10⁶ psi)	Poisson's ratio	Bend radius for thickness (t) over 1.8 mm (0.07 in.)
Commercially pure titanium											
99.5 Ti (ASTM grade 1)	Annealed	240–331 (35–48)	170–241 (25–35)	30	55	…	120 HB	102.7 (14.9)	38.6 (5.6)	0.34	2t
99.2 Ti (ASTM grade 2)	Annealed	340–434 (50–63)	280–345 (40–50)	28	50	34–54 (25–40)	200 HB	102.7 (14.9)	38.6 (5.6)	0.34	2.5t
99.1 Ti (ASTM grade 3)	Annealed	450–517 (65–75)	380–448 (55–65)	25	45	27–54 (20–40)	225 HB	103.4 (15.0)	38.6 (5.6)	0.34	2.5t
99.0 Ti (ASTM grade 4)	Annealed	550–662 (80–96)	480–586 (70–85)	20	40	20 (15)	265 HB	104.1 (15.1)	38.6 (5.6)	0.34	3.0t
99.2 Ti (ASTM grade 7)(b)	Annealed	340–434 (50–63)	280–345 (40–50)	28	50	43 (32)	200 HB	102.7 (14.9)	38.6 (5.6)	0.34	2.5t
98.9 Ti (ASTM grade 12)(c)	Annealed	480–517 (70–75)	380–448 (55–65)	25	42	…	…	…	102.7 (14.9)	…	2.5t
α alloys											
5Al-2.5Sn	Annealed	790–862 (115–125)	760–807 (110–117)	16	40	13.5–20 (10–15)	36 HRC	110.3 (16.0)	…	…	4.5t
5Al-2.5Sn (low O₂)	Annealed	690–807 (100–117)	620–745 (90–108)	16	…	43 (32)	35 HRC	110.3 (16.0)	…	…	…
Near α											
8Al-1Mo-1V	Duplex annealed	900–1000 (130–145)	830–951 (120–138)	15	28	20–34 (15–25)	35 HRC	124.1 (18.0)	46.9 (6.8)	0.32	4.5t
11Sn-1Mo-2.25Al-5.0Zr-1Mo-0.2Si	Duplex annealed	1000–1103 (145–160)	900–993 (130–144)	15	35	…	36 HRC	113.8 (16.5)	…	…	…
6Al-2Sn-4Zr-2Mo	Duplex annealed	900–980 (130–142)	830–895 (120–130)	15	35	…	32 HRC	113.8 (16.5)	…	…	5t
5Al-5Sn-2Zr-2Mo-0.25Si	975 °C (1785 °F) (½ h), AC + 595 °C (1100 °F) (2 h), AC	900–1048 (130–152)	830–965 (120–140)	13	…	…	…	113.8 (16.5)	…	0.326	…
6Al-2Nb-1Ta-1Mo	As-rolled 2.5 cm (1 in.) plate	790–855 (115–124)	690–758 (100–110)	13	34	31 (23)	30 HRC	113.8 (17.5)	…	…	…
6Al-2Sn-1.5Zr-1Mo-0.35Bi-0.1Si	β forge + duplex anneal	1014 (147)	945 (137)	11	…	…	…	…	…	…	…
IMI 685 (Ti-6Al-5Zr-0.5Mo-0.25Si)	β heat treated at 1050 °C (1920 °F), OQ, + aged 24 h at 550 °C (1020 °F)	882–917 (128–133)	758–815 (110–118)	6–11 (on 5D)	15–22	43 (32)	…	~125 (~18)	…	…	…
IMI-829 (Ti-5.5Al-3.5Sn-3Zr-1Nb-0.25Mo-0.3Si)	β heat treated at 1050 °C (1920 °F), AC, + aged 2 h at 625 °C (1155 °F)	930 (min) (35)	820 (min) (119)	9 (min) on 5D	15 (min)	…	…	…	…	…	…
IMI-834 (Ti-5.5Al-4.5Sn-4Zr-0.7Nb-0.5Mo-0.4Si-0.06C)	α-β processed	1030 (min) (149)	910 (min) (132)	6 (min) on 5D	15 (min)	…	…	…	…	…	…

(continued)

(a) If a range is given, the lower value is a minimum; all other values are averages. (b) Also contains 0.2 Pd. (c) Also contains 0.8 Ni and 0.3 Mo. AC, air-cooled; OQ, oil quenched

Table 1.9B (Continued)

Composition, %	Condition	Minimum and average tensile properties(a)						Average or typical properties			
		Ultimate tensile strength, MPa (ksi)	0.2% yield strength, MPa (ksi)	Elongation, %	Reduction in area, %	Charpy impact strength, J (ft · lbf)	Hardness	Modulus of elasticity, GPa (10⁶ psi)	Modulus of rigidity, GPa (10⁶ psi)	Poisson's ratio	Bend radius for thickness (t) over 1.8 mm (0.07 in.)
α-β alloys											
8Mn	Annealed	860–945 (125–137)	760–862 (110–125)	15	32	...	...	113.1 (16.4)	48.3 (7.0)	...	...
3Al-2.5V	Annealed	620–689 (90–100)	520–586 (75–85)	20	...	54 (40)	...	106.9 (15.5)	...	...	...
6Al-4V	Annealed	900–993 (130–144)	830–924 (120–134)	14	30	14–19 (10–14)	36 HRC	113.8 (16.5)	42.1 (6.1)	0.342	5t
	Solution + aging	1172 (170)	1103 (160)	10	25	...	41 HRC	...	...	...	...
6Al-4V (low O₂)	Annealed	830–896 (120–130)	760–827 (110–120)	15	35	24 (18)	35 HRC	113.8 (16.5)	42.1 (6.1)	0.342	...
6Al-6V-2Sn	Annealed	1030–1069 (150–155)	970–1000 (140–145)	14	30	14–19 (10–14)	38 HRC	110.3 (16.0)	...	...	4.5t
	Solution + aging	1276 (185)	1172 (170)	10	20	...	42 HRC	...	...	...	...
7Al-4Mo	Solution + aging	1103 (160)	1034 (150)	16	22	18 (13)	38 HRC	113.8 (16.5)	44.8 (6.5)	...	...
	Annealed	1030 (min) (150)	970 (min) (140)	...	...	...	...	...	...	...	...
6Al-2Sn-4Zr-6Mo	Solution + aging	1269 (189)	1172 (170)	10	23	8–15 (6–11)	36–42 HRC	113.8 (16.5)	...	...	...
6Al-2Sn-2Zr-2Mo-2Cr-0.25Si	Solution + aging	1276 (185)	1138 (165)	11	33	20 (15)	...	122 (17.7)	46.2 (6.7)	0.327	...
	Annealed	1030 (min) (150)	970 (min) (140)	...	...	...	...	...	...	...	...
Corona 5 (Ti-4.5Al-5Mo-1.5Cr)	β annealed plate	910 (132)	817 (118)	...	...	...	...	...	...	...	...
	β worked plate	945 (137)	855 (124)	...	...	...	...	...	...	...	...
	α-β worked	935 (131)	905 (131)	...	...	...	...	...	...	...	...
IMI 550 (Ti-4Al-4Mo-2Sn-0.5Si)	Solution at 900 °C (1650 °F), AC, + aging of 25 mm (1 in.) slice	1100 (160)	940 (136)	7 on 5D	15	23 (17)	...	~115 (~17)	...	...	...

(continued)

(a) If a range is given, the lower value is a minimum; all other values are averages. (b) Also contains 0.2 Pd. (c) Also contains 0.8 Ni and 0.3 Mo. AC, air-cooled; OQ, oil quenched

Table 1.9B (Continued)

| Composition, % | Condition | Minimum and average tensile properties(a) | | | | Charpy impact strength, J (ft·lbf) | Hardness | Average or typical properties | | | |
		Ultimate tensile strength, MPa (ksi)	0.2% yield strength, MPa (ksi)	Elongation, %	Reduction in area, %			Modulus of elasticity, GPa (10⁶ psi)	Modulus of rigidity, GPa (10⁶ psi)	Poisson's ratio	Bend radius for thickness over 1.8 mm (0.07 in.)
β alloys											
13V-11Cr-3Al	Solution + aging	1170–1220 (170–177)	1100–1172 (160–170)	8	...	...	...	101.4 (14.7)	42.7 (6.2)	0.304	...
8Mo-8V-2Fe-3Al	Solution + aging	1276 (185)	1207 (175)	8	...	11 (8)	40 HRC	106.9 (15.5)	...	...	...
	Solution + aging	1170–1310 (170–190)	1100–1241 (160–180)	8	...	...	40 HRC	...	...	...	...
3Al-8V-6Cr-4Mo-4Zr (Beta C)	Solution + aging	1448 (210)	1379 (200)	7	...	10 (7.5)	...	105.5 (15.3)	...	...	...
	Annealed	883 (min) (128 min)	830 (min) (120 min)	15	...	...	...	...	...	...	...
11.5Mo-6Zr-4.5Sn (Beta III)	Solution + aging	1386 (210)	1317 (191)	11	...	...	...	103 (15)	...	...	...
	Annealed	690 (min) (100 min)	620 (min) (90 min)	...	...	...	...	...	...	...	...
10V-2Fe-3Al	Solution + aging	1170–1276 (170–185)	1100–1200 (160–174)	10	19	...	...	111.7 (16.2)	...	...	...
Ti-15V-3Cr-3Al-3Sn (Ti-15-3)	Annealed	785 (114)	773 (112)	22	...	...	...	...	...	...	...
	Aged	1095–1335 (159–194)	985–1245 (143–180)	6–12	...	...	...	...	...	...	...
Ti-5Al-2Sn-2Zr-4Mo-4Cr (Ti-17)	Solution + aging	1105–1240 (160–180)	1305–1075 (150–170)	8–15	20–45	...	...	...	...	...	...
Transage 134 plate	Solution + aging	1055–1380 (153–200)	1000–1310 (145–190)	5–12	10–38	...	...	...	...	...	...
Transage 175 (extruded bar)	Solution + aging	1305 (189)	1250 (180)	10	39	...	...	...	...	...	...
Transage 175 at 425 °C (800 °F)	Solution + aging	1080 (157)	925 (134)	10	56	...	...	...	...	...	...

(a) If a range is given, the lower value is a minimum; all other values are averages. (b) Also contains 0.2 Pd. (c) Also contains 0.8 Ni and 0.3 Mo. AC, air-cooled; OQ, oil quenched

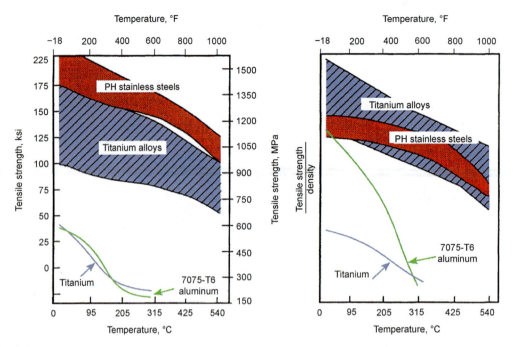

Fig. 1.4 Strength comparison of titanium and titanium alloys and other aerospace alloys. PH, precipitation hardening

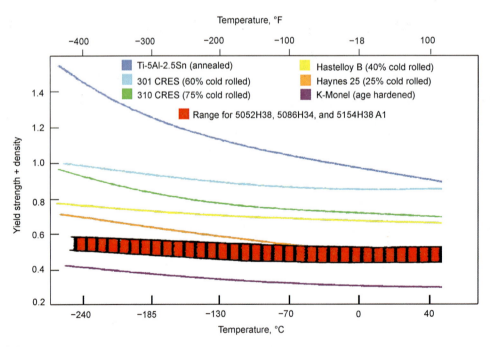

Fig. 1.5 Strength-density ratio comparison. The Ti-5Al-2.5Sn alloy has a significant strength advantage over a wide range of alloys.

Thermal Conductivity and Expansion. The high binding forces between titanium atoms result in a very high melting point and a low thermal expansion. Its thermal conductivity is also low. These properties may be beneficial for close tolerances over a temperature range, or they may be detrimental in applications where they may be paired with iron or nickel alloys.

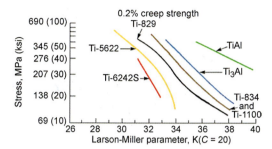

Fig. 1.6 Creep behavior of a number of terminal high-temperature titanium alloys and the intermetallic compounds Ti₃Al and TiAl, showing the enhanced creep behavior of the intermetallics

Table 1.10 Cost of titanium relative to other metals

Item	Contract prices, 2014 U.S. dollars/lb		
	Steel	Aluminum	Titanium
Ore	0.02	0.10	0.22 (rutile)
Metal	0.10	1.10	5.44
Ingot	0.15	1.15	9.07
Sheet	0.30–0.60	1.00–5.00	15.00–50.00

Note: The high cost of titanium compared to aluminum and steel is a result of: (1) High extraction costs. (2) High processing costs. Relatively low processing temperatures are required to control the microstructure and hence the mechanical properties. The low processing temperatures mean high pressures and increased amounts of conditioning between working operations. Conditioning of surface regions contaminated at the processing temperatures and pressures and of surface cracks, both of which must be removed, is done prior to further fabrication. (3) Prices of titanium are contract prices.

Table 1.11 Structural materials

Material	Consumption/year, 10³ metric tons
Titanium	50
Magnesium	320
Aluminum	25,000
Steel	700,000
Wood	400,000

Table 1.12 Cost of titanium precursors

Precursor	Cost, 2014 U.S. dollars/lb	Cost of contained titanium, 2014 U.S. dollars/lb
TiO₂(a)	1.75	2.94
TiCl₄	1.00	4.00
Titanium sponge	5.44	5.44

(a) Metal grade

Electrical Conductivity. Together with its low thermal conductivity, titanium has a correspondingly low electrical conductivity. Its specific resistivity is among the highest of all metallic elements. Therefore, titanium is used in deicing installations and in certain instruments.

Magnetic Susceptibility. Titanium is slightly paramagnetic, which is advantageous in and around instruments. The high fatigue strength and good modulus of elasticity of titanium is also advantageous in instruments. Low magnetic susceptibility plus corrosion resistance are important in certain marine applications.

Economics

Titanium and titanium alloys are relatively expensive materials compared with aluminum and steel (Table 1.10), and this is reflected in worldwide consumption figures (Table 1.11). However, titanium can be the lowest-cost material for many applications due to weight savings, as in engines and airframes, and in maintenance-free, long-life hardware in corrosive environments. Further, the premium performance of vehicles or equipment made of titanium can be attributed to a combination of titanium properties. Thus, in comparing the total cost of alternate materials for hardware, the performance and operating cost of the equipment must be considered in addition to the initial cost. Despite all the recent advances in the production of titanium and its alloys, the cost of production still remains high compared with other metals. Therefore, the need to develop a more cost-effective extraction and powder metallurgy process remains high. (See Chapter 8, "Melting, Casting, and Powder Metallurgy," for more details on powder metallurgy.)

The cost of titanium precursors is shown in Table 1.12, along with the cost of the contained titanium in each of these precursors. The cost of contained titanium in TiO₂ is less than that for the intermediate compound (in the Kroll process), TiCl₄. The price of titanium at various stages of conventionally fabricated components is shown in Fig. 1.7.

The manufacturing cost breakdown for the Boeing 787 side-of-body chord is shown in Fig. 1.8. Note that the machining cost (see Chapter 13, "Machining and Chemical Shaping of Titanium") is almost half the total component cost, thus the attraction of producing near-net shapes.

Process Challenges

The high atomic binding forces that give titanium its desirable characteristics also make it so

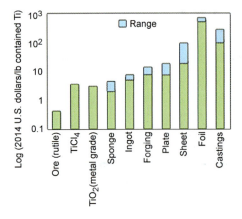

Fig. 1.7 Cost of titanium at various stages of mill product fabrication

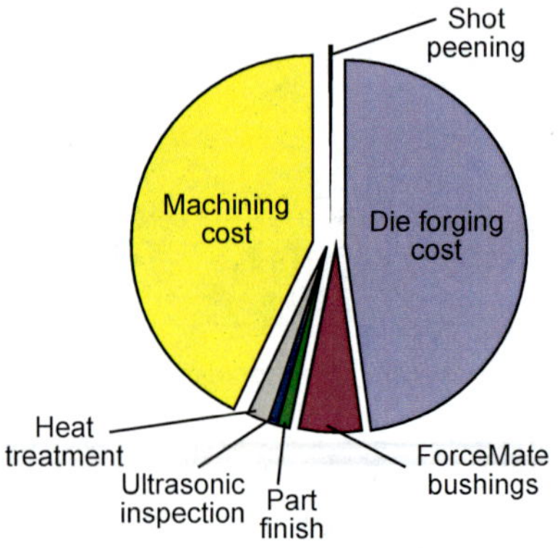

Fig. 1.8 Boeing 787 side-of-body chord manufacturing cost breakdown. Note that the machining cost is almost half the total component cost. Courtesy of R. Boyer and J. Cotton, Boeing Corporation, Sept 2013

reactive that it is difficult to extract from its ores and to maintain in pure form at elevated temperatures. Titanium is so reactive at red heat that it becomes contaminated with nearly every material it contacts. This contamination can lead to embrittlement and to loss of properties. Consequently, finding a suitable material to contain titanium in the molten state and avoid contamination is a challenge. A water-cooled copper crucible freezes molten titanium instantly upon contacting the cold copper; thereafter, the remaining liquid is contained in a shell of solid titanium (referred to as a "skull"), substantially avoiding contamination. Product improvements in processing, such as purification of reactants for winning the metal and better control procedures during melting and fabricating, also minimized problems with contaminants.

In the early 1950s, hydrogen embrittlement threatened the usefulness of titanium. Hydrogen in amounts over 150 parts per million (ppm) embrittles pure titanium under stress. The problem was solved by removing hydrogen in the vacuum-melting process and by maintaining a hydrogen-free material during processing of mill products.

Most titanium alloys are susceptible to stress corrosion when traces of salt (NaCl) are present and when time, elevated temperature, and stress factors exceed critical limits. Residuals from chloride-containing cleaning solvents or from chloride-bearing marking crayons also cause this reaction. Certain grades of nitrogen tetroxide (used as an oxidizer in liquid-fueled rocket motors) and even methyl alcohol can cause stress-corrosion failures of titanium. However, while readily reproduced in the laboratory, stress-corrosion cracking is rarely found in the field.

Certain titanium alloys are susceptible to accelerated crack growth in aqueous solutions of chloride-containing chemicals. This phenomenon, which is also found in other materials, is discussed in Chapter 14, "Corrosion," in this book.

Extractive Metallurgy

Titanium is extremely difficult to extract from its ores. A century and a half passed between the discovery of the element and the development of an economical method for producing titanium commercially. The only two commercial processes for the extraction of titanium from its ores are attributed to Dr. W.J. Kroll and to M.A. Hunter. The major difference in the two processes is that the Kroll process uses magnesium metal for the reduction of $TiCl_4$ while the Hunter process uses sodium. A number of other processes, such as electrowinning and the fluoride process, show promise but have not been developed to the point of being classified as commercial processes. In recent years there has been renewed activity in reducing the cost of extraction.

All commercial methods for reducing oxides to titanium metal have three basic steps:

1. Convert the oxide ore to $TiCl_4$ by chlorination in the presence of carbon or titanium tetrachloride production
2. Reduce $TiCl_4$ to metal using sodium (Na), magnesium (Mg), and, on a much smaller scale, electrolysis

3. Purify titanium by distillation in a vacuum, by an inert gas sweep, and by leaching to remove residual salts and unconsumed reactants

The metal produced using these techniques has an open-pore-type structure and is referred to as titanium sponge due to this porous, spongy appearance (Fig. 1.9). The sponge, which is more than 99% Ti, is blended with alloys, compacted, and melted. Several methods are available for melting titanium, the principal ones being consumable electrode vacuum arc melting and cold hearth melting using either an electron beam or a plasma heat source. Further details on melting and fabrication approaches are presented in Chapter 8, "Melting, Casting, and Powder Metallurgy," in this book.

The first production of titanium sponge offering commercial potential was reported in 1946 by the U.S. Bureau of Mines, where 7 kg (15 lb) batches of good-quality sponge were made using the Kroll process. By 1949, the National Lead pilot plant was operating at 50 kg (100 lb) per day. In January 1950, National Lead formed an equal partnership with Allegheny Ludlum Steel to establish Titanium Metals Corporation of America. By 1952, E.I. du Pont de Nemours Co. at Newport, Delaware; Titanium Metals Corporation of America at Henderson, Nevada; and Crane Company at Chattanooga, Tennessee, were operating small titanium sponge plants. The U.S. Bureau of Mines was also doing research work at its Boulder City, Nevada, pilot plant.

While the Hunter process for production of ductile sponge using sodium metal was first demonstrated in 1910, its introduction on a commercial scale did not occur until August 1955, when the ICI plant in Deeside, United Kingdom, went into production.

Development of electrowinning processes using molten salt electrolysis principles occurred concurrently with the evolution of the thermochemical processes of Kroll and Hunter. These developments continued at varying levels of effort by a decreasing number of companies at increasing scales of operation. By the mid-1980s, U.S. efforts were reduced to two general processes strictly on a pilot/demonstration scale. Because they play a role in future production expansions, they are considered here.

Titanium sponge production is relatively expensive due in part to the high capital and operating cost for production as well as to the cyclic nature of the market, which has been driven primarily by military and commercial aerospace applications. However, in general, there has been an increase in production of sponge, which continued through the end of the 20th century.

Titanium Sponge Production

Titanium Tetrachloride Production. Because all processes use titanium tetrachloride ($TiCl_4$) as a starting material, a few details of manufacture of this product are necessary to understand the extraction story. Titanium tetrachloride is made from rutile by high-temperature reaction with chlorine in the presence of a reducing agent, usually carbon. The reactions are:

$$TiO_2 + 2Cl_2 + C \rightarrow TiCl_4 + CO_2 + Heat \qquad (Eq\ 1.1)$$

$$TiO_2 + 2Cl_2 + 2C \rightarrow TiCl_4 + 2CO + Heat \qquad (Eq\ 1.2)$$

These reactions are exothermic, the first producing $420,000 \times 10^3$ J/kg (101,000 Btu/lb · mol) at 799 °C (1470 °F), and the second, $82,800 \times 10^3$ J/kg (19,800 Btu/lb · mol). They are carried out at 700 to 1000 °C (1290 to 1830 °F).

Two processes used for the chlorination of rutile are fluidized-bed chlorination and static chlorination of briquetted or sintered material. In either instance, intimate contact of carbon and ore is required. The fluidized-bed chlorination technique is most widely used.

There are advantages in operating fluidized chlorinators wherein the particles of coke and rutile are suspended in a fluidlike bed by the upward flow of chlorine and product gases. The sintering or briquetting step is eliminated, because ground coke (or other carbonaceous material) and rutile are fed directly to the bed.

Because most metallic impurities in rutile are as easy to chlorinate as titanium, the product gases contain (in addition to the titanium tetrachloride) iron chloride, vanadyl trichloride, silicon tetrachloride, and other metal chlorides. Also, they contain other impurities, including unreacted rutile and carbon dust, unreacted chlo-

Fig. 1.9 Magnesium-reduced titanium sponge

rine, phosgene, carbon dioxide, and carbon monoxide. A dust collector in the chlorinator off-gas line is used to reduce the amount of solids in the product titanium tetrachloride. From the dust collector, the gas stream goes to a cooling-condensing system where the titanium tetrachloride is recovered. Use of spray towers with circulating, chilled titanium tetrachloride is a satisfactory condensing system. The product titanium tetrachloride is extracted from the condensing system and is transferred to crude storage.

The steps in the purification of titanium tetrachloride are illustrated by Fig. 1.10. The first and simplest step is the removal of solids. This is achieved by simple settling or, more commonly, by vaporizing and condensation of the liquid-phase metal chlorides.

Most of the other impurities, such as stannic chloride, silicon tetrachloride, and other high-boiling-point impurities, are removed by distillation. One of them, vanadium oxytrichloride ($VOCl_3$), poses a special problem because its molecular structure and molecular weight differ slightly from those of titanium tetrachloride. It boils at 126 °C (260 °F) as compared with 136 °C (277.5 °F) for titanium tetrachloride. Thus, it is necessary to treat the product mixture with hydrogen sulfide, copper powder, or other compounds to reduce the contaminant. The reduced vanadium compound is insoluble in $TiCl_4$ and can be separated by simple distillation. Distillation columns are used to remove impurities such as those mentioned previously. It is also common practice in the industry to process and store the purified $TiCl_4$ under an inert gas blanket (argon or helium) to protect it from atmospheric contamination.

The degree of purity required of titanium tetrachloride is very high; for example, the maximum desired total oxygen content is 50 ppm. Metallic impurities in the tetrachloride can be determined readily, but they are not as important in the final quality of the metal as the amounts of carbon, oxygen, and other nonmetallic impurities. The best test for the quality of tetrachloride is the grade of metal it will produce.

Magnesium-Reduction Process (Kroll)

The thermochemical process, first described by Kroll and which now bears his name, was the first production-scale scheme introduced for economically producing acceptable-quality titanium metal on a large scale. The process for making titanium by the Kroll method is best described by the following expression:

$$TiCl_4 (V) + 2Mg (L, V) \rightarrow$$
$$2MgCl_2 (L) + Ti (S) \qquad \text{(Eq 1.3)}$$

This appears to be a straightforward displacement reaction requiring simply the exclusion of species that may inhibit the reaction, such as oxygen or nitrogen. However, in reality, it is not so simple, because there exists a number of other concurrent reactions. These reactions are independently temperature-sensitive and competitive with one another, but all reduce to the expression of Eq 1.3 and are described as follows:

$$TiCl_4 (V) + Mg (L, V) \rightarrow$$
$$2MgCl_2 (L) + TiCl_2 (L) \qquad \text{(Eq 1.4)}$$

$$TiCl_2 (L) + Mg (L, V) \rightarrow MgCl_{2L} + Ti (S) \qquad \text{(Eq 1.5)}$$

$$Ti (S) + TiCl_4 (V) \rightarrow 2TiCl_2 (L) \qquad \text{(Eq 1.6)}$$

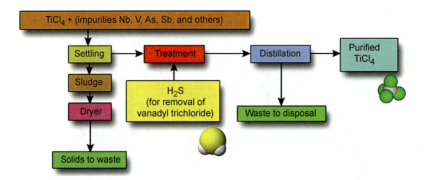

Fig. 1.10 Raw TiCl₄ from the processing of ore contains many impurities, including niobium, vanadium, arsenic, and antimony. The raw mixture is allowed to settle, where undissolved materials are removed as sludge. The liquid, which contains oxytrichloride (vanadyl trichloride, $VOCl_3$) and other chlorides, is treated with hydrogen sulfide (H_2S) and then distilled to further purify the titanium chloride ($TiCl_4$). Pure or nearly pure titanium chloride is clear in color.

Equation 1.6 illustrates how control of the amount of TiCl$_4$ present is necessary to avoid redissolution of titanium. Therefore, magnesium must be omnipresent with the introduction of TiCl$_4$. The magnesium must itself be free of undesirable impurities, because they are more attracted to the titanium reaction product than to the MgCl$_2$ salt.

In the reaction deposit, magnesium, MgCl$_2$, and titanium become mechanically intermingled; the accumulation of MgCl$_2$ tends to choke all three reactions, shielding the magnesium from contact with TiCl$_4$. Therefore, the problem of removing MgCl$_2$ from the reaction zone must be dealt with. During the reduction, molten MgCl$_2$ is removed from the reactor vessel by carefully tapping the reactor, allowing the molten MgCl$_2$ to flow into an awaiting container. Because both magnesium and titanium, when hot, are very reactive metals, air must be excluded from the reaction in a manner so as not to interfere with MgCl$_2$ removal or TiCl$_4$ introduction. Techniques to accomplish this involve manipulation of the tapping schedule and the TiCl$_4$ feed program, and by maintaining special conditions at the reactor tapping point. The details of these methods are often considered proprietary and are therefore not discussed. The reduction vessel is typically maintained under an inert gas atmosphere throughout the reduction cycle.

A schematic of the Kroll process system is shown in Fig. 1.11. This diagram contains all the necessary elements of the Kroll reaction, which include a means of generating and/or recycling magnesium and chlorine. Not all producers are fully integrated; some purchase materials for portions of the process.

The process of harvesting finished product from the reaction vessel involves various combinations of mechanical methods, such as boring and extrusion. The details depend somewhat on the method by which residual magnesium and MgCl$_2$ are separated from the reaction product. Figure 1.12 illustrates the two chief methods whereby this separation is accomplished with Kroll reduction product: leaching and vacuum distillation (vapor-phase gas sweep). There are, in production, various hybrids of leaching and distillation that are presumed to be more cost-effective. One such practice is an inert gas sweep over the reduction product to carry the magnesium and MgCl$_2$ to condensation chambers. A final leaching operation completes the salt removal.

The process of leaching the Kroll reduction product begins with extracting the product from the reactor vessel, usually by boring. The chips at this point contain residual magnesium and MgCl$_2$. The latter salt is hygroscopic and accumulates water of hydration. Therefore, magnesium chloride salts must be reduced to very low levels prior to melting so as not to introduce oxygen to the ingot. The bored chips are usually crushed and then leached in acid solutions to accomplish removal of residues from the reduction. This is followed by washing and drying by thermal, mechanical centrifuge, or vacuum, or some hybrid of these.

Those Kroll reduction processes that incorporate vacuum distillation to remove the magnesium and MgCl$_2$ residue of the reduction process can, for example, invert and seal the original reactor over a similar chilled vessel. Heat is then applied to the reaction mass while maintaining a vacuum in the chamber in a manner similar to that shown in Fig. 1.1. The combined actions of gravity and vacuum aid in transferring magnesium

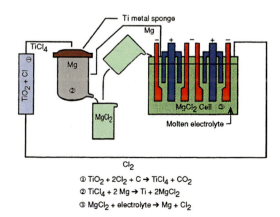

Fig. 1.11 Schematic of Kroll process using magnesium as the reacting metal, illustrating the closed loop or recycling of magnesium and Cl$_2$

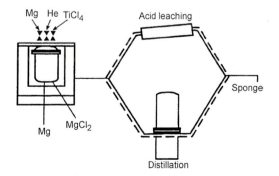

Fig. 1.12 Schematic of the magnesium reduction process for producing titanium sponge, illustrating alternate means of removing magnesium and MgCl$_2$ from the sponge

and $MgCl_2$ from the reactor mass to the condensing vessel. The condensing vessel is then prepared for separation and/or the beginning of the next Kroll reduction. The portion of original reactor mass that is pure titanium is generally removed via mechanical means, possibly aided by using a removable reactor vessel liner, which is easily stripped from the distilled titanium.

Recovery of magnesium and $MgCl_2$ is more complete using the vacuum distillation process than using the Kroll leach process. Therefore, it should affect the economics of plant operation in a positive manner. On the other hand, more expensive capital equipment and greater total energy use for this process erode some of the advantage.

Sodium-Reduction Process (Hunter)

The sodium-reduction process, developed first by M.A. Hunter at the General Electric Company in 1906, is described by the following reaction:

$$TiCl_4\,(V) + 4Na\,(L) \rightarrow 4NaCl\,(L, S) + Ti\,(S) \quad (Eq\ 1.7)$$

However, there are steps that occur simultaneously and sometimes competitively:

$$TiCl_4\,(V) + 2Na\,(L) \rightarrow TiCl_2\,(L) + 2NaCl\,(S) \quad (Eq\ 1.8)$$

$$TiCl_2\,(L) + 2Na\,(L) \rightarrow Ti\,(S) + 2NaCl\,(S, L) \quad (Eq\ 1.9)$$

$$Ti\,(S) + TiCl_4\,(V) \rightarrow 2TiCl_2\,(L) \quad\quad (Eq\ 1.10)$$

The United States, United Kingdom, and Japanese companies used the sodium-reduction process. Although differing somewhat in detail between companies, RMI, the only major titanium producer in the United States to use the sodium-reduction process, separated the reaction into two distinct operations (Fig. 1.13) incorporating the following steps:

- Near-stoichiometric portions of sodium and $TiCl_4$ are reacted at a temperature of approximately 230 °C (450 °F) under positive pressure of an inert gas to produce $TiCl_2$ and NaCl.
- Discharge this free-flowing product mixture using a screw conveyor into another vessel called a sintering pot.
- Charge mixed salts and more sodium to near-stoichiometric portions in the sintering pot for the reaction of Eq 1.9.
- Take the reaction to completion at a temperature below 1065 °C (1950 °F) under a positive pressure of inert gas.

- Chip out the product, a mixture of titanium and NaCl in proportions described by Eq 1.7, followed by crushing, leaching, and washing it.
- Dry the product by heating in air or, due to a high portion of reactive fines, by using a centrifuge to remove the bulk of the liquid, followed by vacuum drying.

In the United Kingdom, it is believed that the sodium-reduction process for producing titanium sponge involved a single-vessel operation. This accomplishes the entire reaction of Eq 1.7 in one vessel. It is not known if sodium is all introduced at the beginning, or if it is fed in a program associated with the $TiCl_4$ feed program.

An advantage of the two-step process is the continuous nature of the first step with its low operating temperature, where equipment construction materials offer fewer problems.

Neither of the RMI or U.K. sodium processes is in operation.

Electrolytic Winning of Titanium

Although electrochemical processes have been under investigation for nearly as long as the previously discussed thermochemical processes, early acceptance of the latter as production practices drew attention away from the electrochemical method to the point that only two versions of electrochemical reduction received funding activity in the United States. The early days of titanium production were burdened with significant technical problems, commanding attention of the industry's best investigators. Once the sodium and magnesium production practices were estab-

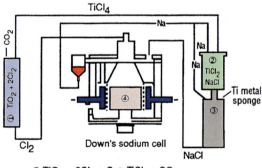

Fig. 1.13 Schematic description of the sodium reduction scheme for producing titanium sponge evolved from the process developed by Hunter

lished, there was little driving force for change, there was no driving need for the superpurity promised by the electrowinning process, and the cyclic nature of the market made investors less than enthusiastic about funding the development of a new process. Therefore, electrowinning development work suffered on-again, off-again cycles, never receiving sustained efforts for scaleup to production-sized operations.

The picture could change somewhat, because evidence indicates that many titanium alloys of the future will require use of high-purity titanium sponge in their manufacture. This, coupled with a projected need to expand the industry's capacity in the future, may rekindle interest in various versions of the electrowinning process.

The D.H. Titanium Company piloted a small plant based on modifications of the original U.S. Bureau of Mines patent, wherein electrowinning is carried out in the fused-salt electrolyte in two distinct steps described by the following cathodic reactions:

$$Ti^{4+} + 2e^- \rightarrow Ti^{2+} \qquad (Eq\ 1.11)$$

$$Ti^{2+} + 2e \rightarrow Ti^0 \qquad (Eq\ 1.12)$$

The anodic reaction is:

$$2Cl^- \rightarrow Cl_2\ (g) + 2e^- \qquad (Eq\ 1.13)$$

The schematic drawing in Fig. 1.14 shows this electrolytic cell with two cathodes and one anode. The cathode at the center in the figure depicts the dichloride generator wherein the electrolytic reduction of the tetrachloride occurs, producing titanium dichloride, which is soluble in the molten salt electrolyte (catholyte). The $TiCl_2$ is then

further reduced to titanium, as in Eq 1.12. This reaction occurs at the second cathode, where titanium attaches for later harvesting using batch or semicontinuous schemes. Note that at the anode there is maintained a diaphragm transparent to Cl^- ions but little else, to ensure separation of anions and cations. An important ingredient to any molten salt electrolysis of titanium is the establishment and maintenance of this diaphragm.

The TIMET Division of Titanium Metals Corporation of America piloted its own version of the molten salt electrowinning process for titanium production. The TIMET electrolytic cell is schematically portrayed at the left in Fig. 1.14. The cathodic reactions combining Eq 1.11 and 1.12 are carried out in a single-cathode volume surrounded by a diaphragm, outside of which are the anodes. Cathodes hanging in the volume established by the diaphragm basket collect the final product of Eq 1.12 and are harvested by withdrawal and disassembly of the entire cathode diaphragm structure.

The harvested crystal, containing residuals of the electrolyte salts from all electrowinning processes, is leached, washed, and dried.

Both of the electrowinning processes demonstrated the capability of producing a large fraction of their product titanium with 90 HB or better.

A characteristic of electrowinning processes for titanium reduction is that there is no need to be associated with a source of sodium or magnesium. The reducing metals are produced, in situ, electrolytically. These processes are inherently more simple operations than the Kroll and sodium-reduction practices and are generally capable of producing higher-quality titanium.

Costs for Production

Plant capital costs for producing ductile titanium sponge are very high. According to government figures, there are only 11 commercial producers of titanium sponge worldwide. Production cost per ton of capacity per year is between $10,000 and $15,000 (approximately 1985 dollars). Initial capital investments for a plant with an annual capacity of 5000 tons (4.5 ´ 10^6 kg) would range from $70 to $100 million dollars (approximately 1985 dollars). Typical requirements and costs for materials and supplies to produce 1 kg (2.2 lb) of titanium sponge metal are shown in Table 1.13.

Although cost of money has increased by a factor of 2.3 (based on the inflation rate of 3% per year), the price of titanium sponge remained relatively stable between 1982 and 2014. This is pri-

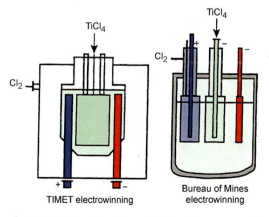

Fig. 1.14 Schematic of electrowinning cells as operated by TIMET and the U.S. Bureau of Mines

marily due to advances made in titanium production technologies and innovations. Relative production cost of titanium production from ore to the mill product is shown in Fig. 1.15.

If the magnesium chloride formed in the reduction reaction in Eq 1.3 is recycled, approximately 0.4 to 1 kg (0.2 to 0.5 lb) of magnesium and 0.5 to 1 kg (1 to 2 lb) of chlorine are required to make up for losses. When sodium is used, recycling is not practical. Power requirements range from 7 to 22 kW · h per 0.5 kg (1 lb) of sponge. This depends on how accountability for lower costs to produce sodium or magnesium is handled. The larger power requirements include power consumed for recycling of the magnesium chloride and for production of magnesium or sodium. These costs are, in general, nonprocess specific, whereas labor and overhead become sensitive to specific processes and can range from $1 to $4 per 0.5 kg (1 lb) of titanium sponge. Therefore, typical direct price for sponge ranged from $3.50 to $6.50 per 0.5 kg (1 lb) in 2014.

Table 1.14 reflects an assessment of the qualitative cost differences among the various sponge production practices.

Table 1.15 highlights some of the areas where electrowinning impacts the cost of titanium production. Particularly noteworthy is the cost for magnesium or sodium, which may be produced in situ by electrowinning processes. Unpublished work at TIMET indicated that there may be operating cost advantages with the electrowinning processes over the common thermochemical processes, but these are thought to be offset somewhat by higher initial capital costs.

When considering a process selection, interpretation of Table 1.15 must take into consideration factors such as availability of low-cost power, availability of affordable labor, and proximity to sources of, and disposal sites for, sodium, NaCl, magnesium, $MgCl_2$, coke, Cl_2, and ore. Local conditions may prove very persuasive.

While the intrinsic capabilities of each of the reviewed processes to produce high-quality titanium are adequate to meet current standards of alloy manufacture, there remain some differences that can influence downstream processing. Table 1.15 offers qualitative comparisons assuming optimized downstream processes for each product.

From a quality standpoint, as indicated by Brinell hardness of melted buttons (which reflects impurity content), the thermochemical processes are potentially identical, with downstream processing, such as melting, requiring the most adjustments. The electrochemical process, on the other hand, offers some potential quality advantage, but this remains unproven on a very large production scale. This latter fact, along with high initial capital requirements, poses distinct barriers to acceptance of the technology.

Extraction Processes Under Development

During the 2000s and early 2010s, a number of innovative extraction processes were explored. A summary of a number of the extraction techniques being developed is shown in Table 1.16 (Ref 1.7,

Table 1.13 Costs to produce 1 kg (2.2 lb) of titanium sponge in 1982 to 2014

2.2 kg (4.8 lb) rutile at $0.18/kg ($0.08/lb)	$0.40
3.5 kg (7.7 lb) chlorine at $0.08/kg ($0.04/lb)	$0.28
1.25 kg (2.75 lb) magnesium at $1.36/kg ($0.61/lb) (or 2.1 kg, or 4.6 lb, sodium)	$1.69
0.30 kg (0.66 lb) petroleum coke at $0.09/kg ($0.05/lb)	$0.03
Miscellaneous	$0.10
Total	$2.50

Table 1.14 Cost differences among sponge production processes

Process	Operating cost	Capital cost
Kroll—vacuum distillation	Moderate	High
Kroll—leach	Moderate	Medium
Sodium	Moderate	Medium
Electrolytic	Low	High

Table 1.15 Qualitative comparison of titanium production processes

Process	Quality, HB	Residual volatiles	Fines content
Kroll—vacuum distillation	Good	Low	Low
Kroll—leach	Good	High	Moderate
Sodium	Good	Moderate	High
Electrolytic	Excellent	Moderate	High

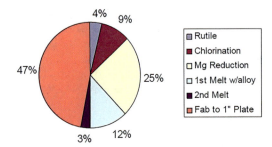

Fig. 1.15 Relative cost of titanium plate production from rutile. Source: Ref 1.6

- 4%
- 9%
- 47%
- 25%
- 3%
- 12%
- ☐ Rutile
- ■ Chlorination
- ☐ Mg Reduction
- ☐ 1st Melt w/alloy
- ■ 2nd Melt
- ☐ Fab to 1" Plate

1.8). Further details of a number of these processes and other innovative processes are given in Fig. 1.16 to 1.24. In many cases, the processes produce a powder that can be used to dramatically reduce the number of steps required to fabricate a titanium shape. (See Chapter 8, "Melting, Casting, and Powder Metallurgy," for more details.)

Table 1.16 Titanium extraction processes

Techniques	Comments
FFC	Oxide, electrolytic molten $CaCl_2$
MER	Oxide, electrolytic
SRI	Fluidized bed H_2 reduction of $TiCl_4$
BHP (Billiton, Australia)	Oxide electrolytic, prepilot plant
Idaho Ti	Plasma quench, chloride
Ginatta, Italy	Electrolytic, chloride
OS (Ono, Japan)	Electrolytic/calciothermic oxide
MIR, Germany	Iodide reduction
CSIR, South Africa	Electrolysis of oxide
Okabe-1 (Tokyo, Japan)	Oxide, reduction by calcium
Okabe-11 (Tokyo, Japan)	Oxide, calcium vapor reduction
Vartech, Idaho	Oxide, calcium vapor reduction
Northwest Institute for Non-Ferrous Metals	Innovative hydride-dehydride
CSIRO, Australia	Chloride, fluidized bed, sodium
Armstrong/1TP	Chloride, continuous reduction with sodium
DMR	Aluminothermic rutile feedstock
MIT	Oxide, electrolysis
QIT/Rio Tinto	Slag, electrolysis
Tresis	Argon plasma, chloride
Dynamet Technology	Low-cost feedstock

The Armstrong/International Titanium Powder method (Fig. 1.16, 1.17) is continuous and uses molten sodium to reduce titanium tetrachloride, which is injected as a vapor. The resultant powder does not need further purification and can be used directly in the conventional ingot approach. The powder is most efficiently used in the powder metallurgy technique. A range of alloys can be produced (including Ti-6Al-4V) as a high-quality, homogeneous product suitable for use in many applications. International Titanium Powder currently operates a research and development facility in Lockport, Illinois, and has broken

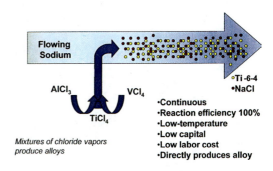

Fig. 1.16 The Armstrong/International Titanium Powder process. The $TiCl_4$ is directly injected in a vapor form, resulting in the reduction of $TiCl_4$ to commercially pure titanium. Courtesy of K. Akhtar, Armstrong/Crystal, Sept 2013

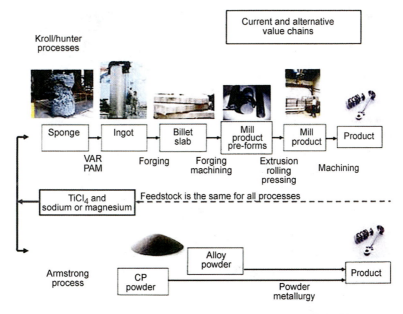

Fig. 1.17 Schematic of how the Armstrong/International Titanium Powder process can simplify the fabrication of titanium shapes. The Armstrong process results in commercially pure (CP) titanium, which can be combined or directly used in powder metallurgy processes to produce a final product. VAR, vacuum arc remelting; PAM, plasma arc melting. Courtesy of K. Akhtar, Armstrong/Crystal, Sept 2013

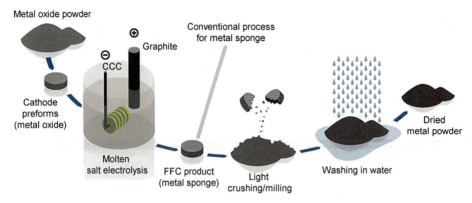

Fig. 1.18 The FCC Cambridge process. Courtesy of D. Vaughn, Metalysis Corporation, Oct 2013

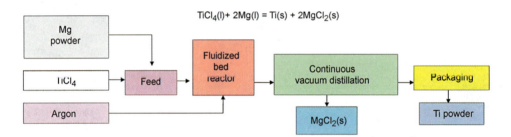

Fig. 1.19 The CSIRO TiRO process. Courtesy of J. Barnes, CSIRO, Oct 2013

ground on a 4 million pound per year expansion in Ottawa, Illinois, to produce both commercially pure titanium and Ti-6Al-4V alloy powder.

In the FFC Cambridge approach (Fig. 1.18), titanium metal is produced at the cathode in an electrolyte (generally $CaCl_2$) by the removal of oxygen from the cathode. This technique enables the direct production of alloys such as Ti-6Al-4V at a cost that could be less than the product of the conventional Kroll process. The process is being developed by Metalysis in South Yorkshire, United Kingdom.

The CSIRO technique (Fig. 1.19, 1.20) builds on the fact that Australia has some of the largest mineral and sand deposits in the world. In this approach, cost-effective commercially pure titanium is produced in a continuous fluidized bed in which titanium tetrachloride is reacted with molten magnesium (the TiRO process). Continuous production of a wide range of alloys, including aluminides and Ti-6Al-4V, has been demonstrated on a large laboratory scale. The commercially pure titanium powder produced has been used to fabricate extrusions, thin sheet by continuous roll consolidation, and cold-sprayed

Fig. 1.20 Titanium metal powder suitable for use in near-net shape manufacturing, which produces components that are close to the finished size and shape. Courtesy of J. Barnes, CSIRO, Oct 2013

complex shapes including ball valves and seamless tubing. CSIRO built a pilot reactor with production capacity of 2 kg/h (4.5 lb/h) of titanium,

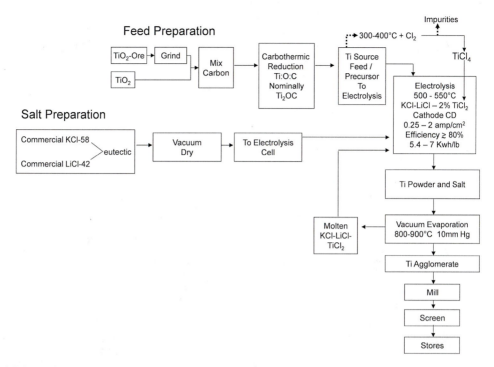

Fig. 1.21 The MER electrolytic process. Courtesy of J. Withers, MER, Nov 2013

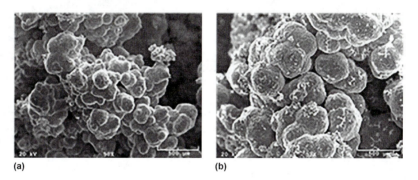

(a)   (b)

Fig. 1.22 Examples of electrowon titanium particulate by the MER process. Courtesy of J. Withers, MER, Nov 2013

the design of which was suitable for scaleup to commercial scale.

The MER approach (Fig. 1.21, 1.22) is an electrolytic method that uses a composite anode made of TiO_2, a reducing agent, and an electrolyte, mixed with fused halides. Projections are for titanium production at a significantly lower cost than the conventional Kroll process.

The CSIR method (Fig. 1.23) essentially aims to directly produce titanium powder in a continuous metallothermic $TiCl_4$-reduction process in a molten salt reaction medium. In principle, it is possible

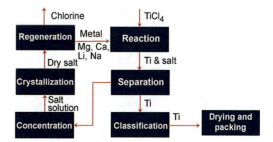

Fig. 1.23 Block flow diagram of the CSIR process. Courtesy of D. Van Vuuren, CSIR, Sept 2013

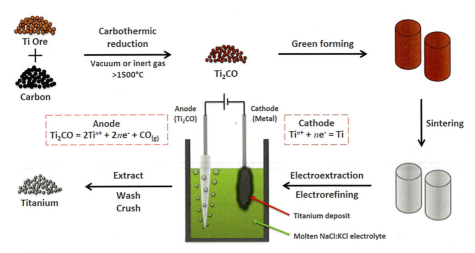

Fig. 1.24 The Chinuka process refines as well as reduces. Courtesy of D. Fray, University of Cambridge, Sept 2013

to use any one of the alkali or alkali earth metals as the reducing metal, but there are advantages and disadvantages associated with each. This is in contrast to what the MER Corporation is doing by integrating the $TiCl_4$ production at the anode and reducing metal production at the cathode. The CSIR method uses a more conventional approach of producing the $TiCl_4$ and reducing metal prior to introducing the reactants in a reactor system.

The Chinuka technique (Fig. 1.24), unlike the FFC process, refines as well as reduces. The process was devised to treat ores with a few percent of other oxides, containing fines and calcium oxide. Some of the impurities remain with the anode or form a sludge, while others build up in the electrolyte or evaporate.

ADMA (Fig. 1.25) produces non-Kroll-process titanium sponge cooled in a hydrogen atmosphere rather than conventional inert gas. The hydrogenated sponge is easily crushed and, in the hydrogenated condition, is compacted to a higher density than conventional low-hydrogen sponge. Subsequent hydrogen removal is easily accomplished with a simple vacuum anneal. The remnant chloride content of the hydrogenated sponge is reportedly at low levels (helping to avoid porosity and enhancing weldability). The sponge is easily ground into powder for the production of titanium components by way of blended elemental powder metallurgy. Titanium alloys produced by this approach are effectively heat treated to the strength levels achieved in ingot metallurgy solution-treated and aged alloys of identical composi-

Fig. 1.25 Hydrogenated titanium sponge produced by the ADMA Products non-Kroll process. Courtesy of V. Moxson, ADMA Products, Sept 2013.

tion. ADMA hydrogenated titanium powder (TiH_2) has been produced in a laboratory-scale titanium powder manufacturing unit (for more details see Chapter 8 "Melting, Casting, and Powder Metallurgy").

Trends in Sponge Production

Figures 1.26 and 1.27 show recent trends in worldwide titanium sponge production, with a major increase in Chinese production for nonaerospace applications. The U.S. capabilities are mainly for aerospace domestic use, with TIMET producing 12.6 kt/year and ATI at 21 kt/year in 2013.

Traditionally, sponge production numbers and aircraft deliveries have followed similar trend lines. A comparison of sponge production and air-

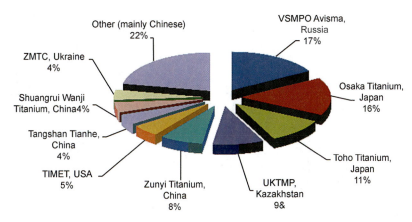

Fig. 1.26 Pie chart of worldwide major titanium sponge manufacturers. Courtesy of P. Dewhurst, Roskill Information Services, Oct 2013

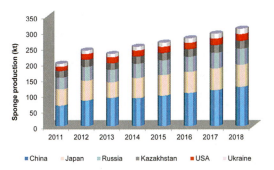

Fig. 1.27 Current and predicted global sponge production. Courtesy of P. Dewhurst, Roskill Information Services, Oct 2013

craft deliveries (Fig. 1.28) does show that the linkage is being maintained.

Trends in Mill Products Shipments

Projected mill products shipments are shown in Tables 1.17 and 1.18 and Fig. 1.29. (The global output is predicted to reach 310 kt/year in 2018, with aerospace-grade output at the 100 kt/year level.)

Summary

Titanium was first discovered in the 1790s, but pure titanium was not produced until the early 1900s. The U.S. Bureau of Mines began successful production of titanium in 1946 using the magnesium-reduction process developed by W.J. Kroll. From 1947 through the 1950s, many titanium-production ventures were launched in the United States, the United Kingdom, continental Europe, the former Soviet Union, and Japan. The titanium industry market, although cyclic in nature, has grown overall.

The most economical ore for production of titanium metal is rutile, because it contains the highest content of TiO_2. However, the mineral ilmenite (iron titanate) is found on every continent and is the most abundant source of the minerals that contain titanium.

All commercial methods for reducing titanium oxides to metal have three basic steps: chlorination (to produce titanium tetrachloride), reduction, and purification. The two basic methods for reducing titanium tetrachloride are thermochemical and thermoelectrical. Two thermochemical processes (Kroll, which uses magnesium reduction, and Hunter, which uses sodium reduction) are used commercially because of their relative cost advantages. Electrowinning has been done in demonstration/pilot-scale projects since the late 1980s but has not achieved commercial success. The purification that follows reduction involves removal of residual salts and unconsumed reactants. The result is titanium sponge, so called because of its open-pore-type structure. A number of promising developmental processes for production of titanium are being evaluated.

Titanium has a density between that of aluminum and steel. It is a highly reactive metal and depends on a thin protective oxide film for its corrosion resistance. Other properties include a high melting point, low thermal expansion, low electrical conductivity, high fatigue strength, and low magnetic susceptibility.

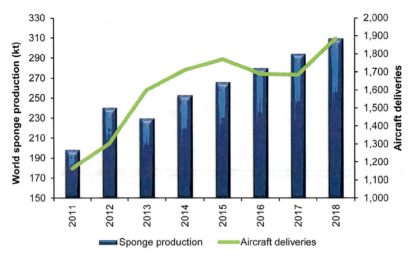

Fig. 1.28 Comparison of forecasts for sponge production and aircraft deliveries. Courtesy of Roskill Information Services .

Table 1.17   U.S. supply and disposal of titanium mill products, 2003 to 2012 (in tons)

	2003	2004	2005	2006	2007	2008	2009	2010	2011	2012
Production	21.3	26.3	30.9	36.1	38.2	39.7	31.9	36.3	40.5	39.8
Exports	6.5	8.3	11	13.5	15.8	19.2	12.6	14.9	24.9	23.5
Imports	3.6	3.6	3.9	5.8	5.3	7.2	5.1	7.1	5.6	6.9
Reported net shipments: plate, sheet, strip	5.2	7.9	10	8.9	14.7	15.6	11.4	...	...	...
Billet	5.4	6.4	8.1	8.5	12.4	12.3	9.5	19.7	16.8	16.7
Rod and bar	4.5	4.6	5.2	4.2	5.4	...	...	...	...	...
Other	0.6	0.2	0.6	0.5	0.7	6.9	6.7	18.5	28.7	23.0
Castings	0.5	...	...	...	1.8	...	...	...	...	...
Total	16.1	19.1	23.8	22.1	35	34.8	27.6	38.2	45.5	39.7

Table 1.18   World forecast demand for titanium mill products in 2018 (in kilotons)

	European Union and North America	China	Rest of world	Total
2012				
Industrial applications	16	44	27	87
Aerospace	45	5	10	59
Consumer and other	12	4	4	19
Total	72	53	40	165
Average annual growth rate, %				
Industrial applications	2.5	8.0	4.5	6.0
Aerospace	2.5	5.0	4.0	3.0
Consumer and other	2.0	5.0	2.0	2.7
Total	2.4	7.5	4.2	4.6
2018				
Industrial applications	19	70	35	123
Aerospace	52	7	12	71
Consumer and other	13	5	4	22
Total	83	82	51	216

Courtesy of Roskill Information Services

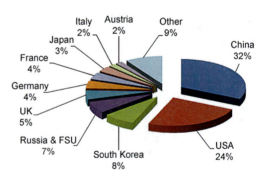

Fig. 1.29 World estimated division of consumption of titanium mill products by principal countries in 2012. Courtesy of P. Dewhurst, Roskill Information Services, Oct 2013

REFERENCES

1.1   F.H. Froes et al., Ed., *Titanium Technology: Present Status and Future Trends*, Titanium Development Association, 1985

1.2 S. Abkowitz, The Emergence of the Titanium Industry and the Development of the Ti-6Al-4V Alloy, *JOM Monograph Series,* Vol 1, TMS, Warrendale, PA, 1999

1.3 "Titanium Statistics and Information," U.S. Geological Survey, U.S. Department of the Interior, http://minerals.usgs.gov/minerals/pubs/commodity/titanium/

1.4 M.A. Imam, K. Housley, and F.H. Froes, *Titanium and Titanium Alloys,* Kirk-Othmer, 2011

1.5 *ASM Metals Reference Book,* ASM International, 1993

1.6 "Summary of Emerging Titanium Cost Reduction Technologies," a study performed for U.S. Department of Energy and Oak Ridge National Laboratory, Subcontract 4000023694

1.7 E.H. Kraft, "Opportunities for Low Cost Titanium in Reduced Fuel Consumption, Improved Emissions, and Enhanced Durability Heavy-Duty Vehicles," Oak Ridge National Laboratory, July 2002

1.8 E.H. Kraft, "Summary of Emerging Titanium Cost Reduction Technologies," Oak Ridge National Laboratory, Dec 2003

SELECTED REFERENCES

- R. Boyer, E.W. Collings, and G. Welsch, Ed., *Materials Properties Handbook: Titanium Alloys,* ASM International, 1994
- E.W. Collings, *The Physical Metallurgy of Titanium Alloys,* American Society for Metals, Metals Park, OH, 1984
- M.J. Donachi, *Titanium: A Technical Guide,* 2nd ed., ASM International, 2000
- H. Kuhn and D. Medlin, Ed., *Mechanical Testing and Evaluation,* Vol 8, *ASM Handbook,* ASM International, 2000
- P. Lacombe, R. Tricot, and G. Beranger, Ed., *Proceedings of the Sixth International Conference on Titanium* (Nice, France), Metallurgical Society of AIME, Warrendale, PA, 1988
- G. Lutjering and J.C. Williams, *Titanium,* Springer, 2003
- G. Lutjering, U. Zwicker, and W. Bunk, *Proceedings of the Fifth International Conference on Titanium,* Metallurgical Society of AIME, Warrendale, PA, 1984
- *Proceedings of the Seventh International Conference on Titanium* (San Diego, CA), Metallurgical Society of AIME, Warrendale, PA, 1992
- *Proceedings of the Eighth International Conference on Titanium* (San Birmingham, U.K.), Metallurgical Society of AIME, Warrendale, PA, 1995
- *Proceedings of the Ninth International Conference on Titanium,* Metallurgical Society of AIME, Warrendale, PA, 2000
- *Proceedings of the Tenth International Conference on Titanium,* Metallurgical Society of AIME, Warrendale, PA, 2004
- *Proceedings of the Eleventh International Conference on Titanium,* Metallurgical Society of AIME, Warrendale, PA, 2008
- *Proceedings of the Twelfth International Conference on Titanium,* Metallurgical Society of AIME, Warrendale, PA, 2012

CHAPTER 2

Introduction to Solidification and Phase Diagrams*

THIS CHAPTER DISCUSSES the structures, phases, and phase transformations observed in metals and alloys as they solidify and cool to lower temperatures (Ref 2.1–2.3). It also shows how this information is presented diagrammatically and is used in practical applications. This chapter introduces common terminology, elementary structures, general theories on solidification and alloying, and the construction and practical application of binary and ternary phase diagrams. Specific Ti-*X* phase diagrams (Ref 2.4) are presented and discussed in Chapter 3, "Principles of Alloying Titanium," in this book.

Atoms

Elementary Particles. All materials are composed of one or more of the 118 known elements (Ref 2.3). All of these elements have uniquely different structures and properties. The smallest complete unit of an element is called an atom. It consists of a central core or nucleus (consisting of the elementary particles neutrons and protons; the proton number is the same number as the number of electrons orbiting the nucleus), which is surrounded by a cloud of one or more rapidly moving electrons. Electrons are extremely small atomic particles having a specific negative charge. The total charge of the electrons in a complete atom is equal to but opposite that of the nucleus. For example, a titanium atom has 22 electrons with negative charges in orbit around a central nucleus that contains 22 protons having a positive charge.

Electrons, and especially the outermost electrons, determine the nature of the bonds between similar and dissimilar atoms. As a consequence, they affect crystal structures and the basic chemical, mechanical, electrical, thermal, optical, and certain other properties. Thus, the electronic structure accounts for most of the different characteristics observed among the various elements and compounds (molecules) formed with the elements.

The smallest atom is that of hydrogen. The complete or neutral atom has a single electron, an atomic number of one (1), and an atomic mass of 1.0080 amu. Atomic mass unit (amu), formerly referred to as atomic weight, is usually defined as a mass one-twelfth ($\frac{1}{2}$) of the atomic mass of carbon-12, which has an amu of 12. Larger atoms have correspondingly more electrons, larger nuclei, and greater masses. For example, the neutral titanium atom has 22 electrons, and therefore, it has an atomic number of 22. Its atomic mass is 47.90 amu, with most of the mass concentrated in the relatively large nucleus.

States of Matter. Atoms interact with one another to form different states or phases. The most common states, or forms of matter, are solids, liquids, and gases, depending on the temperatures and pressure to which the material is exposed. Thus, if a solid is heated, its atoms vibrate more energetically, atomic bonds are weakened, and the material expands as interatomic distances increase. When materials are heated to their melting points, solid-state bond strengths are greatly reduced, atomic arrangements become almost completely random, and a liquid phase appears. With additional heating, bonds are weakened further and a greater percentage of the atoms and molecules have sufficient energy to escape to the vapor phase. At the

---

*Adapted and revised from Howard B. Bomberger and Francis H. Froes, *Titanium and Its Alloys*, ASM International.

higher temperatures, significant vapor pressures can be measured. When boiling points are reached, all atoms and stable molecules have sufficient energy to escape to the vapor phase. Atomic and molecular arrangements are completely random in the vapor phase. Examples of melting and boiling points of some metals are given in Table 2.1 (Ref 2.1).

On cooling, energy is removed, vapors condense to the liquid phase, and, at lower temperatures, liquids transform to the solid phase. These processes are completely reversible for stable materials.

Solidification of Metals

Nucleation. Liquid metals behave like other homogeneous liquids. As the temperature of a liquid is lowered close to its melting (also freezing) point, interatomic forces become stronger and small crystalline units, or clusters of atoms, form. These small units or nuclei, consisting of several to thousands of atoms, are not stable above the melting point. However, as the temperature of the liquid is lowered below the equilibrium freezing temperature, the probability increases that atoms (and molecules) will form clusters greater than the critical, or stable, size (Ref 2.1, 2.3). (A molecule consists of two or more atoms chemically combined and represents the smallest possible unit of a chemical compound.) Such initial undercooling or supercooling is an important step for the formation of stable nuclei in high-purity liquids. For example, in some high-purity metals the undercooling may be as great as a few hundred degrees Celsius. However, initiation of freezing can be greatly assisted by naturally occurring foreign particles and by container surfaces that serve as stable nuclei. Thus, in most commercial operations the presence of solid nucleating agents can limit undercooling to only a few degrees.

Crystal Growth. After stable, solid nuclei form, crystal or grain growth occurs as more atoms from the liquid become attached to the solid (Ref 2.1). However, the nature of growth depends on how the heat is removed. First, the specific heat (the heat required to change the temperature of a unit weight one degree) must be removed. This is then followed by the removal of the latent heat of fusion (the heat released as the liquid phase changes to a solid without a temperature change). Most of the heat is removed from the liquid and solid phases by radiation and conduction. The manner in which heat is removed affects the growth mechanism of crystals.

If a nucleated liquid metal is cooled slowly, but in a manner in which the liquid is actually at a higher temperature than the solid-liquid interface, the temperature profile will be as indicated in Fig. 2.1. In this case, a smooth, planar interface surface will develop (Ref 2.1). Planar growth is favored because rapid growth of individual grains or protrusions is limited by the higher temperatures within the liquid.

However, if a liquid metal has little or no nucleation, the liquid tends to undercool below the freezing temperature. In this case, rapid growth of large dendritic grains normal to the interface is encouraged, as indicated by Fig. 2.2. Treelike dendrites grow as the latent heat of fusion is conducted into the undercooled liquid and raises its temperature toward the freezing temperature. This process also encourages the formation of dendrite "arms," which assist in removing heat from the liquid. Dendritic grain growth proceeds in this manner until the undercooled liquid warms to the freezing temperature. The remaining liquid solidifies by planar growth.

Table 2.1 Melting and boiling points of some metals

Metal	Melting point		Boiling point	
	°C	°F	°C	°F
Aluminum	660	1220	2467	4473
Iron	1536	2797	3000	5432
Nickel	1453	2647	2730	4946
Titanium	1668	3034	3260	5900
Vanadium	1900	3452	3000	5432
Zirconium	1852	3366	3580	6476

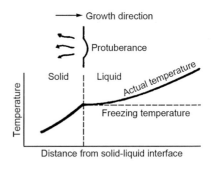

Fig. 2.1 Smooth, planar crystal growth is observed at the solid-liquid interface when liquid metals are not undercooled. Growth of long individual grains and protuberances into the liquid are prevented whenever the liquid is above the freezing point. Arrows indicate the direction of heat flow. Reprinted with permission from Ref 2.5

Cast Structures. Solidification of a cast ingot is depicted in Fig. 2.3 (Ref 2.1, 2.3). The mold wall consists of a heat sink and also numerous stable nuclei from which grains can grow. The largely dendritic grain growth is perpendicular to the mold wall, which is the direction of heat removal. Growth is most rapid in those grains having favorable orientations. As a result, the long, parallel columnar grains generally have the same crystal orientation, and such materials have anisotropic (directional) properties.

After the specific heat is removed from the remaining liquid, stable nuclei form and grow in the liquid. This accounts for the smaller, but equiaxed, uniform-sized grains located in the center of the solid ingot cross section in Fig. 2.3.

The amount of undercooling in a liquid increases with increasing cooling rates. Because of this, the number of stable nuclei increases and they produce more, but finer, grains. Thus, dendritic grain sizes decrease with increasing cooling rates. Rapid solidification is used to obtain extremely fine structures, and, in special cases,

crystallization of some alloys can be avoided. Such noncrystalline alloys have a glasslike, or amorphous (random), structure.

Alloy Solidification. The solidification steps for alloys are similar to those described previously except that as most alloys freeze, temperatures and compositions vary over a wide range (Ref 2.1, 2.3). Thus, the first materials to solidify, including stable nuclei and the central cores of dendrites, have higher solidification temperatures and different compositions than the last liquid to freeze. The last material to freeze has the lowest melting temperature and is found between the dendrite branches and in grain boundaries. The solidification steps involving nucleation and grain growth of a uniformly cooled alloy are depicted in Fig. 2.4.

Crystal Structures. As liquid metals and many nonmetals solidify, atoms are arranged spontaneously in a three-dimensional lattice structure in which the atoms occupy specific positions in the

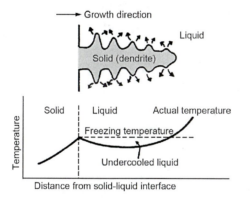

Fig. 2.2 Rapid dendritic growth is observed in liquids that have been undercooled below the freezing temperature. The latent heat of fusion, released by the solidificatio process, raises the temperature of the interface, as shown.

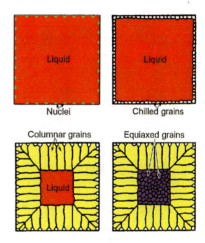

Fig. 2.3 Solidification of ingots and large castings involves nucleation, growth of small surface grains, preferred growth of columnar grains, and finally, growth of smaller equiaxed grains. Reprinted with permission from Ref 2.5

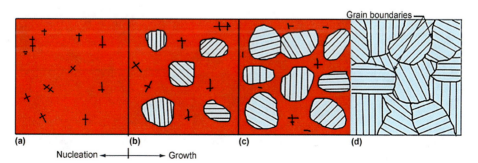

Fig. 2.4 Freezing of a uniformly cooled alloy liquid involves nucleation, grain growth, and variable composition within each grain. Each dendrite develops branches along three sets of axes, each at 90° to the other. Only two sets of axes are shown here; the third set is at right angles to the illustration.

geometric pattern. The dimensions and the crystalline forms of the unit cells formed are characteristic of the given material (Ref 2.3). These structures are imposed by the bond energies and angles of the atoms involved.

All crystal structures fall within seven basic systems and 14 lattice types. The most common crystal structures found among metals are face-centered cubic (fcc), body-centered cubic (bcc),

and hexagonal close-packed (hcp). Figure 2.5 illustrates unit cells of these structures and the lattice positions occupied by atoms. Table 2.2 lists common metals and their crystal forms.

A number of metals and nonmetals can exist in more than one solid phase, depending largely on the temperature and composition. For example, as unalloyed titanium solidifies at approximately 1668 °C (3034 °F), the atoms arrange themselves

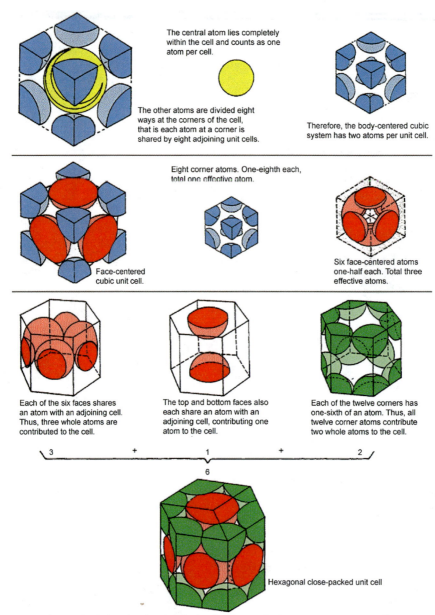

The central atom lies completely within the cell and counts as one atom per cell.

The other atoms are divided eight ways at the corners of the cell, that is each atom at a corner is shared by eight adjoining unit cells.

Therefore, the body-centered cubic system has two atoms per unit cell.

Eight corner atoms. One-eighth each, total one effective atom.

Face-centered cubic unit cell.

Six face-centered atoms one-half each. Total three effective atoms.

Each of the six faces shares an atom with an adjoining cell. Thus, three whole atoms are contributed to the cell.

The top and bottom faces also each share an atom with an adjoining cell, contributing one atom to the cell.

Each of the twelve corners has one-sixth of an atom. Thus, all twelve corner atoms contribute two whole atoms to the cell.

$$\frac{3 \quad + \quad 1 \quad + \quad 2}{6}$$

Hexagonal close-packed unit cell

The hcp cell is further refined by subdividing into 3 sections, so that the total atoms per unit cell is 6/3 or 2

Fig. 2.5 Unit cells of the most common crystal structures found in metals: body-centered cubic (top), face-centered cubic (middle), and hexagonal close-packed (hcp) (bottom)

in the bcc structure, which persists until the metal is cooled to 885 °C (1625 °F), where the atoms are rearranged to form the hcp structure. Such transformations occur spontaneously on heating and cooling as the atoms move to their lowest possible energy state at a given temperature. The temperatures at which phase transformations occur can vary greatly with alloy content.

Grains (crystals) and grain boundaries contain a large number of defects in the form of atom vacancies and dislocations. Such defects have very important effects on the deformation, strength, and solid-state diffusion of metals.

Alloying

Properties of most metals can be improved by alloying with one or more other elements. Such properties include strength, hardness, toughness, and corrosion resistance. It is important to know how alloying elements affect the microstructures of materials, because properties are influenced by the structures obtained. An important variable is solid solubility.

Solid solutions are obtained if two or more elements are dispersed throughout the same crystal structure, forming a single phase (Ref 2.3). They are similar to liquid solutions in that the different atoms can be distributed quite uniformly throughout the same solid phase. For example, if liquid copper and liquid nickel are mixed, a single liquid phase is formed. If this uniform mixture is solidified, the crystals will consist of just one solid phase in which the copper and nickel atoms are fairly well distributed throughout the same lattice structure. Copper and nickel are completely soluble in each other and therefore form a complete series of solid solutions. In some cases, solid solubility is limited or is incomplete. An example of this is salt in water. When such liquid solutions are frozen, almost all of the salt is rejected, forming a second solid phase, because salt has very little solubility in ice. Many examples of limited solid solubility are observed in metal alloys and are discussed later in this chapter.

Types of Solid Solubility. Two types of solid solubility are substitutional and interstitial. Extensive substitutional solid solubility occurs if atomic sizes, crystal structures, and chemical factors are very similar. In this case, solute atoms occupy the same lattice positions used by the solvent. However, if solute atoms are very small, they locate in the interstices (lattice spaces) between the larger solvent atoms and form limited interstitial solid solutions. The relatively small hydrogen, oxygen, nitrogen, and carbon atoms dissolve interstitially in titanium and are referred to as interstitial elements. Examples of the two forms of solid solubility are given in Fig. 2.6.

Alloying Factors. Extensive substitutional solid solutions are possible when certain conditions (the Hume-Rothery rules, Ref 2.3) are met. The solute and solvent atoms must have similar sizes with variations of no more than 15%, they

Table 2.2 Solid phases of some metals

Metal	Solid phase(a)	Temperature ranges for phase stability	
		°C	°F
Aluminum	fcc	<660	<1220
Iron	bcc	1400–1535	2552–2795
	fcc	912–1400	1674–2552
	bcc	<912	<1674
Nickel	fcc	<1455	<2651
Titanium	bcc	885–1668	1625–3034
	hcp	<885	<1625
Vanadium	bcc	<1890	<3434
Zirconium	bcc	865–1850	1589–3362
	hcp	<865	<1589

(a) fcc, face-centered cubic; bcc, body-centered cubic; hcp, hexagonal close-packed

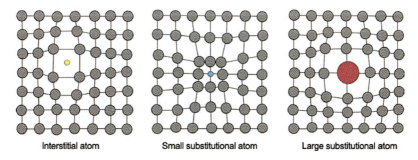

Interstitial atom   Small substitutional atom   Large substitutional atom

Fig. 2.6 Types of solid solution. An interstitial atom occupies a space between the atoms of the crystal lattice. Substitutional atoms replace or substitute for an atom in the crystal structure.

must have the same crystal structures, and they must have similar chemical characteristics, including valence and electronegativity. Valence refers to the number of electrons present in, or absent from, an atom that can participate in chemical reactions and atomic bonding. Electronegativity describes the relative tendency of an atom to gain an electron. Differences in this property between elements are an indication of their tendency to combine chemically. Interstitial solubility is possible when solute atoms are relatively small, with diameters less than 60% that of the solvent atoms.

Figure 2.7 gives information on elements that have favorable sizes for solubility in solid titanium. Table 2.3 provides other important data on factors that also affect solid solubility of elements in titanium.

Intermetallic Phases. When alloying additions exceed the solid solubility, a second phase appears. If the solute and solvent have grossly dissimilar chemical characteristics, solubility is restricted, and the precipitates tend to be intermetallic compounds. Most intermetallic compounds have high hardness and strength and are not ductile. Controlled precipitation of such phases can

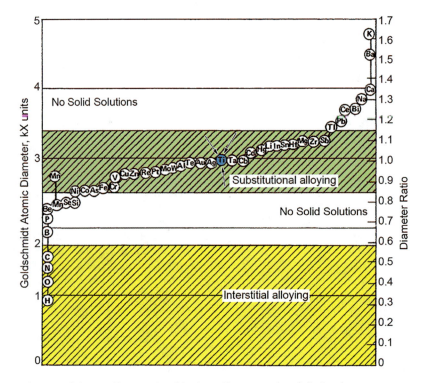

Fig. 2.7 Atomic diameters of elements. The atom size of titanium with respect to that of alloying elements

Table 2.3 Data useful for predicting solid solubility of metals in titanium metal

Metal	Atomic diameter(a), Å	Crystal form(b)	Common valences	Electronegativity factor, X	Approximate maximum solid solubility, wt%	
					Alpha	Beta
Aluminum	2.82	fcc	3	1.5	14	2
Iron	2.52	fcc, bcc	2, 3	1.7	0.2	25
Molybdenum	2.80	bcc	6, 3, 5	1.6	0.8	100
Silicon	2.82	dc	4	1.8	<0.1	3
Titanium	2.93	bcc, hcp	3, 4	1.6	...	...
Vanadium	2.71	bcc	4, 5	1.6	2	100
Zirconium	3.19	bcc, hcp	4	1.4	100	100

(a) An angstrom is a small unit of length frequently used for atomic diameters; it is equivalent to 10^{-8} cm. (b) fcc, face-centered cubic; bcc, body-centered cubic; dc, diamond cubic (dc structures are similar to the fcc structure, with additional atoms located on the tetrahedral positions [1/4, 1/4, 1/4] of the cube); hcp, hexagonal close-packed

provide useful strengthening effects in some alloy systems. The compounds Ti_3Al and $TiAl$ have been extensively evaluated as "stand-alone" materials for structural use (see discussions of these materials in subsequent chapters). Examples of intermetallic phases or compounds found in titanium alloys are given in Table 2.4.

Phase Diagrams

Alloying (solute) elements influence solid-to-liquid and solid-to-solid transformations. These effects (solubility limits) and the distinctly different structures (phases) observed can be shown conveniently in a graphic representation of entire alloy systems (Ref 2.4). In these diagrams, it is customary to show only the effect of temperature and composition. Pressure is rarely shown in metal systems because it is considered to have little or no influence.

These displays are usually referred to as phase diagrams, constitution diagrams, and sometimes as equilibrium diagrams, because an effort is

made to show the phases that prevail under equilibrium or near-equilibrium conditions. *Equilibrium* is a term that implies a state of complete rest and stability in which no further changes can occur without the addition of energy from outside the system. Thus, a system at equilibrium is at its lowest energy level and is said to be stable. However, some special diagrams do show intermediate and metastable phases.

Most phase diagrams involve only two (binary) or three (ternary) components. Compositions in binary systems are frequently shown in both weight and atomic percents, and temperatures are usually given on the Celsius and Fahrenheit scales.

Alloys Completely Soluble in the Solid State. Figure 2.8 illustrates a simple binary system composed of similar metals, A and B, having complete solid solubility. Therefore, it is referred to as an isomorphous system. The upper curved line is called the liquidus and the lower one is the solidus. Liquid and solid phases coexist between these two lines. Many alloy systems have diagrams similar to the one in Fig. 2.8. A number of titanium binary alloy systems are similar, except they also show solid-state transformations (which is discussed later in this chapter).

Solidification of liquid alloys of this type to achieve solid-state equilibrium involves nucleation and growth of alloy crystals in the liquid and diffusion (atomic movement), resulting in a uniform (homogeneous) composition in each crystal if sufficient time at temperature is allowed.

Table 2.4 Some intermetallic compounds found in titanium alloys

Ti_3Al, $TiAl$, $TiAl_3$	Ti_2Co	Ti_2Ni
$TiBe$	Ti_2Cu	Ti_5Si_3
$TiCr_2$	$TiFe$	Ti_3Sn

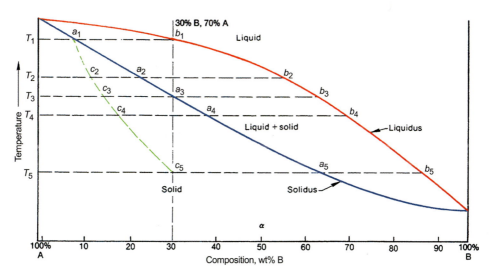

Fig. 2.8 Binary alloy phase diagram of metals A and B that are completely soluble in each other in the solid state. Liquidus line (red) shows the liquid composition at various temperatures as the alloy is slowly cooled (equilibrium is maintained). Solidus line (blue) shows the composition of alloys that crystalize as temperatures are slowly cooled. Dashed green line shows crystal composition if the melt is cooled rapidly (nonequilibrium conditions).

Consider, for example, an alloy of 30% B and 70% A under equilibrium conditions (Fig. 2.8). If the alloy is cooled very slowly from its liquid state to temperature T_1, which is just slightly under the liquidus, solid crystals of composition a_1 form. Liquid at temperature T_1 has the composition b_1. Thus, for this alloy at T_1, under equilibrium conditions, two phases exist as a liquid of composition b_1 and a solid of composition a_1.

If the melt is cooled to temperature T_2, the solid has a composition a_2 and liquid of composition b_2 remains. This process occurs continuously as the melt is cooled. Finally, at temperature T_3 the last bit of liquid to solidify has the composition b_3 and the solid has the overall composition a_3 (70% A and 30% B).

Thus, as the melt is cooled under equilibrium conditions, the liquid composition follows the liquidus line and the solid follows the solidus line.

However, if the melt is cooled at a moderate rate, equilibrium conditions do not exist. Consider the same melt of composition 30% B and 70% A but cooled under nonequilibrium conditions from the liquid state. At temperature T_1, solid precipitating from the liquid has a composition slightly richer in metal A than it would be if equilibrium conditions were maintained. Upon cooling to temperature T_2, the average composition of the solidifying solid is that of composition c_2 instead of equilibrium composition a_2. Because the solid is now richer in metal A, the liquid is richer in metal B, which means it has a lower-than-equilibrium solidus temperature. Further cooling to temperature T_3 produces a solid of composition c_3, and the liquid which should be converted to solid does not solidify until a much lower temperature is reached. Cooling to temperature T_4 produces a solid of composition c_4. Finally, the melt solidifies completely at temperature T_5.

The gradual stepwise growth of alloy crystals under nonequilibrium conditions is illustrated in Fig. 2.9. Successive "shells" of different, but gradual, solid composition form when there is not sufficient time for diffusion to make the composition uniform. In such cored structures, the composition of the grains varies from the center to the edge.

Cored structures melt at lower temperatures than the same alloy cooled slowly. In the example, if originally cooled under equilibrium conditions (very slow cooling), the alloy 30% B and 70% A on reheating begins to melt at temperature T_3. If the same alloy originally were cooled rapidly, when reheated, it will start to melt at some lower temperature, T_5.

Cored structures often exhibit low strength in the center and low ductility in the grain boundaries. Thus, cored structures are undesirable in most cases. Coring can be removed by holding the metal for some time at a temperature below the solidus but high enough for diffusion to homogenize the composition. In the case of the cored 70% A and 30% B alloy, a temperature slightly below T_5 would be selected to avoid melting at temperature T_5. Such heat treatments are called homogenization treatments.

Alloys Partially Soluble in the Solid State. Alloys in which there is partial solubility in the solid state are the most common. The typical binary diagram (Ref 2.4) for this behavior is shown as Fig. 2.10 (referred to as a eutectic phase diagram). Because regions of solid solubility in this system are usually at the ends of the diagram, they are called terminal solid solutions or limited solid solutions. They are generally designated with the Greek letters α (alpha) and β (beta), although some systems are now in use that denote the terminal elements by enclosing the elemental chemical symbol in parentheses.

All Alpha. Consider cooling an alloy "J" of composition 10% B and 90% A from the liquid state in Fig. 2.10. Until the liquidus temperature is reached, the alloy exists as a homogeneous liquid of 10% B in 90% A. At a temperature slightly below the liquidus, crystals of solid-solution alpha solidify. They are rich in metal A and have the composition represented by X. If the melt is

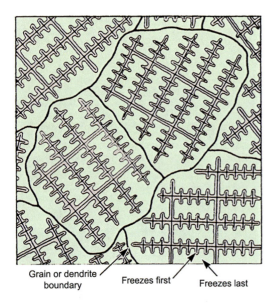

Grain or dendrite boundary — Freezes first — Freezes last

Fig. 2.9 Microstructure of a binary alloy exhibiting a cored structure

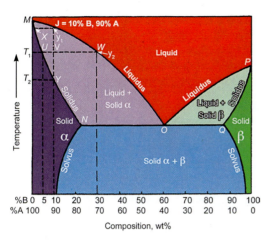

Fig. 2.10 Binary phase diagram of metals A and B, which are partially soluble in the solid state

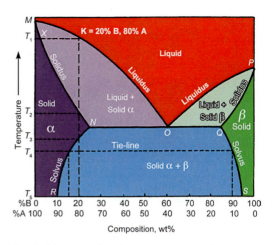

Fig. 2.11 Binary phase diagram of metals A and B, which are partially soluble in the solid state

cooled under equilibrium conditions, in the two-phase region, the composition of the solid-solution alpha being precipitated follows the solidus curve MN and the composition of the liquid remaining follows the liquidus curve MO.

In addition to showing the compositions of the liquid and solid, phase diagrams can also be used to calculate the amount of each phase present at any given temperature. For example, at temperature T_1, a horizontal tie-line (UVW) is drawn. At this temperature, solid-solution alpha has the composition U and the liquid remaining in the melt has the composition W. The amount of liquid and solid present at temperature T_1 is given by the Lever rule (Ref 2.3, 2.5):

% liquid = (Distance UV)/(Distance UW) × 100

$$= (10 - 5)/(30 - 5) \times 100$$

$$= 20\% \qquad \text{(Eq 2.1)}$$

% solid = (Distance VW)/(Distance UW) × 100

$$= (30 - 10)/(30 - 5) \times 100$$

$$= 80\% \qquad \text{(Eq 2.2)}$$

The tie-line distances needed for the calculations can be obtained from the horizontal composition scale or by measuring them on the diagram with a ruler, if the scale is uniform.

As slow cooling proceeds, the proportions change (UV shrinking and VW growing) as more solid is formed. Finally, at temperature T_2, the melt is completely solidified and has the composition Y which is the same as that of the initial alloy "J." If alloy "J" of 10% B and 90% A is cooled rapidly so that equilibrium is not reached, a cored structure results. However, this solid will

be a single-phase structure, all-alpha solid solution, at low temperatures.

Alpha-Beta. Consider cooling alloy "K" having 20% B and 80% A from the liquid state, as shown in Fig. 2.11. Above the liquidus, the alloy exists as a homogeneous liquid of composition 20% B and 80% A. Just slightly under the liquidus temperature, solid alpha crystals of composition X solidify. On further cooling under equilibrium conditions, solid-solution alpha precipitates, and its composition follows the solidus MN, while the composition of the liquid follows the liquidus MO, as shown previously for alloy "J." The all-alpha phase exists only between T_2 and T_3.

After solidification is completed at N, the solid cools to T_3 as solid-solution alpha of the composition 20% B and 80% A. At temperature T_3, the cooling solid-solution alpha intersects the solvus line NR. To the right of NR, the solid solubility of metal B in metal A is exceeded, and a second phase, beta, forms.

After the alloy is cooled slightly under temperature T_3, solid beta begins to precipitate from solid-solution alpha. Thus, below T_3, two phases are present: solid-solution alpha and tiny particles of solid-solution beta. Because this alloy now consists of two different crystal forms, it has two phases and is said to be heterogeneous. The amount of each phase at a given temperature can be calculated by means of a tie-line and the Lever rule.

For example, at temperature T_4, the following amounts of each phase are present:

% solid-solution α = (90 − 20)/(90 − 15) × 100

$$= (70/75) \times 100$$

$$= 93\% \qquad \text{(Eq 2.3)}$$

% solid-solution $\beta = (20 - 15)/(90 - 15) \times 100$

$= (5/75) \times 100$

$= 7\%$ (Eq 2.4)

At temperature T_4, solid-solution alpha has the composition 15% B and 85% A, and solid-solution beta contains 10% A and 90% B.

On cooling to room temperature, more beta phase precipitates according to the solvus lines QS and NR, and the percentage of primary alpha decreases accordingly. Such precipitates are useful in some alloy systems for strengthening.

At room temperature, T_5, the following amounts of each phase are present:

% solid-solution $\alpha = (95 - 20)/(95 - 10) \times 100$

$= (75/85) \times 100$

$= 88\%$ (Eq 2.5)

% solid-solution $\beta = (20 - 10)/(95 - 10) \times 100$

$= (10/85) \times 100$

$- 12\%$ (Eq 2.6)

At room temperature, T_5, solid-solution alpha has the composition 10% B and 90% A, and the solid-solution beta has the composition 5% A and 95% B under equilibrium conditions.

Eutectic Reactions. Consider cooling alloy "F" of composition 30% B and 70% A from the liquid state, as shown in Fig. 2.12. Above the liquidus, the alloy exists as a homogeneous liquid of 30% B and 70% A. Just slightly under the liquidus temperature, solid crystals of alpha (5% B) form. On slow cooling to temperature T_3, the eutectic temperature, the liquid composition follows the liquidus MO and the solid composition follows the solidus MN.

For example, at temperature T_2, the following amounts of liquid and solid phases are present:

% solid $\alpha = (45 - 30)/(45 - 10) \times 100$

$= (15/35) \times 100$

$= 43\%$ (Eq 2.7)

% liquid $= (30 - 10)/(45 - 10) \times 100$

$= (20/35) \times 100$

$= 57\%$ (Eq 2.8)

The composition of the liquid at temperature T_2 is 45% B and 55% A, and the composition of the solid is 10% B and 90% A. Slightly above the eutectic temperature T_3, the following amounts of solid alpha and liquid of eutectic composition (point O on Fig. 2.12) are present:

% solid $\alpha = (60 - 30)/(60 - 20) \times 100$

$= (30/40) \times 100$

$= 75\%$ (Eq 2.9)

% eutectic liquid $= (30 - 20)/(60 - 20) \times 100$

$= (10/40) \times 100$

$= 25\%$ (Eq 2.10)

At the eutectic temperature, the solid-solution alpha is 80% A and 20% B. The eutectic liquid is 60% B and 40% A.

When the alloy is cooled to the eutectic temperature, all of the remaining liquid of eutectic composition solidifies at a fixed temperature, similar to a pure metal. The difference is that now two phases (alpha and beta) solidify simultaneously from the liquid, forming the eutectic structure.

To further illustrate the point, if an alloy of 60% B and 40% A (eutectic composition) is cooled from the liquid, it does not solidify until it reaches temperature T_3. The entire melt solidifies at that temperature to form 100% eutectic. The eutectic is a finely divided structure composed of the two alternate solid phases, alpha and beta, having the compositions N and Q, respectively. Pearlite in carbon steel is a similar lamellar but is a eutectoid (solid states; $S_1 \geq S_2 + S_3$) structure (Ref 2.3).

Below the eutectic temperature, two solid phases exist in this alloy. These phases are solid-solution alpha and solid-solution beta. As the temperature is lowered below the eutectic temperature, the composition of alpha follows the solvus NR and that of beta follows the solvus QS, if equilibrium is maintained.

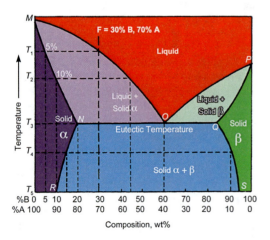

Fig. 2.12 Binary phase diagram of metals A and B, which are partially soluble in the solid state

At temperature T_4, the amounts of alpha and beta in alloy "F" are:

$$\% \; \alpha \; (\text{part in eutectic}) = (90 - 30)/(90 - 15) \times 100$$
$$= (60/75) \times 100$$
$$= 80\% \qquad \text{(Eq 2.11)}$$

$$\% \; \beta \; (\text{all in eutectic}) = (30 - 15)/(90 - 15) \times 100$$
$$= (15/75) \times 100$$
$$= 20\% \qquad \text{(Eq 2.12)}$$

At temperature T_4, the composition of alpha is 15% B and 85% A and that of solid-solution beta is 90% B and 10% A.

Monotectic Reactions. A monotectic reaction is defined as one in which a liquid phase (L_1) decomposes upon cooling into a solid phase (S) and a second liquid phase (L_2). The following reversible equation can be written for monotectic reactions:

$$L_1 \xleftarrow[\text{Heating}]{\text{Cooling}} S + L_2 \qquad \text{(Eq 2.13)}$$

A sketch of a binary diagram of metals A and B having a monotectic reaction at point M is given in Fig. 2.13. In this alloy system, the two liquids are insoluble (like oil and water) over a certain composition range and above the monotectic reaction temperature, T_M. This behavior is represented by $L_1 + L_2$ for the two-liquid region in the diagram.

Consider the equilibrium cooling of alloy "X" in Fig. 2.13. Above the monotectic line, the alloy is composed of a single-phase liquid, L_1. At the monotectic temperature, this liquid decomposes into a solid phase, alpha, and a liquid phase, L_2, of composition R. This reaction is analogous to the eutectic decomposition in which the liquid phase

decomposes into two solid phases. However, in the monotectic reaction, one liquid and one solid phase are produced from the decomposition of a liquid. Aluminum-lead, aluminum-bismuth, and copper-lead systems are of this type, for example.

Peritectic Reactions. In the peritectic reaction, on cooling, the liquid phase (L) reacts with one solid phase (S_1) to form another solid phase (S_2). The peritectic reaction can be summarized by the following equation:

$$L + S_1 \xleftarrow[\text{Heating}]{\text{Cooling}} S_2 \qquad \text{(Eq 2.14)}$$

A phase diagram of metals A and B with a peritectic reaction is shown in Fig. 2.14. Three types of alloys are used to help understand this diagram.

Type 1. Consider the equilibrium cooling of alloy "X" shown in Fig. 2.14. This is the peritectic composition. Above the liquidus, the alloy exists as a homogeneous liquid solution of composition X. After cooling to slightly below the liquidus, at temperature T_1, solid-solution alpha (of composition α_1) starts to precipitate. From temperature T_1 at the liquidus to slightly above the peritectic reaction temperature, solid-solution alpha continues to precipitate, with the liquid composition going from L_1 to L_2 and the solid composition going from α_1 to α_2.

Just slightly above temperature T_2, the following amount of each phase is present:

$$\% \; \text{solid-solution} \; \alpha = (\beta_2 - L_2)/(\alpha_2 - L_2) \times 100$$
$$= (70 - 55)/(70 - 25) \times 100$$
$$= (15/45) \times 100$$
$$= 33\% \qquad \text{(Eq 2.15)}$$

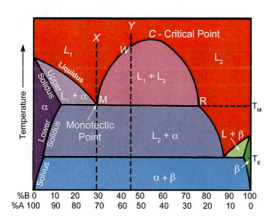

Fig. 2.13 Phase diagram with a monotectic reaction

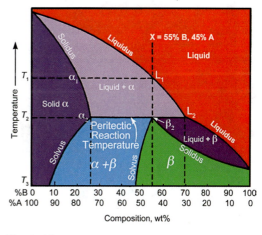

Fig. 2.14 Phase diagram with a peritectic reaction

% liquid = $(\alpha_2 - \beta_2)/(\alpha_2 - L_2) \times 100$

$\qquad = (55 - 25)/(70 - 25) \times 100$

$\qquad = (30/45) \times 100$

$\qquad = 67\%$ (Eq 2.16)

At temperature T_2, liquid (of composition L_2) reacts with the solid solution of composition α_2 to produce a new solid phase, solid-solution β of composition β_2. From temperature T_2 to room temperature, T_3, the solid cools as solid-solution β.

Type 2. Consider the equilibrium cooling of alloy "Y" shown in Fig. 2.15. This is a hypoperitectic composition (less B than that required for 100% peritectic). Above the liquidus, the alloy exists as a homogeneous liquid solution of composition Y. On cooling to slightly below the liquidus to temperature T_1, solid-solution α of composition α_1 starts to precipitate. From temperature T_1 at the liquidus to slightly above T_2 (the peritectic reaction temperature), solid solution continues to precipitate, with the liquid composition going from L_1 to L_2 and the solid composition changing from α_1 to α_2.

Just slightly above temperature T_2, the following amount of each phase is present:

% solid-solution α = $(Y - L_2)/(\alpha_2 - L_2) \times 100$

$\qquad = (70 - 40)/(70 - 27) \times 100$

$\qquad = (30/43) \times 100$

$\qquad = 70\%$ (Eq 2.17)

% liquid = $(\alpha_2 - Y)/(\alpha_2 - L_2) \times 100$

$\qquad = (40 - 27)/(70 - 27) \times 100$

$\qquad = (13/43) \times 100$

$\qquad = 30\%$ (Eq 2.18)

At temperature T_2, all the remaining liquid of composition L_2 reacts with the solid-solution α (of composition α_2) to produce solid-solution β. However, because there is not enough liquid to react with all of the solid-solution α, there is an excess of the solid-solution α after β forms.

Therefore, just slightly below T_2 (the peritectic reaction temperature), a mixture of the two solid-solutions α and β exists. The following amounts are present:

% solid-solution α = $(Y - \beta_2)/(\alpha_2 - \beta_2) \times 100$

$\qquad = (55 - 40)/(55 - 27) \times 100$

$\qquad = (15/28) \times 100$

$\qquad = 54\%$ (Eq 2.19)

% solid-solution β = $(Y - \alpha_2)/(\alpha_2 - \beta_2) \times 100$

$\qquad = (40 - 27)/(55 - 27) \times 100$

$\qquad = (13/28) \times 100$

$\qquad = 46\%$ (Eq 2.20)

On cooling to room temperature, T_3, the composition of solid-solution alpha follows the solvus α_2 to α_3 and that of solid-solution beta follows the solvus β_2 to β_3, if equilibrium is maintained.

Type 3. Consider the equilibrium cooling of alloy "Z" shown in Fig. 2.16. Above the liquidus, the alloy exists as a liquid alloy solution of composition Z. After cooling to slightly below the liquidus at temperature T_1, solid-solution alpha (of composition α_1) precipitates. From temperature T_1 to just slightly above T_2 (the peritectic reaction temperature), solid-solution alpha continues to precipitate, with the liquid composition changing from L_1 to L_2 and the solid composition going from α_1 to α_2. Just slightly above tempera-

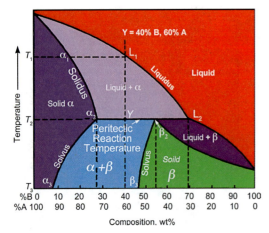

Fig. 2.15 Constitutional diagram with peritectic reaction (Y is hypoperitectic in composition)

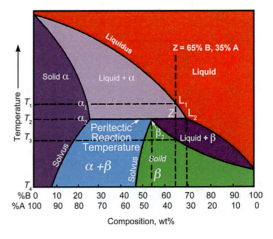

Fig. 2.16 Constitutional diagram with peritectic reaction (Z is hyperperitectic in composition)

ture T_2, the following amounts of each phase are present:

% solid-solution $\alpha = (Z - L_2)/(\alpha_2 - L_2) \times 100$
$$= (70 - 65)/(70 - 25) \times 100$$
$$= (5/45) \times 100 = 11\% \quad \text{(Eq 2.21)}$$

% liquid $= (\alpha_2 - Z)/(\alpha_2 - L_2) \times 100$
$$= (65 - 25)/(70 - 25) \times 100$$
$$= (40/45) \times 100 = 89\% \quad \text{(Eq 2.22)}$$

At temperature T_2, all the solid-solution alpha reacts with the liquid to produce solid-solution beta. However, because there is not enough solid-solution alpha to react with all the liquid, the excess liquid remains.

Therefore, just slightly below T_2 (the peritectic reaction temperature), liquid and solid-solution beta exist. The following amounts of each phase are present:

% solid-solution $\beta = (Z - L_2)/(\beta_2 - L_2) \times 100$
$$= (70 - 65)/(70 - 55) \times 100$$
$$= (5/15) \times 100 = 33\% \quad \text{(Eq 2.23)}$$

% liquid $= (\beta_2 - Z)/(\beta_2 - L_2) \times 100$
$$= (65 - 55)/(70 - 55) \times 100$$
$$= (10/15) \times 100$$
$$= 67\% \quad \text{(Eq 2.24)}$$

On cooling from temperature T_2 to temperature T_3, solid-solution beta continues to precipitate from the liquid. Finally, at temperature T_3, all the liquid is converted to solid-solution beta of composition Z. From temperature T_3 to room temperature, T_4, the alloy cools simply as solid-solution beta.

Intermediate Phases in Binary Alloys. If two pure elements have similar atomic diameters, chemical natures, and the same crystal structure, they usually form solid solutions of unlimited solid solubility (Fig. 2.8). Nickel-copper and gold-silver are two binary alloy systems that show this complete solid solubility. Titanium exhibits unlimited solid solubility in beta phase (bcc structure) with hafnium (Hf), molybdenum (Mo), niobium (Nb), tantalum (Ta), vanadium (V), and zirconium (Zr) (Fig. 2.17).

When differences between the two metals in an alloy system increase, the equilibrium diagram usually shows a eutectic or a peritectic with ter-

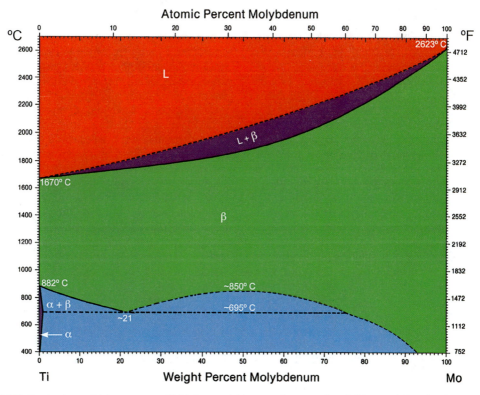

Fig. 2.17 The titanium-molybdenum system. Molybdenum, niobium, tantalum, vanadium, hafnium, and zirconium form a complete series of beta solid solutions with titanium; hafnium and zirconium also form a complete series of alpha solid solutions.

minal solid solutions (Fig. 2.10). Still greater differences between the elements in a binary alloy, especially with crystal structure, lead to one or more intermediate phases and solid-state transformation. Most titanium systems contain such phases and transformations.

Solid-State Transformations. Discussions thus far have been concerned mainly with the construction and use of fairly simple phase diagrams. Little has been said about solid-state reactions except where there is a decrease in solubility in the solid state and precipitates form, as, for example, in Fig. 2.11 and 2.12.

As discussed previously, some metals and nonmetals exist at different temperatures in more than one solid state. For example, when liquid titanium solidifies, the atoms are arranged in a bcc (beta) structure, but at a lower temperature they are rearranged to the hcp (alpha) lattice form. Such solid-state transformations can have important consequences and are used in some strengthening heat treatments and for grain refinement.

Diagrams with solid-state transformations, intermediate, and intermetallic phases are interpreted and used with the guidelines noted previously. A brief discussion and a few examples should clarify these important features common to titanium and the other metal systems.

Isomorphous Alloys. The simplest binary diagram was shown in Fig. 2.8 in which all of the alloying factors are favorable and there is no solid-state transformation (Ref 2.4).

Figure 2.17 is typical of the titanium isomorphous alloy system with molybdenum, niobium, tantalum, vanadium, hafnium, and zirconium. The first four of these elements have only bcc structures in the solid state. They are completely soluble in bcc titanium, but they have only a little solubility in hcp titanium. These elements also stabilize the bcc structure when added in large amounts, but they are less effective at low concentrations. For example, an alloy containing 10 wt% Mo solidifies as the beta phase, but on cooling slowly below its beta transus (at 780 °C, or 1435 °F, in this case), a portion of the unstable beta transforms to alpha; alpha precipitates appear in the beta matrix. The beta is enriched and stabilized by the molybdenum not soluble in the alpha phase.

However, if the alloy contains only 1% Mo or less, none of the beta phase is stable, and it transforms completely to alpha as the alpha transus is crossed. The term *transus* refers to the temperature at the appropriate solvus lines. Above the beta transus temperature, all the material has the beta structure, and below the alpha transus temperature, the structure is all alpha. Between these lines or transus temperatures, the alloy consists of two phases, alpha and beta.

The titanium-hafnium and titanium-zirconium systems are special cases. These elements have very similar alloying factors and form a complete series of solid solutions in both the alpha and beta phases.

Eutectoid Alloys. Although no two alloy diagrams are identical, Fig. 2.18 for the titanium-iron system is fairly representative of those having a eutectoid reaction and intermetallic phases (Ref 2.4). The eutectoid temperature is the temperature at which the eutectoid reaction occurs. The eutectoid temperature in Fig. 2.18 is at 595 °C (1105 °F), and the eutectoid is shown in this figure at 17% Fe. The diagram also shows two eutectics and a peritectic.

At the eutectoid in Fig. 2.18, the beta phase transforms to two other solid phases of alpha and the intermetallic compound, TiFe. This reversible reaction can be shown as:

$$\beta \xleftrightarrow[\text{Heating}]{\text{Cooling}} \alpha + \text{TiFe} \qquad \text{(Eq 2.25)}$$

Other titanium alloy systems of this type include those of beryllium, chromium, cobalt, copper, nickel, and silicon.

Eutectoid reactions are similar to eutectic reactions except they involve the transformation of a solid phase rather than a liquid to form two new solid phases. In most practical alloys, solute concentrations of such alloys are limited to only a few percent or less to avoid serious instability and embrittlement problems.

Peritectoid Alloys. Peritectoid reactions are found in binary titanium systems with aluminum, carbon, cerium, germanium, nitrogen, oxygen, and tin (Ref 2.4). The peritectoid reaction is similar to the peritectic reaction except two solid phases, instead of a solid and liquid phase, combine on cooling to form one solid phase. An example of a peritectoid is found in the titanium-aluminum diagram in Fig. 2.19. In this case, beta and gamma combine on cooling to form alpha at 1285 °C (2345 °F):

$$\beta + \gamma \xleftrightarrow[\text{Heating}]{\text{Cooling}} \alpha \qquad \text{(Eq 2.26)}$$

The diagram also shows that alpha is not stable at high aluminum levels (e.g., more than 6% Al at 600 °C, or 1110 °F) and that it can form the ordered structure, α_2. Other interesting features are

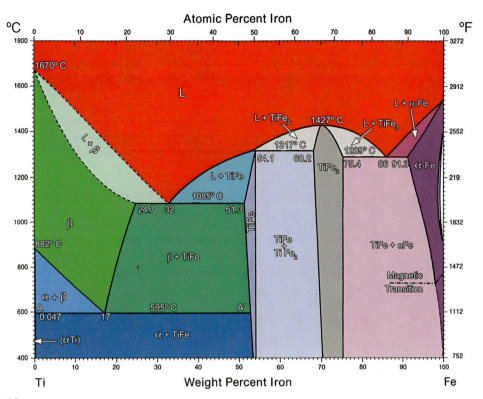

Fig. 2.18 The titanium-iron system

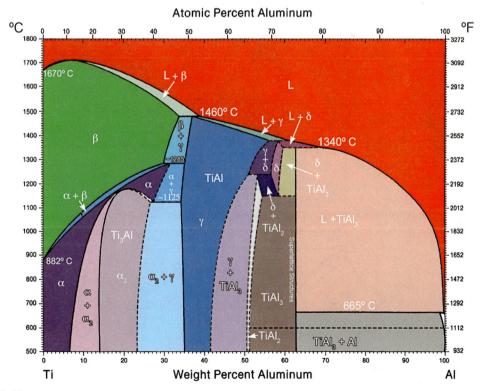

Fig. 2.19 The titanium-aluminum system

the intermetallic compound TiAl$_3$ and two peritectics consisting of L + β ↔ γ and L + γ × δ.

Ternary Phase Diagrams. Only binary alloy diagrams have been discussed to this point. Attention is now given to diagrams of three elements, or ternary alloys. Ternary alloy diagrams are usually constructed with an equilateral triangle as a base. Alloys are represented on this base, with the pure metals located at each corner of the triangle.

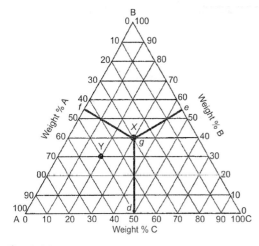

Fig. 2.20 Ternary equilibrium diagram

Binary and ternary alloys of metals A, B, and C are represented in Fig. 2.20.

Binary alloys appear on the three sides of the triangle, and ternary alloys are within the triangle. Usually, only one temperature is portrayed (called an isothermal section). To show a range of temperatures, a three-dimensional figure is necessary (like the topography of a mountain range), although contour lines showing temperatures can be superimposed, as shown in Fig. 2.21.

The composition of any combination is easily determined. In Fig. 2.20, 100% metal A is at one corner, 100% metal B is at another, and 100% metal C is at the remaining corner. The amount of each metal in the ternary alloy is given by the perpendicular distance from the side opposite the metal in question to the point representing the ternary alloy. For example, alloy "X" in Fig. 2.20 is composed of e minus g = 30% A, d minus g = 40% B, and f minus g = 30% C. The composition of alloy "Y" in Fig. 2.20 is 50% A, 30% B, and 20% C.

Temperatures are represented by height measurements. Thus, a three-dimensional model is needed to show a complete ternary equilibrium diagram. However, because it is impractical to draw such a diagram on a flat page, the temperatures of simple surfaces are plotted as a series of contours on the triangle base in the same way that elevations are

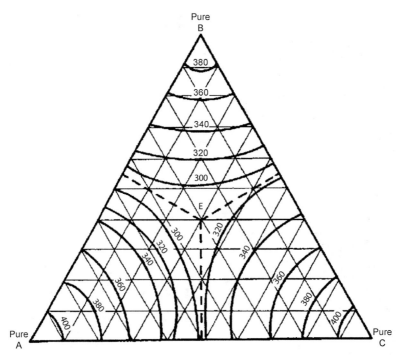

Fig. 2.21 Liquidus surface of ternary alloy A, B, and C. Temperature, °F

shown on a land map. As an example, the liquidus surface of this imaginary ternary eutectic is shown in Fig. 2.21. The eutectic point is at *E*. Solidus surfaces are more complex, and an entire ternary phase diagram is almost impossible to illustrate.

The liquidus surface of a ternary system can be shown by a series of contours. However, to show solid phases, a definite temperature is selected and the phases that exist at that temperature are given on the triangle. If a three-dimensional solid diagram were sliced horizontally across the triangle at the desired temperature and the top removed, then looking down on the triangle, one would see the phases present at that temperature. This form of representation is most common.

An illustration of the Ti-Al-V ternary diagram at 1100 °C (2012 °F) is shown in Fig. 2.22. The percentages of each phase at any location can be calculated from the diagram.

Summary

The common states of matter are solid, liquid, and gas. Metallurgists are concerned mainly with the liquid and solid states. However, gases in liquid and solid metals are also important.

The three most common crystal structures in metals are fcc, bcc, and hcp. Titanium crystallizes in the bcc structure and transforms to the hcp structure at a lower temperature.

Metals solidify by forming nuclei (in the liquid state) that grow into crystals. These processes are called nucleation and growth. Most crystals form as dendrites. Growing crystals, or grains, meet at grain boundaries. If the grains are almost the same size in all directions, they are called equiaxed grains; if they are elongated, they are called columnar grains.

As alloys solidify, the composition of crystals tends to change according to the solidus, and the liquid changes according to the liquidus. However, if equilibrium conditions are maintained, or if a homogenizing treatment is used, uniform alloy distribution can be achieved by atom diffusion.

Two distinct phases observed in alloys are solid solutions and intermetallic compounds. Intermetallic compounds are usually hard and strong with a more or less definite composition. Solid solutions can be of the interstitial type or the substitutional type. To be of the interstitial

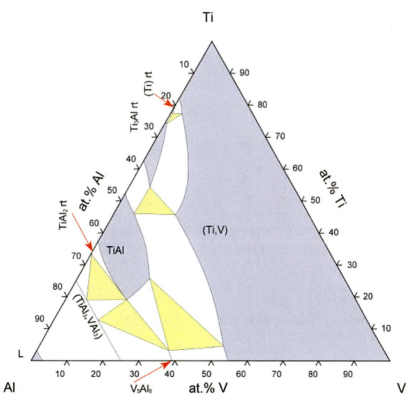

Fig. 2.22 Ternary isothermal diagram for Ti-Al-V at 1100 °C (1376 K, 2012 °F). Source: Ref 2.6

type, solute atoms must be sufficiently small to fit in between the atoms in the lattice of the solvent. In the substitutional type, solute atoms substitute in the place of solvent atoms. Solid solutions have the structure of the solvent.

Extensive solid solubility (the Hume-Rothery rules) between two or more elements requires that the elements have compatible sizes, structures, and chemical characteristics.

Alloys can contain one phase only; such alloys under equilibrium conditions are called homogeneous alloys. Alloys containing more than one phase are called heterogeneous or two- or multiphase alloys.

Binary equilibrium diagrams are useful for understanding alloys because they help:

- Describe the freezing and melting of alloys, the melting points of different phases, and what kinds of phases exist at each temperature.
- Identify the composition and the amount of each phase in a two-phase region. By drawing a tie-line and using the Lever rule, it is possible to calculate the percentage and composition of each phase in two-phase regions.
- Predict the microstructure of alloys when equilibrium or near-equilibrium conditions exist.

A binary eutectic-type alloy forms when the liquidus lines of two elements, which are partially insoluble in each other, intersect to form the lowest-melting composition of the alloy system. The temperature at which the intersection occurs is called the eutectic temperature, and the composition is called the eutectic composition. The following equation sums up the binary eutectic reaction:

$$\text{Liquid} \xleftrightarrow[\text{Heating}]{\text{Cooling}} \text{Solid}_1 + \text{Solid}_2 \qquad \text{(Eq 2.27)}$$

A ternary eutectic would be similarly illustrated:

$$\text{Liquid} \xleftrightarrow[\text{Heating}]{\text{Cooling}} \text{Solid}_1 + \text{Solid}_2 + \text{Solid}_3 \qquad \text{(Eq 2.28)}$$

The binary peritectic reaction involves (on cooling) the reaction of liquid with the solid phase to form a new solid phase. This reaction can be summarized by the equation:

$$\text{Liquid} + \text{Solid}_1 \xleftrightarrow[\text{Heating}]{\text{Cooling}} + \text{Solid}_2 \qquad \text{(Eq 2.29)}$$

Phase transformations occur as atoms relocate to states or positions having lower energy. Solid-state reactions include the precipitation of a second phase as solubilities decrease with decreasing temperature. Important new phases also appear with decreasing temperature by eutectoid and peritectoid reactions. These two reactions are similar to the eutectic and peritectic reactions.

Glossary

alpha transus. The temperature below which an alloy consists entirely of the alpha phase. By heating the alloy above this temperature, beta phase appears and the two phases exist together until the beta transus is exceeded.

beta transus. A term used in titanium technology to express the temperature above which a solid alloy consists entirely of the beta phase. Just below this temperature, the alloy consists of two phases (alpha and beta) in equilibrium.

compound. A substance produced by the chemical combination of two or more chemical elements.

constitution diagram. See *equilibrium diagram*.

dendrite. A crystal formed by solidification and characterized by a treelike pattern composed of many branches.

equilibrium. A dynamic condition of balance between atomic movements, where the resultant is zero and the condition appears to be one of rest rather than change.

equilibrium diagram. A graphical representation of the equilibrium temperature and composition limits of phase fields and phase reactions in an alloy system.

eutectic. (1) The isothermal (constant temperature) reversible reaction of a liquid that forms two different solid phases (in a binary alloy system) during cooling. (2) The alloy composition that freezes at constant temperature, undergoing the eutectic reaction completely. (3) The alloy structure of two (or more) solid phases formed from the liquid eutectically.

eutectoid. Same as eutectic except that a solid rather than a liquid decomposes into two other solids.

liquidus. A line on a binary phase diagram or a surface on a ternary equilibrium diagram, representing the temperatures at which freezing begins during cooling or melting ends during heating under equilibrium conditions.

miscibility gap. A heterogeneous area in an equilibrium diagram where two liquid phases are not soluble in each other. The $L_1 + L_2$ semicircular area of Fig. 2.13 represents an area of liquid-phase immiscibility.

monotectic reaction. Reversible reaction in a binary system where a liquid, during cooling across a specific isothermal reaction temperature, forms a solid and a second liquid of different composition.

monotectoid reaction. Similar to the monotectic, except that above the reaction temperature, a solid phase decomposes to similar and dissimilar solids.

peritectic reaction. An isothermal reversible reaction in binary alloy systems in which a solid and a liquid phase react during cooling to form a second solid phase.

peritectoid reaction. Same as the peritectic except that the liquid phase is replaced by a solid phase.

phase diagram. See *equilibrium diagram.*

solid solutions. There are two types of solid solutions: (1) Interstitial solid solutions in which atoms of the alloying element occupy spaces between atoms of the solvent element, which retain their original lattice positions. (2) Substitutional solid solutions in which atoms of the alloying element occupy atom sites previously filled by other atoms.

solidus. A line in a binary phase diagram, or a surface on a ternary phase diagram, representing the temperatures at which freezing ends during cooling or melting begins during heating, under equilibrium conditions.

solvus. A name applied to solid solubility curves (maximum solubility of one atomic species in another) of alloy equilibrium diagrams.

REFERENCES

2.1 M.C. Flemings, *Solidification Processing,* McGraw-Hill, New York, 1973

2.2 D.A. Porter, K.E. Easterling, and M. Sherif, *Phase Transformations in Metals and Alloys,* 3rd ed. (revised), Chapman and Hall, 2009

2.3 R.E. Reed-Hill, *Physical Metallurgy Principles,* D. Van Nostrand, New York, 1973

2.4 J. Murray, *Phase Diagrams of Binary Titanium Alloys,* ASM International, 1987

2.5 R. Askeland, *The Science and Engineering of Materials,* Brooks/Cole Engineering Division, Wadsworth, Inc., 1984, p 157, 163

2.6 F.H. Hayes, The Al-Ti-V (Aluminum-Titanium-Vanadium) System, *J. Phase Equilib.,* Vol 16, 1995, p 163–176

SELECTED REFERENCES

- R. Boyer, E.W. Collings, and G. Welsch, Ed., *Materials Properties Handbook: Titanium Alloys,* ASM International, 1994
- M.J. Donachi, *Titanium: A Technical Guide,* 2nd ed., ASM International, 2000
- H. Kuhn and D. Medlin, Ed., *Mechanical Testing and Evaluation,* Vol 8, *ASM Handbook,* ASM International, 2000
- P. Lacombe, R. Tricot, and G. Beranger, Ed., *Proceedings of the Sixth International Conference on Titanium* (Nice, France), Metallurgical Society of AIME, Warrendale, PA, 1988
- G. Lutjering and J.C. Williams, *Titanium,* Springer, 2003
- G. Lutjering, U. Zwicker, and W. Bunk, *Proceedings of the Fifth International Conference on Titanium,* Metallurgical Society of AIME, Warrendale, PA, 1984
- *Proceedings of the Seventh International Conference on Titanium* (San Diego, CA), Metallurgical Society of AIME, Warrendale, PA, 1992
- *Proceedings of the Eighth International Conference on Titanium* (San Birmingham, U.K.), Metallurgical Society of AIME, Warrendale, PA, 1995
- *Proceedings of the Ninth International Conference on Titanium,* Metallurgical Society of AIME, Warrendale, PA, 2000
- *Proceedings of the Tenth International Conference on Titanium,* Metallurgical Society of AIME, Warrendale, PA, 2004
- *Proceedings of the Eleventh International Conference on Titanium,* Metallurgical Society of AIME, Warrendale, PA, 2008
- *Proceedings of the Twelfth International Conference on Titanium,* Metallurgical Society of AIME, Warrendale, PA, 2012

CHAPTER 3

Principles of Alloying Titanium*

TITANIUM IS A MEMBER of the group of elements called the transition elements. These metals have several important characteristics, including high strength and allotropic (i.e., different solid-state crystal structures in different temperature ranges) behavior.

The titanium atom has a median diameter compared with other elements, and it is relatively light, resulting in a metal with an intermediate density. This relatively low density, when combined with high strength, results in a useful engineering metal (Ref 3.1). The average atomic diameter indicates favorable alloying behavior.

The titanium atoms are aligned in the solid state in either a hexagonal close-packed (alpha) or body-centered cubic (beta) structure. The addition of alloying elements either stabilizes the alpha phase to higher temperatures (alpha stabilizers) or the beta phase to lower temperatures (beta stabilizers). One additional group, neutral additives, has only a minor influence on stabilization of these phases.

The addition of alloying elements results in three classes of terminal titanium alloys. The alloy classes are described by the crystal structures that exist at or near room temperature. They are alpha (α), alpha-beta (α-β), and beta (β) alloys. Each class of alloys has its own distinctive properties. In addition to these three classes of terminal alloys, intermetallic titanium aluminides (Ti_3Al, $TiAl$, and $TiAl_3$) are of significance.

In this chapter, the titanium alloy systems are described along with the alloying mechanisms involved and the characteristics and properties that can be developed.

Atomic Structure of Titanium

The characteristics of metals depend on the atomic structure. Some properties can be estimated with a knowledge of atom structure and lattice imperfections. Because the usefulness of titanium is due to a unique combination of physical and mechanical properties (Ref 3.1), an examination of the titanium atom is desirable.

The atom is a fundamental unit composed of many parts (Ref 3.2). For our purposes, we need be concerned with only the electrons. Figure 3.1 shows a schematic view of the titanium atom. This representation, although not entirely accurate, is useful in understanding the nature of the element. Almost the entire mass of this atom is concentrated in the nucleus, which has a diameter only 1/100,000 of that of the whole atom.

The electrons revolve within this atom in orbits, or shells. Each shell, referred to as an energy level, can accept only a limited number of electrons. This number varies according to well-defined physical laws. For example, the 1s and 2s energy levels contain up to two electrons each, while the 2p level can accept up to six electrons (Ref 3.2).

The outer electrons in the unfilled shells are loosely bound to the atom and are called valence electrons. They largely determine the chemical and physical properties of the metal. The remaining electrons make up the stable core.

Each element has a different number of electrons. For example, titanium has 22, iron has 26, and hydrogen has 1. Several elements have over 100 electrons. Table 3.1 shows the electronic configuration of the first 29 elements in the

---

*Adapted and revised from Walter E. Herman, originally from Stan R. Seagle and Patrick A. Russo, *Titanium and Its Alloys*, ASM International.

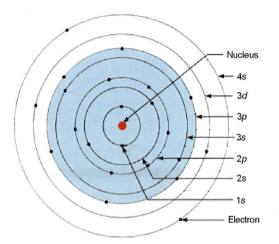

Fig. 3.1 Schematic of a titanium atom. The shaded area is the inner electron core; the outer electrons are the valence electrons.

Table 3.1 Electronic configuration of first 29 elements in atomic series

Element	Atomic number(a)	Number of electrons in indicated energy level						
		1s	2s	2p	3s	3p	3d	4s
Hydrogen (H)	1	1	...	...	...	...	...	...
Helium (He)(b)	2	2	...	...	...	...	...	...
Lithium (Li)	3	2	1	...	...	...	...	...
Beryllium (Be)	4	2	2	...	...	...	...	...
Boron (B)	5	2	2	1	...	...	...	...
Carbon (C)	6	2	2	2	...	...	...	...
Nitrogen (N)	7	2	2	3	...	...	...	...
Oxygen (O)	8	2	2	4	...	...	...	...
Fluorine (F)	9	2	2	5	...	...	...	...
Neon (Ne)(b)	10	2	2	6	...	...	...	...
Sodium (Na)	11	2	2	6	1	...	...	...
Magnesium (Mg)	12	2	2	6	2	...	...	...
Aluminum (Al)	13	2	2	6	2	1	...	...
Silicon (Si)	14	2	2	6	2	2	...	...
Phosphorous (P)	15	2	2	6	2	3	...	...
Sulfur (S)	16	2	2	6	2	4	...	...
Chlorine (Cl)	17	2	2	6	2	5	...	...
Argon (A)(b)	18	2	2	6	2	6	...	...
Potassium (K)	19	2	2	6	2	6	0	1
Calcium (Ca)	20	2	2	6	2	6	0	2
Scandium (Sc)(c)	21	2	2	6	2	6	1	2
Titanium (Ti)	22	2	2	6	2	6	2	2
Vanadium (V)	23	2	2	6	2	6	3	2
Chromium (Cr)	24	2	2	6	2	6	5	1
Manganese (Mn)	25	2	2	6	2	6	5	2
Iron (Fe)	26	2	2	6	2	6	6	2
Cobalt (Co)	27	2	2	6	2	6	7	2
Nickel (Ni)	28	2	2	6	2	6	8	2
Copper (Cu)	29	2	2	6	2	6	10	1

(a) Also number of electrons. (b) Inert. (c) Scandium (Sc) through nickel (Ni) are transition elements.

atomic series. As one progresses through the series to higher atomic numbers, the number of electrons and their relation to the nucleus increase in a regular manner. For example, the hydrogen atom has one electron in the 1s energy level.

For helium, a second electron is added to complete this energy level. Because no more electrons enter this energy level, it is now a stable core with no valence electrons. Because of the lack of valence electrons, this element will not react with others; it is referred to as inert. Similar situations are also found when the 2p (neon) and 3p (argon) energy levels are filled. Again, these elements are inert and do not react with other elements. Argon is used in some areas in the manufacture of titanium products due to this lack of reactivity.

The regularity in this sequence as the number of electrons increases is interrupted at scandium, element number 21. Instead of systematically adding a third electron in an outer energy level, the electron is added to the 3d energy level. This is the start of the first transition series, in which titanium is located. Also included in this series are several important industrial metals, including iron, chromium, cobalt, and nickel.

In this transition series, after one or two electrons enter an outer orbit (4s), additional electrons enter an inner orbit (3d). This occurs because the energy of the 4s electrons is less than, but close to, the 3d electrons. This not only causes a variable valence in the transition elements but also affects properties.

The transition metals have several important characteristics. They have high cohesive strength, which results in high tensile strength, low thermal

expansion, and relatively high melting points. Several metals in the transition series can exist in more than one crystalline form. It is shown later that this is an extremely important asset for titanium (as well as iron-base alloys), because this is the major basis for strengthening through heat treatment and processing.

Atom Diameter. As one of the early elements in the periodic system, the titanium atom has a relatively light nucleus. This, combined with an average atomic diameter, results in a favorable density midway between aluminum and iron (Table 3.2) (Ref 3.1, 3.2).

The average atomic diameter is also very beneficial for alloying. Favorable substitutional alloying (high solubility) can be expected when the diameter of the alloying element does not differ more than 15% from that of the parent metal (Hume-Rothery rules discussed in Chapter 2, "Solidification and Phase Diagrams").

Figure 3.2 shows the atomic diameters of many of the elements. Substitutional alloying can occur when the diameter ratio is between 0.85 and 1.15

Table 3.2 Comparison of physical properties of titanium alloys with those of other metal alloys

Properties	Ti	Ti-5Al-2.5Sn	Ti-6Al-4V	Ti-3Al-8V-6Cr-4Mo-4Zr	7075 aluminum	17-7PH steel	4340 steel
Density, g/cm³ (lb/in.³)	4.540 (0.164)	4.484 (0.162)	4.429 (0.160)	4.816 (0.174)	2.796 (0.101)	7.640 (0.276)	7.833 (0.283)
Thermal conductivity, W/m · K (Btu/h · ft² · ft · °F)	17.0 (9.8)	7.8 (4.5)	6.7 (3.9)	6.9 (4.0)	121.1 (70.0)	16.6 (9.6)	37.5 (21.7)
Electrical resistivity, μΩ · m at 21 °C (μΩ · in. at 70 °F)	0.61 (24)	1.57 (62)	1.71 (67.4)	1.52 (60)	0.06 (2.3)	0.86 (34.0)	0.22 (8.8)
Coefficient of thermal expansion, m/m · °C (10⁻⁶ in./in. · °F)	10.1 (5.6) (20–650 °C, or 70–1200 °F)	9.7 (5.4) (20–650 °C, or 70–1200 °F)	11.0 (6.09) (20–650 °C, or 70–1200 °F)	8.8 (4.9)	26.4 (14.4) (20–300 °C, or 70–572 °F)	12.5 (6.9) (20–425 °C, or 70–800 °F)	14.8 (8.1) (20–650 °C, or 70–1200 °F)
Specific heat, J/kg at 21 °C (Btu/lb · °F at 70 °F)	540 (0.129)	523 (0.125)	565 (0.135)	515 (0.123)	962 (0.23)	502 (0.12)	448 (0.107)
Melting range, °C (°F)	1670 (3038)	1600 (2910)	1605–1670 (2920–3040)	1650 (3000)	475–640 (890–1180)	1400–1455 (2550–2650)	1505 (2740)
Alloy type	α	α	α + β	β	...	...	...

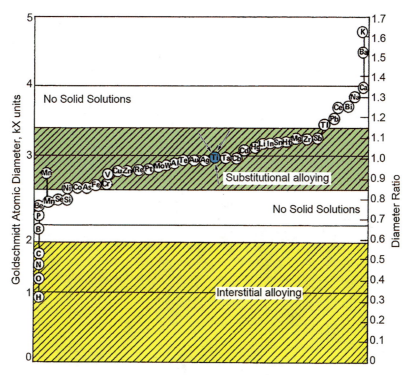

Fig. 3.2 Atomic diameters of elements. The chart compares the atom size of titanium with other potential alloying elements.

(15% of the atomic diameter of titanium). (See Fig. 2.3 to 2.5 in Chapter 2, "Introduction to Solidification and Phase Diagrams," in this book.) Many commercially important alloying metals fall in this shaded area, although they are not all substitutional alloying elements.

Manganese, iron, vanadium, molybdenum, aluminum, tin, and zirconium are some impor-

tant alloying elements used in titanium (Ref 3.3). Because many common elements fall in the favorable area for alloying, numerous combinations exist for altering the properties of titanium. This favorable alloying situation, in combination with the formation of two crystallographic structures, greatly enhances the usefulness of titanium.

When the atomic diameters of alloying elements are less than 0.6 times the diameter of titanium, alloy elements can enter the metal interstitially. The elements carbon, oxygen, nitrogen, and hydrogen are included in this group (Ref 3.3). The importance of interstitial elements to the properties of titanium is shown later.

Lattice Structure. Atoms in the liquid and solid states attract each other. If they did not, all matter would be gaseous. The reasons for the formation of certain crystallographic patterns are rather complicated and are not discussed here. However, the type of crystallographic pattern that forms has an important influence on the properties. Figure 3.3 lists typical metals of three of the most common of the 14 basic structures (Ref 3.4).

Most metals have a cubic crystal structure, either body-centered or face-centered (Ref 3.4). Only a few of the commercially important metals have the hexagonal type of crystal structure. Of the three types listed, face-centered cubic metals are the most ductile. Silver and gold, two of the most malleable and ductile metals, are in this category. The metals in the body-centered group are less ductile, and the hexagonal metals are the least ductile of the three groups.

Titanium is allotropic. At low temperatures it exists as a hexagonal close-packed (hcp) structure (Fig. 3.4) and is termed alpha phase. This structure is present in pure titanium at temperatures up to the beta transus point of 885 °C (1625 °F). At temperatures above the beta transus but below the melting point of 1670 °C (3038 °F), titanium exists in the beta phase as a body-centered cubic (bcc) structure.

The atoms in the bcc crystal structure are not as closely packed as in the hcp structure; thus, a volume expansion during transformation is expected. This transformation of alpha to beta in pure titanium results in slight expansion and thus a decrease in density, as illustrated in Fig. 3.5.

The temperature at which the transformation of alpha (hcp) to beta (bcc) occurs in pure titanium is referred to as the beta transus. The addition of alloying elements alters this temperature. Beta transus is the lowest equilibrium temperature at which the alloy consists of 100% beta phase.

The bcc crystal structure of the beta phase exists up to the melting point, which for pure titanium is 1670 °C (3038 °F).

The dimensions between atom sites influence the properties of a crystal. In the hexagonal system, two dimensions are necessary to define the crystal structure. The short distance is defined as the "a" parameter, while the height is the "c" parameter. The ratio c/a for an ideal hcp system is 1.633. As shown in the compilation of physical properties in Table 3.3, the ratio of 1.5873 for titanium is slightly less than ideal. The ratio, although less than ideal, is nearer 1.633 than, for example, the 1.568 value for the less ductile metal beryllium.

Physical Properties. Table 3.3 presents the physical properties of pure titanium (Ref 3.5). Alloying titanium changes many of these properties. Table 3.2 lists some of the more important physical properties of titanium and titanium alloys. The titanium alloys are representative of the three alloy types: alpha, alpha-beta, and beta. Included for comparison with titanium are data on two iron-base alloys and an aluminum-base alloy.

The densities of current commercial titanium alloys range from 4.318 to 4.872 g/cm³ (0.156 to 0.176 lb/in.³), depending on the type and amount of alloying elements. The density of titanium is

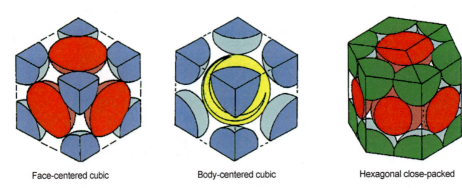

Face-centered cubic   Body-centered cubic   Hexagonal close-packed

Fig. 3.3 The crystal structures representative of most metals are the face-centered cubic (fcc), body-centered cubic (bcc), and hexagonal close-packed (hcp). Common fcc metals include aluminum, iron (above 910 °C, or 1670 °F), copper, stainless steel (18Cr-8Ni), nickel, lead, silver, and gold. The bcc metals include titanium (above 885 °C, or 1625 °F), iron (below 910 °C, or 1670 °F), tantalum, niobium, molybdenum, and tungsten. The hcp metals include titanium (below 885 °C, or 1625 °F), magnesium, beryllium, zinc, and cadmium.

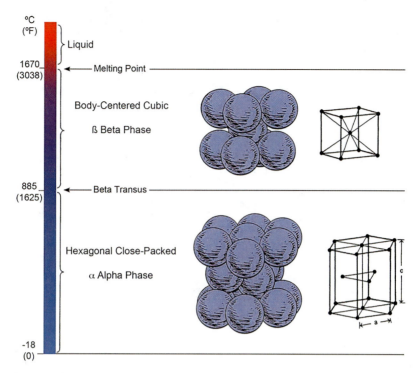

Fig. 3.4 Crystal structure of titanium. Titanium is allotropic: hexagonal close-packed (alpha) up to 885 °C (1625 °F) and body-centered cubic (beta) from 885 to 1670 °C (1625 to 3038 °F).

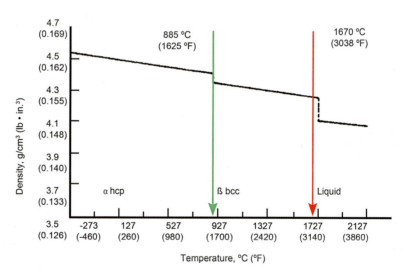

Fig. 3.5 Density of pure titanium as a function of temperature. The volume increases as the hexagonal close-packed (hcp) alpha phase transforms to body-centered cubic (bcc) beta at 885 °C (1625 °F), the beta transus.

56% that of steel and 40% greater than aluminum. This property, in combination with other desirable characteristics, accounts for the usefulness of the metal. Because the strength levels of titanium-base alloys are similar to those of steels, the lower den-sity of titanium results in a more efficient design due to lower weight.

The thermal conductivity (measure of heat flow) of titanium at room temperature is near 17.3 W/m · K (10 Btu/h · ft · °F). This value is low

Table 3.3 Physical properties of titanium (at room temperature unless otherwise noted)

Atomic number: 22
Atomic weight: 47.90 g/mol
Electronic structure, neutral state:
  $1s^2, 2s^2, 2p^6, 3s^2, 3p^6, 3d^2, 4s^2$
Absorption cross section for thermal neutrons: 5.6 barns

Crystal structure:

  α: to 885 °C (1625 °F) hexagonal
    close-packed
    $c = 4.6832$ Å, $a = 2.9504$ Å, $c/a = 1.5873$
  β: 885 °C (1625 °F) to melting point: body-centered cubic
    $a = 3.3065$ Å, closest atom distance, 2.860 Å

Density:

  α: 4.505 g/cm³ (0.163 lb/in.³), 4.35 g/cm³ (0.157 lb/in.³)
    at 870 °C (1600 °F)
  β: 4.32 g/cm³ (0.156 lb/in.³) at 900 °C (1650 °F)

Elastic moduli, polycrystalline:

  Young's: 109 GPa (15.8×10^6 psi)
  Shear: 37 GPa (5.6×10^6 psi)
  Bulk: 123 GPa (17.8×10^6 psi)
Poisson's ratio: 0.34
Transformation temperature (hexagonal close-packed to
  body-centered cubic): 885 °C (1625 °F)
Melting temperature: 1670 °C (3038 °F)
Boiling temperature: 3260 °C (5900 °F)
Heat of transformation: 4.351 kJ/mol (1.050 kcal/mol)
Heat of fusion: 20.9 kJ/mol (5 kcal/mol)

Heat of vaporization: 470.7 kJ/mol (112.5 kcal/mol)
Heat of oxide formation: 914.6 kJ/mol (218.6 kcal/mol)
Surface tension (for liquid-state temperatures):
  1200 dynes/cm (6.852×10^{-3} lb/in.)

Thermal expansion coefficient:

  α: 9.9 μcm/cm · °C (5.5 μin./in. · °F)⊥ to C,
  12.1 μcm/cm · °C (6.7 μin./in. · °F) II to C,
  10.3 μcm/cm · °C (5.7 μin./in. · °F) for random
    polycrystalline,
  11.5 μcm/cm · °C (6.4 μin./in. · °F) at 871 °C (1600 °F)
  β: 11.5 μcm/cm · °C (6.4 μin./in. · °F) at 900 °C (1650 °F)
Thermal conductivity: 17.3 W/m · K (10 Btu/h · ft² · ft · °F)
Specific heat: 5.976 cal/mol · °C
Electrical resistivity: 42 μΩ · cm
Magnetic susceptibility: 3.2×10^{-6} emu/g
Thermoelectric force (Ti-Pt couple): 0.2 mV

Emissivity:

  α: 810–870 °C (1490–1598 °F): 0.459
  β: 1400 °C (2550 °F): 0.42
Work function: 4.17 eV

X-ray spectra:

  Kα: 2.750 Å
  Kβ: 2.514 Å
  K absorption edge: 2.496 Å

relative to many other metallic elements, and the addition of alloying elements further reduces thermal conductivity. The thermal conductivity of titanium alloys is one-half that of stainless steels and one-tenth that of aluminum alloys.

The electrical resistivity (resistance to electrical flow) of titanium is lower than that of 17-7PH precipitation-hardening stainless steel and 10 times greater than that of aluminum alloys. Alloying elements substantially increase the resistivity of titanium and all other metals.

The linear coefficient of thermal expansion (change in dimensions due to heating) of titanium is lower than that of the aluminum alloys and steels. The value is near those of ceramics and glass.

The specific heat defines the amount of energy required to change the temperature of a unit weight one degree. In general, the specific heats of titanium and titanium alloys are nearly the same as that of steels and approximately 60% of aluminum alloys.

The high-temperature performance of a metal is partially related to the melting point. Before 1910, the melting point of titanium was thought to be greater than 5500 °C (10,000 °F). This led to much optimism for its potential high-temperature use. However, as better measuring techniques and higher-purity titanium became available, the reported melting point decreased over the years and now is at 1670 °C (3038 °F). This melting point is

150 to 200 °C (270 to 360 °F) higher than that of steel and approximately 980 °C (1760 °F) higher than that of pure aluminum. Alloying elements can either increase or decrease the melting range.

Alloying Elements

Pure titanium, although very ductile, has relatively low strength. The metal is more useful when its strength is increased by alloying. The addition of other elements to titanium results in three classes of terminal alloys (Ref 3.3):

- Alpha (α) alloys
- Alpha-beta (α-β) alloys
- Beta (β) alloys

They are characterized by the phases (crystal structures) that exist in the alloy near room temperature.

The temperatures at which the alpha and beta phases can exist are altered as alloying elements are added to pure titanium. Based on their influence on the proportions of the alpha and beta phases below the beta transus, the alloying elements are divided into three groups:

- Alpha stabilizers
- Beta stabilizers
- Neutral additions

Some examples of the strengthening effect of alloying elements from each group are shown in Table 3.4. The following sections cover the behavior of each group of alloying elements.

Alpha Stabilizers. Some elements, when added to titanium, increase the temperature at which the alpha phase can exist (Ref 3.3). These elements preferentially dissolve in the alpha phase and are known as alpha stabilizers. The titanium-aluminum phase diagram (Fig. 3.6) is a typical example.

There exists an intermediate, ordered phase called Ti_3Al (α_2) and ordered TiAl (γ) at higher aluminum content (Ref 3.6, 3.7). As aluminum is added, these atoms replace titanium atoms while still maintaining the hexagonal structure until the solubility limit is reached. For example, at 650 °C (1200 °F), the hexagonal phase can accept only approximately 8 wt% Al. Further additions form the intermediate phase, Ti_3Al.

The influence of aluminum content on the ambient temperature strength and ductility of titanium is shown in Fig. 3.7. The strength level of titanium is greatly increased by aluminum. Ductility is excellent up to 8% Al, where a sufficient amount of Ti_3Al can cause embrittlement. Neutral

Table 3.4 Effect of alloy additions on strength of titanium

Element	Addition to base, %	Yield strength Annealed(a) MPa	Annealed(a) ksi	Heat treated(a) MPa	Heat treated(a) ksi
Base titanium	...	241	35	...	...
Alpha stabilizer					
Nitrogen	0.1	483	70	...	...
Oxygen	0.1	365	53	...	...
Carbon	0.1	324	47	...	...
Aluminum	4	496	72	...	...
Neutral stabilizer					
Zirconium	4	331	48	...	...
Tin	4	310	45	...	...
Beta stabilizer					
Iron	4	593	86	703	102
Chromium	4	510	74	655	95
Manganese	4	503	73	634	92
Molybdenum	4	490	71	620	90
Tungsten	4	483	70	572	83
Vanadium	4	400	58	496	72
Niobium	4	310	45	324	47
Hydrogen	0.1	241	35(b)	...	...
Silicon	1	448	65	...	...

(a) Annealed (beta transus: 93 °C, or 200 °F), furnace cooled to 482 °C (900 °F), air cooled, heat treated (beta transus: 38 °C, or 100 °F), ½ h water quenched 538 °C (1000 °F), 2 h air cooled. (b) Strengthening has been noted in α + β and some α alloys.

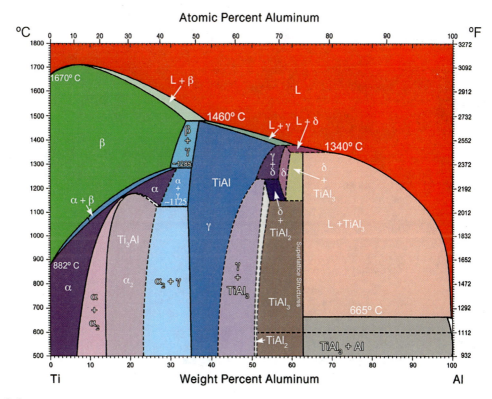

Fig. 3.6 The titanium-aluminum phase diagram

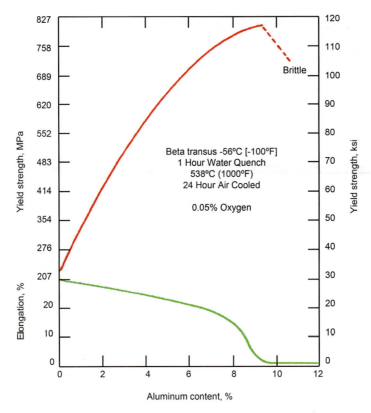

Fig. 3.7 Influence of aluminum on embrittlement in the titanium-aluminum system. Aluminum (up to approximately 8%) sharply increases the strength level of titanium, but it lowers ductility.

elements tin and zirconium, in combination with aluminum, also cause embrittlement or lack of ductility after creep exposure. The maximum amount of these elements that can be tolerated in titanium is expressed by the following equation:

$$\%Al + \%Sn/3 + \%Zr/6 \leq 8 \qquad \text{(Eq 3.1)}$$

Many commercial titanium alloys use the greatest possible amount of these elements. Some examples are Ti-5Al-2.5Sn, Ti-6Al-4V, Ti-6Al-2Sn-4Zr-2Mo, Ti-6Al-2Nb-1Ta-0.8Mo, and Ti-8Al-1Mo-1V.

Alloys referred to as α_2 aluminides are formulated based on the Ti_3Al (Ti-15.8wt%Al) compound. This compound, which has been shown to be inherently brittle, offers the potential for large improvements in elevated-temperature properties (see Figure 1.7, Chapter 1, "History and Extractive Metallurgy," and later in the present chapter). Additions of beta stabilizers to this system produce α_2 aluminides and TiAl aluminides having useful engineering properties. Other elements also stabilize the alpha phase (Table 3.5). Of par-

Table 3.5 Classification of major alloying elements in titanium

Alpha stabilizing	Beta isomorphous	Beta eutectoid	Neutral
Aluminum	Vanadium	Copper	Zirconium
Gallium	Niobium	Silver	Hafnium
Germanium	Tantalum	Gold	Tin
Lanthanum	Molybdenum	Indium	...
Cerium	Rhenium	Lead	...
Oxygen	...	Bismuth	...
Nitrogen	...	Chromium	...
Carbon	...	Tungsten	...
...	...	Manganese	...
...	...	Iron	...
...	...	Cobalt	...
...	...	Nickel	...
...	...	Uranium	...
...	...	Hydrogen	...
...	...	Silicon	...

ticular interest are the interstitials oxygen, nitrogen, and carbon (Ref 3.3). Instead of replacing a titanium atom on a lattice site, these elements, because of their small diameter, are found between the titanium atoms in the interstices. At small concentrations and properly controlled, interstitials improve strength. At high levels, dramatic decreases in ductility occur.

Small quantities of each of the interstitial elements are found in the metal after the reduction of the ore. During the late 1950s and early 1960s, the interstitial content of titanium gradually decreased with improvements in the reduction and melting processes. Because the concentrations of all the interstitials have decreased, oxygen is intentionally added to compensate for the loss in strength. Consequently, oxygen is an important alloying element. Carbon and nitrogen are not used as extensively, because oxygen additions are easier to make and are highly effective.

Although carbon, oxygen, and nitrogen stabilize the alpha phase and raise the beta transus (Ref 3.3), important differences exist in the titanium-phase equilibriums for these elements at dilute concentrations. The titanium-rich sections of the phase diagrams for each of these interstitials are shown in Fig. 3.8. Carbon has only limited solubility in both the alpha and beta phase. Nitrogen and oxygen have extensive solubility in the alpha phase, with oxygen having the greater. A re-view of Fig. 3.2 (atomic diameters) shows that of these three elements, oxygen has the smallest atomic diameter and thus the greatest solubility in alpha, while carbon has the largest atomic diameter and therefore the least solubility in alpha titanium.

The alpha-stabilizing interstitial elements carbon, oxygen, and nitrogen increase the strength and decrease the ductility of titanium. Nitrogen is the most potent. Strengthening by addition of oxygen and nitrogen is not a linear function with composition but follows a parabolic curve. However, the strengthening effect can be considered linear for the small composition ranges. In pure titanium, the strengthening effect of 0.03% C, 0.02% O, and 0.01% N additions is nearly identical (Fig. 3.9). This approximation is often expressed as an oxygen equivalency, where $\frac{2}{3}(\%C)$ + 2(%N) + (%O) equals the oxygen equivalent interstitial content in titanium.

Oxygen and nitrogen strengthen alpha titanium alloys but not on the same equivalency noted for pure titanium. Although nitrogen is more potent than oxygen, the actual strengthening effect of these elements appears related to the alloy con-

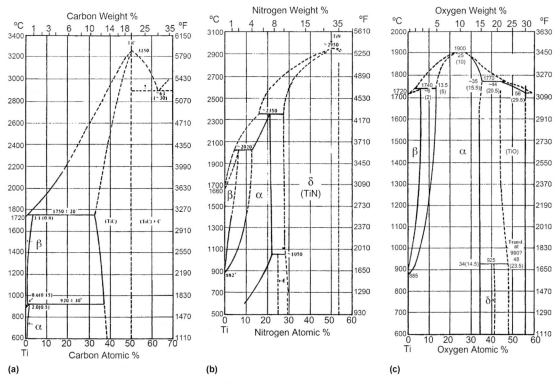

Fig. 3.8 Partial phase diagrams for (a) carbon, (b) nitrogen, and (c) oxygen systems. Of these three elements, oxygen has the smallest atomic diameter and hence the greatest solubility in alpha titanium; carbon has the largest atomic diameter and the lowest solubility in alpha titanium.

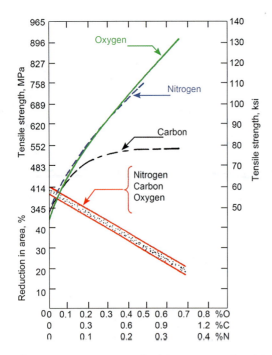

Fig. 3.9 Tensile properties of iodide titanium as affected by carbon, oxygen, and nitrogen. Generally, these alpha-stabilizing interstitial elements increase strength and decrease ductility of titanium.

tent. In studies of alpha alloy Ti-5Al-2.5Sn, the influence of oxygen and nitrogen on strength was substantially less than noted for unalloyed titanium.

The influence of these elements on the properties of alpha-beta titanium alloys is more complex, because most of the oxygen and nitrogen is present in the alpha phase. In the predominantly alpha-matrix alloys, such as Ti-6Al-4V, oxygen and nitrogen have a strengthening influence of approximately 8.3 MPa (1200 lb/in.²) per 0.01% O and 9.6 MPa (1400 lb/in.²) per 0.01% N. This is less than the reported values for 0.01% additions to pure titanium of 12.1 MPa (1750 lb/in.²) for oxygen and 24.1 MPa (3500 lb/in.²) for nitrogen. Based on the available data, the following oxygen equivalency formulas are indicated:

%Oe = (%O) + 1.2(%N) + 0.67(%C)

for Ti-6Al-4V (Eq 3.2)

%Oe = (%O) + 2.0(%N) + 0.67(%C)

for unalloyed titanium (Eq 3.3)

Although the interstitials have a favorable influence on strength, they are detrimental to fracture toughness as measured by a K_{IC} test. As a result, certain alloys designed for use in high-fracture-

toughness applications contain extralow interstitials. Commercially, these alloys are referred to as extralow interstitial alloys.

Beta Stabilizers. While alpha stabilizers increase the transformation temperature of titanium, beta stabilizers depress it and stabilize the beta phase to lower temperatures (Ref 3.3). The titanium-niobium system (Fig. 3.10) is an example of a beta-stabilized system.

Important features in this diagram include:

- The titanium beta phase is isomorphous with niobium; both have the bcc crystal structure. As a result, this type of system is referred to as beta isomorphous. Other binary systems of this type are titanium-molybdenum, titanium-vanadium, and titanium-tantalum. These elements, particularly vanadium and molybdenum, are frequently used in titanium alloys.

- A 100% Ti beta phase can exist at low temperatures, provided a sufficient amount of alloying element is added. For the titanium-niobium example in Fig. 3.10, nearly 56 wt% Nb is required to stabilize the beta phase to 400 °C (750 °F).

- Both alpha and beta phases can coexist at temperatures where most titanium alloys are used (up to 600 °C, or 1110 °F). This depends on the total amount of beta-stabilizing elements present. For example, at 800 °C (1470 °F), both phases coexist at niobium contents between 2 and 7 wt%. At 600 °C (1110 °F), the range is expanded to between 4 and 28% Nb. By selecting the composition and temperature in this two-phase area, the percent of each phase present and the alloy content (niobium, in this example) in each phase can be determined. This principle forms the basis for much of the heat treatment of titanium and is discussed in detail in Chapter 4, "Principles of Beta Transformation and Heat Treatment of Titanium Alloys," in this book.

- In addition to the beta-isomorphous system, some beta stabilizers form beta-eutectoid systems (Ref 3.3). An example, the titanium-iron phase diagram, is shown in Fig. 3.11. Other elements that form this type of system are listed in Table 3.5. A similarity exists between the eutectoid system and the isomorphous system just discussed.

The beta transus is depressed in both systems. However, the similarity ends at this point. In the eutectoid system, a decomposition or transformation of the beta phase occurs at the eutectoid temperature, which is represented as A'-A in Fig.

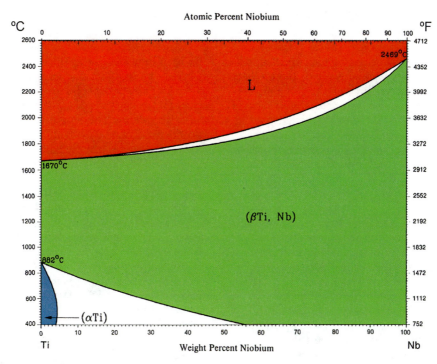

Fig. 3.10 The titanium-niobium phase diagram. This beta-stabilized system is typical of the beta-isomorphous type. Both titanium and niobium have a body-centered cubic crystal structure.

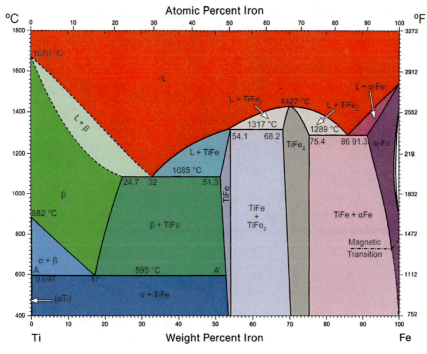

Fig. 3.11 The titanium-iron phase diagram, a typical beta-eutectoid system

3.11. At this temperature, the beta phase decomposes, and at equilibrium it forms the alpha phase plus an intermediate phase, typically an intermetallic compound. In the example shown in Fig. 3.11, the intermediate phase is the compound TiFe.

There are many titanium eutectoid systems, and their eutectoid temperatures vary from near 538 °C (1000 °F) to above 815 °C (1500 °F). At higher temperatures below the eutectoid temperature, the eutectoid reaction is rapid, and the beta phase transforms readily into the components alpha plus an intermediate phase. However, at lower temperatures sufficient thermal energy is not available to readily decompose the beta phase; thus, the eutectoid reaction proceeds sluggishly. Alloys in which this transformation to eutectoid is sluggish can often be treated like alloys in the beta-isomorphous system.

When the transformation is rapid and occurs during normal heat treatment cycles, the constitution system is known as active eutectoid. The degree of eutectoid transformation occurring during heat treatment depends on the eutectoid temperature and the amount of beta stabilizer in relation to the composition at the eutectoid point.

Eutectoid transformation is favored by high eutectoid temperatures and high alloy contents. Table 3.6 lists the eutectoid temperature and composition for several titanium binaries. These systems are listed by decreasing eutectoid temperatures. The active eutectoid systems encompass those between silicon and cobalt, while the sluggish eutectoids are normally those below cobalt in Table 3.6.

The active eutectoid elements are used sparingly in formulating titanium alloys because of possible thermal instability or embrittlement. Alloys developed in the United States and Great Britain use the active eutectoid element silicon to enhance creep performance. Small additions of silicon are quite effective in improving creep strength. Figure 3.12 shows the dramatic effect of silicon on the creep strength of the Ti-6Al-2Sn-4Zr-2Mo (Ti-6242) alloy. An optimum addition

exists near 0.1% Si in the Ti-6Al-2Sn-4Zr-2Mo alloy. This coincides with maximum silicon solubility for the heat treatment applied. Alloys have also been formulated with silicon content as high as 0.5%. The optimum amount of silicon depends on the beta content of the alloy (solubility of silicon is greater in beta phase than in alpha phase) and on the presence of alloying elements such as zirconium that have a great affinity for silicon.

Table 3.4 shows that the additions of beta stabilizers can substantially increase the strength of annealed titanium. The strength of these systems can be further improved by heat treatment. This is also illustrated in Table 3.4 and is discussed in the next chapter.

Not all beta stabilizers improve the properties of titanium. One notable example is the interstitial element hydrogen. In the early 1950s, hydrogen embrittlement threatened to disrupt the titanium industry. However, the problem was isolated and solved by minimizing the hydrogen content through vacuum arc melting, vacuum annealing, better atmospheric control while heating, and use of strongly oxidizing pickling media during processing. However, hydrogen can be used in a positive way to enhance the mechanical properties of titanium by using it as a temporary alloying element (see the section "Thermohydrogen Processing" in Chapter 8, "Melting, Casting, and Powder Metallurgy," in this book) (Ref 3.8).

Table 3.6 Eutectoid temperature and composition in binary titanium alloys

Alloying element	Eutectoid temperature °C	Eutectoid temperature °F	Eutectoid composition, wt%	Eutectoid behavior
Silicon	860	1580	0.9	
Silver	850	1565	19.8	
Gold	830	1530	15.9	
Copper	790	1450	8.1	Active eutectoids
Nickel	770	1415	5.5	
Tungsten	715	1320	28	
Cobalt	685	1265	9	
Chromium	670	1240	15	
Iron	595	1100	16	Sluggish eutectoids
Manganese	550	1020	20	

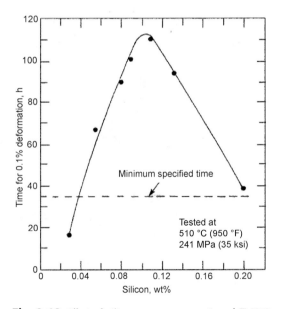

Fig. 3.12 Effect of silicon on creep properties of Ti-6242 (15.9 mm, or 5/8 in. bar). Heat treatment: (β transus: −142 °C, or −255 °F), 1 h air cool + 595 °C (1100 °F), 8 h air cool

Nevertheless, hydrogen typically is still present in titanium in quantities to 200 ppm (0.02 wt%). Under certain circumstances, small amounts of dissolved hydrogen can affect several properties of titanium. Hydrogen is often suspected of being involved with hot salt corrosion, aqueous stress corrosion, thermal instability, and poor notch toughness; these properties are discussed in subsequent chapters.

Four principal sources of hydrogen in titanium and its alloys are an impurity in titanium sponge and alloy additions, pickup from acid pickling and descaling baths, atmospheres in heating furnaces, and intentional additions (up to 2 wt%) in special equipment for temporary alloying with hydrogen.

The effects of hydrogen are complex. Hydrogen embrittlement is observed under varying conditions of stress, time, and temperature. It is most pronounced at room and lower temperature, where embrittlement can cause a loss in ductility. At these low temperatures, a titanium hydride phase can precipitate.

The titanium-hydrogen phase diagram (Fig. 3.13) shows several important features (Ref 3.3). The addition of hydrogen lowers the beta transus to the eutectoid temperature of approximately

300 °C (570 °F). Most important is the solubility of hydrogen in titanium. At low temperatures, the phase diagram shows no solubility of hydrogen in alpha; thus, the hydride phase delta (δ) is present at low hydrogen concentrations. Conversely, hydrogen has appreciable solubility in the beta phase.

Based on these observations, an alloy containing both alpha and beta phases should have the hydrogen content partitioned: the beta phase containing substantially all the hydrogen (because of high solubility) and the alpha phase containing very little.

Two different types of hydrogen embrittlement are observed in titanium and titanium alloys. Alloys containing predominantly the alpha phase are susceptible to hydrogen embrittlement in a rapid strain-rate test such as impact testing. Figure 3.14 shows an example of this type of embrittlement.

For pure titanium, the breaking energy in a notched impact specimen decreases rapidly as the hydrogen content increases. Because impact embrittlement results from the presence of a hydride, embrittlement is lessened by alloying to increase the hydrogen solubility in alpha phase. As shown in Fig. 3.14, aluminum is quite effective in this

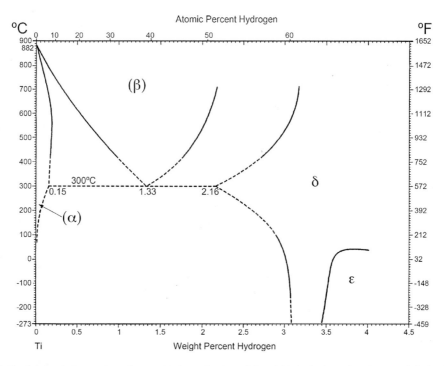

Fig. 3.13 The titanium-hydrogen phase diagram. Hydrogen is substantially soluble in the beta phase but essentially insoluble in the alpha phase at room temperature.

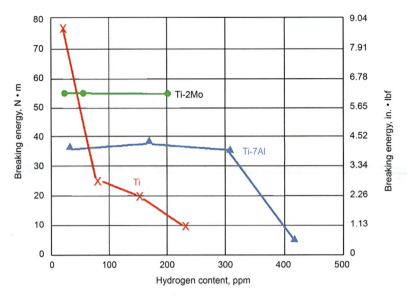

Fig. 3.14 Effect of hydrogen on impact strength. Small additions of the beta stabilizer molybdenum and the alpha stabilizer aluminum increase the tolerance for hydrogen at room temperature. Note that unalloyed titanium is embrittled severely by hydrogen.

respect. The addition of 7% Al raises the hydrogen content necessary to cause the embrittlement from approximately 55 ppm to more than 300 ppm.

Small additions of beta stabilizers also increase the hydrogen tolerance of an alpha alloy. A 2% Mo addition (Fig. 3.14) appreciably increases the tolerance for hydrogen. This improvement is attributed to the high solubility of hydrogen in the beta phase of the Ti-2%Mo alloy.

Alpha-beta-type alloys may be susceptible to both the impact type and a strain-aging type of hydrogen embrittlement. The strain-aging type of embrittlement is most severe at slow testing speeds, which indicates that hydrogen diffusion is a significant factor. An example of this type of behavior is shown in Fig. 3.15. In this example, notched tensile specimens of Ti-8Mn sheet were loaded to stresses below the typical failure stress (100%).

Samples containing 260 ppm hydrogen failed after 10 h at a stress as low as 50% of the typical failure strength. Sheet containing 169 and 20 ppm showed no indication of this severe hydrogen embrittlement. As a result of numerous tests similar to this example, hydrogen specifications for titanium were established at 125 to 200 ppm maximum, depending on the alloy. Fortunately, most alpha-beta alloys are not as susceptible to this embrittlement as the Ti-8Mn alloy.

Hydrogen embrittlement in alpha-beta alloys is attributed to the direct diffusion of hydrogen atoms to the alpha-beta phase interface during

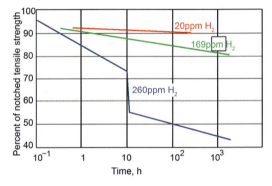

Fig. 3.15 Notched stress-rupture properties under sustained loading. The Ti-8Mn sheet alloy is susceptible to the strain-aging type of hydrogen embrittlement. Notched tensile specimens were loaded to stresses below normal failure stress.

plastic deformation. During deformation, the hydrogen concentration increases in the beta phase to progressively higher values. When the concentration finally reaches the solubility limit for hydrogen in beta, a hydride precipitates at the interface. The hydride is relatively brittle, and microcracks can readily form and propagate, particularly in zones of high stress concentration, such as the apex of stressed notches.

Two methods are recognized for increasing the tolerance of hydrogen in titanium. One method consists of increasing the solubility of hydrogen in alpha titanium by aluminum additions. A second method is to add a beta-stabilizing element such as vanadium or molybdenum to stabilize a small amount of beta phase to room temperature. Because the solubility of hydrogen in the beta phase is greater than that in the alpha phase, hydrogen remains in solution in the beta particles, and, consequently, tolerance for hydrogen increases. Due to this high solubility of hydrogen in beta phase, the metastable beta titanium alloys have a much greater tolerance for hydrogen than alpha alloys.

Several hydrogen treatments were introduced to modify and refine the microstructure of titanium alloy products and net-shape parts such as those produced by powder metallurgy and casting. The low beta transus temperature (Fig. 3.13) allows beta solute treatments to be performed on Ti-6Al-4V at temperatures as low as 815 °C (1500 °F). Several experimental and commercial treatments based on temporary alloying with hydrogen were developed, including constitutional solution treatment and thermohydrogen processing; the latter was developed by the U.S. Air Force Materials Laboratory (Ref 3.8).

Neutral Additions. The last group of alloying elements is known as neutral additions (Ref 3.3).

The titanium-zirconium phase diagram, typical of this type of addition, is shown in Fig. 3.16. The titanium-tin and titanium-hafnium systems are also included in this group. The latter two elements are sometimes classified as beta stabilizers because they depress the beta transus temperature. However, they can take on the characteristics of alpha stabilizers such as aluminum when added to alpha-matrix alloys.

Figure 3.17 illustrates the effect of hafnium, zirconium, and aluminum additions on tensile strength and strain after a high-temperature creep exposure. The alloy addition is expressed in this figure as atomic percent (at.%) rather than weight percent (wt%). The ultimate strength increases as the atomic percent of the combined elements increases. This improvement in strength is accompanied by greater creep strength (less plastic strain). In this example, neutral additions zirconium and hafnium improved the properties in a similar manner as the alpha stabilizer aluminum.

Neutral additions are used successfully to formulate near-alpha alloys such as Ti-6Al-2Sn-4Zr-2Mo and alpha alloys such as Ti-5Al-2.5Sn.

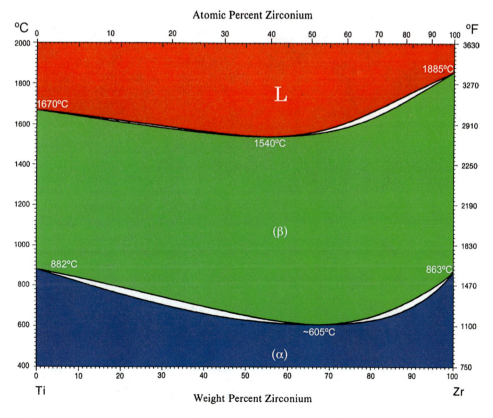

Fig. 3.16 The titanium-zirconium phase diagram. This system is typical of neutral addition elements such as zirconium, tin, and hafnium.

Other examples include their use in the alpha-beta alloy Ti-6Al-6V-2Sn and in the beta alloy Ti-3Al-8V-6Cr-4Mo-4Zr.

The characteristics of the neutral alloying additions include:

- Depression of the beta transus
- Substantial solubility in both the alpha and beta phases
- Presence of alpha phase at room temperature over a wide range of compositions

Effect of Alloying Elements on the Beta Transus Temperature. Based on a multiple linear regression analysis of experimental data, the effect of a number of alloying elements on the beta transus temperature is shown in Table 3.7 (Ref 3.9). The beta stabilizers molybdenum, vanadium, chromium, and iron all decrease the temperature, the alpha stabilizer aluminum increases the temperature, and the neutral element zirconium has a considerably smaller effect on the temperature.

Titanium Alloys

Three classes of terminal alloys that emerge as a result of alloy additions are alpha (or near-) alpha alloys, alpha-beta alloys, and beta alloys (Fig. 3.18). These basic alloy groups are defined by the phases predominant in their microstructure near room temperature.

In addition to these basic groups, advanced alpha-stabilized systems based on the intermetallic Ti_3Al and TiAl are emerging as important alloys. Each group has distinctive characteristics.

Alpha (or Near-Alpha) Alloys. Alloys that predominantly consist of the hexagonal alpha crystal structure at room temperature are classed as alpha alloys, such as Ti-5Al-2.5Sn. They sometimes contain small amounts of beta stabilizers (1% Mo plus 1% V in Ti-8Al-1Mo-1V, for example). The active beta-eutectoid element silicon is also added to alpha-matrix alloys to improve creep strength. Examples of alloys of this type include Ti-6Al-2Sn-4Zr-2Mo-0.08Si and Ti-5Al-6Sn-2Zr-1Mo-0.25Si.

Although not an alloy, commercially pure titanium is placed into this group. Small additions of iron and oxygen are sometimes added to commercially pure titanium to increase its strength.

The predominant alloying element in alpha alloys is aluminum. As discussed previously, aluminum has substantial solubility in titanium and strengthens the alpha phase. This results in a moderate strength level at room temperature, which is maintained at high temperatures. The data in Fig. 3.19 illustrate this point. The yield and creep strengths are shown as a function of testing temperature for alpha alloy Ti-5Al-2.5Sn and alpha-beta alloy Ti-8Mn. Both alloys have identical yield strengths from room temperature to 425 °C (800 °F). Upon increasing the temperature further, the Ti-5Al-2.5Sn alloy exhibits superior yield strength. The same conclusion is made for the stress at 1% plastic deformation. The excellent high-temperature strength of the alpha alloy is the most important characteristic of this group.

Alpha alloys consist predominantly of the alpha phase regardless of cooling rate from high temperatures; thus, these alloys do not respond to strengthening heat treatments. The alloys are most often used in an annealed condition. Compared with the other titanium alloy groups, they have the lowest tensile strengths at room temperature.

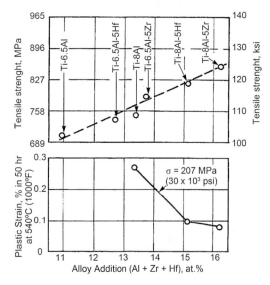

Fig. 3.17 Tensile strength and creep resistance of Ti-Al, Ti-Al-Zr, and Ti-Al-Hf. Ultimate strength rises as the atomic percent of the combined elements increases; creep strength also rises.

Table 3.7 Effect of alloying element on beta transus temperature 870 °C (1600 °F)

Element	Coefficient (for 1 wt%)	Standard error of the coefficient
Molybdenum	−14	2
Aluminum	+42	3
Zirconium	−8	3
Vanadium	−22	2
Chromium	−26	3
Iron	−15	12

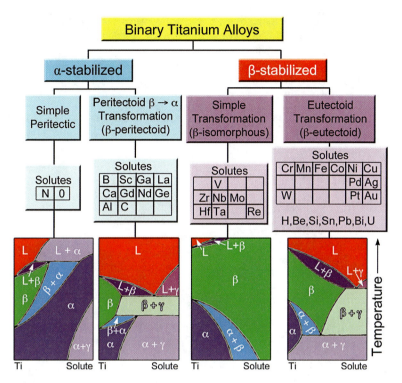

Fig. 3.18 Classification of terminal titanium alloys. Adapted from Ref 3.10

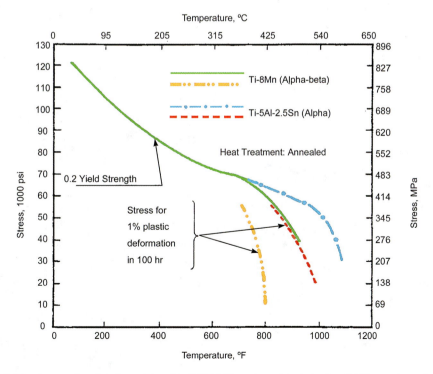

Fig. 3.19 Yield and creep strengths of an alpha alloy Ti-5Al-2.5Sn and an alpha-beta alloy Ti-8Mn are shown for a range of temperatures.

Because the alpha alloys do not respond to heat treatment when cooled through the transformation range, good ductility is usually achieved in fusion weldments.

Certain forming characteristics (as a group) are not as favorable in the alpha alloys as in other types of titanium alloys. Using the same examples (Ti-8Mn and Ti-5Al-2.5Sn), the following minimum bend radii are possible for sheets under 1.78 mm (0.070 in.) thick:

Ti-8Mn (α-β alloy): 3.0 × thickness

Ti-5Al-2.5Sn (α alloy): 4.0 × thickness

At identical strength levels, the predominantly alpha alloy is not as formable as the alpha-beta alloy. As discussed previously, the hexagonal-type crystal structure has less ductility. The specific formability and ductility of an alloy depends not only on crystal structure but also on the alloying elements and strength. In the preceding example, the strengths were identical. Commercially pure titanium with low strength levels and nearly all alpha has excellent formability. A forged compressor disc or wheel made from the near-alpha alloy IMI 685 is shown in Fig. 3.20.

In summary, characteristics of alpha alloys include:

- Excellent high-temperature strength
- Good weldability
- Nonresponsive to heat treatment
- Fair fabricability
- Good fracture toughness and fatigue crack growth rate (slow), particularly at low temperatures
- Fully alpha alloys, including commercially pure titanium and Ti-5Al-2.5Sn, have medium strength, good creep resistance, and good weldability.
- Near-alpha alloys were developed for good high-temperature performance.

Alloys used in jet-engine compressor components, including discs, blades, vanes, cases, and so on, generally contain 5 to 8% Al, some zirconium and tin, and some beta stabilizer (molybdenum, vanadium, niobium), plus silicon, which gives solid-solution strengthening and sometimes (at higher levels of >0.1) precipitation strengthening. Lenticular alpha promotes creep; equiaxed alpha enhances low-cycle fatigue and ductility. Examples of near-alpha alloys are Ti-6Al-2Sn-4Zr-2Mo-0.1Si, Ti-8Al-1Mo-1V, and IMI 685, 829, and 834.

Advanced Alpha Alloys. Titanium aluminides are alloys of titanium and aluminum based on the ordered compounds Ti_3Al ($α_2$) and TiAl ($γ$) (Ref 3.6, 3.7). These alloys offer a significant step forward in high-temperature property improvements compared with conventional titanium alloys. They have relatively light weight and good oxidation resistance. The improvements in Young's

Fig. 3.20 Forged compressor disc or wheel made from the near-alpha alloy IMI 685

modulus for Ti$_3$Al compared with the Ti-5Al-1Mo-1V alloy are shown in Fig. 3.21. Ductility at room temperature for Ti$_3$Al is quite low, as shown in Fig. 3.22. Improvements in ductility are not substantial until temperatures exceed 540 °C (1000 °F). Alloy additions of transition metals such as vanadium and niobium have been explored to improve the properties of these alloys. These additions stabilize the beta phase, resulting in ductility improvements with somewhat increased density.

The relatively low ductility and accompanying thermal cracking sensitivity of these materials makes processing difficult. The TiAl-type alloys have basically been used as castings, whereas wrought products have been produced from the Ti$_3$Al-type alloys.

The characteristics of these alloys include excellent high-temperature properties and low room-temperature ductility.

These materials are discussed further in the section "Intermetallic Compounds Ti$_3$Al and TiAl" later in this chapter.

Alpha-Beta Alloys. This group contains the largest number of alloys (Ref 3.3). The alloys are formulated so both the hexagonal alpha phase and bcc beta phase exist at room temperature. These alloys, to a certain degree, compromise some of the characteristics of both alpha and beta alloys.

Many alpha-beta alloys contain aluminum. The influence of this alpha stabilizer on elevated-temperature tensile strength is shown in Fig. 3.23. The strength of the alpha-beta alloy containing aluminum is substantially greater than either an alpha alloy containing aluminum or an aluminum-free alpha-beta alloy.

The properties that can be attained in alpha-beta alloys result from their response to heat treatment. This response is achieved by altering both the phase composition and amounts of the alpha and beta phases. This area is covered in detail in Chapter 4, "Principles of Beta Transformation and Heat Treatment of Titanium Alloys," in this book.

Most alpha-beta alloys contain substantial amounts of the beta-isomorphous elements molybdenum or vanadium. This results in excellent stability of the properties after exposure to high temperatures and stresses. The addition of beta-eutectoid elements, although they are excellent contributors to strength, causes instability due to the formation of intermetallic compounds.

The weldability of the alpha-beta alloys is related to the amount of beta phase and the way it

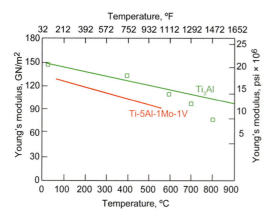

Fig. 3.21 Modulus of Ti$_3$Al

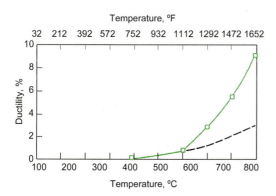

Fig. 3.22 Ductility of Ti$_3$Al. Note: Dashed line thought to be more accurate

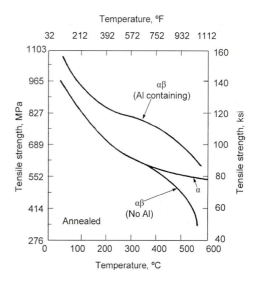

Fig. 3.23 Short-time tensile properties of three types of titanium alloys are compared.

transforms on cooling through the transformation range. Generally, the weldability of an alloy is satisfactory if the total beta-stabilizer content is low or the beta-stabilizing elements are weak. The characteristics of this class of alloys are summarized as follows and in Tables 3.8 and 3.9:

- Heat treatable to moderate to high strength levels
- Good thermal stability (if active beta-eutectoid content is low)
- Poor weldability except when the beta-stabilizing content is low (Ti-6Al-4V is weldable)
- Fair to good fabricability
- Generally good toughness and fatigue crack growth rate (slow)
- Contain elements to stabilize and strengthen the alpha and beta phases, for example, Ti-6Al-4V (IMI 318), Ti-6Al-2Sn-4Zr-6Mo, and Ti-4.5Al-5Mo-1.5Cr (Corona 5)
- Generally used as-annealed but can be strengthened considerably (to 1.2 GPa, or 180 ksi) by precipitation of the α phase
- For the α alloys, the morphology (lenticular versus equiaxed) of the α phase strongly influences the mechanical properties (see Chapter 6, "Mechanical Properties and Testing of Titanium Alloys").

Beta Alloys. This class of alloys is the smallest and consists of alloys that contain nearly 100% beta phase after air cooling from an annealing temperature (Ref 3.3). Although the group classification indicates that these are single-phase alloys, they depend on the partial transformation of the beta phase to alpha or an intermediate phase as a means of achieving high strength levels. Thus, they are sometimes referred to as metastable beta alloys and are discussed in detail in the next chapter.

Heat treating these alloys produces strengths exceeding 1380 MPa (200,000 psi). However, the strength-to-weight ratio values are somewhat compromised by the fact that the alloys are denser than most other titanium alloys. The high density is a result of the large additions of the heavy beta stabilizers (such as molybdenum, vanadium, and chromium) that are necessary to stabilize the beta phase.

The highly ductile beta phase has a large capacity for cold work; thus, these alloys can be fabricated cold in the annealed condition. The alloys can be heat treated to high strength levels after fabrication.

This class of alloys is weldable in the annealed condition. After heat treating, the ductility of the weldment decreases, although some combination of heat treatments, as well as metal deforming and heat treatment, improves weldment ductility. Following is a summary of beta-alloy characteristics:

- Heat treatable to high strength levels
- Excellent fabricability
- Excellent tensile strength up to 370 °C (700 °F)
- Poor creep strength above 370 °C (700 °F)
- Good weldability as solution treated
- Good toughness and fatigue crack growth rate (slow)
- Contain sufficient beta stabilizer to retain a fully beta structure to room temperature; the bcc structure increases ductility
- Cold formable/ageable, for example, Ti-13V-11Cr-3Al (Beta 1), Ti-11.5Mo-6Zr-4.5Sn (Beta III), Ti-15V-3Sn-3Cr-3Al (Ti-15-3), Ti-10V-2Fe-3Al (Ti-10-2-3), Ti-15Mo-2.7Nb-3Al-0.25Si (β21S)
- Beta 1 alloy used successfully on SR-71, Mach 3 aeroplane
- Solution treatment and aging similar to aluminum alloys except that generally α phase is precipitated, for example, Beta III

Table 3.8 Properties of annealed Ti-6Al-4V forgings

Annealed 2 h at 705 °C (1300 °F), air cooled after forging

Property	Forging treatment(a)	
	α + β phase field	β phase field
Tensile ultimate, MPa (ksi)	978 (142)	991 (144)
Tensile yield, MPa (ksi)	940 (136)	912 (132)
Tensile elongation, %	16	12
Reduction in area, %	45	22
Fracture toughness, MPa · m1/2 (ksi · in.1/2)	52 (47)	79 (72)
10⁷ fatigue limit(b), MPa (ksi)	±494 (±72)	±744 (±108)

(a) α/β transus 1005 °C (1840 °F). (b) Axial loading: smooth specimens, $K_t = 1.0$

Table 3.9 Solid-solution strengthening and β-stabilizing alloying elements

Property	Element							
	V	Cr	Mn	Fe	Co	Ni	Cu	Mo
Solid-solution strengthening, MPa · wt%⁻¹	19	21	34	46	48	35	14	27
Minimum alloy content to retain β on quenching, %	14.9	6.3	6.4	3.5	7	9	13	10

Source: Ref 3.11

Terminal Alloy Formulation

Terminal alloys can be compared in many ways, based on their composition (e.g., their type and effect of alloying element). One method of grouping titanium alloys is illustrated in Fig. 3.24. This simplified type of presentation illustrates the relative position of alloy compositions in relation to properties. The "alloy line" starts at the left and represents alloys containing 100% alpha at room temperature. Moving toward the right, it represents increasing amounts of beta phase, due to the addition of beta-stabilizing elements.

Finally, on the extreme right are the metastable beta alloys that retain 100% beta at room temperature. The upper portion of the figure illustrates how the indicated property can be improved by altering the amount of alpha and beta phases. For example, response to heat treatment improves by increasing the percent beta phase. Several other important properties are included in this figure.

The relative positions of some commercial titanium alloys are shown in the lower portion of Fig. 3.24. The position of these alloys on this schematic diagram can be determined easily with the knowledge of the relative beta-stabilizing effect of the alloy additions. The beta-stabilizing effect is closely related to the strengthening effect of the element in titanium, as discussed in the next chapter.

Table 3.4 presents information on the strengthening influence of many of the important alloying elements. Iron is the most potent strengthener in the beta-stabilizing group, while niobium is one of the least effective. On this basis, considerably more niobium than iron would be required to stabilize equal amounts of the beta phase.

Using the information in Table 3.4, the position of an alloy on this alloy line can be estimated in Fig. 3.24. For example, an alloy of Ti-6Al contains only the strong alpha stabilizer aluminum. This composition falls at the 100% alpha point. The addition of 4% Nb to this alloy (Ti-6Al-4Nb) shifts it to the right, because niobium is a beta stabilizer.

However, niobium is a weak beta stabilizer (Table 3.4); thus, the alloy falls between the Ti-8Al-1Mo-1V and Ti-6Al-2Sn-4Zr-2Mo alloys. If 4% Cr is substituted for the niobium (Ti-6Al-4Cr), and because the chromium is a strong beta stabilizer and more potent than 4% Mo, this alloy falls to the right of Ti-7Al-4Mo and to the left of Ti-8Mn.

These examples illustrate how titanium alloys can be classified and compared when the composition is known. A good understanding of the effects of alloying elements is a valuable asset when dealing with the numerous titanium alloys. Using the approach just discussed, an alloy can be quickly assessed based on the chemical composition.

Intermetallic compounds Ti$_3$Al and TiAl

Two titanium-aluminum intermetallics, Ti$_x$Al (x = 1 or 3), have been studied extensively (see the titanium-aluminum phase diagram in Fig. 3.25),

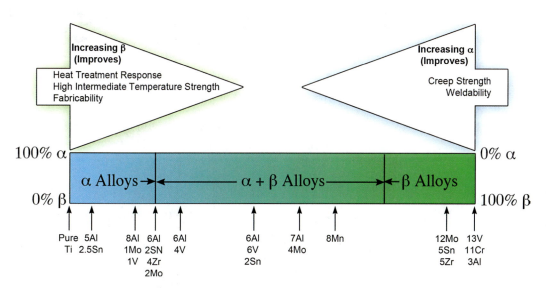

Fig. 3.24 Compositional relationship of several terminal titanium alloys by increasing or decreasing the percentage of alpha and beta phases of an alloy

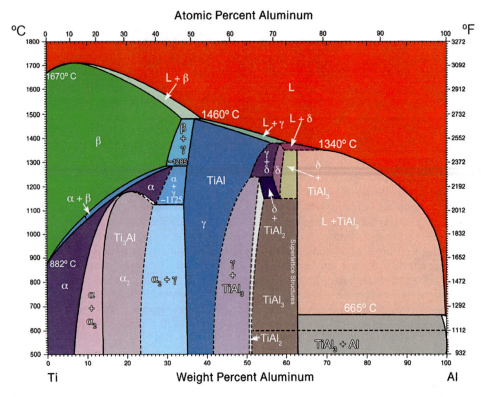

Fig. 3.25 The titanium-aluminum phase diagram showing the temperature-composition in which Ti$_3$Al and TiAl are stable

with the equiatomic TiAl showing the greatest potential for commercialization (Fig. 3.26) (Ref 3.6, 3.7). TiAl$_3$ is not considered here, because it has not shown sufficient promise for commercialization and has not been as extensively researched.

The crystal structure of Ti$_3$Al (α_2) phase and possible slip planes and slip vectors in the structure are shown in Fig. 3.27.

Crystal structure and Burgers vectors in the TiAl (γ) phase are shown in Fig. 3.28. The difficulty in dislocation motion in both the intermetallics (particularly TiAl) contributes to their good high-temperature creep behavior, but, at the same time, they exhibit low room-temperature ductility. This lack of ductility requires a change in philosophy by designers who generally use a fracture mechanics approach in structural design, assuming propagation of a preexisting crack. They traditionally also demand a minimum level of ductility (4 to 5% elongation) as a further "comfort" factor. Further discussion of the mechanical properties of the intermetallics can be found in Chapter 6, "Mechanical Properties and Testing of Titanium Alloys," in this book.

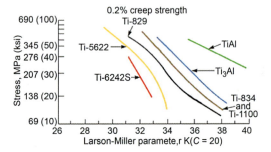

Fig. 3.26 Creep behavior of a number of terminal high-temperature titanium alloys and the intermetallic compounds Ti$_3$Al and TiAl, showing the enhanced creep behavior of the intermetallics, particularly the equiatomic TiAl

Summary

- Titanium is a typical transition element; it has high strength and allotropic behavior. The titanium atom has a median diameter compared with other elements. Thus, it has low density and favorable alloying characteristics.

- Alloying other elements with titanium results in three classes of alloys: alpha, alpha-beta, and beta, each having its own distinctive properties.

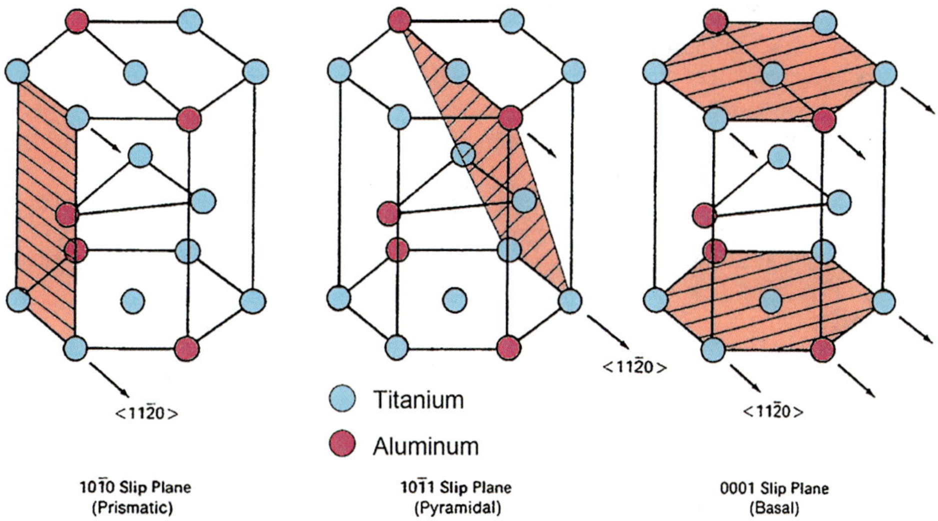

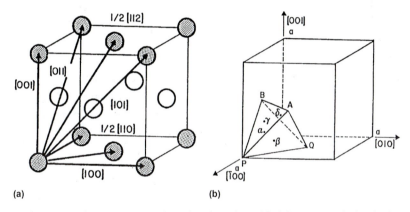

Fig. 3.27 Crystal structure of Ti₃Al (α2) phase and possible slip planes and slip vectors in the structure

Fig. 3.28 (a) Crystal structure and Burgers vectors in the TiAl (γ) phase. (b) Modified Thompson tetrahedron for the L1 structure

- The atomic diameter of titanium makes it suitable for alloying with several elements, including manganese, iron, vanadium, molybdenum, aluminum, tin, and zirconium.
- Substitution alloying is possible when the alloying element is within ±15% of the diameter of the titanium atom. Interstitial alloying is possible when the alloying element is less than 0.6 times the diameter of the titanium atom.
- Titanium is an allotropic element. At room temperature, it exists as an hcp structure. When heated to a temperature exceeding 885 °C (1625 °F), the structure changes to bcc.
- Density of titanium ranges from 4.318 to 4.872 g/cm³ (0.156 to 0.176 lb/in.³), depending on the amount and type of alloying elements. Thus, the density of titanium and titanium alloys is 56% that of steel and 40% greater than most aluminum alloys.

- Alloying elements are divided into three classes: alpha stabilizers, beta stabilizers, and neutral additions.
- Some elements, when added to titanium, increase the temperature at which the alpha phase can exist and are thus known as alpha stabilizers. Aluminum is a notable example.
- As opposed to alpha stabilizers, beta stabilizers depress the transformation temperature of alpha to beta. The titanium-niobium system exemplifies beta stabilization. It is possible for 100% beta phase to exist at room temperature if sufficient amounts of beta-stabilizing elements are added.
- A third group of alloying elements is known as neutral additions. Zirconium, tin, and hafnium are typical examples. Characteristics of neutral alloying additions include some depression of the beta transus temperature, substan-

tial solubility in both alpha and beta phases, and alpha being present at room temperature over a wide range of conditions.

- Alpha alloys have good strength at elevated temperature, good weldability, and fair fabricability, but they do not respond to heat treatment.

- The alpha-beta alloy group represents the greatest number of titanium alloys. These alloys are characterized by good heat treating response, good thermal stability, generally poor weldability, and fair to good fabricability.

- The beta alloys represent the smallest group of titanium alloys. Their characteristics include good response to heat treatment, excellent fabricability, poor creep strength, and good weldability in the annealed condition.

- The phase diagram for the titanium-aluminum system shows the Ti_xAl (where $x = 1$ or 3) intermetallics. These compounds feature creep performance better than the terminal titanium alloys but low room-temperature ductility. These characteristics require a change in philosophy by design engineers who typically require 4 to 5% elongation.

REFERENCES

3.1 F.H. Froes, D. Eylon, and H.B. Bomberger, Ed., *Titanium Technology: Present Status and Future Trends,* Titanium Development Association, Dayton, OH, 1985

3.2 W.D. Callister, Jr. and D.G. Rethwisch, *Materials Science and Engineering: An Introduction,* John Wiley and Sons, Hoboken, NJ, 2010

3.3 J.L. Murray, *Phase Diagrams of Binary Titanium Alloys,* ASM International, Metals Park, OH, 1987

3.4 R.E. Reed-Hill, *Physical Metallurgy Principles,* Van Nostrand, New York, 1973

3.5 M.A. Imam, K.L. Housley, and F.H. Froes, *Titanium and Titanium Alloys,* Kirk-Othmer, 2011

3.6 H.A. Lipsitt et al., The Deformation and Fracture of Ti_3Al at Elevated Temperatures, *Metall. Trans. A,* Vol 11, 1980, p 1369–1375

3.7 Y.-K. Kim, *Acta Metall. Mater.,* Vol 40 (No. 6), 1992, p 1121–1134

3.8 F.H. Froes, O.N. Senkov, and J.I. Qazi, Hydrogen as a Temporary Alloying Element in Titanium Alloys: Thermohydrogen Processing, *Int. Mater. Rev.,* Vol 49 (No. 3–4), 2004, p 227–245

3.9 F.H. Froes, private communication, 2013

3.10 D.H. Herring, Heat Treatment of Titanium Alloys, *Ind. Heat.,* Feb 7, 2013

3.11 C. Hammond and J. Nutting, *Met. Sci.,* Vol 11, 1977, p 474

SELECTED REFERENCES

- R. Boyer, E.W. Collings, and G. Welsch, Ed., *Materials Properties Handbook: Titanium Alloys,* ASM International, 1994

- E.W. Collings, *The Physical Metallurgy of Titanium Alloys,* American Society for Metals, Metals Park, OH, 1984

- M.J. Donachi, *Titanium: A Technical Guide,* 2nd ed., ASM International, 2000

- H. Kuhn and D. Medlin, Ed., *Mechanical Testing and Evaluation,* Vol 8, *ASM Handbook,* ASM International, 2000

- P. Lacombe, R. Tricot, and G. Beranger, Ed., *Proceedings of the Sixth International Conference on Titanium* (Nice, France), Metallurgical Society of AIME, Warrendale, PA, 1988

- G. Lutjering and J.C. Williams, *Titanium,* Springer, 2003

- G. Lutjering, U. Zwicker, and W. Bunk, *Proceedings of the Fifth International Conference on Titanium,* Metallurgical Society of AIME, Warrendale, PA, 1984

- *Proceedings of the Seventh International Conference on Titanium* (San Diego, CA), Metallurgical Society of AIME, Warrendale, PA, 1992

- *Proceedings of the Eighth International Conference on Titanium* (San Birmingham, U.K.), Metallurgical Society of AIME, Warrendale, PA, 1995

- *Proceedings of the Ninth International Conference on Titanium,* Metallurgical Society of AIME, Warrendale, PA, 2000

- *Proceedings of the Tenth International Conference on Titanium,* Metallurgical Society of AIME, Warrendale, PA, 2004

- *Proceedings of the Eleventh International Conference on Titanium,* Metallurgical Society of AIME, Warrendale, PA, 2008

- *Proceedings of the Twelfth International Conference on Titanium,* Metallurgical Society of AIME, Warrendale, PA, 2012

CHAPTER 4

Principles of Beta Transformation and Heat Treatment of Titanium Alloys*

THE PROPERTIES OF TITANIUM ALLOYS are governed by chemistry and micro/macrostructure, all of which are interrelated. Thermomechanical processing, which includes heat treatment, plays a major role in establishing both microstructure and macrostructure for a given alloy. Therefore, for a given chemistry, the structure can be varied dramatically by heat treatment. In general, the heat treatment of titanium alloys is analogous to that of other precipitation-hardening systems. For example, when heat treating for increased strength, the material is heated (solution treated) to dissolve much or all of the precipitated phase (in this case, alpha) and cooled (quenched) at a rate sufficiently fast to retain much or all of this phase in metastable solution. The material is then heated again to a lower temperature (aged) to precipitate the strengthening phase. The lower the age temperature, the higher is the amount of precipitate and the finer the precipitates, and therefore, the higher the strength. Thus, heat treatment of titanium alloys normally involves the decomposition or transformation of the high-temperature, body-centered cubic beta phase to precipitate alpha phase in the beta matrix. However, a variety of intermediate phases are possible.

To fully understand the principles of heat treatment of titanium alloys, it is necessary to first understand the transformations that can occur. For example, the high-temperature beta phase can undergo a thermal decomposition to titanium martensites or omega. Also, the beta phase can decompose by a nucleation-and-growth process

to form equilibrium alpha phase or isothermal omega, beta prime, or eutectoid compound plus alpha. Although less important commercially, the alpha phase can also decompose to form ordered alpha-two phase or compound. Therefore, by controlling beta transformation and a variety of microstructures, thus mechanical properties can be developed in titanium alloys.

In this chapter, the various transformation products that result from heat treatment of titanium are discussed. The practical uses of such transformations in commercial heat treat practice are also presented.

Beta Transformation

Equilibrium Phase Relationships. Figure 4.1 shows a schematic of a partial beta-isomorphous equilibrium phase diagram typical of the binary alloy systems such as titanium molybdenum (Ti-Mo), titanium-vanadium (Ti-V), titanium-niobium (Ti-Nb), and titanium-tantalum (Ti-Ta) (Ref 4.1). The following examples illustrate several characteristics of this diagram that are important for understanding the heat treatment of titanium alloys.

Consider an alloy of composition A (Fig. 4.1) containing 6% alloying element and 94% Ti. Heating the alloy to temperature T_3 and holding at this temperature results in 100% beta phase with an alloying element content in the beta phase of 6%. When the temperature drops to T_2, the phase diagram shows that alpha and beta phases coexist.

---

*Adapted and revised from Paul J. Bania, originally from Stan R. Seagle, *Titanium and Its Alloys*, ASM International.

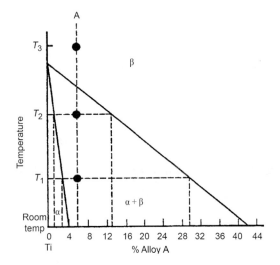

Fig. 4.1 Partial beta-isomorphous phase diagram. Raising the temperature of alloy A from T_1 to T_2 increases the volume percent of the beta phase and correspondingly decreases the amount of alpha phase. This is accompanied by a decrease in the alloy content in the beta phase.

The Lever Rule. The volume percentage and composition of each of the two phases (alpha and beta) can be calculated by applying the Lever rule (Ref 4.2). In Fig. 4.1, a line is extended horizontally (isothermally) from T_2 until the alpha and beta phase boundaries (transi) are intersected, which occurs at 1.5 and 13%, respectively.

To calculate using the Lever rule:

$$\% \text{ alpha } (\alpha) = [(13 - 6)/(13 - 1.5)] \times 100$$
$$= (7/11.5) \times 100 = 61\% \qquad \text{(Eq 4.1)}$$

$$\% \text{ beta } (\beta) = [(6 - 1.5)/(13 - 1.5) \times 100$$
$$= (4.5/11.5) \times 100 = 39\% \qquad \text{(Eq 4.2)}$$

In this example, 61% alpha and 39% beta are present at temperature T_2, and the alloying element content in the alpha phase is 1.5% and in the beta phase is 13%.

A similar analysis can be made at temperature T_1, which is summarized as follows:

Temperature	Phase present, vol%		Alloying element content, wt%	
	Alpha	Beta	Alpha	Beta
T_3	0	100	...	6
T_2	61	39	1.5	13
T_1	89	11	3.0	30

These examples illustrate important characteristics of this type of alloy system. As the temperature decreases from T_3 to T_1:

- The volume percent beta decreases.
- The alloying element content in the beta phases increases rapidly.
- The alloying element content in the alpha phase increases slightly.

These phase relationships in the binary system can be altered substantially by the addition of a third element to form a ternary alloy system. Figure 4.2 shows a change in phase relationships in the titanium-vanadium system as a result of a 6% Al addition. This addition raises the beta transus temperature, because aluminum is an alpha stabilizer. Furthermore, at a constant vanadium content and temperature, the aluminum increases the percent alpha phase present and increases the vanadium content in the remaining beta phase.

This is illustrated by the phase relationships for Ti-4%V and Ti-6%Al-4%V alloys at 760 °C (1400 °F) in Fig. 4.2. The former alloy contains approximately 66% alpha and 34% beta phases. The beta phase contains 10% V, while the alpha phase has only 1%. The Ti-6Al-4V alloy consists of nearly 86% alpha phase. However, the small amount of beta phase is extremely rich in vanadium content (nearly 16%), while the vanadium content in the alpha phase is similar to that of the binary Ti-4V.

In summary, the addition of an alpha stabilizer to a binary beta-stabilized alloy results in:

- An increase in the beta transus temperature
- An increase in the volume of alpha phase and therefore a corresponding decrease in the volume of beta phase at any given temperature in the alpha + beta phase field
- An increase in the alloy content in the remaining beta phase at any given temperature in the alpha + beta phase field

Many alloying elements used in titanium are the beta-eutectoid type. The amount of these elements added to titanium is generally small and less than that in the eutectoid composition. Thus, these are known as hypoeutectoid alloys.

The phase relationships for a beta-eutectoid diagram can be predicted by extrapolating the beta phase boundaries below the eutectoid temperature. This is illustrated in Fig. 4.3, where the beta phase transforms into alpha and an intermetallic compound designated as gamma, γ (e.g., $TiCr_2$, Ti_2Ni, Ti_2Cu).

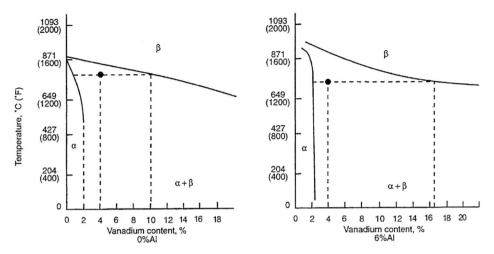

Fig. 4.2 Partial titanium-vanadium phase diagrams at 0 and 6% Al. The addition of aluminum, an alpha stabilizer, raises the beta transus temperature. It also increases the amount of alpha phase and the vanadium content in the remaining beta phase.

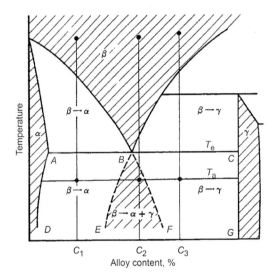

Fig. 4.3 Beta transformation in a eutectoid system. Phase relationships can be predicted by extrapolating the beta phase boundaries below the eutectoid temperature. The beta phase transforms into alpha and an intermetallic phase, gamma.

When an alloy of composition C_1 (hypoeutectoid alloy in Fig. 4.3) is quenched from the beta field to the aging temperature T_α, alpha of the composition indicated by line AD begins to precipitate. As this occurs, the composition of the beta phase is shifted from C_1 toward line BE. When the beta composition reaches line BE, gamma begins to precipitate. The decomposition of beta to alpha and gamma continues until the beta phase no longer exists.

The time required for the compound gamma to precipitate depends on temperature and alloy composition. If the alloy composition C_1 is close

to line BE, the composition of the beta phase reaches line BE more quickly than a leaner alloy composition, which would be further to the left. The transformation, or decomposition, of beta to alpha and intermetallic phase is a diffusion-controlled process. It requires the diffusion of atoms, and therefore, it proceeds faster at high temperatures where diffusion is more rapid. Thus, high alloy contents and high temperatures approaching, but below, the eutectoid temperature favor intermetallic compound formation.

Next, consider an alloy of composition C_2 (eutectoid alloy) quenched from a temperature in the beta field to T_α. In this instance, the beta is unstable with respect to both alpha and gamma, and both phases immediately form as a fine eutectoid structure similar to pearlite in steel.

Now consider an alloy of composition C_3 (hypereutectoid alloy), which is very high in alloy content. When this alloy is quenched to temperature T_α, the gamma phase precipitates until the beta phase composition reaches line BF; then both alpha and gamma precipitate.

Metastable Phases and Metastable Phase Diagrams

Phase diagrams such as the one shown in Fig. 4.1 are known as equilibrium phase diagrams, because they indicate the types and amounts of phases expected upon cooling a given composition from a known temperature (Ref 4.1, 4.3). The example cited previously for Fig. 4.1 shows that for alloy A at temperature T_2, there is 61% alpha phase containing 1.5% element A and 39% beta phase containing 13% element A. A cooling rate slow enough to permit diffusion to occur results in equilibrium conditions, that is, 95% alpha at 4% alloy content A and 5% beta at 42% alloy content A at room temperature.

However, what happens when alloy A in Fig. 4.1 is rapidly quenched from temperature T_2? If rapid enough, the quench could restrict diffusion, resulting in metastable phases. A metastable phase is one that can exist in transition to an equilibrium, or more stable, state. Metastable phases encountered in titanium alloys are martensite, metastable beta, omega, and beta prime. To indicate such phases on phase diagrams, dashed lines are often used to delineate metastable phase boundaries to differentiate them from equilibrium phase boundaries.

In summary, metastable phases are nonequilibrium phases that result from a cooling rate too rapid to permit diffusion to occur and therefore too rapid to allow formation of equilibrium phases. Metastable phases are transient phases that form equilibrium phases upon further heat treatment. Metastable phases are delineated by dashed lines on phase diagrams.

Titanium Martensites. The name *martensite* is taken from steel terminology. Some of the general characteristics of the martensitic reaction are:

- Transformation to martensite is independent of time and depends for its progress only on decreasing temperature. This is termed athermal martensite.

- Transformation is diffusionless and involves no change in chemical composition. It is a nucleation-and-shear-type transformation.

- The temperature range of martensite formation is characteristic of a given alloy and is not lowered by increasing the cooling rate.

Figure 4.4(a) is a schematic binary phase diagram showing the metastable phase boundaries for alpha prime (α'), martensite. The M_s indicates the martensite-start temperature, and the M_f indicates the martensite-finish temperature. The transformation of beta to martensite begins at the M_s temperature, and all quenched beta phase is transformed to martensite when the M_s is reached. Upon reaching the M_s temperature, beta suddenly starts to transform by a shear displacement process. The transformation is so sudden (i.e., diffusionless) that the high alloy content of the beta phase is retained. If the alloy content is greater than that at the alpha transus, the result is a supersaturated alpha phase, which is titanium martensite. Figure 4.4(b) describes the phases that exist at room temperature on quenching from the indicated area. Composition C_1 is the intersection of the M_f temperature with room temperature, and C_2 is the M_s intersection. Alloys with alloy content less than C_1 when quenched from the beta

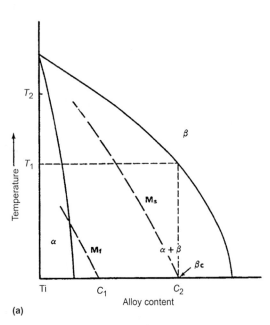

(a)

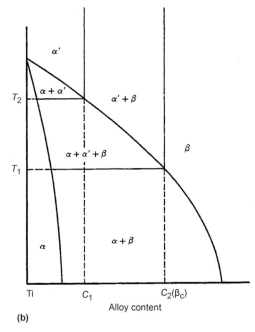

(b)

Fig. 4.4 Partial phase diagram of beta-isomorphous system. The metastable phase boundaries are shown in (a), while the metastable phases present at room temperature after quenching from the indicated temperature are shown in (b). The martensite phase is indicated by α'.

region contain only the martensitic alpha phase (α'). At concentrations from C_1 to C_2, the martensite transformation does not go to completion. Thus, the structure consists of α' and β. Finally, at concentrations greater than C_2, only the beta phase is present after quenching from the beta region.

In the two-phase $\alpha + \beta$ region at temperatures below T_1, the alloy concentration in the beta phase is greater than C_2, and therefore, beta is retained during a quench. These alloys contain $\alpha + \beta$ phases.

In the temperature range between T_1 and T_2, the alloy content of the beta phase is less than the critical C_2 and greater than C_1. This results in transformation of a portion of the beta to martensite. The final structure of alloys quenched from

this region contain α, α', and β. In the two-phase region above T_2, the alloy content of the beta phase is less than C_1, and this phase transforms to martensitic alpha during a quench. As a result, the phases existing at room temperature are α and α'. When alloys are heated in the all-alpha region and quenched, α phase is retained.

The M_s and M_f temperatures indicated in Fig. 4.4(a) vary with alloy content, but they are not believed to be altered by quenching rate. This is illustrated in Fig. 4.5. As molybdenum content is increased, the M_s temperature is depressed from approximately 840 °C (1545 °F) for unalloyed titanium to approximately 580 °C (1075 °F) for a Ti-7.1%Mo alloy. The rate of cooling does not alter this transformation temperature.

However, the M_s temperature is a function of alloy content. The example in Fig. 4.4(a) shows that the M_s temperature drops with increased alloy content at nearly the same rate as the beta transus. The intersection of the M_s curve at room temperature is the critical, or minimum, composition for retaining the beta phase by rapid cooling. Examples of M_s curves for several binary systems are shown in Fig. 4.6. Iron (Fe) depresses the M_s temperature at the greatest rate, while tantalum has the least effect of the elements illustrated. The rate of decrease in M_s temperature relates to observed strengthening in beta-stabilized systems. Iron is one of the more potent strengtheners in titanium, while tantalum has only a minor influence on strength. Elements having the greatest beta-stabilizing ability (i.e., rapidly decreasing M_s temperature) have the greatest strengthening effect in titanium.

Two types of athermal titanium martensite are hexagonal α' (alpha prime) and orthorhombic α'' (alpha double-prime), as shown in Fig. 4.7. The most prevalent is the α', which is formed mostly

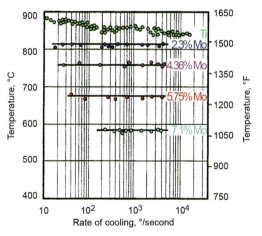

Fig. 4.5 Effect of cooling rate on martensite-start (M_s) transformation temperature. As molybdenum content increases, the M_s drops below 600 °C (1110 °F) at 7.1% Mo in the titanium-molybdenum alloy system. The M_s transformation temperature is not affected by quenching rate.

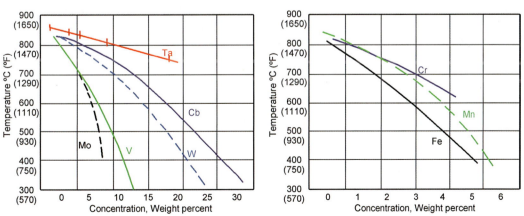

Fig. 4.6 Effect of alloy content on martensite-start (M_s) temperature of binary alloys. Tantalum has the least effect, while iron depresses the M_s at the greatest rate. These rates correlate with strengthening in beta-stabilized systems.

in low-alloy-content titanium alloys such as Ti-6Al-4V. The α" usually occurs in more highly alloyed material, such as Ti-6Al-2Sn-4Zr-6Mo, particularly alloys containing molybdenum. It also forms when a retained beta phase just to the right of β_c in Fig. 4.7 is mechanically deformed, as discussed in the next section.

Tempering of martensite (especially orthorhombic martensite) can lead to low ductility. However, structures containing as-quenched martensite are typically not brittle.

The deliberate formation of martensite by heat treatment is not often used in commercial titanium heat treating practice. It often poses a prob-

lem when an alloy such as Ti-6Al-2Sn-4Zr-6Mo is welded and undergoes a rapid cooling rate in the fusion and heat-affected zones. In such cases, stress relieving tempers the α", causing embrittlement. Special precautions are typically taken to prevent or minimize such occurrences.

Metastable Beta Phase. Figure 4.4 shows that when alloy content exceeds a critical value (β_c), quenching from above the beta transus into the two-phase alpha + beta phase results in a retained-beta structure. The structure is termed metastable beta, because under equilibrium conditions, alpha phase forms. However, by cooling fast enough, alpha phase formation is suppressed.

The critical alloy content, β_c, for retaining 100% beta is different for each alloy system. Table 4.1 lists the minimum alloy content necessary to retain 100% beta after quenching for several binary systems. Values obtained in the former USSR are presented together with those quite frequently referenced in U.S. literature. There is generally good agreement between the two sources. These data show appreciable differences in β_c between binary systems. For example, only approximately 4% Fe added to titanium is required to retain the beta phase, while nearly 50% Ta is required before 100% beta can be retained at room temperature.

The alloy composition β_c for retention of beta can be estimated by the average concentration of outer, or valence, electrons for each atom. The β_c values in weight percent for species A are converted to atomic percent using the formula:

$$\text{at.\% A} = (\text{wt\% A})(\text{atomic weight A})/[(\text{wt\% A})$$
$$(\text{atomic weight A}) +$$
$$(\text{wt\% B})(\text{atomic weight B})]$$

Then, using the assumed number of valence electrons, the average electron density per atom can be calculated using the equation:

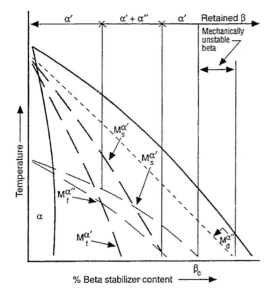

Fig. 4.7 Schematic representation depicting martensites formed after a quench from the beta field. The alpha double-prime martensite forms in more heavily stabilized alloys and can be formed by deformation of unstable beta.

Table 4.1 Influence of alloying on alloy concentration necessary to retain 100% beta

Alloying element	Assumed valence electrons	Minimum alloy content to retain 100% beta after quenching (β_c)				Calculated average valence electrons per atom at β_c	
		U.S. data		Former USSR data			
		wt%	at.%	wt%	at.%	U.S. data	Former USSR data
Mn	7	6.5	5.6	5.3	5.0	4.17	4.15
Fe	8	3.5	3.0	5.1	4.7	4.12	4.19
Cr	6	6.3	5.8	9.0	8.4	4.12	4.17
Co	9	7.0	5.8	6.0	4.9	4.29	4.20
W	6	22.5	6.7	26.8	8.7	4.13	4.17
Ni	10	9	7.5	7.2	5.9	4.45	4.36
Mo	6	10.0	5.3	11.0	5.8	4.11	4.12
V	5	15.0	14.2	19.4	18.4	4.14	4.18
Cb	5	36.0	22.5	36.7	23.0	4.22	4.23
Ta	5	45.0	17.8	50.2	21.0	4.18	4.21

$$\text{el}_{\text{avg}} = (\text{at.\% A}/100)\ \text{el}_A + (\text{at.\%B}/100)\ \text{el}_B \quad (\text{Eq 4.3})$$

where el_{avg} is the average valence electrons per atom; el_A and el_B are valence electrons of A and B, respectively; and at.% is atomic percent of the element (A, B, etc.).

In Table 4.1 it appears that an average valence electron density exceeding approximately 4:1 is necessary to retain the beta phase upon quenching from above the beta transus. This approach can be used to estimate the degree of beta stability of commercial alloys, although it is only an approximation. An alloy such as Ti-15V-3Cr-3Sn-3Al converted to atomic percent, 0.765Ti-0.142V-0.028Cr-0.012Sn-0.054Al, shows the electron density from Eq 4.3 as:

$$0.765(4) + 0.142(5) + 0.028(6) + 0.012(4)$$
$$+ 0.054(3) = 4.148\ \text{el/atom}$$

Table 4.2 lists the results of similar calculations for a variety of commercial alloys and whether beta is retained upon quenching. The values given in the table are consistent with the trend noted earlier. The alpha-beta alloys (such as Ti-6Al-4V with el_{avg} values less than 4.0) result in martensite upon quenching, whereas the beta alloys (such as Ti-15-3 with el_{avg} values exceeding 4.1) readily retain the beta phase upon quenching.

When referring to beta phase stability, note that if the solution treatment is carried out below the beta transus and the material is quenched from that point, the beta phase is enriched in beta stabilizer. If enriched sufficiently, beta is retained upon quenching. Referring to Fig. 4.1, by solutionizing at T_2, the beta phase is enriched to 13% alloying

element. If the alloying element is molybdenum, then according to Table 4.1, the beta phase (which is at a 39% volume fraction at T_2) is retained as beta upon quenching. On the other hand, quenching from above the beta transus, T_3, the 100% beta structure at T_3 contains only 6% Mo, which, according to Table 4.1, is insufficient to retain beta.

Another factor must be considered when quenching an alloy that is marginal in terms of retention of the beta phase. In some cases, such as noted for Ti-10V-2Fe-3Al in Table 4.2, even though the beta phase is retained upon quenching, the beta phase can be transformed to martensite (α'') with the addition of a small amount of mechanical energy. Stresses encountered during quenching could be sufficient to cause the beta phase to transform to martensite. Figure 4.8 shows stress-strain curves for Ti-10V-2Fe-3Al alloy quenched from two temperatures. When quenched from above the beta transus, the beta phase is unstable and a transformation occurs at a relatively low stress. When quenched from below the transus, the beta phase is stable and no transformation occurs. A partial binary phase diagram shown in Fig. 4.9 depicts the relationship between M_s and M_D (deformation-induced martensite). In the figure, composition C represents an alloy such as Ti-10V-2Fe-3Al, which can be quenched to retain beta but will transform to martensite upon application of stress.

Table 4.2   Comparison of valence electrons per atom for various commercial alloys

Alloy	el_{avg}	Quenched structure(a)
Ti-15V-3Cr-3Sn-3Al (Ti-15-3)	4.15	Beta
Ti-3Al-8V-6Cr-4Mo-4Zr (Beta C or Ti-3-8-6-4-4)	4.17	Beta
Ti-13V-11Cr-3Al (Ti-13-11-3)	4.26	Beta
Ti-10V-2Fe-3Al (Ti-10-2-3)	4.16	Beta(b)
Ti-6Al-4V (Ti-6-4)	3.80	Martensite (+ β)
Ti-6Al-6V-2Sn-0.5Fe-0.5Cu (Ti-6-6-2)	3.96	Martensite (+ β)
Ti-6Al-2Sn-4Zr-6Mo (Ti-6-2-4-6)	3.95	Martensite (+ β)
Ti-6Al-2Sn-4Zr-2Mo (Ti-6-2-4-2)	3.91	Martensite
Ti-8Al-1Mo-1V (Ti-8-1-1)	3.89	Martensite

(a) Quenched from above beta transus. (b) Beta phase is mechanically unstable and can transform to martensite with application of stress. Also, a thermal omega may be present.

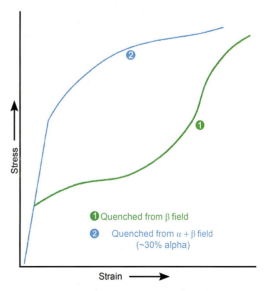

Fig. 4.8 Stress-strain curves for Ti-10V-2Fe-3Al alloy solution treated above (curve 1) and below (curve 2) the beta transus. The lower yield point in curve 1 is caused by a stress-induced transformation of beta to martensite (alpha double-prime).

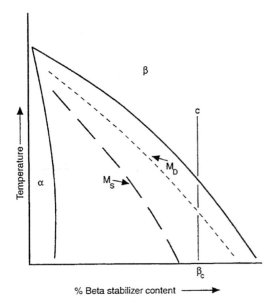

Fig. 4.9 Partial binary phase diagram depicting quenched martensite start (M_s) and deformation-induced martensite (M_D). Alloy C will retain beta upon quenching from above the beta transus but will transform to martensite upon application of a critical strain at room temperature.

Omega. Another metastable phase that forms in titanium alloys is the omega (ω) phase. It was first identified in early titanium studies using x-ray diffraction and later by transmission electron microscopy. The omega phase is extremely fine and cannot be resolved optically. The early investigations of omega were prompted by an embrittlement noted during aging studies.

Omega phase can be formed athermally or isothermally. Athermal omega (ω_a) is believed to form without a change in composition, analogous to martensite. It can occur in alloys with beta stabilization near or below β_c but with an M_f below room temperature. The volume fraction formed depends on alloy chemical composition, but it is usually a low volume fraction and therefore has little effect on as-quenched mechanical properties. An alloy such as Ti-10V-2Fe-3Al, thought to be all beta upon quenching, can actually have ω_a present. In such cases, the formation of ω_a is believed to compete with α'' formation. Isothermal omega (ω_{iso}) is generally formed by aging a retained beta (or $\beta + \omega_a$) structure in a temperature range of 200 to 500 °C (390 to 930 °F). The amount of ω_{iso} formed depends primarily on alloy composition and aging time, although the kinetics is quite rapid.

Extended aging time in the omega aging-temperature range eventually results in equilibrium alpha formation. The structure of omega is generally believed to be hexagonal. Isothermal omega has a composition intermediate between alpha and beta. Alloying additions such as aluminum, tin, and zirconium reduce the amount of omega that can form. Aluminum and oxygen do so by promoting alpha phase formation, and tin and zirconium stabilize the beta phase. Conflicting data on the presence or lack of omega in certain alloys are explained by small differences in such elements.

The effect of omega on tensile properties has been studied since the early embrittlement problem surfaced. Generally, embrittlement occurs when the volume fraction of omega phase exceeds roughly 50%. At intermediate volume fractions (roughly in the 25 to 45% range), omega is an effective strengthener, and useful ductility can be achieved. However, no commercial alloy is currently being used in an omega-aged condition. At low volume fractions (in the neighborhood of 20%), omega has only a minor effect on properties.

Several commercial alloys exhibit omega phase. When quenched from above the beta transus with minimal quenching strains, Ti-10V-2Fe-3Al forms a $\beta + \omega_a$ structure, which is quite ductile. However, aging at a temperature of approximately 300 °C (570 °F) promotes ω_{iso}, and embrittlement will ensue. If strained either during or after the quench, orthorhombic martensite (α'') forms.

Beta Prime. If sufficiently stabilized to preclude martensite and omega, the retained metastable beta phase undergoes a phase separation to $\beta + \beta'$ upon aging in the temperature range of 200 to 500 °C (390 to 930 °F). Both phases have a body-centered cubic structure, and the β' phase is the solute-lean phase. The β' precipitates are very fine and can only be detected by thin foil electron microscopy. Commercial Ti-15V-3Cr-3Sn-3Al alloy forms β' upon aging at 315 °C (600 °F) for 10 h from a starting structure of 100% retained beta. No hardness increase is noted for the $\beta + \beta'$ structure over the all-β structure. Continued aging results in the precipitation of alpha phase (presumably at the β' sites) and a rapid increase in hardness.

A partial binary phase diagram depicting the $\beta + \beta'$ and omega regions is shown in Fig. 4.10.

Examples of Phases Present after Quenching. The metastable phases and principles governing their formation play an important role in the heat treatment of commercial titanium alloys. Figure 4.11 shows a variety of standard commercial alloys and the phases present after quenching from the indicated temperatures to highlight the

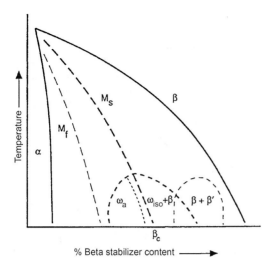

Fig. 4.10 Partial binary phase diagram depicting the omega phase and beta prime (beta phase separation)

effects of alloy content and specific chemistry. The alloys are arranged in ascending order of beta stabilization. This is evidenced by the decreasing temperature at which primary alpha is present. The beta transus is the temperature above which no primary alpha exists.

The Ti-8Al-1Mo-1V alloy shows that when quenched from above or just below the beta transus, the beta phase transforms to α'. However, from lower temperatures the beta phase is enriched and some α'' begins to form. At approximately 870 °C (1600 °F) and below, the beta phase is enriched sufficiently to be retained upon quenching. The situation is somewhat similar in the case of Ti-6Al-4V except that α'' does not form upon quenching. In the Ti-6-2-4-6 case, no α' is noted, but α» is formed upon quenching from near or above the beta transus. Both alloys exhibiting α'' formation contain molybdenum.

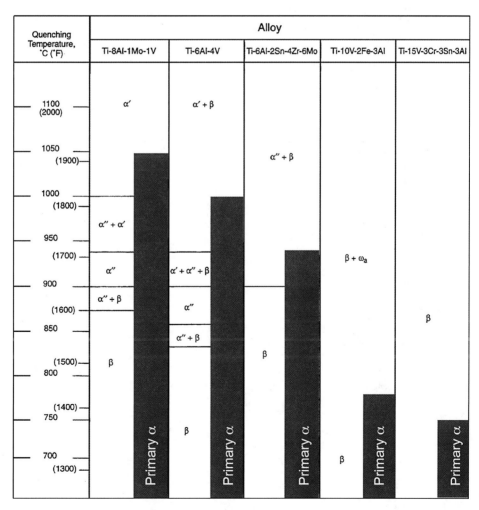

Fig. 4.11 Phases present in standard commercial alloys at various quenching temperatures

The Ti-10-2-3 alloy is a near-beta alloy. Although no martensites are reported upon quenching, a thermal omega is present when the alloy is quenched from near or above the beta transus. It should also be noted that the retained beta phase resulting from a quench near the beta transus is mechanically unstable. An applied strain, even as small as quenching strain, results in the retained beta phase transforming to α''. As with the previous examples, quenching from sufficiently below the beta transus enriches the beta phase sufficiently to retain beta upon quenching.

The Ti-15-3 alloy is a beta alloy that readily retains the beta phase even with a quench from above the beta transus. Alloys such as Beta-C (Ti-3Al-8V-6Cr-4Zr-4Mo) and Beta 1 (Ti-13V-11Cr-3Al) fall into this category.

Transformation Kinetics

In the previous discussions, equilibrium phases and metastable phases produced by a quench were considered. In the equilibrium case, cooling rates were assumed to be very slow, and therefore, sufficient time at temperature enabled the proper diffusion to occur. In the metastable case, rapidly quenched material severely restricted diffusion. However, a specific cooling rate was not considered in either case. For example, how fast does a quench have to be to retain the beta phase present at a higher temperature? This is a very important and practical consideration for the reason indicated in Fig. 4.12, which shows the time/temperature profile for center portions of quenched 152, 76, and 25 mm (6, 3, and 1 in.) diameter sections. The center of the thinner sections reaches the quench temperature faster than the thicker sections. In heat treatable alloys, it is important to cool below a critical temperature (T_c) in a critical time (t_c). Figure 4.12 shows that a 25 mm (1 in.) thick part can be heat treated properly, a 76 mm (3 in.) thick part would be marginal, and a 152 mm (6 in.) thick part cannot be heat treated. Note also that the outer regions of the 152 mm thick part probably would cool fast enough so the part would be nonuniform in heat treatment response.

This example shows that it is necessary to know the T_c and t_c relationships for heat treatable alloys. The best source of such information is a time-temperature-transformation (TTT) diagram, which relates the transformation of the beta phase to the time and temperature conditions to which it is subjected. Each diagram describes the decomposition of beta in a given titanium alloy. Thus, each titanium alloy has a different TTT diagram.

Most TTT curves in the literature were obtained by quenching samples from a temperature in the beta field directly to the transformation temperature, holding for various times, and rapidly cooling to room temperature. The final cooling stops the transformation reaction. Some portions of the transformation can be observed microstructurally, although other techniques, including resistivity, x-ray, and thermal expansion, are used for the harder-to-de`tect transformations.

Several common types of TTT curves for titanium alloys are shown in Fig. 4.13 and 4.14. Figure 4.13 shows the curves for two alloy compositions (A and B) from a beta-isomorphous system. The first C-curve in the TTT diagram represents the start of the transformation of beta to alpha. The final curve indicates the transformation is 95% complete and equilibrium has nearly been achieved.

In Fig. 4.13, the alloy content of composition A is sufficiently high (exceeds β_c) so that cooling rapidly misses the knee of the C-curve and 100% beta is retained at room temperature. Composition B has less alloy content and completely transforms to martensite during the quench.

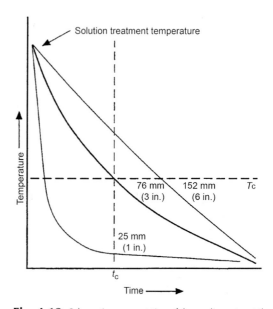

Fig. 4.12 Schematic representation of the cooling rates at the center of 152, 76, and 25 mm (6, 3, and 1 in.) thick sections that have been solution treated and quenched. The 25 mm (1 in.) section can be quenched below the critical temperature (T_c) before the critical time (t_c) has elapsed. The 76 mm (3 in.) section is marginal, and the 152 mm (6 in.) section cannot be properly solution treated.

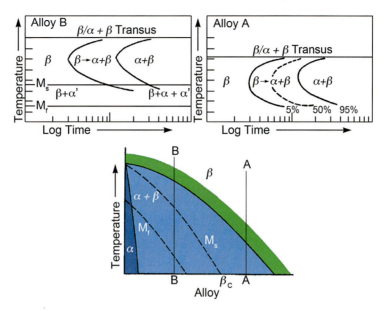

Fig. 4.13 Time-temperature-transformation curves for two alloys of a beta-isomorphous system. Start of transformation of beta to alpha and its completion are indicated by the C-curves.

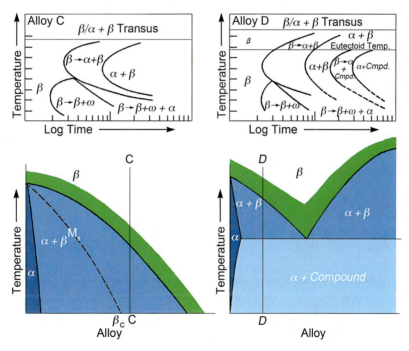

Fig. 4.14 Time-temperature-transformation curves for beta-isomorphous and beta-eutectoid systems. The curves show omega forming at low temperatures and eventually forming the equilibrium products of alpha plus beta.

The TTT curves for a composition C in Fig. 4.14 illustrate the transition omega reaction. The curves show omega forming at low temperatures and eventually forming equilibrium products

alpha plus beta. Figure 4.14 also shows typical TTT curves for a composition D in a beta-eutectoid system. This diagram illustrates the transformation of a hypoeutectoid alloy discussed previ-

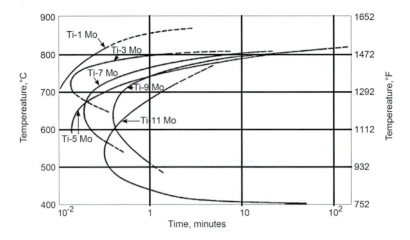

Fig. 4.15 Effect of molybdenum on start of beta-to-alpha transformation. Increasing the molybdenum content in titanium-molybdenum alloys shifts the initial transformation of beta to alpha to the right. Hence, beta is more readily retained.

ously. The beta initially transforms to alpha plus beta, and eventually, equilibrium phases alpha plus intermetallic develop.

The retention of the beta phase by a rapid cool from high temperatures is achieved more readily in alloys that have their initial C-curves shifted to the right. This can be done by increasing the beta-stabilizing content, as illustrated in Fig. 4.15. In this example, increased molybdenum content shifts the initial transformation of beta to alpha to longer times, or farther to the right, in the TTT diagram. Alpha stabilizers have the opposite effect. Figure 4.16 illustrates the influence of oxygen on the initial transformation of beta to alpha in a Ti-11Mo alloy. The knee of the curve shifted from 3 min at low oxygen content to less than 0.1 min at high oxygen content. Oxygen additions also increase the temperature at which the knee occurs, probably as a result of increasing the beta-transus temperature.

Figure 4.17 illustrates the effect of an addition of the alpha stabilizer aluminum on the TTT characteristics of a Ti-15%V alloy. Aluminum, like oxygen, also decreases the time for initial transformation of beta to alpha. However, the aluminum addition also increases the time for the initial beta transformation to omega. A sample of Ti-15V cannot be quenched from a temperature in the beta field to room temperature without passing through the beta-to-omega transformation range. However, the addition of 2.75% Al shifts the knee of the omega C-curve to 1 min, making it possible to quench past this transformation.

Effect of Alloy Additions on Transformation Nose Time. A multilinear regression analysis of experimental data resulted in the data shown in Table 4.3 (Ref 4.4). Beta stabilizers such as mo-

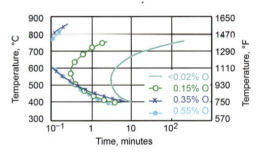

Fig. 4.16 Effect of oxygen on start of beta-to-alpha transformation. Oxygen, an alpha stabilizer, shifts the transformation curve to the left, decreasing the time associated with the nose of the C-curve.

lybdenum, vanadium, chromium, manganese, and iron all increase the time for decomposition of the beta to alpha, while the neutral element tin has little effect on the time. The alpha stabilizer aluminum shortens the time for transformation.

Heat Treatment

Titanium and titanium alloys are heat treated to:

- Increase strength (age hardening)
- Produce an optimum condition of ductility, machinability, and dimensional and structural stability (annealing)
- Reduce residual stresses developed during fabrication (stress relieving)

Annealing and age-hardening treatments are intended to alter the mechanical properties. Stress

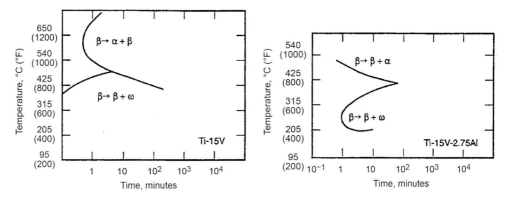

Fig. 4.17 Effect of aluminum on initial transformation of beta to alpha. Aluminum, like oxygen, decreases the time for initial transformation of beta.

Table 4.3 Effect of alloy element on nose time

	Coefficient (for 1 wt%)	Element	Standard error of coefficient
ln(nose time [t_n], s) = −1.74	+0.34	Mo	0.03
	−0.25	Al	0.06
	+0.03	Sn	0.03
	+0.26	V	0.03
	+0.63	Cr	0.04
	+0.59	Mn	0.28
	+0.50	Fe	0.20

relieving is used chiefly at temperatures between 450 and 800 °C (840 and 1470 °F) to prevent distortion and to condition the metal for subsequent forming and fabrication operations. Stress-relief treatments are discussed in detail in Chapter 5, "Deformation and Recrystallization of Titanium and Its Alloys," in this book. Typical heat treatment temperature ranges are illustrated in Fig. 4.18 for an alpha-beta alloy and a metastable beta alloy. The salient difference is that beta alloys are often solution treated, stress relieved, and annealed above the beta transus, while all these treatments are usually carried out below the beta transus for alpha-beta alloys.

The final properties achieved depend not only on the heat treatment cycle but also on the alloy composition. Figure 4.19 shows the strength trends that can occur through different heat treatment procedures in a beta-isomorphous system. The strength of fully annealed material increases as the alloy content or percent beta phase increases. However, for material cooled rapidly from a temperature in the beta field, a more complex strength-composition relationship exists. This relationship depends on the martensite transformation of beta to alpha. The temperature at

which this reaction is initiated (M_s) and completed (M_f) during cooling is shown on the phase diagram. Some strengthening can be obtained by this martensite transformation when the alloy content is low. However, this increase in strength is not as great as that achieved in ferrous alloys.

The maximum strength obtainable from martensitic alpha occurs at the composition where the M_f temperature is at room temperature. If the omega phase is not formed during the quench, the yield strength decreases to a minimum at the composition where the M_s temperature is at room temperature (β_c). This is the lowest alloy content at which 100% beta can be retained by a rapid quench. The beta phase is mechanically unstable, and, under strain, it transforms to martensitic alpha; as the alloy content is raised further, the strength increases, because less mechanically unstable beta is present.

The maximum strength for this system is achieved by an age-hardening process. This involves rapidly quenching from a high temperature (solution treatment) and then aging at an intermediate temperature. Figure 4.19 shows that for a given age cycle, maximum strength should be achieved at the composition where the M_s temperature coincides with room temperature (β_c).

The two heat treatments most commonly used for titanium are age hardening and annealing. Age hardening significantly increases the strength of beta- and alpha-beta-type alloys, while annealing is used for all titanium alloys to impart a good balance of strength, ductility, formability, and thermal stability. Each of these heat treatment processes is discussed here.

Age Hardening. The procedure for age hardening titanium alloys is a two-step cycle (Ref 4.2). The alloy is first heated to a temperature high in the

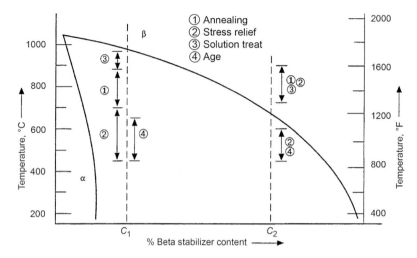

Fig. 4.18 Examples of heat treatment temperatures for alpha-beta alloys such as Ti-6Al-4V (C_1) and beta alloys such as Ti-15V-3Cr-3Sn-3Al (C_2)

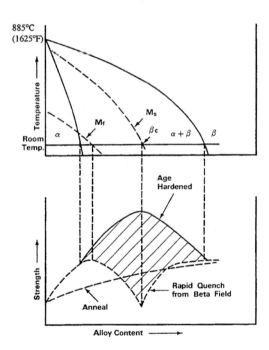

Fig. 4.19 Heat treatment of beta-isomorphous alloys. Curves indicate the strength trends that occur through different heat treatments.

alpha-beta field or in the beta field, held for 30 to 60 min, and then quickly quenched (usually in water) to room temperature. Rapid cooling is necessary to retain a metastable beta phase. This portion of the cycle is called solution treatment.

After quenching, the alloy is reheated to a lower temperature in the alpha-beta field. Various temperatures and times can be used. Most treatments are conducted at a temperature between 480 and 620 °C (895 and 1150 °F) for 2 to 16 h. Temperatures below 480 °C are not used, to avoid formation of the brittle omega transition phase. The alloy is then cooled in air. This second heat treatment step is called aging. It usually results in the precipitation of a fine dispersion of alpha in the beta phase and therefore increases strength. The total process of solution treating and aging is called age hardening.

The increased strength depends on the transformation of a retained metastable phase, such as the beta phase, to precipitated alpha phase in beta phase. The finer the precipitate (thus a greater number of precipitates), the greater the strength increase (see Figure 4.24). A fine precipitate is favored by low aging temperatures. However, ductility decreases as strength increases, so a trade-off must be made.

The specific reaction occurring during the solution treating and aging process can be examined using the information developed earlier from Fig. 4.1: solution treat the Ti-6%Al alloy at temperature T_2 and subsequently age it at temperature T_1. At T_2, 61% alpha and 39% beta are present, while the alloy contents of the alpha and beta phases are 1.5 and 13%, respectively; at T_1, 89% alpha and 11% beta are present if equilibrium is achieved, and the alloy contents are 3.0 and 30%, respectively. The reaction can be written as:

$$\underline{\text{Afer solution treatment at } (T_2)}$$
$$39\% \ \beta_{13} + 61\% \ \alpha_{1.5}$$

$$\underline{\text{Afer aging } (T_1)}$$
$$11\% \ \beta_{30} + 89\% \ \alpha_{3.0}$$

This reaction describes the age-hardening process. The prefix in the aforementioned reaction denotes the volume percent of each phase, while the subscripts indicate the approximate alloy content in each phase. The solution treatment retains the high-temperature equilibrium form of alpha and beta at room temperature. Reheating at low temperatures (aging) causes readjustments in the phases in an attempt to achieve equilibrium at the new temperature, T_1. As a result, the volume percent of beta decreases from 39 to 11%, while a corresponding increase in the amount of the alpha phase occurs. The change in volume percent of phases during aging is accompanied by a rather large increase in the alloy content of the beta phase (from 13 to 30%).

The aforementioned transformation reaction is often written in a condensed form as:

$$\beta_o + \underline{\alpha} \rightarrow \underline{\alpha} + \alpha + \beta_u$$

where β_o is the original beta, with an alloy content of 13%; $\underline{\alpha}$ is primary alpha (i.e., alpha existing at the solution-treatment temperature); α (no underline) is the new α that precipitated in the beta phase during the aging; and β_u is the enriched beta (up to 30% alloy content).

Because there is little change in the primary alpha during the age, the reaction can be condensed further to:

$$\beta_o \rightarrow \alpha + \beta_u$$

The beta retained after solution treating (β_o) transforms during aging to alpha and an enriched beta, β_u.

The strengthening during aging is due to the coherency strains between the precipitating alpha and the matrix.

This example illustrates, in a simple manner, the heat treatment of an ideal alpha-beta alloy. In a weakly beta-stabilized system, martensitic alpha or omega is present after the quench. The following aging reaction applies in these instances:

$$\alpha' \rightarrow \alpha + \beta_u$$

or

$$\beta_O \rightarrow \beta_r + \omega \rightarrow \beta_r + \alpha + \omega \rightarrow \beta_u + \alpha$$

where β_r is a slightly enriched beta phase. Although the metastable phase complicates the aging reaction, the equilibrium phases alpha and beta form eventually.

Solution Treatment. For alpha-beta alloys, solution treatment at temperatures in the alpha +

beta field is most commonly used. Figure 4.20 shows the effect of solution-treating temperature on the properties of Ti-6Al-4V after solution treating. The ultimate strength increases with increasing solution temperature. The minimum yield strength occurs after solution treating at the beta transus of the β_o composition. This occurs, as shown in Fig. 4.11, at 830 °C (1525 °F) for Ti-6Al-4V. At this temperature, the beta phase is retained during the quench, but it is mechanically unstable. This results in low yield strengths and a large spread between yield strength and ultimate strength. Lower ductility is encountered in Ti-6Al-4V when the solution treatment temperature exceeds the beta transus. This is due largely to a coarsening in grain size. This low ductility also persists after aging. Thus, most solution treatments are done below the beta transus. An exception is the heat treatment of the metastable beta alloys such as Ti-15V-3Cr-3Sn-3Al. These alloys have high beta-stabilizing content and therefore low beta transus. Heating above the beta transus for short periods of time is not detrimental, because grain growth is not as rapid at these lower temperatures.

Solution treating conditions the material for the subsequent strengthening aging process. Figure 4.21 illustrates the influence of solution temperature on the properties of Ti-6Al-4V after aging at 540 °C (1000 °F) for 8 h. As the solution temperature increases, the strength after aging increases because of the increased amount of

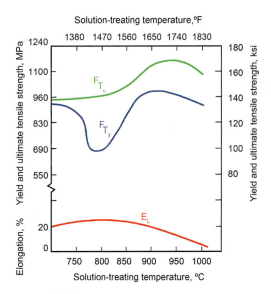

Fig. 4.20 Effect of solution-treating temperature on the tensile properties of Ti-6Al-4V

metastable (beta) phase available for aging (by precipitation of alpha).

Cooling rate from the solution temperature has an important effect on the aging response of titanium alloys. This is even more critical in weakly beta-stabilized systems where transformation of the beta phase occurs more readily during a quench. The effect of quench delay is shown in Fig. 4.22 for Ti-6Al-4V bar that was subsequently aged at 480 °C (900 °F) for 6 h. The strength decreases rapidly as the delay time increases.

The example illustrates that Ti-6Al-4V is a weakly beta-stabilized alloy and thus shows a dramatic effect of quench delay. More highly beta-stabilized alloys are not as sensitive. Not only do delays from the furnace to the quenching medium lower the strengths after aging, but the interior areas of large, solution-treated sections show similar effects. The interior of quenched large sections cool less rapidly (similar to the slower cooling that occurs during delays in processing) and thus do not respond as well to heat treatment.

The effect of section size refers to the hardenability of an alloy. For example, Ti-6Al-4V (a weakly beta-stabilized alloy) can only be fully hardened at thicknesses up to roughly 25 mm (1 in.). On the other hand, a more heavily stabilized alloy such as Ti-10V-2Fe-3Al can be fully hardened in sections up to approximately 127 mm (5 in.) thick.

The type of quenching medium used depends on the severity of quench necessary to obtain aging response. In weakly beta-stabilized alpha-beta alloys, water quenching is used. However, in thin sections, oil (and other less severe quenching media) sometimes is used to minimize distortion. In the heavily beta-stabilized alloys, such as Ti-15V-3Cr-3Sn-3Al, air cooling is sufficiently rapid to retain the beta phase. As section size increases, air cooling may not suffice.

During solution treatment of large parts, some growth or change in dimensions occurs, which could remain after the solution treatment. A growth of 1% has been noted in Ti-6Al-4V by using a slow heating rate of 3 °C (6 °F) per minute and holding for 2 h at the solution temperature. The amount of growth decreases by rapid heating and holding at the solution temperature for as short a time as possible.

Aging. The final step in heat treating titanium alloys to increase strength is aging. Aging is typically carried out in a temperature range of 480 to 600 °C (900 to 1110 °F). Aging results in the precipitation of alpha phase in the retained beta. Figure 4.23 shows some general features of the aging process. Strength increases to a maximum, then gradually decreases. In addition, there is the attendant trade between strength and ductility. The maximum strength is also called the peak aged condition. When material is aged for times less than that which produces peak strength, it is considered underaged. When the aging time passes the peak condition, the material is overaged. Typically, it is desirable to age to the peak or slightly beyond (i.e., overaged) for optimum tensile properties.

Figure 4.23 also shows that as aging temperature increases, peak strengths decrease and time to peak condition decreases. As aging tempera-

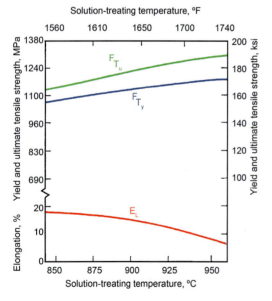

Fig. 4.21 Effect of solution-treating temperature on the solution-treated and aged properties of Ti-6Al-4V. Age cycle: 540 °C (1000 °F) for 8 h

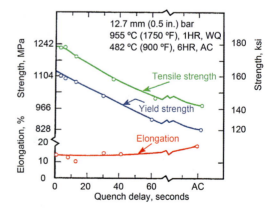

Fig. 4.22 Effect of quench delay on tensile properties of aged Ti-6Al-4V bar. WQ, water quenched; AC, air cooled

ture increases, the precipitation reaction occurs more readily, thus the shorter time to the peak condition. Also, at higher aging temperatures, precipitated alpha particles are fewer in number but coarser in size. Strengthening by precipitation is directly related to particle size and distribution. The greater number of finer particles formed at lower aging temperatures (Fig. 4.24) are a more potent strengthener than fewer, coarser particles.

There are other important variables besides chemistry that affect aging response. One is the heating rate to the aging temperature. If the heating rate is very slow, enough time may be spent

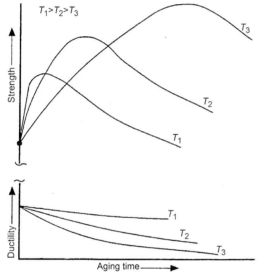

Fig. 4.23 Schematic representation of the effects of aging time and temperature (T)

just below the desired aging temperature such that aging may begin at the lower temperature. In some cases, the time at lower temperatures could precipitate omega phase (a finer precipitate than alpha), which leads to embrittlement. Whether it is omega, or simply a finer dispersion of alpha, lower-temperature aging ultimately leads to higher strengths and lower ductility. This effect can contribute to scatter in property data for a given material. Table 4.4 illustrates the effect in Ti-15V-3Cr-3Sn-3Al sheet product.

Another significant variable in the aging process is the degree of residual work in the product before aging. For example, solution-treated Ti-15V-3Cr-3Sn-3Al is often cold formed prior to aging. The stored energy (residual cold work) assists the aging process and results in increased strengths. Although solution-treated Ti-15-3 would not age measurably at 205 °C (400 °F) in several hundred hours, the addition of 50% cold work forces an aging response. A similar effect is seen for other alloys and residual hot work as well as cold work.

Annealing is performed to impart a "soft," or low-strength, condition to the metal. In the annealed condition, the material is generally most suitable for forming and machining and is stable (i.e., not age hardenable). The following annealing terms are discussed with respect to Ti-6Al-4V with a beta transus of approximately 995 °C (1825 °F).

Mill Anneal. Heat at 735 °C (1350 °F) for 2 to 4 h and air cool to room temperature (AC to RT). This generally results in a very fine alpha grain size.

Recrystallize Anneal. "In material deformed approximately 50% or more," heat at 925 to

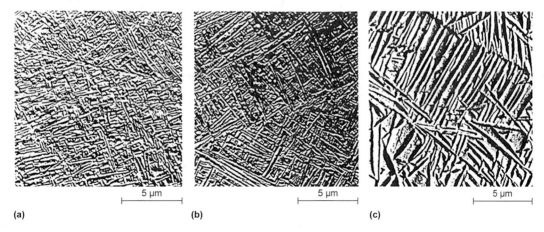

Fig. 4.24 Effect of aging temperature on the dispersion of alpha precipitates in the Beta III alloy (Ti-11.5Mo-6Zr-4.5Sn). The size and spacing of the alpha increases as the aging temperature is increased from (a) 480 °C (900 °F), (b) 535 °C (1000 °F), and (c) 595 °C (1100 °F) after solution treating above the beta transus temperature, water quenching, and aging.

955 °C (1700 to 1750 °F) for 4 h or more, furnace cool to 760 °C (1400 °F), and AC to RT. This results in a structure consisting of equiaxed alpha grains and grain-boundary beta (Ref 4.3, 4.4). See Chapter 6, "Mechanical Properties and Testing of Titanium Alloys," for further details on how this structure develops.

Beta Anneal. Heat at 1035 °C (1900 °F) for 30 min, AC, reheat to 735 °C (1350 °F) for 2 h, and AC. This results in the transformed beta structure containing platelet alpha.

Duplex Anneal. Heat at 955 °C (1750 °F) for 10 to 30 min, AC, reheat to 675 to 735 °C (1250 to 1350 °F) for 4 h, and AC to RT. This generally results in a structure with both equiaxed primary alpha and transformed beta (platelet alpha) regions.

Stress-Relief Anneal. Heat at 595 °C (1100 °F) for 2 to 4 h and AC. This treatment does not change the incoming microstructure and is only used to relieve stored deformations.

The temperatures cited previously are, in many cases, tied to the beta transus. For example, Ti-6Al-2Sn-4Zr-6Mo with a typical beta transus of 930 °C (1710 °F) would be beta annealed at 980 °C (1800 °F) to minimize grain growth. Similarly, the recrystallize anneal and duplex anneal would be from 890 °C (1630 °F).

Each of the first four anneals produces different microstructures, which, in turn, affect resultant properties. Figure 4.11 provides a summary of the general trends noted for each of these microstructures. These trends are general in nature and are not strictly obeyed for all alloys. There are trade-offs associated with each microstructure. For example, the beta-annealed microstructure results in the highest creep resistance, fatigue crack growth resistance, and toughness. However, this microstructure generates the lowest ductility and smooth fatigue resistance.

Further discussion regarding the effects of microstructure on properties in titanium alloys can be found in Ref 4.5 and 4.6.

Titanium-Aluminum Alloy. In alpha-titanium alloys, the beta phase transforms on cooling to the alpha phase, and little, if any, response to heat

treatment occurs as a result of the beta transformation. However, in alpha alloys containing 5% or more aluminum, another type of heat treatment possibility exists. The titanium-aluminum phase diagram in Fig. 4.25 (Ref 4.1) shows that the ordered intermediate phase α_2 (Ti$_3$Al) can be present when the aluminum content is 5% and greater.

This ordered phase, although difficult to distinguish by optical microscopy, can have a profound influence on mechanical properties and stress-corrosion resistance. The α_2 phase develops more readily in the higher-aluminum-containing (7 to 8% Al) alloys and by heat treating at temperatures in the $\alpha + \alpha_2$ phase field for long time periods. The effect of this phase on the properties of the Ti-8Al-1Mo-1V alloy is shown in Fig. 4.26. In this example, all specimens were heat treated at 790 °C (1450 °F) and cooled at various rates through the $\alpha + \alpha_2$ phase region.

The slowest cooling (15 °C, or 25 °F, per hour) promotes the greatest amount of α_2. This figure shows that yield strength strongly depends on cooling rate: slower cooling rates produce higher strength levels. Ultimate strength is affected in a similar manner but to a lesser extent. Elongation is slightly greater for the slowly cooled material. A significant increase in the modulus also occurs with the development of the α_2 phase. The most detrimental characteristic of the α_2 ordered intermediate phase is the low fracture toughness associated with its presence. The notched tensile strength is appreciably less for the slowly cooled material. Other tests also confirm this degradation of fracture toughness with the presence of α_2.

The effects of the α_2 phase can be changed by alloying. Additions of beta-isomorphous elements (molybdenum, vanadium, niobium, and tantalum) slow down the rate of formation of α_2. Without the presence of 1% Mo and 1% V, the Ti-8Al-1Mo-1V is more sensitive to heat treatment in the $\alpha + \alpha_2$ phase field. To avoid the deleterious effect of α_2, heat treatments and fabrication of alloys containing 5% or more aluminum should be designed to avoid slowly cooling, or dwelling for long times, at temperatures below approximately 735 °C (1350 °F).

Contamination during Heat Treatment. Titanium is chemically active at temperatures higher than 650 °C (1200 °F) and readily reacts with the oxygen in air. Nitrogen also reacts with titanium but at a much slower rate than oxygen; thus, it presents no serious contamination problem. Oxygen pickup during heat treatment results in a predominantly alpha structure at the surface (alpha case) in addition to the formation of scale.

Table 4.4 Effect of aging in Ti-15V-3Cr-3Sn-3Al

Material: 1.8 mm (0.07 in.) thick sheet; age cycle: 540 °C (1000 °F) for 8 h

Heating rate, ambient to 540 °C (1000 °F)	Yield strength		Ultimate tensile strength		Elongation, %
	MPa	ksi	MPa	ksi	
1.25 h	1120	162	1190	173	13
16 h	1230	178	1300	186	7

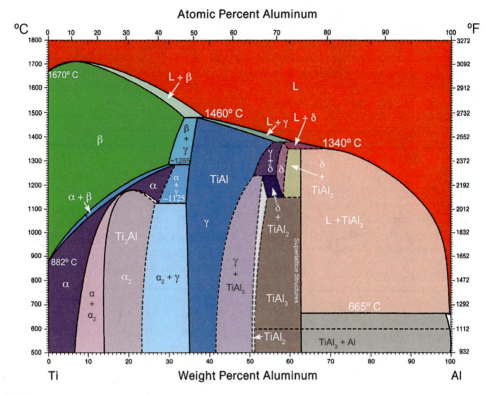

Fig. 4.25 The titanium-aluminum phase diagram

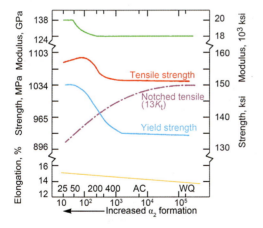

Fig. 4.26 Effect of cooling rate from annealing temperature on tensile properties. Tests on Ti-8Al-1Mo-1V 2.3 mm (0.090 in.) sheet indicate that tensile and yield strengths are higher with slower cooling rates. Elongation is also slightly greater, but notched strength is lower. AC, air cool; WQ, water quench

The alpha case (see Chapter 7, "Metallography of Titanium and Its Alloys," in this book) is extremely brittle and can extend to depths exceeding 0.25 mm (0.010 in.), depending on the alloy and heat treating temperature and time. Several

methods are used to retard oxidation and remove both scale and the alpha case.

Hydrogen pickup also occurs from furnace atmospheres during heat treatment. Hydrocarbon fuels produce hydrogen as a by-product of incomplete combustion, and electric furnaces with air atmospheres contain hydrogen from the breakdown of water vapor. Small amounts of hydrogen pickup can be tolerated. Most titanium heat treating operations are performed in conventional furnaces using oxidizing (air) atmospheres. If excessive hydrogen pickup occurs, vacuum annealing treatments at temperatures in the range of 600 to 760 °C (1110 to 1400 °F) are used to reduce hydrogen content.

Summary

In summary, titanium alloys are heat treated by age hardening and annealing. The age-hardening treatment is a two-step process consisting of solution treating followed by aging. Solution treating and quenching conditions the material for the subsequent aging treatment. After solution treatment and quenching, metastable phases are retained at room temperature. These can be metastable beta, omega, and martensite. The amount of each phase present after solution treatment and quenching depends on the solution temperature and alloy composition.

Retention of metastable beta is favored by low solution temperatures and highly beta-stabilized alloys. Omega and martensite form in low-alloy-content beta-stabilized alloys and by using high solution temperatures. Strengthening occurs from transformation of the metastable phases during aging to the equilibrium phases, generally alpha and an enriched-alloy-content beta. The embrittling omega phase is avoided during aging by selecting temperatures higher than approximately 425 °C (800 °F) and aging for 2 h or longer. High strength is achieved by using high solution temperatures in combination with low aging temperatures.

The time for initial transformation of beta during aging is changed by the alloy content. Higher alpha-stabilizing content decreases the time for transformation of beta to alpha, while higher beta stabilizers increase it. Aluminum extends the time for initial transformation of beta to omega.

Annealing temperatures are selected to obtain the best combination of mechanical properties and to ensure adequate thermal stability. High annealing temperatures alter the grain morphology, resulting in improved creep strength and fracture toughness (due to an increased amount of lenticular alpha). This is accompanied by some loss in both ductility and tensile strength. High thermal stability is achieved in highly beta-stabilized alpha-beta alloys by annealing at low temperatures to enrich the beta phase in alloy content. This is done by a low-temperature stabilization anneal or by furnace cooling from the annealing temperature.

REFERENCES

4.1 J.L. Murray, *Phase Diagrams of Binary Titanium Alloys,* ASM International, Metals Park, OH, 1987

4.2 R.E. Reed-Hill, *Physical Metallurgy Principles,* Van Nostrand, New York, 1973

4.3 *Titanium Technology: Present Status and Future Trends,* Titanium Development Association, Dayton, OH, 1985

4.4 F.H. Froes, private communication, 2013

4.5 G. Lutjering and J.C. Williams, *Titanium,* Springer, 2003

4.6 R. Boyer, E.W. Collings, and G. Welsch, Ed., *Materials Properties Handbook: Titanium Alloys,* ASM International, 1994

SELECTED REFERENCES

• E.W. Collings, *The Physical Metallurgy of Titanium Alloys,* American Society for Metals, Metals Park, OH, 1984

• M.J. Donachi, *Titanium: A Technical Guide,* 2nd ed., ASM International, 2000

• H. Kuhn and D. Medlin, Ed., *Mechanical Testing and Evaluation,* Vol 8, *ASM Handbook,* ASM International, 2000

• P. Lacombe, R. Tricot, and G. Beranger, Ed., *Proceedings of the Sixth International Conference on Titanium* (Nice, France), Metallurgical Society of AIME, Warrendale, PA, 1988

• G. Lutjering, U. Zwicker, and W. Bunk, *Proceedings of the Fifth International Conference on Titanium,* Metallurgical Society of AIME, Warrendale, PA, 1984

• *Proceedings of the Seventh International Conference on Titanium* (San Diego, CA), Metallurgical Society of AIME, Warrendale, PA, 1992

• *Proceedings of the Eighth International Conference on Titanium* (Birmingham, U.K.), Metallurgical Society of AIME, Warrendale, PA, 1995

• *Proceedings of the Ninth International Conference on Titanium,* Metallurgical Society of AIME, Warrendale, PA, 2000

• *Proceedings of the Tenth International Conference on Titanium,* Metallurgical Society of AIME, Warrendale, PA, 2004

• *Proceedings of the Eleventh International Conference on Titanium,* Metallurgical Society of AIME, Warrendale, PA, 2008

• *Proceedings of the Twelfth International Conference on Titanium,* Metallurgical Society of AIME, Warrendale, PA, 2012

CHAPTER 5

Deformation and Recrystallization of Titanium and Its Alloys*

NEARLY ALL METALS AND METAL ALLOYS can be strengthened by cold working, and titanium and its alloys are no exception (Ref 5.1). The mode of plastic deformation that occurs varies considerably among different metals and metal alloys. Metal properties altered by cold deformation can be restored by recrystallization. Recrystallization temperatures vary over a wide range, depending on the metal and alloy.

This chapter provides treatment detailed discussion on deformation and recrystallization of titanium alloys and factors that influence the results.

The predominant mode of plastic deformation in titanium is slip. The most common crystallographic slip planes in hexagonal (alpha) titanium are basal, prismatic, and pyramidal. The planes involved during deformation depend on alloy composition, temperature, grain size, and crystal orientation. Deformation also occurs by twinning, which is favored over slip by coarse grain size, high purity, and low-temperature deformation. With more slip systems available, the body centered cubic structure (beta phase) is more ductile than the hexagonal close packed structure (alpha phase).

The hexagonal close-packed (hcp) crystal structure is less symmetrical than the body-centered cubic (bcc) lattice. As a result, directionality in certain properties (anisotropy) occurs. When anisotropy is controlled, improved strength can be achieved under biaxial stress conditions. Increased strength as a result of crystallographic texture is called texture strengthening (Ref 5.2, 5.3). It is most pronounced in alpha-titanium alloys and offers a potential method for improved design in biaxial stress applications such as rocket cases and pressure bottles.

Resistance to additional amounts of cold deformation due to strain hardening is greatest in alpha-titanium alloys. Increasing amounts of the beta phase decrease the strain-hardening rate. As a result, beta alloys have the best cold formability.

After appreciable plastic deformation, heat treatments are generally necessary to restore the original, or more useful, combination of properties. Titanium alloys, like most common metals, are stress relieved to remove residual stresses remaining from plastic deformation. Higher-temperature treatments are used to fully recrystallize the worked structure. However, heating to extremely high temperatures through the beta transus temperature causes extreme grain growth, which is undesirable as it degrades the mechanical properties.

Alpha-beta titanium alloys display superplastic behavior (Ref 5.4); that is, they exhibit both high elongation and high strain-rate sensitivity at elevated temperatures. This characteristic allows complex parts to be formed at temperatures in the alpha-beta phase field. See Chapter 11, "Forming of Titanium Plate, Sheet, Strip, and Tubing" for further discussion on superplastic forming.

Deformation

Titanium, like other metals, is processed into standard shapes, tube, billets, bar, sheet, and wire. These processes require deformation of the metal to produce the desired form or finished product. This ability of a solid to be permanently deformed without rupture is characteristic of metals and a major reason for their wide use. The manner in which metals respond to the

---

*Adapted and revised from Patrick A. Russo and Stan R. Seagle, *Titanium and Its Alloys*, ASM International.

external stresses determines their mechanical properties.

Slip. Atoms in metals are spaced at definite intervals in a repeating pattern. The pattern is determined by the closest packing of atoms that will yield the lowest energy in the metal (Ref 5.5). In pure titanium, the pattern or crystal structure is hcp below 885 °C (1625 °F) and bcc above 885 °C (1625 °F) to the melting point. Atoms are bonded in position by the equilibrium of the cohesive forces acting between them. These bonds function like springs that resist tension and compression forces.

When a crystal structure (Fig. 5.1a) is subjected to minor stresses, the atomic bonds are stretched or contracted (Ref 5.1, 5.5), behaving like springs. Crystals are deformed while force is applied (Fig. 5.1b) and regain their original shape as the force is removed. This deformation, which exists only when the metal is stressed, is called elastic deformation. The maximum stress that permits only this type of deformation is called the elastic limit. When this limit is exceeded, the atoms relocate, resulting in permanently deformed metal.

Atoms of the hcp and bcc crystal structures are grouped into natural planes, which are numbered according to a standard reference system. For example, the base plane of the hexagon is identified as (0001). The density of the atoms on these crystallographic planes is greater than the average density of the crystal. As a result, cohesive forces

holding the atoms of a given plane in position are stronger than the forces holding two planes together. Thus, when a crystal is subjected to a stress greater than the elastic limit, the bonds holding the planes together are the first to yield, and deformation occurs by slipping between two adjacent planes. This type of deformation is called slip.

Figure 5.1 shows a simplified mechanism of slip. The atom planes are in equilibrium in Fig. 5.1(a). The application of a force (Fig. 5.1c) shifts the plane of atoms so they are displaced by a distance of one atom. In this simplified example, slip occurs when the force is great enough to break all atom bonds along the slip plane.

The stress necessary to initiate this type of slip is calculated using information available on interatomic forces. Calculated results indicate that metals should be much stronger than they are (a thousandfold greater). This discrepancy results from imperfections in the crystal structure called dislocations (Ref 5.1, 5.5). A model of such an imperfection is shown in Fig. 5.1(d).

Dislocations exist in large numbers in engineering metals. Grain boundaries are actually an array of dislocations. The number and distribution of dislocations are affected by chemistry, mechanical working, and heat treatment. Dislocations reduce the stress necessary for initiating slip. This is often compared to the movement of a large rug across a floor. The rug is difficult to move if shifted as one piece, like the layer of atoms in Fig. 5.1(c). However, if a wrinkle is

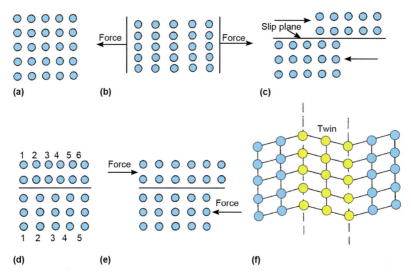

Fig. 5.1 Deformation in a metal crystal. When a crystal structure is stressed, the atomic bonds stretch or contract as shown. (a) Portion of unstrained lattice crystal. (b) Lattice deformed elastically. (c) Slip deformation. (d) Example of dislocation with extra row of atoms above the slip plane. (e) Slip by dislocation movement. (f) Twinning deformation. Source: Ref 5.6

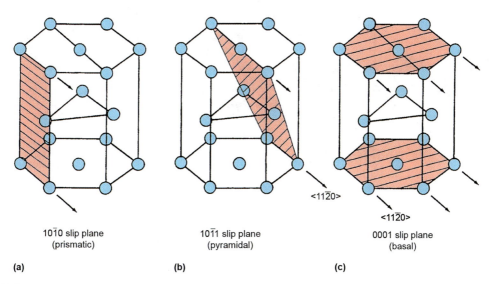

1010 slip plane
(prismatic)

(a)

1011 slip plane
(pyramidal)

(b)

<1120>

<1120>

0001 slip plane
(basal)

(c)

Fig. 5.2 Principal slip planes in alpha-titanium

made in the rug, similar to a dislocation, and then moved a bit at a time by pushing the wrinkle, the rug can be moved easily. This process is analogous to slip-by-dislocation movement.

All slip planes are crystallographic planes, but not all crystallographic planes are slip planes. Slip usually occurs on the planes with the highest density of atomic population. Because of their high symmetry, cubic crystal structures have high-density planes running in many directions and therefore can slip in many directions. This has an important effect on ductility.

By slipping simultaneously in several directions, a crystal can change into any shape that has the same volume. It can adjust its shape to fit its neighboring grains without having to open up holes or cracks. Thus, the entire mass can be reshaped without breaking up. The hexagonal structure, because of its lower symmetry, does not have this property to the same extent as cubic structures. Thus, hexagonal metals are generally more difficult to work mechanically.

Slip does not occur with equal ease in all crystallographic directions. Each slip plane has a required (critical) shear force before slip occurs. Although a high stress may be applied to a metal, the critical resolved shear stress must be exceeded before slip occurs on a given plane. Thus, the orientation of the slip planes in relation to the applied stress direction must be favorable for permanent deformation to occur.

The most common slip planes in alpha-titanium are shown in Fig. 5.2. The (1010) is the prismatic plane, the (1011) is the pyramidal plane,

Table 5.1 Resolved shear stress for slip in titanium as a function of purity

Oxygen plus nitrogen, %	Slip plane	Resolved shear stress	
		MPa	psi
0.01	1010	14	2,000
	0001	62	9,000
0.10	1010	90	13,075
	1011	97	14,075
	0001	107	15,500

while the (0001) is the basal plane. Planes operating during deformation depend on the composition, temperature, grain size, and crystal orientation. Table 5.1 shows the influence of interstitial elements on the resolved shear stress necessary to induce slip on each plane.

In high-purity titanium, (1010) slip predominates. Slip occurs on the (0001) plane only when favorable orientation (high resolved stresses) occurs with the applied stress. Slip is not observed on the 1011 plane in low-interstitial titanium. However, at high interstitial levels, all three slip planes are nearly equally favored.

Although several slip planes may be operative in alpha-titanium, they all slip in the <1120> direction. This is illustrated in Fig. 5.2.

Slip modes in unalloyed beta-phase titanium have not yet been established. It is expected that they would be the same as for bcc metals, which are (110), (112), and (123) planes with a <111> direction. These systems are operative in the beta alloys titanium-vanadium and titanium-molybdenum (Fig. 5.3).

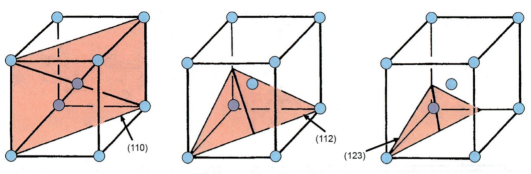

Fig. 5.3 Typical slip planes and directions in the body-centered cubic crystal structure

Fig. 5.4 Twinning in high-purity titanium. The twins are the needlelike bands in the grains. In some instances, the twins extend entirely across a grain. Etchant: 10%HF-5%HNO₃. Original magnification: 250x

Twinning. Permanent deformation also occurs in titanium as a result of twinning. This process, like slip, occurs on characteristic crystal planes called the twinning planes. This is illustrated in Fig. 5.1(f). Twinning occurs when the energy of elastic deformation exceeds a critical value or when a critical shear stress is acting. A group of atoms then function cooperatively to form the twinned region in the crystal. In the process, each atom moves only slightly with respect to its neighboring atoms. However, the final orientation of the twinned region is such that the twin lattice is the mirror image of the untwinned lattice across the twinning plane.

An example of twinning in high-purity titanium is shown in Fig. 5.4. Note that twins form on parallel planes that are crystallographic in nature. Twinning is favored over slip deformation in high-purity, coarse-grained material at low temperatures. High-purity, unalloyed titanium twins easily. Often, the mechanical work performed in cutting, grinding, and polishing metallographic specimens is sufficient to induce twins. Titanium twins on a number of crystallographic planes (Fig.

5.5). Most twinning is observed in the (1102) plane, although several other planes have been noted for room-temperature deformation.

The preferred mode of deformation of the hexagonal structure depends on temperature. Twinning is favored when the temperature during deformation drops below room temperature; increasing numbers of twin systems are observed. Although not favored by low temperatures, reorientation can occur on the prismatic plane. In addition to temperature, strain rate influences modes of deformation. Twinning is the favored mode of slip during high-velocity deformation.

Development of Texture in Titanium

Of the three most common metal crystal structures, the hexagonal type displays the most polycrystalline anisotropy. That is, its properties can vary significantly with direction of measurement. Polycrystalline face-centered and body-centered cubic metals are usually nearly isotropic in their plastic-flow characteristics, even though they may be strongly textured.

The anisotropy of fracture behavior, such as fracture toughness or reduction of area, is largely due to directionality of microstructure, grain shape, secondary phases, and weak interfaces. However, anisotropy of plastic behavior is the result of preferred orientation of the crystal structure, referred to as crystallographic texture. Such texture influences material behavior such as yield strength, elastic modulus, and strain hardening. This plastic anisotropy is a consequence of the fact that the deformation mechanisms, slip and twinning, are crystallographic in nature. The effect of test direction on the elastic moduli of a single crystal of alpha-titanium is shown in Fig. 5.6. The modulus is 145 GPa (21.0 × 10⁶ psi) when tested perpendicular to the basal planes,

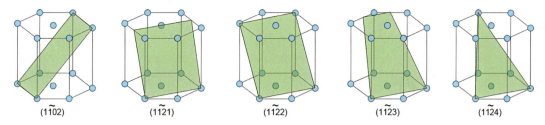

Fig. 5.5 Twinning planes in titanium. Although most twinning occurs along the (1102) plane, deformation at room temperature also takes place along other planes.

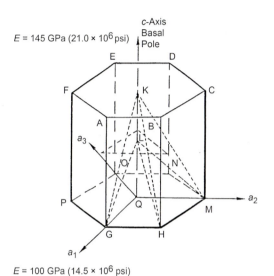

E = 145 GPa (21.0 × 10^6 psi)

E = 100 GPa (14.5 × 10^6 psi)

Fig. 5.6 Variation of modulus (E) with test direction for alpha-titanium

compared with 100 GPa (14.5 × 10^6 psi) in the direction of the a-axis. Strength also varies substantially in these test directions.

Titanium mill products are not single crystals but are made up of millions of unit cells and many grains. The unit cells can be oriented in a somewhat random fashion or aligned so a crystallographic texture is developed. The development of texture in titanium products is the result of thermomechanical processing. Early rolling practice called for straight-away rolling. Later, cross rolling was used to minimize the anisotropy in mechanical properties. However, it was recognized that crystallographic anisotropy could be advantageous for certain applications.

The most prominent textures in the widely used alpha-beta titanium alloy Ti-6Al-4V are the basal texture, where basal planes are parallel to the rolling planes, and the transverse texture, where the basal planes are perpendicular to the rolling plane and parallel to the rolling direction

(Ref 5.2, 5.3). Variations of these, such as basal/transverse texture, where both basal and transverse portions are present, are commonly encountered. Schematic views of basal and transverse textures are shown in Fig. 5.7. Unidirectional rolling of Ti-6Al-4V at temperatures below approximately 900 °C (1650 °F) gives a basal/transverse texture, whereas cross rolling below 900 °C (1650 °F) results in a basal texture. A transverse texture is developed by unidirectional rolling above 900 °C (1650 °F). Tensile properties of solution-treated and aged Ti-6Al-4V plate processed to develop these different textures are shown in Table 5.2. Significant changes in modulus and strength are apparent, depending on texture and test direction. The highest modulus and highest yield strengths were achieved when the testing direction was perpendicular (T-TD) or nearly perpendicular (B/T-TD) to the basal planes.

Crystallographic texture can also have an important influence on fatigue properties. Figure 5.8 shows the influence of texture on the high-cycle fatigue properties of fine-grained Ti-6Al-4V tested in a vacuum. A pronounced influence of texture is apparent.

Texture Strengthening

Even when the tensile properties of titanium sheet and plate are balanced in the longitudinal and transverse directions, a significant increase in strength may be found if a tensile test through the thickness of the material was performed. The term *texture strengthening* or *hardening* refers to increased resistance to yielding under various conditions of loading caused by crystallographic texture. A useful form of texture strengthening is the increased resistance to yielding under biaxial tension (Ref 5.5), as shown in Fig. 5.9. These special stress-field conditions exist in such applications as spherical pressure bottles and cylindrical rocket-motor cases. Material loaded in this

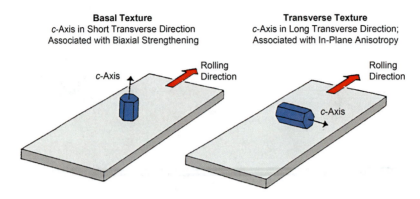

Fig. 5.7 Schematic view of alpha-phase textured plates

Table 5.2 Effect of texture and test direction on solution-treated and aged Ti-6Al-4V plate

Texture and test direction(a)	Elastic modulus		0.2% offset yield strength		Fracture strength		Fracture strain
	GPa	10^3 ksi	MPa	ksi	MPa	ksi	
B/T-RD	107	15.5	1120	162	1650	239	0.62
B/T-45°	113	16.4	1055	153	1560	226	0.76
B/T-TD	123	17.8	1170	170	1515	220	0.55
T-RD	113	16.4	1105	160	1540	223	0.57
T-45°	120	17.4	1085	157	1610	234	0.76
T-TD	126	18.3	1170	170	1665	241	0.70
B-RD	109	15.8	1120	162	1505	218	0.70

(a) B, basal texture; T, transverse texture; RD, rolling direction; TD, transverse direction

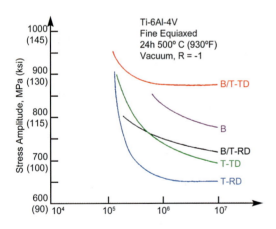

Fig. 5.8 Influence of texture and test direction on high-cycle fatigue strength. B, basal texture; T, transverse texture; TD, transverse direction; RD, rolling direction. Reprinted with permission from Ref 5.7

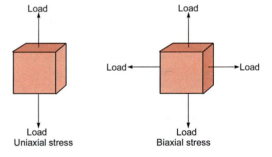

Fig. 5.9 Uniaxial and biaxial stresses. The uniaxial material can contract in the thickness and width directions, while the biaxial material can contract only in the thickness direction.

manner cannot contract freely in either width or length; thus, yielding must proceed by thinning (reduction in thickness). Yielding in a two-to-one biaxial stress field (cylinder loading) occurs at 1.15 times the uniaxial yield strength for isotropic material, while strengths up to 1.6 times the yield strength can occur in anisotropic material.

Even when tensile specimens cut from sheets at 0° (longitudinal), 45°, and 90° (transverse) to the rolling directions have similar yield and tensile strengths, a significant increase in strength may be found if a tensile test across the thickness was performed.

The yield strength and macroscopic stress-strain behavior of an anisotropic material varies with the direction of measuring, as shown previously. One criterion for evaluating sheet material is the comparison of the two principal transverse-contraction strains during a tensile test. This is illustrated in Fig. 5.10. When a load, P, is applied in a tensile test, the specimen elongates. This elongated strain is referred to as ε_l. Because equal volume must exist before and after stress, contractual strains must occur as ε_w and ε_t (width and thickness). The anisotropy parameter, R, can be calculated from the ratio of ε_w to ε_t.

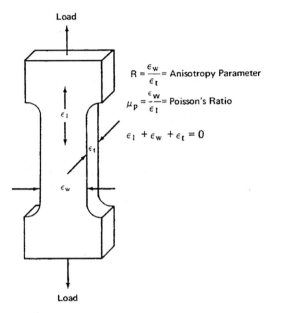

Fig. 5.10 Three principal directions of strain result from applying a tension load, P, to a sheet tensile specimen.

Because it is difficult to accurately measure strain in the thickness direction of sheet, R is calculated using the strain in the length direction, ε_l. When ε_l and ε_w are measured, ε_t can be calculated from the following relationship:

$$\varepsilon_l + \varepsilon_w + \varepsilon_t = 0$$

The ratio of ε_w to ε_l is the plastic Poisson's ratio, μ_p. The R-value can also be calculated with Poisson's ratio, as follows:

$$R = \varepsilon_w / \varepsilon_t = \mu_p / 1 - \mu_p$$

High values of R indicate anisotropy and high resistance to thinning in a biaxial stress field.

Table 5.3 contains uniaxial tensile, biaxial tensile, and R-values for several titanium alloys. The ratio of the biaxial tensile ultimate strength to the standard uniaxial tensile strength (BTU/UTS) is also included. Alpha alloys Ti-5Al-2.5Sn, Ti-7Al-12Zr, and Ti-8Al-1Mo-1V have high BTU/UTS ratios and high R-values because they contain predominantly hexagonal crystal structures. Less texture strengthening is noted as the amount of beta stabilizing is increased. For example, the Ti-4Al-3Mo-1V and Ti-6Al-6V-2Sn alloys have low R-values and low BTU/UTS ratios.

One potential application of an intentionally textured titanium alloy is in spherical pressure vessels for cryogenic temperatures. Thus, the effect of low temperatures on the anisotropy parameter becomes important. Table 5.4 contains data on the effect of low temperatures on the biaxial yield properties of two alpha alloys. The Ti-4Al-0.20 alloy was developed solely for biaxial loading conditions. This alloy and Ti-5Al-2.5Sn

Table 5.3 Room-temperature uniaxial and biaxial tensile properties of several titanium alloys

Alloy(a)	Testing direction(b)	Tensile strength		Yield strength		Elongation, %	BTU(c)		Ratio	R(d)
		MPa	ksi	MPa	ksi		MPa	ksi		
Ti-5Al-2.5Sn	L	1110	161	952	138	8	1455	211	1.31	2.4
	T	1076	156	938	136	8	1531	222	1.42	2.1
Ti-7Al-12Zr	L	1069	155	882	128	11	1372	199	1.28	7.0
	T	1007	146	862	125	12	1434	208	1.42	7.7
Ti-8Al-1Mo-1V	L	1200	174	1020	148	7	1517	220	1.26	2.0
	T	1200	174	1034	150	5	1427	207	1.19	1.8
Ti-6Al-4V	L	1172	170	993	144	8	1476	214	1.26	1.6
	T	1186	172	1055	153	5	1503	218	1.27	2.0
Ti-4Al-3Mo-1V	L	1034	150	903	131	9	1289	187	1.25	1.3
	T	1034	150	917	133	6	1310	190	1.27	2.3
Ti-6Al-6V-2Sn	L	1193	173	1062	154	6	1379	200	1.15	0.6
	T	1262	183	1124	163	6	1379	200	1.09	0.6

(a) After 60% reduction at 650 °C (1200 °F), except Ti-6Al-4V, which was rolled 60% at 482 °C (900 °F). (b) L, longitudinal; T, transverse. (c) BTU, biaxial tensile ultimate strength, UTS = standard uniaxial tensile strength. (d) Anisotropy parameter

develop biaxial yield strengths substantially greater than those predicted for isotropic materials. Strengthening ranges from 1.25 to 1.50 times the uniaxial yield strength. Strengthening is greater for the more heavily textured Ti-4Al-0.20 alloy than for Ti-5Al-2.5Sn extra low interstitial (ELI) at all temperatures. Consequently, although Ti-4Al-0.20 has lower uniaxial yield strength than Ti-5Al-2.5Sn ELI at the three testing temperatures, the biaxial yield strength for both alloys at 21 and −196 °C (70 and −320 °F) are nearly identical. At −253 °C (−423 °F), the biaxial yield strength of Ti-4Al-0.20 is approximately 9% greater than that of Ti-5Al-2.5Sn ELI.

The amount of anisotropy decreases at lower testing temperatures. The R-value for Ti-4Al-0.20 dropped from 4.70 to 2.49 when the temperature was lowered from room temperature to −253 °C. Above room temperature, the R-value does not change, because Poisson's ratio is constant to at least 540 °C (1000 °F). Because R is related to Poisson's ratio (μ_p), no change in R is anticipated up to 540 °C.

Texture strengthening is applied in the manufacture of 51 mm (2 in.) diameter cylindrical pressure vessels. All alloys were shear spun to 50% reduction in wall thickness at 540 °C (1000

°F) and machined to a uniform 1 mm (0.040 in.) wall thickness after annealing and before testing. Results of pressure vessel tests are given in Table 5.5. These alloys show some degree of texture strengthening. However, the degree of texture strengthening is most significant in the Ti-4Al alloy. The biaxial strength for all three alloys increases as the stress-relieving temperature drops from 760 to 540 °C (1400 to 1000 °F).

The preceding results indicate that texture strengthening (high R-values) is a reality for titanium alloys under proper conditions of alloying and processing. Of immediate practical importance are the strengthening effects that can be obtained in biaxial stress fields.

Strain Hardening

Cold working strengthens metals (Ref 5.1, 5.5). With increasingly greater deformation, the resistance of the metal to further working constantly increases due to strain hardening. Simultaneously, ductility decreases. The effect of cold reduction on the yield strength and elongation of an alpha alloy and a beta alloy is illustrated in Fig. 5.11.

Table 5.4   Average uniaxial and biaxial yield properties of Ti-4Al-0.20 and Ti-5Al-2.5Sn extra low interstitial (ELI) sheet

Material	Testing temperature		Uniaxial yield strength (UYS)		1:2 biaxial yield strength (BYS)		BYS/UYS ratio	Strain ratio (R)	Plastic Poisson's ratio (μ_p)
	°C	°F	MPa	ksi	MPa	ksi			
Ti-4Al-0.20	21	70	651	94.4	997	144.6	1.53	4.70	0.845
	−196	−320	1083	157.1	1534	222.5	1.42	3.06	0.697
	−253	−423	1387	201.1	1889	274.0	1.36	2.49	...
Ti-5Al-2.5Sn ELI	21	70	793	104.9	992	143.8	1.37	2.55	0.732
	−196	−320	1191	172.8	1531	222.0	1.28	1.83	0.617
	−253	−423	1403	203.5	1740	252.4	1.24	1.53	0.557

Table 5.5   Results of pressure vessel tests on three titanium alloys shear spun at 538 °C (1000 °F)

Alloy	Anneal or stress relieve		0.2% yield strength				Calculated R
			Uniaxial		1:2 stress ratio		
	°C	°F	MPa	ksi	MPa	ksi	
Ti-5Al-2.5Sn	760	1400	876	127	1117	162	1.70
	650	1200	979	142	1324	192	2.34
	538	1000	1007	146	1386	201	2.63
Ti-4Al	760	1400	669	97	1014	147	4.30
	650	1200	731	106	1214	176	7.98
	538	1000	772	112	1324	192	9.85
Ti-6Al-4V	460	1400	910	132	1117	162	1.43
	650	1200	1007	146	1324	192	2.08
	538	1000	1041	151	1413	205	2.44

Increased cold reduction raises the yield strength and lowers ductility. The strength of alpha alloy Ti-5Al-2.5Sn increases more rapidly with cold reduction than the strength of beta alloy Ti-12Mo-6Sn. Alpha alloys strain harden more rapidly than beta alloys. The low rate of strain hardening for beta alloys makes them more amenable to cold working. Alloys of this type can be given greater reductions before annealing is necessary than equivalent-strength alpha alloys.

Another, more common method of measuring the strain-hardening characteristics of an alloy is by analyzing the tensile stress-strain relationship. (True stress is calculated from the actual cross-sectional area of the specimen rather than using the original cross-sectional area throughout the tensile test, as is the case in calculating the engineering stress.) Figure 5.12(a) shows a typical true-stress/true-strain diagram obtained by tensile testing a specimen. The test bar is elongated and the load necessary to cause this elongation is measured. While the test bar is pulled until it fractures, data on changes in length and diameter are noted with load data. True-stress and true-strain values (calculated from these data) are plotted to obtain the true-stress/true-strain curve.

Most stress-strain curves are of the engineering type, where the stress at each point of deformation is determined from the cross-sectional area of the original test bar. However, the strain-hardening characteristics are determined from a true-stress/true-strain curve.

True strain in the elastic range is nil, and the stress-strain curve in Fig. 5.12(a) rises directly to the stress axis until the range for plastic deformation is reached. Strain then increases rapidly with increasing stress until fracture occurs. The slope of the straight-line portion of the plastic range is

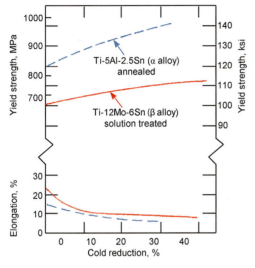

Fig. 5.11 Effect of cold reduction on yield strength and elongation (ductility) in alpha and beta alloys

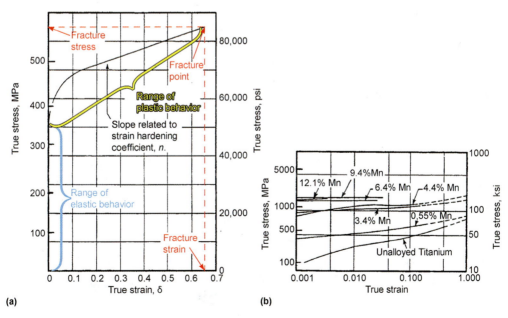

Fig. 5.12 Stress-strain relationship. (a) True stress/true strain derived from tensile testing a specimen and measuring the actual area at each strain. (b) Strengthening effect of manganese on flow strength of titanium-manganese alloys annealed at 750 °C (1380 °F)

related to the strain-hardening exponent. A large slope, or a rapid increase in stress with increase in strain, indicates a high strain-hardening rate.

Strain hardening of alpha-titanium is greater than that of beta-titanium, as shown earlier in Fig. 5.11. Small amounts of beta phase in alpha-matrix alloys decrease the strain-hardening rate. This is illustrated in Fig. 5.12(b) for high-purity titanium-manganese alloys quenched from 750 °C (1380 °F). An increasing percentage of manganese, a beta stabilizer, raises the amount of retained beta phase. The flow curves shown give the true stress needed to initiate plastic flow for various amounts of true strain. The slopes of these flow curves decrease with manganese content and the corresponding amount of beta phase.

Strain Effects

The ability to readily form a part is important. However, the effect of such forming (plastic deformation) on the properties of the workpiece must be considered. The most often analyzed effect of forming is the Bauschinger effect (Ref 5.1, 5.5). It is a phenomenon by which plastic deformation of polycrystalline metal, caused by stress applied in one direction, reduces the yield strength when the stress is applied in the opposite direction.

An example of this phenomenon is shown in Fig. 5.13 for Ti-6Al-4V. A sample of Ti-6Al-4V was solution treated and strained in tension. Tensile strain results in an appreciable decrease in the compressive yield strength. The loss in compressive yield strength that occurs in the solution-treated condition can be recovered by aging. Therefore, compressive properties can be improved, if desired, by heat treatment after straining.

Superplasticity

Fine-grained metals are said to be superplastic when they exhibit both high elongation and high strain-rate sensitivities (Ref 5.4). These characteristics allow complex parts to be formed to near-net shape at substantial reductions of yield loss and machining costs. Rate of strain, temperature, and grain size are the major factors that influence the superplasticity of titanium alloys. Strain-rate sensitivity, m, is defined as:

$$m \approx \frac{\Delta(\log \sigma)}{\Delta(\log \varepsilon')}$$

where σ is flow stress, and ε' is strain rate. The Δ signifies a change, and m is thus the change in log σ with respect to a change in log $\dot{\varepsilon}$. The m-value can be obtained from the slope of a plot of log σ versus log $\dot{\varepsilon}$.

High values of strain-rate sensitivity ($m > 0.5$) are required for superplastic behavior. The relationship between m-values and strain rate at 925 °C (1700 °F) for Ti-6Al-4V having grain sizes from 6.4 to 20 μm is shown in Fig. 5.14. An optimum strain rate to produce high m-values exists for each grain size. Finer grain size enhances superplastic behavior.

A summary of superplastic characteristics for various titanium alloys is shown in Table 5.6. Elongations and m-values are shown for various strain rates and temperatures. Elongations ap-

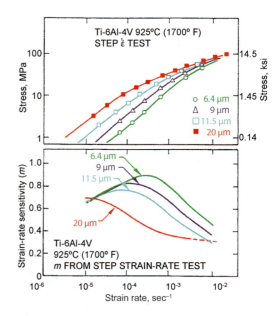

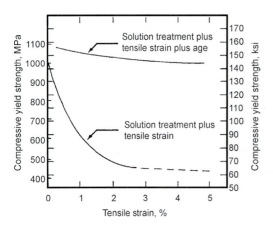

Fig. 5.13 Effect of tensile strain on the compressive yield strength (Bauschinger effect) of solution-treated plus strained Ti-6Al-4V sheet and after subsequent aging

Fig. 5.14 Effect of average grain size on the strain-rate sensitivity of flow stress and m for Ti-6Al-4V at 925 °C (1700 °F)

proaching 1200% are possible for Ti-6Al-4V under certain conditions. Generally, alpha-beta alloys have the best superplastic-forming potential. This is due to the development of a two-phase structure (alpha and beta) at the forming temperature. Beta alloys, such as Ti-13V-11Cr-3Al, which are one phase at the forming temperature, do not display superplastic characteristics to any degree. Alpha alloys such as Ti-5Al-2.5Sn, which consist essentially of the alpha phase at the superplastic-forming temperature, are also poor superplastic-forming candidates. Information on the effect of microstructure on superplasticity is shown in Fig. 5.15. These data show that approximately 65% alpha at the forming temperature appears optimum for superplasticity, as measured by *m*-values. Further discussion of superplastic forming

and the related diffusion bonding can be found in Chapter 11, "Forming of Titanium Plate, Sheet, Strip, and Tubing," in this book.

Internal Changes

Plastic deformation produces extensive changes in the external appearance of a metal. Substantial internal changes also occur. The effects of both hot and cold working should be considered when evaluating the structural effects of plastic deformation.

A metal is considered to have been cold worked if its grains are in a distorted condition when plastic deformation is completed (Fig. 5.16a). Changes occur in almost all of its physical and

Table 5.6 Summary of superplastic characteristics for titanium alloys

Alloy	Test temperature °C	Test temperature °F	Strain rate, s⁻¹	Strain-rate sensitivity(*m*)	Elongation, %
Ti-6Al-4V	840–870	1544–1600	1.3×10^{-4} to 10^{-3}	0.75	750–1170
Ti-6Al-5V	850	1565	8×10^{-4}	0.70	700–1100
Ti-6Al-2Sn-4Zr-2Mo	900	1650	2×10^{-4}	0.67	538
Ti-4.5Al-5Mo-1.5Cr	870	1600	2×10^{-4}	0.63–0.81	>510
Ti-6Al-4V-2Ni	815	1500	2×10^{-4}	0.85	720
Ti-6Al-4V-2Co	815	1500	2×10^{-4}	0.53	670
Ti-6Al-4V-2Fe	815	1500	2×10^{-4}	0.54	650
Ti-5Al-2.5Sn	1000	1832	2×10^{-4}	0.49	420
Ti-15V-3Cr-3Sn-3Al	815	1500	2×10^{-4}	0.50	229
Ti-13V-11Cr-3Al	800	1472	...	...	<150
Ti-8Mn	750	1380	...	0.43	150
Ti-15Mo	800	1472	...	0.60	100
Commercially pure Ti	850	1565	1.7×10^{-4}	...	115

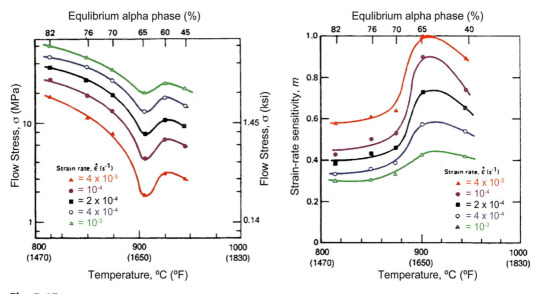

Fig. 5.15 Effects of temperature and alpha-phase content on flow stress and strain rate for Ti-6Al-4V

mechanical properties during cold work. Strength, hardness, and electrical resistance increase, whereas ductility decreases. When a cold-worked metal is heated at a sufficiently high temperature, the deformed grains change into strain-free grains by a process of nucleation and growth called recrystallization (Fig. 5.16b) (see also Chapter 7, "Metallography of Titanium and Its Alloys," in this book). If the working is performed at temperatures substantially above the recrystallization temperature, recrystallization occurs almost simultaneously. This process is referred to as hot working.

Annealing

The thermal treatment given to a cold-worked metal to return it to its strain-free state is referred to as annealing. Three stages of annealing are recovery, recrystallization, and grain growth.

Recovery. In the recovery process, the physical and mechanical properties that changed during cold working recover to their original values. The changes occur without significant changes in microstructure. The recovery process is shown in relation to the recrystallization and grain-growth processes in Fig. 5.17. During recovery, internal (residual) stresses decrease, whereas strength is only slightly affected.

The heat treatment by which the residual stresses are relieved is referred to as stress-relief annealing. The relief of residual stress is a function of both time and temperature (Fig. 5.18). Data on commercially pure titanium, Ti-6Al-4V, and Ti-5Al-2.5Sn are included in the figure. The Larson-Miller parameter is used to combine both temperature and time effects into a single parameter.

Examination of Fig. 5.18 reveals several important features. Both increasing time and temperature reduce residual-stress levels. Commercially pure titanium, stress relieved at 480 °C (900 °F) for 1 h, has approximately 35% of the original stress remaining. Increasing the time to 2 h at 480 °C reduces the stress remaining to 30% (use Larson-Miller parameter to locate position on plot). However, increasing the temperature to

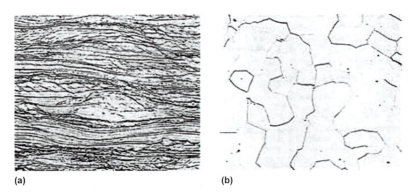

(a)   (b)

Fig. 5.16 Grain shape of titanium. (a) Distorted grains are evidence of cold work. (b) Deformed grains in (a) changed to strain-free grains by recrystallization at 790 °C (1450 °F). Etchant: 10% HF-5%HNO$_3$. Original magnification: 250x

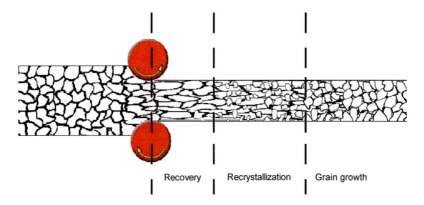

Recovery   Recrystallization   Grain growth

Fig. 5.17 Recovery, recrystallization, and grain growth occur after cold working operations such as cold rolling followed by annealing.

540 °C (1000 °F) for 1 h for commercially pure titanium decreases the remaining residual stress to approximately 15% of the original value. This shows that the stress-relieving process is much more sensitive to changes in temperature than to changes in time. The curves shown also depend on initial stress level and should only be used as approximations of stress-relieving behavior.

Figure 5.18 also shows that stress-relief behavior is strongly dependent on alloy composition. Commercially pure titanium can be stress relieved at lower temperatures than Ti-6Al-4V. The alloy Ti-5Al-2.5Sn requires the highest stress-relief temperature of the three alloys shown. The alloy effects on stress relief are similar to alloy effects on creep strength. Alloys that have good high-temperature creep strength, such as alpha and near-alpha alloys, require higher stress-relief temperatures and longer times than alloys having poorer creep strength, such as alpha-beta alloys. Table 5.7 shows stress-relief temperatures used for various titanium alloys. Because the alpha-beta and beta-titanium alloys are heat treatable, stress-relief treatments must be placed in context with the final thermal condition of the material. This is important so that heat treating response and subsequent mechanical properties are not adversely affected by the stress relief.

Recrystallization is defined as the nucleation and growth of strain-free grains in a cold-worked metal. As mentioned previously, the recovery process precedes recrystallization to some degree. The properties that a metal has after recrystallization are nearly the same as the metal had

before it was cold worked. Some differences in properties can occur due to differences in grain size and crystallographic texture.

Because recrystallization is a nucleation-and-growth process, it begins slowly, accelerates to a maximum rate, and finishes slowly during isothermal annealing. An example of a typical isothermal annealing curve is shown in Fig. 5.19. The rate at which the recrystallization process occurs depends on several factors, including temperature, amount of cold work, grain size prior to cold work, and purity of the metal. Higher annealing temperatures, increasing amounts of cold work, smaller grain size prior to cold working,

Table 5.7 Stress-relief treatments for titanium alloys

Alloy	Temperature		Time
	°C	°F	
Commercially pure Ti (all grades)	480–595	900–1100	15 min–4 h
α or near-α alloys			
Ti-5Al-2.5Sn	540–650	1000–1200	15 min–4 h
Ti-8Al-1Mo-1V	595–705	1100–1300	15 min–4 h
Ti-6Al-2Sn-4Zr-2Mo	595–705	1100–1300	15 min–4 h
Ti-6Al-2Nb-1Ta-0.8Mo	595–650	1100–1200	15 min–4 h
Ti-0.3Mo-0.8Ni (ASTM grade 12)	480–595	900–1100	15 min–4 h
α-β alloys			
Ti-6Al-4V	480–650	900–1200	1–4 h
Ti-6Al-6V-2Sn (Cu + Fe)	480–650	900–1200	1–4 h
Ti-3Al-2.5V	540–650	1100–1200	30 min–2 h
Ti-6Al-2Sn-4Zr-6Mo	595–705	1100–1300	15 min–4 h
β or near-β alloys			
Ti-13V-11Cr-3Al	705–730	1300–1350	5–15 min
Ti-11.5Mo-6Zr-4.5Sn (Beta III)	720–730	1325–1350	5–15 min
Ti-3Al-8V-6Cr-4Zr-4Mo (Beta C)	705–760	1300–1400	10–30 min
Ti-10V-2Fe-3Al	675–705	1250–1300	30 min–2 h
Ti-15V-3Al-3Cr-3Sn	790–815	1450–1500	5–15 min

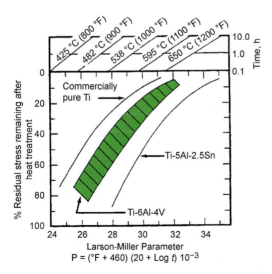

Fig. 5.18 Effect of stress-relief time and temperature on the percent of original residual stresses remaining

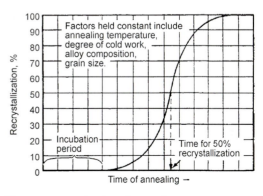

Fig. 5.19 Typical isothermal recrystallization curve. Reprinted with permission from Ref 5.8

and higher purity increase the rate at which the recrystallization process proceeds.

Recrystallization of unalloyed titanium occurs well below the beta transus. Figure 5.20 shows the influence of cold work on the recrystallization range for pure titanium. This range is the temperature region necessary for the formation of new grains until the structure is composed of 100% recrystallized grains. The lower line represents the temperature at which recrystallization is initiated (in 30 min), while the upper line represents the completion of recrystallization. The temperature for both initiation and completion of recrystallization is lowered by increased cold work.

Alloy additions can affect recrystallization of titanium, although their effects are not completely consistent. Additions of beta stabilizers can either lower or raise the recrystallization temperature, while the alpha additions aluminum, oxygen, and nitrogen can substantially increase the recrystallization temperature. The additions of beta stabilizers to form alpha-beta alloys can also retard recrystallization in some instances. Chromium has been particularly beneficial in this respect. The addition of compound formers such as boron and carbon also inhibits recrystallization.

Grain Growth. Once new grains have recrystallized from the cold-worked metal, their growth depends on time and temperature. Figure 5.21 shows the average grain size of recrystallized grains as a function of annealing time for various annealing temperatures for iodide titanium that was cold worked 94%. The average grain diameter follows the relationship:

$$D = Kt^n$$

where D is the average grain diameter, t is the annealing time, and n and K are grain-growth parameters.

For the data shown, n equals approximately 0.33. This value varies with alloy, purity, and annealing temperature.

The addition of elements that form second phases can retard grain growth. Combinations of boron and carbon, beryllium and carbon, and boron and silicon are effective in this respect. Sulfur and carbon also refine grain size. The effect of carbon on the grain size of Ti-11Mo alloy is shown in Fig. 5.22. Generally, the compound-formers adversely affect ductility.

Iron, which is present at some level in all commercially pure titanium products, can have a significant effect on grain growth. Figure 5.23 shows the effect of iron content on the grain growth of commercially pure titanium. Increasing amounts of iron retard the rate of grain growth. Iron is only soluble to a limited degree in alpha-titanium. Generally, iron levels in excess of 0.1% result in the formation of beta spheroids, which are responsible for retarding grain growth (Fig. 5.24). In alpha-beta alloys, small amounts of alpha phase retard grain growth at temperatures below the beta transus.

Neocrystallization

Metals that undergo a high-temperature allotropic transformation on heating or cooling (such as alpha to beta in titanium) experience reformation of their grain structure very similar to recrystallization.

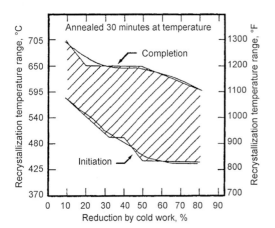

Fig. 5.20 Effect of cold reduction on the temperature for initiation and completion of recrystallization for unalloyed titanium

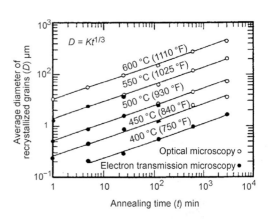

Fig. 5.21 Effects of annealing time and temperature on the average diameter of recrystallized grains of iodide titanium cold rolled 94%

In recrystallization, the initial phase is made unstable as a result of cold working, and it reverts during annealing to the stable, non-cold-worked condition. The recrystallization temperature is merely one at which recrystallization proceeds at a fairly high rate.

By comparison, neocrystallization requires no cold working, because the beta phase is stable above the beta transus temperature and a different phase, alpha, is stable below this temperature. Therefore, neocrystallization can occur only when a given phase is held at a temperature outside the temperature range in which it is stable. Transformation to the stable phase at this temperature proceeds by a process of nucleation and growth that is essentially the same as encountered in recrystallization.

Heating through the transformation range in steel results in nucleation of many small grains.

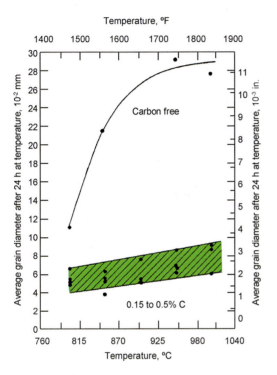

Fig. 5.22 Effect of carbon and annealing temperature (for 24 h) on grain growth in Ti-11Mo alloy

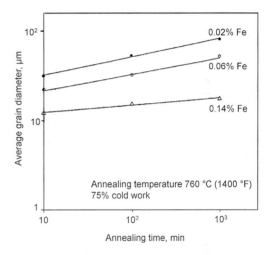

Fig. 5.23 Effect of iron and annealing time on grain growth of titanium annealed at 760 °C (1400 °F) after 75% cold work

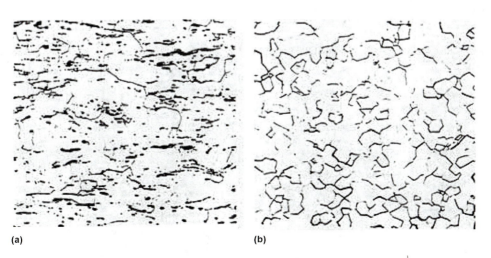

(a)   (b)

Fig. 5.24 Typical microstructures of commercially pure titanium plate with (a) normal (0.17%) and (b) low (0.03%) iron content. The beta-spheroids in the high-iron material tend to restrict grain growth. Original magnification: 100x. Courtesy of RMI Co.

However, in titanium this recrystallization is not apparent when heating through the alpha-to-beta transition (beta transus). The reason is the excellent conformity, or accommodation, of the titanium alpha- and beta-phase lattices and the small difference in their specific volumes. The orientation relations of the alpha- and beta-phase lattices are retained during the transformation. Lattice arrangement is such that the (0001) alpha plane is parallel to the (110) beta plane, and the <1120> slip direction in alpha is parallel to the <111> slip direction in beta.

Because of this high degree of conformity and similarity of crystal volume, few new grains are formed. In actual practice, extreme grain growth occurs. This is illustrated in Fig. 5.25, using the alpha alloy Ti-5Al-2.5Sn as an example. This alloy consists of the alpha phase up to 955 °C (1750 °F), where some alpha transforms to beta. By the time the temperature reaches 1015 °C (1860 °F), all the alpha has transformed and only the beta phase is present. Little grain growth occurs until the alpha begins to transform to beta. At the beta transus temperature, the grain growth is extremely rapid, and this growth continues appreciably at temperatures greater than the beta transus.

The development of this large grain size by neocrystallization is attributed to lower ductility found in certain alloys heated or processed above the beta transus. The alloys most sensitive to this treatment are those that contain both alpha and beta phases after cooling from a temperature in the beta field. However, near-alpha and the near-beta alloys have excellent ductility after a beta-recrystallization anneal. The advantages of this type of processing treatment are covered in Chapter 4, "Principles of Beta Transformation and Heat Treatment of Titanium Alloys," in this book.

Gamma Titanium Aluminide

Gamma titanium aluminide (TiAl) is of interest for use in aerospace and high-temperature automotive engine applications (Ref 5.9). Gamma TiAl has excellent mechanical and oxidation- and corrosion-resistance properties at elevated temperatures (over 600 °C, or 1110 °F). General Electric made the first commercial use of TiAl in its stage 6 and 7 compressor blades for the new GEnx (General Electric Next-Generation) engine, which powers the Boeing 787 and 747-8 aircraft. The replacement of the nearly twice-as-dense nickel-base superalloy traditional compressor blades with TiAl resulted in greater thrust-to-weight ratios. Gamma TiAl can be isothermally forged using hot dies under a constant strain rate ($<10^{-2}$/s) or at an initial strain rate of 1×10^{-3}/s at 1150 °C (2100 °F). In the latter case, the forging of medium-sized ingot billets that have undergone hot isostatic pressing can be as much as 80% reduction.

Summary

Slip is the predominant mode of plastic deformation in hexagonal (alpha) titanium. Common crystallographic slip planes are basal, prismatic, and pyramidal. The resistance to cold deformation is greatest in the alpha alloys. Strain-hardening rate is decreased as the amount of beta phase increases.

The hcp crystal structure, which characterizes titanium, is less symmetrical than the bcc lattice. The hexagonal type displays more crystalline anisotropy; thus, the properties can vary significantly with the direction of measurement.

In view of the anisotropy that can prevail in titanium, processes are generally designed to minimize texture. However, certain applications can take advantage of texture produced by special processing.

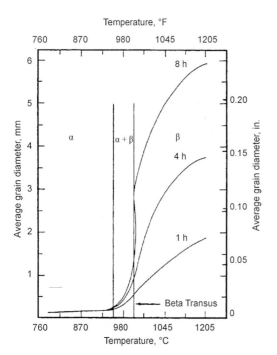

Fig. 5.25 Effect of annealing temperature on grain size of Ti-5Al-2.5Sn. Grain growth is very rapid at the beta transus temperature (1015 °C, or 1860 °F) and higher.

As is true for most metals, cold working results in strain hardening of titanium alloys. Increased cold reduction raises yield strength and decreases ductility. The most common method of measuring the strain-hardening characteristics of titanium alloys is by analysis of the tensile stress-strain relationship.

Fine-grained titanium alloys are considered to be superplastic when they exhibit both high elongation and high strain-rate sensitivities. These characteristics allow forming to near-net shape at substantial reductions in machining costs. Rate of strain, temperature, and grain size are the major factors that affect superplasticity.

Annealing is the term applied to the thermal treatment given to cold-worked titanium alloys to return them to their original strain-free state. Annealing consists of three processes: recovery, recrystallization, and grain growth. These processes occur at increasing temperature; that is, recovery occurs at the lowest temperature and grain growth at the highest temperature.

Stress-relief behavior of titanium and titanium alloys depends greatly on alloy composition. For example, commercially pure titanium can be stress relieved at a lower temperature than Ti-6Al-4V.

In recrystallization, the initial phase is made unstable as a result of cold working. On the other hand, neocrystallization requires no cold working because the beta phase is stable above the beta transus temperature, whereas a different phase, alpha, is stable below this temperature.

Gamma titanium aluminides are generally isothermally forged using hot dies at a temperature in excess of 1100 °C (2010 °F), at which temperature the forging reduction of medium-sized ingot billets that have undergone hot isostatic pressing can be as much as 80%.

ACKNOWLEDGMENTS

The assistance of Dr. Y.-W. Kim of Gamtech in compiling the section on gamma titanium aluminides is greatly appreciated.

REFERENCES

5.1 R.E. Reed-Hill, *Physical Metallurgy Principles,* D. Van Nostrand, Hoboken, NJ, 1973

5.2 M. Peters and G. Lutjering, Control of Microstructure and Texture in Ti-6Al-4V, *Titanium Science and Technology '80, Proc. of the Fourth International Conference on Titanium* (Kyoto, Japan), AIME, 1980, p 933

5.3 A.W. Sommer and M. Creager, "Research toward Developing an Understanding of Crystallographic Texture on Mechanical Properties of Titanium Alloys," AFML-TR-76-222, Jan 1977

5.4 C.H. Hamilton, Superplasticity in Titanium Alloys, *Superplastic Forming,* American Society for Metals, Metals Park, OH, 1985, p 13–22

5.5 W.D. Callister, Jr. and D.G. Rethwisch, *Materials Science and Engineering: An Introduction,* John Wiley and Sons, Hoboken, NJ, 2010

5.6 A.C. Reardon, Ed., *Metallurgy for the Non-Metallurgist,* 2nd ed., ASM International, 2011

5.7 G. Lutjering and A. Gysler, Critical Review—Fatigue, *Titanium Science and Technology, Proceedings of the Fifth International Conference on Titanium,* Deutsche Gesellschaft für Metallkunde e.V., Federal Republic of Germany, 1984, p 2071

5.8 A.G. Guy, *Elements of Physical Metallurgy,* Addison-Wesley Publishing Co., Inc., 1974, p 429

5.9 Y.-K. Kim, *Acta Metall. Mater.,* Vol 40 (No. 6), 1992, p 1121–1134

SELECTED REFERENCES

• R. Boyer, E.W. Collings, and G. Welsch, Ed., *Materials Properties Handbook: Titanium Alloys,* ASM International, 1994

• K.G. Budinski and M.K. Budinski, *Engineering Materials: Properties and Selection,* 7th ed., Prentice Hall, 2002

• M.J. Donachi, *Titanium: A Technical Guide,* 2nd ed., ASM International, 2000

• P. Lacombe, R. Tricot, and G. Beranger, Ed., *Proceedings of the Sixth International Conference on Titanium* (Nice, France), Metallurgical Society of AIME, Warrendale, PA, 1988

• G. Lutjering and J.C. Williams, *Titanium,* Springer, 2003

• G. Lutjering, U. Zwicker, and W. Bunk, *Proceedings of the Fifth International Conference on Titanium,* Metallurgical Society of AIME, Warrendale, PA, 1984

• *Proceedings of the Seventh International Conference on Titanium* (San Diego, CA), Metal-

lurgical Society of AIME, Warrendale, PA, 1992
- *Proceedings of the Eighth International Conference on Titanium* (Birmingham, U.K.), Metallurgical Society of AIME, Warrendale, PA, 1995
- *Proceedings of the Ninth International Conference on Titanium,* Metallurgical Society of AIME, Warrendale, PA, 2000
- *Proceedings of the Tenth International Conference on Titanium,* Metallurgical Society of AIME, Warrendale, PA, 2004
- *Proceedings of the Eleventh International Conference on Titanium,* Metallurgical Society of AIME, Warrendale, PA, 2008
- *Proceedings of the Twelfth International Conference on Titanium,* Metallurgical Society of AIME, Warrendale, PA, 2012

CHAPTER 6

Mechanical Properties and Testing of Titanium Alloys*

IN THIS CHAPTER, the effects of titanium characteristics are discussed as they relate to mechanical properties. Briefly, alloy composition establishes the alloy types, designated by the predominant phases present: alpha, alpha-beta, and beta. The amount and type of thermomechanical reduction and the heat treatment are the two most important factors that influence microstructure and thereby the mechanical properties of titanium alloys (Ref 6.1, 6.2).

A range of properties is available to designers and engineers. However, this chapter emphasizes the more basic properties of terminal alloys: hardness; tensile strength; ductility; creep, stress-rupture, and fatigue strengths; toughness; and fatigue crack growth rate (FCGR). The fracture mechanics approach to lifing of a part (by assuming a pre-existing crack) using the FCGR and the fracture toughness of the material is presented. The intermetallics Ti_xAl (where $x = 1$ or 3) are discussed (Ref 6.3, 6.4) and their characteristics summarized. The properties of titanium metal-matrix composites (both continuously and discontinuously reinforced) are also presented (Ref 6.5).

Effect of Alpha Morphology on Titanium Alloy Behavior

The morphology (shape) of the alpha phase has a significant effect on the mechanical properties of titanium alloys (Ref 6.6). Figure 6.1 shows a phase diagram of the alpha-beta alloy Ti-6Al-4V. Figure 6.2 illustrates the effect of working temperature on the morphology of the alpha phase after a subsequent anneal in the al-

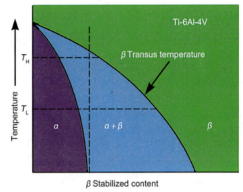

T_H: High $\alpha + \beta$ annealing temperature
T_L: Low $\alpha + \beta$ annealing temperature

Fig. 6.1 Phase diagram of the alpha-beta alloy Ti-6Al-4V

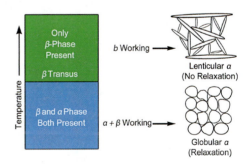

Fig. 6.2 Effect of working temperature on the morphology of the alpha phase after a subsequent anneal in the alpha-beta phase field

pha-beta phase field. Beta working is conducted with no alpha phase present, so the subsequent anneal in the alpha-beta phase field results in no

---

*Adapted and revised from R. Terrence Webster, originally from O. Bertea, *Titanium and Its Alloys*, ASM International.

"relaxation" of the alpha phase, which assumes its "unrelaxed" lenticular morphology. In contrast, working in the alpha-beta phase field occurs with alpha phase present and a relaxation of the alpha phase (with a sufficient amount of deformation, approximately 30 to 50%) on subsequent annealing in the alpha-beta phase field. This relaxation results in the alpha phase assuming an equiaxed morphology (a maximization of the volume-to-surface-area ratio).

Figure 6.3 illustrates the effect of the time of alpha-beta annealing after alpha-beta working on the morphology of the alpha phase. Short annealing times result in a cusping of the lenticular alpha phase, while longer annealing times (with sufficient working) result in the alpha phase assuming an equiaxed morphology. Cooling rate from the alpha-beta annealing temperature also affects the microstructure (Fig. 6.4). On annealing (after sufficient working) the alpha phase present (in accordance with the phase diagram in Fig. 6.1) assumes an equiaxed morphology (top left of Fig. 6.4). On subsequent slow cooling (approximately 30 °C, or 55 °F, per hour), the diffusion time allows for the alpha regions to grow through expansion of the spherical regions. In contrast, with a fast cooling rate, there is insufficient time for diffusion to the pre-existing equiaxed alpha regions to occur; the resultant microstructure consists of the small re-

gions of alpha previously formed during the alpha-beta anneal, with martensite formation in the regions between the alpha phase.

Figure 6.5 illustrates the effect of a duplex alpha-beta anneal on the microstructure of the Ti-6Al-4V alloy. Equiaxed alpha is formed at the higher alpha-beta annealing temperature, T_H (with sufficient alpha-beta working). Following a fast cool and a subsequent lower-temperature (T_L) alpha-beta anneal, lenticular alpha (because this alpha has not been worked) forms in accordance with the phase diagram.

The general effect of the alpha morphology is shown in Table 6.1, with upward-facing arrows indicating an enhancement in that mechanical property. These mechanical properties are discussed in more detail in subsequent sections of this chapter. The effect of alpha morphology on ductility and toughness is rationalized in Fig. 6.6. The lenticular morphology causes the crack path to be more tortuous, giving higher toughness, while equiaxed (or globular) alpha results in larger plastic zones and higher ductility.

The variation in fracture toughness with alpha morphology in the alpha-beta alloy Corona 5 (Ti-4.5Al-5Mo-1.5Cr) is shown in Fig. 6.7. The ductility shows a trend in the opposite direction (higher ductility with an equiaxed alpha morphology and lower with a lenticular morphology).

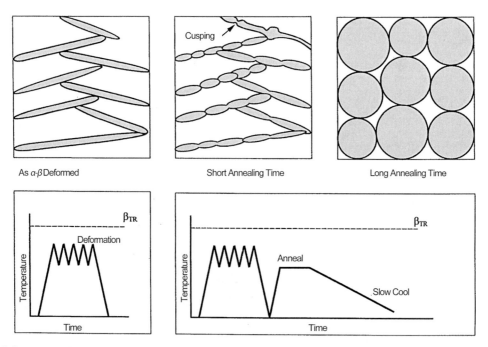

As α-β Deformed   Short Annealing Time   Long Annealing Time

Fig. 6.3 Effect of the time of alpha-beta annealing after alpha-beta working on the morphology of the alpha phase in Ti-6Al-4V

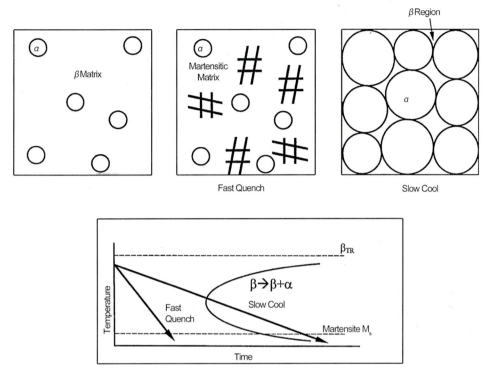

Fig. 6.4 Effect of cooling rate from the alpha-beta annealing temperature on microstructure

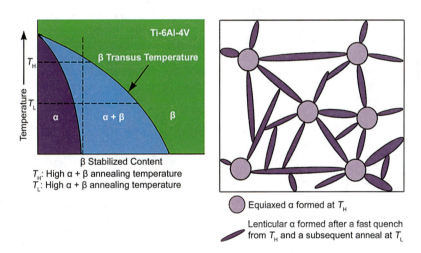

Fig. 6.5 Effect of a duplex alpha-beta anneal on the microstructure of the Ti-6Al-4V alloy

Table 6.1 Effect of alpha morphology on titanium alloy behavior

Lenticular morphology	Globular morphology
↑ Fracture toughness	↑ Ductility
↑ Fatigue crack growth rate	↑ Low-cycle fatigue behavior (initiation)
↑ Creep behavior	↑ Superplastic forming/diffusion bonding
	↗ Strength

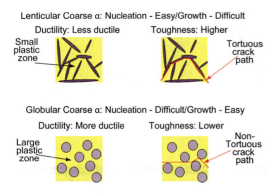

Fig. 6.6 Effect of alpha morphology on ductility, toughness, and crack formation and growth

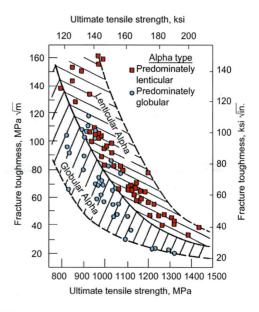

Fig. 6.7 Effects of alpha morphology and strength on fracture toughness in the alpha-beta alloy Corona 5 (Ti-4.5Al-5Mo-1.5Cr)

Hardness

Hardness is a measure of the resistance of an indenter to penetration of the surface of a material (Ref 6.7). Because of its nature, it is the simplest and easiest method of estimating the strength level of a material. The hardness of a material is measured by the size of the indentation made by the indenter; a small indentation represents a high hardness, while a large indentation represents a low hardness. There are a number of hardness testing methods; the most often used is the Vickers method (providing a Vickers hardness number, or VHN), which has a wide range of applications.

Titanium alloys in general exhibit a much higher hardness than aluminum alloys, approaching the hardness of the heat treated alloy steels. High-purity titanium has a hardness of 90 VHN; unalloyed commercially pure titanium has a hardness of approximately 160 VHN; and, when alloyed and heat treated, titanium can attain a hardness in the range of 250 to 500 VHN. The workhorse Ti-6Al-4V alloy with a 930 MPa (130 ksi) yield strength has a hardness of approximately 320 VHN.

Tensile Strength

The tensile strength of titanium is tested using the same standard ASTM International test specimens, procedures, and equipment as for other metallic materials (Ref 6.8, 6.9). Assuming that satisfactory specimens are made, the most important factors during such testing are:

- Strain rate
- Testing temperature
- Section size

To obtain accurate, reproducible results with titanium specimens, the strain rate must be closely controlled during testing. Strain rate is the rate at which the gage length of the test specimen is deformed. Although most metals show some increase in apparent strength with increasing strain rate, this effect is usually small enough to be negligible. However, the strength characteristics of titanium are strongly rate-sensitive. Therefore, in testing titanium, a uniform, slow strain rate must be maintained if results are to show realistic differences in the strength of the material. This effect is shown in Table 6.2.

Specifications recommend a strain rate of 0.076 to 0.178 mm (0.003 to 0.007 in.) per minute in tensile testing to obtain uniform results. This speed must be maintained through the 0.2% offset yield. A strain rate of 1.3 to 25.4 mm (0.05 to 1.0 in.) per minute is used from yield to fracture. The effect of this increased testing speed on elongation is relatively small in the region of plastic deformation, and it can be ignored. Tensile hardness testing equipment has either manual or automatic control of strain rate.

Tensile properties are also affected by the testing temperature. Both yield strength and ultimate strength of all titanium alloys change with temperature. As testing temperature decreases below 21 °C (70 °F), strength increases rapidly. Conversely, as the temperature is raised above 21 °C

(70 °F), strength drops. Initially, the decrease is relatively rapid. However, strength levels off in an intermediate temperature range of 150 to 320 °C (300 to 600 °F). The length of this plateau depends on alloy type. As the temperature increases further, strength drops rapidly. The useful, long-time, maximum temperature limit for titanium alloys exposed to high stress levels is 540 °C (1000 °F). Alloys such as Ti-6Al-2Sn-4Zr-2Mo are being tested and used at higher temperatures (Table 6.3, Fig. 6.8). Alloy Ti-13V-11Cr-3Al was the first beta alloy developed (and used extensively on the high-speed, high-flying SR71) and is used in this chapter as representative of the beta alloy class of alloys.

Section size is also important in considering the strength characteristics of titanium. In flat

Table 6.2 Effect of testing speed on observed yield strength

Material	Strength		0.2% offset yield, % of ultimate		Reduction in yield(b), %
	MPa	ksi	1 to 2 min(a)	2 h(a)	
Unalloyed titanium	578	83.8	72	59	17
Ti-8Mn	896	130.0	83	76	9
302 stainless	667	96.8	43	43	None
7SST aluminum	525	76.2	87	86	<1

Note: All tests were made at room temperature. (a) Time required to rupture specimen. (b) As a result of slower crosshead speed

Table 6.3 Typical elevated-temperature properties of some titanium alloys

Temperature		Ti-0.2O₂ (sheet)	Ti-8Mn (sheet)	Ti-5Al-2.5Sn (bar)	Ti-6Al-4V (bar)	Ti-6Al-2Sn-4Zr-2Mo (forged wheel)	Ti-5Al-6Sn-2Zr-1Mo-0.25Si (bar)	Ti-6Al-6V-2Sn (bar, solution treated and aged)
°C	°F							
Tensile strength, MPa (ksi)								
−54	−65	...	1132 (164)	1090 (158)	...	...	...	...
21	70	586 (85)	959 (139)	918 (133)	1063 (154)	979 (142)	1021 (148)	1289 (187)
93	200	504 (73)	849 (123)	814 (118)	966 (140)	883 (128)	952 (138)	1241 (180)
204	400	359 (52)	752 (109)	690 (100)	869 (126)	814 (118)	849 (123)	1180 (171)
315	600	283 (41)	676 (98)	607 (88)	814 (118)	766 (111)	773 (112)	1042 (151)
427	800	228 (33)	600 (87)	566 (82)	752 (109)	724 (105)	711 (103)	959 (139)
538	1000	124 (18)	310 (45)	518 (75)	635 (92)	655 (95)	689 (100)	689 (100)
649	1200	...	269 (39)	...	...	...	...	...
0.2% offset yield, MPa (ksi)								
−54	−65	...	993 (144)	925 (134)	...	...	...	...
21	70	517 (75)	827 120)	849 (123)	973 (141)	890 (129)	899 (129)	1256 (182)
93	200	393 (57)	697 (101)	725 (105)	849 (123)	745 (108)	800 (116)	1187 (172)
204	400	241 (35)	573 (83)	573 (83)	759 (110)	655 (95)	676 (98)	1104 (160)
315	600	165 (24)	469 (68)	462 (67)	725 (105)	780 (84)	586 (85)	890 (129)
427	800	124 (18)	414 (60)	435 (63)	614 (89)	552 (80)	531 (77)	835 (121)
538	1000	79 (11.5)	228 (33)	400 (58)	531 (77)	524 (76)	497 (72)	510 (74)
Elongation, % in 50.8 mm (2 in.)								
−54	−65	...	15	18	...	...	...	...
21	70	23	15	16	15	14	12	11
93	200	24	15	15	15	12	12	12
204	400	30	14	17	15	12	15	15
315	600	30	14	17	15	15	15	17
427	800	22	17	16	15	18	15	19
538	1000	80	28	19	20	23	15	36
649	1200	...	...	40	...	...	...	...
Reduction of area,%								
−54	−65	...	...	36	...	...	...	...
21	70	...	...	40	35	...	...	31
93	200	...	...	42	36	...	...	44
204	400	...	...	42	37	...	...	51
315	600	...	...	41	42	...	...	55
427	800	...	...	41	49	...	...	63
538	1000	...	...	42	64	...	...	83
649	1200	...	...	58	...	...	...	...

Note: Annealed condition unless otherwise noted. Courtesy of Crucible Steel Co. and RMI, Inc.

products, for example, sheet exhibits higher strength than plate. This results from the thermomechanical work imparted to the material. The same relationship applies to rolled and forged bar and billet products, and it becomes quite striking as billet sizes are extended beyond the 254 mm (10 in.) diameter range.

Processing of large sections usually requires higher temperatures to prevent surface rupturing and to stay within the maximum permissible loads on processing equipment. Therefore, in many instances, the mass of metal being worked and the resultant finished size dictate the property levels of the material being processed. Basically, the rule of thumb is that the lower the processing temperature and the smaller the section size, the higher the strength levels that can be achieved.

Also, the effects of solution treating and aging become less pronounced as section size increases. This is especially applicable to plate thickness over 25 mm (1 in.) and to bar and billet products from 50 to 76 mm (2 to 3 in.) in diameter. This is illustrated in Fig. 6.9.

When comparing properties of titanium alloys, the content of interstitial elements of the materials must be known, because interstitial elements have a marked effect on the strength and ductility of titanium. This must be considered when assessing the effects of metallic alloying, thermomechanical work, and heat treatment on the properties of titanium alloys.

Ductility

Two common measures of ductility determined in the tensile test are elongation and reduction in area (Ref 6.8, 6.9). Elongation is the amount that the gage length of a tensile specimen elongates before fracture. Expressed as a percentage, it is determined by dividing the elongation that produces fracture by the original gage length of the specimen. Generally, there is localized necking. Thus, the elongation of each elemental length within the gage length of the test specimen is not uniform. Because calculated elongations change with the gage length, it is important to compare elongations of specimens of similar gage lengths.

Reduction in area is the ratio (expressed in percent) of the amount the cross-sectional area of the specimen contracted at failure to the original cross-sectional area. For most titanium applications, the material exhibits good ductility if elongation is 8 to 10% or more and reduction in area is 15 to 20% or higher. This depends on alloy type, as well as the application and service environment. Many metallurgical factors that affect ductility relate to alloying, processing, and heat treatment.

Generally, as strength increases in all alloy types, there is a reduction in ductility (Table 6.3). In addition, exposure to fast strain rates and strain aging lowers ductility. Excessive amounts of interstitially soluble elements (nitrogen, oxygen,

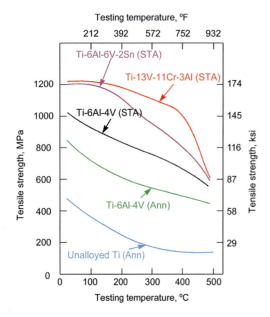

Fig. 6.8 Effect of testing temperature on tensile strength of several titanium alloys. STA, solution treated and aged; Ann, annealed

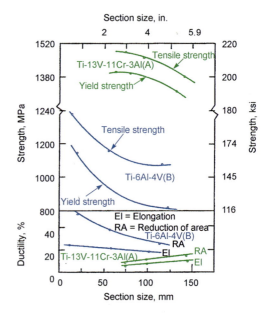

Fig. 6.9 Effect of section size on tensile properties of Ti-6Al-4V and Ti-13V-11Cr-3Al. A is 788 °C (1450 °F) 30 min water quenched; 482 °C (900 °F) 48 h air cooled. B is 927 °C (1700 °F) 1 h water quenched; 482 °C (900 °F) 6 h air cooled. Courtesy of Titanium Metal Corp. of America

hydrogen, and carbon) lower ductility. For a given interstitial content, nitrogen is the most effective of the group. Although these interstitials increase the strength of titanium, they drastically lower ductility.

Hydrogen also reduces ductility. Therefore, most specifications limit hydrogen to a range of 100 to 200 ppm. Above this level, hydrides sometimes form and lower ductility. The formation of hydrides varies depending on the alloy and the temperature. In applications requiring high ductility and toughness at the expense of maximum strength, extralow interstitial titanium alloys should be specified. Hot working the material at temperatures near or above the beta transus temperature can produce microstructures that have lower ductility. However, these microstructures, when properly controlled, can enhance other desirable properties, such as fracture toughness, as discussed previously.

Creep and Stress Rupture

Creep strength and stress-rupture (Ref 6.7) properties determine the ability of material to carry loads at elevated temperatures for long periods of time. These properties are determined by the stress-rupture test. In the test, a specimen is held under constant load and temperature until failure occurs. Creep strength is the constant stress that causes a specified amount of creep in a given time at constant temperature.

These properties are very important in determining the usefulness of titanium in high-temperature applications. Of the three alloy types, alpha alloys exhibit the best creep and stress-rupture characteristics. Generally, the larger the amount of alpha phase within the microstructure, the higher the creep strength. Many alpha alloys contain small amounts (1 to 2%) of beta stabilizers and are more accurately called alpha-beta-lean alloys. For simplicity, here they are classified as alpha alloys.

Aluminum, tin, and zirconium enhance the creep strength of the alpha phase (Table 6.4). Alloys of this type are Ti-5Al-2.5Sn and Ti-6Al-2Sn-4Zr-2Mo. Further improvements in creep strength have been achieved by adding silicon to the alpha-base alloys. Typical alloys of this class (Table 6.3) include Ti-6Al-5Zr-1W-0.25Si and Ti-5Al-6Sn-2Zr-1Mo-0.25Si.

The improved creep strength of the silicon-containing alloys is due to solid-solution strengthening and, in some instances, by a dispersed silicide compound. The solubility of silicon in these alloys is believed to be between 0.2 and 0.3%. Additional silicon beyond the solubility limit precipitates as a complex titanium-zirconium silicide compound.

In studying the creep performance of a number of structural metals including titanium, Larson and Miller offered an analysis of the data based on the theory that the rate at which the creep process proceeds depends on time and temperature, among other factors. They concluded that time and temperature are interchangeable in their effect on strength. The Larson-Miller parameter, which is constant for a given stress, is expressed as:

$$LM = T(C + \log 10\ t) \qquad \text{(Eq 6.1)}$$

where T is absolute temperature, t is time, and C is a constant dependent on the material. The value of C is 20 for titanium. Figure 6.10 compares the creep resistance of several of the more important high-temperature alloys on the basis of the Larson-Miller parameter, P.

The use of the Larson-Miller parameter can be explained by referring to Fig. 6.10. To illustrate: At a stress of approximately 200 MPa (29 ksi), Ti-8Al-1Mo-1V alloy deforms 0.1% under all combinations of time and temperature that yield a value of the Larson-Miller parameter of 30×10^3. There are an infinite number of possible combinations, including 515 °C (960 °F)/10 h, 480 °C

Table 6.4 Creep-resistant titanium alloys

Nominal compositions	Oxygen, %	Type alloy	Stress, 6.89 MPa (1 ksi) for 0.1% deformation in 100 h at indicated temperature			Products available	
			426 °C (800 °F)	482 °C (900 °F)	510 °C (950 °F)	Sheet, strip, plate	Bar, forging, fastener
Ti-5Al-2.5Sn	0.08–0.20	α	5	...	...	X	X
Ti-6Al-4V	0.08–0.20	α + β	34	...	...	X	X
Ti-8Al-1 Mo-1 V	0.07–0.12	α	50	27	...	X	X
Ti-6Al-2Sn-4Zr-2Mo	0.08–0.12	α	70	45	...	X	X
Ti-2.25Al-11Sn-15Zr-1Mo-0.2Si	0.11	α + Compound	70	45	...	...	X
Ti-6Al-5Zr-1W-0.2Si	0.11	α + Compound	70	62	55	...	X
Ti-5Al-6Sn-2Zr-1Mo-0.25Si	0.11	α + Compound	70	70	58	...	X

(900 °F)/100 h, and 450 °C (840 °F)/1000 h (refer to the top of Fig. 6.10).

For convenience, the time-temperature scale is added to the top of the figure. The highest creep strength is achieved with Ti-5Al-6Sn-2Zr-1Mo-0.25Si alloy. All the silicon-containing alloys are more creep resistant than either Ti-6Al-2Sn-4Zr-2Mo or Ti-8Al-1Mo-1V, a high-temperature alloy.

Aluminum is an effective alpha stabilizer and primary contributor to high-temperature titanium alloys. Early research work in titanium concentrated on breaking the aluminum "barrier." Aluminum additions to titanium-base alloys do not exceed 8%. A Ti_3Al phase can precipitate from high-aluminum-containing alpha phase when alloys are exposed for long times between 480 and

705 °C (900 and 1300 °F). This phase results in the loss of toughness and ductility. Additions of molybdenum, niobium, tantalum, and vanadium stabilize a small amount of beta phase, producing stable alloys (Table 6.5).

The thermal stability of a material is its ability to exhibit no significant change in strength or loss in ductility after exposure to heat and stress. Alloying is very critical to obtain optimum stability. The strengthening produced by aluminum additions and the tendency toward embrittlement (low ductility) in binary alloys with more than 8% Al after exposure at 540 °C (1000 °F) are discussed in other chapters. It is important to note here that thermal stability can be a limiting factor in the application of alloys.

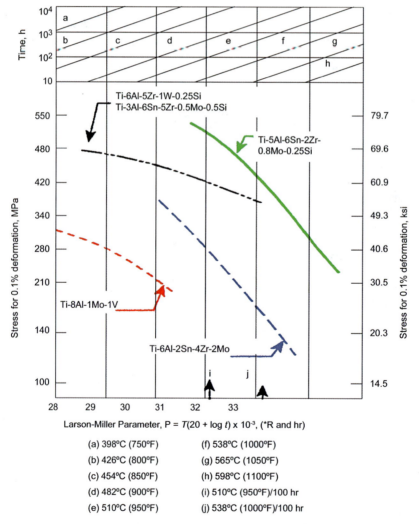

Fig. 6.10 Creep strength of several titanium alloys

Table 6.5 Thermal stability of titanium alloys

Alloy composition	Exposure						Before exposure			After exposure		
	Temperature		Stress		Time,	Deformation,	Tensile strength		Reduction	Tensile strength		Reduction
	°C	°F	MPa	ksi	h	%	MPa	ksi	of area, %	MPa	ksi	of area, %
Ti-5Al-6Sn-2Zr-1Mo-0.2Si	510	950	415	60	100	0.12	1028	149	21	1020	148	21
Ti-6Al-4V	370	700	485	70	100	0.11	1221	177	39	1180	171	32
Ti-6Al-2Sn-4Zr-2Mo	480	900	310	45	150	0.12	973	141	25	938	136	28

Another important consideration in alloying titanium is the limiting amounts of both the alpha stabilizer (aluminum) and the neutral additions (tin and zirconium). Alloy stability, as related to the precipitation of Ti_3Al, is a function of the total alpha solid-solution alloy and oxygen content. The aluminum equivalency factor (AEF) is calculated by the following formula:

$$AEF = \%Al + (\%Sn/3 + \%Zr/6) \qquad (Eq\ 6.2)$$

For AEF between 8 and 9, oxygen content must be below 0.1%. This limits the development of titanium-base alloys for use above 540 °C (1000 °F), because the substitutionally soluble alpha stabilizers aluminum, tin, and zirconium are the important elements for high-temperature strengthening.

The type and amount of beta-stabilizing element present is also important, because thermal stability can occur due to transformation of the beta phase. Table 6.5 contains examples of the stability of several experimental and commercial alloys. An addition of beta-eutectoid stabilizer (7% Cr) severely embrittles this experimental alloy after creep exposure.

Large additions of beta-eutectoid-type alloying elements are detrimental to thermal stability. However, if properly controlled and used at low concentration, good stability can be achieved, as illustrated by the 0.25% Si addition to the Ti-5Al-6Sn-2Zr-1Mo-0.25Si alloy. Alloys containing moderate additions of the beta-isomorphous elements (such as molybdenum and vanadium) are generally very stable. This is illustrated by Ti-3%Mo, Ti-6Al-4V, and Ti-6Al-2Sn-4Zr-2Mo alloys in Table 6.5.

Hydrogen can cause poor thermal stability. However, this problem can be easily remedied by vacuum treatment to lower the hydrogen content.

Fatigue Strength

Fatigue characteristics of the materials must be considered in the design of parts subjected to dynamic stresses (Ref 6.7). Dynamic loads can cause failure at stresses below the yield strength or elastic limit of a material. Fatigue failure of this type occurs in all structural metals; titanium is no exception.

Fatigue strength is the maximum level of alternating stress, that is, reversing tension and compression, which can be applied for a specified number of cycles without specimen failure. It can be defined for any number of cycles required by the design in question. However, the fatigue strength of titanium is generally specified at 10^7 or 10^8 cycles.

Endurance limit also is used in discussing fatigue. This term implies a cyclic stress level below which no failure occurs regardless of the number of cycles imposed. There is some question whether metals really do have an endurance limit, or whether fatigue strength decreases continuously with increasing number of stress cycles. For many metals, fatigue strength does continue to decrease as fatigue testing time increases. However, titanium and titanium alloys have an actual endurance limit.

Fatigue strength is commonly shown graphically in an *S-N* curve, where maximum stress is plotted against the number of cycles to failure. Figure 6.11 shows two typical *S-N* curves for Ti-6Al-4V. As noted earlier, fatigue strength is usually defined for pure dynamic stresses where the stress cycle includes both tension and compression, that is, complete stress reversal.

Under such conditions, the mean stress is zero. The lower curve in Fig. 6.11 illustrates the fatigue strengths for this condition. Tests run on reversed-bending or rotating-beam fatigue machines impose stresses of this type. In many applications, stress conditions consist of a dynamic stress superimposed on a static stress. In such instances, the mean stress is the static stress.

The upper curve in Fig. 6.11 illustrates the results obtained when a static tensile stress is also imposed. Here, the maximum tensile stress (static plus dynamic, or tension-tension fatigue) is plotted versus cycles to failure. Tension-tension fatigue data are preferred by engineers and designers concerned with the application of titanium.

The endurance limit under combined static and dynamic stresses depends on the level of each stress. As the static stress increases, the dynamic stress that can be superimposed decreases. However, the maximum stress (static plus dynamic) is increased. The usual method for showing this relationship for a specific alloy is by means of the Goodman diagram or modified Goodman diagram (alternating stress versus mean stress). Examples of Goodman-type diagrams for Ti-6Al-4V, Ti-5Al-2.5Sn, and Ti-13V-11Cr-3Al are shown in Fig. 6.12 to 6.15.

In addition to the effect of the combination of static and dynamic stress conditions, fatigue strength is also affected by and related to four factors:

- Tensile strength
- Surface condition (finish and stress condition)
- Notches and notch sensitivity
- Heat treatment

The basic good fatigue characteristics of titanium alloys are illustrated in Table 6.6. The endurance ratios indicate that the fatigue performance of titanium is in fairly constant ratio with tensile strength. Thus, alloys that have high tensile strength also have high fatigue strength. This relationship between tensile strength and fatigue strength also is valid as strength is changed by temperature, heat treatment, and alloying.

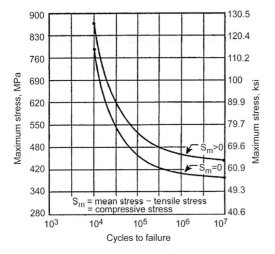

Fig. 6.11 Fatigue (*S-N*) curves for Ti-6Al-4V alloy in solution-treated and aged condition

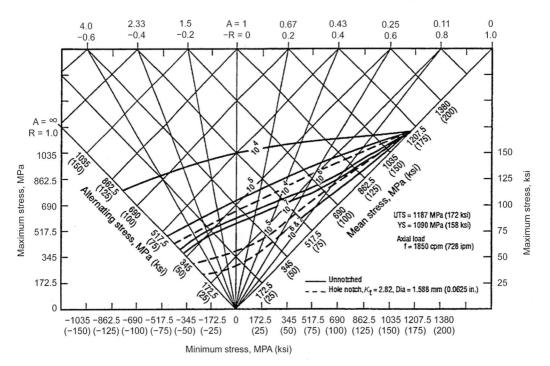

Fig. 6.12 Goodman-type diagram for fatigue behavior of solution-treated and aged Ti-6Al-4V sheet at room temperature. UTS, ultimate tensile strength; YS, yield strength

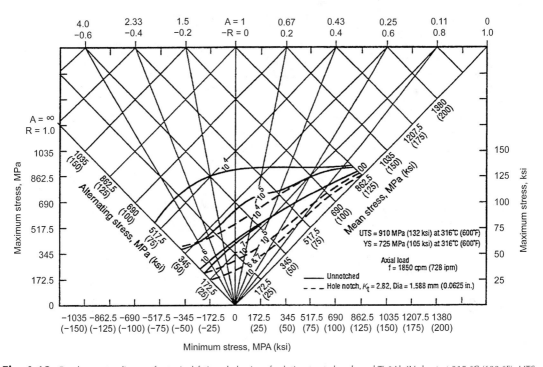

Fig. 6.13 Goodman-type diagram for typical fatigue behavior of solution-treated and aged Ti-6Al-4V sheet at 315 °C (600 °F). UTS, ultimate tensile strength; YS, yield strength

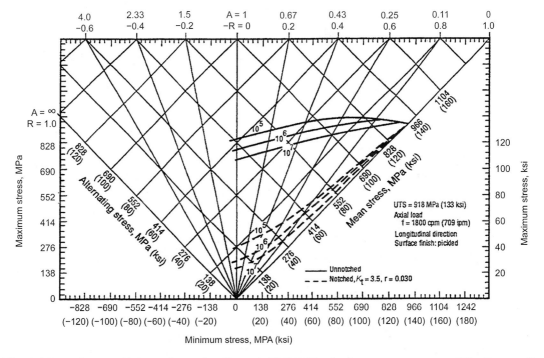

Fig. 6.14 Partial constant-life fatigue diagram for mill-annealed Ti-5Al-2.5Sn alloy sheet at room temperature. UTS, ultimate tensile strength

The fatigue strength of titanium is altered by the presence of notches and the sensitivity of a given material to stress concentrations produced by these notches. In a smooth test specimen stressed in tension, the stress is uniform over the entire cross section. In notched specimens, the stresses are not uniform.

The constraint imposed by the notch concentrates the stress at the root of the notch. The degree to which the stresses are so concentrated depends on the geometry, physical dimensions of the notch, and whether the stress imposed is in tension, compression, or bending. The degree of stress concentration for any notch is expressed by the theoretical

stress-concentration factor, K_t. The factor is the ratio of the maximum stress that results from the presence of the notch to the nominal stress that is obtained by the simple relation P/A (load per unit cross-sectional area at the base of the notch).

The effect of notches on the fatigue strength is that, as notch severity increases, fatigue strength decreases. There is no relation between notch sensitivity and alloy type. This effect is shown in Fig. 6.16 for Ti-6Al-4V.

Although normal endurance ratios are determined on specimens that are machined and polished, certain surface finishes may have a marked effect on fatigue performance. Generally, fatigue

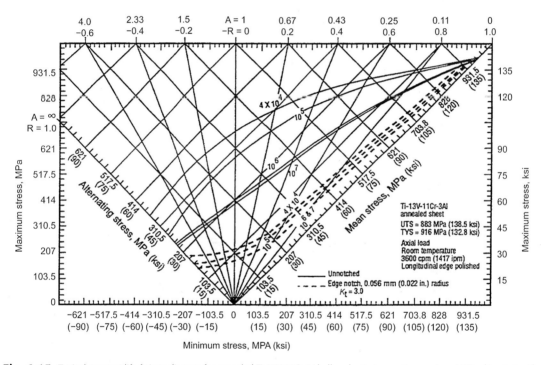

Fig. 6.15 Typical constant-life fatigue diagram for annealed Ti-13V-11Cr-3Al alloy sheet at room temperature. UTS, ultimate tensile strength; TYS, tensile yield strength

Table 6.6 Fatigue properties of titanium alloys

Nominal composition	Tensile strength		Fatigue strength 10^7 cycles		Endurance ratio(a)
	MPa	ksi	MPa	ksi	
Unalloyed titanium	623	90.3	317	46	0.51
Ti-5Al-2.5Sn	918	133	524	76	0.57
Ti-8Mn	959	139	483	70	0.50
Ti-6Al-4V	1062	154	565	82	0.53
Ti-6Al-6V-2.5Sn	1083	157	455	66	0.42
Ti-13V-11Cr-3Al	952	138	228	33	0.24
Ti-6Al-2Nb-1Ta-1Mo	911	132	379	55	0.42

(a) Fatigue strength at 10^7 cycles/tensile strength. Source: Crucible Steel Co. and RMI, Inc.

tests on specimens having rough surfaces yield lower endurance limits than tests on specimens having smooth surfaces. This is expected in view of the previously discussed effect of notches inadvertently present on rough surfaces.

The effect of surface preparation on the development or removal of surface stress is also important in determining the endurance limit of an alloy. Surface compressive stresses are beneficial, whereas surface tensile stresses are detrimental. This effect can be seen in Fig. 6.17, where shot-peened specimens of Ti-5Al-2.5Sn have a higher endurance limit than ground specimens. This is because shot peening imparts residual compressive stresses, and grinding imparts residual tensile stresses.

Because oxygen, nitrogen, and hydrogen embrittle titanium, any treatment that promotes the diffusion of these interstitials into the surface lowers the endurance limit. An example of this is shown in Fig. 6.17, where electrical-discharge machining produces an oxidized surface, resulting in a drastic reduction in endurance limit. Chemical milling is expected to increase the endurance limit of as-rolled material due to the removal of surface layers. However, chemical milling decreases the endurance limits slightly due to the increase in hydrogen content. However, when the hydrogen content is reduced by vacuum annealing, the endurance limit increases greatly.

Heat treatment of alpha-beta titanium alloys improves strength characteristics. This is true of fatigue and tensile strength. Heat treatment must be selected carefully so sufficient ductility and fracture toughness is retained for useful application of the improved strength.

Low-cycle fatigue is becoming increasingly important in component design, including jet-engine components. Many components, such as rotating disks, experience very few stress cycles per

flight; thus, the low-cycle region of S-N curves is of particular interest in considering alloys for use as rotating disks. Figure 6.18 shows the low-cycle region (10^3 to 10^5 cycles) for notched specimens of Ti-6Al-4V. The stress-to-cycles relation is shown for the appearance of the first crack and also for failure of the specimen.

If small flaws are initially present in the material, the rate at which these flaws propagate under cyclic stresses inevitably determines the life of the material. The rate of propagation can be accelerated in certain environments, such as seawater and even distilled water, in some titanium alloys. Forging conditions and heat treatment variables can greatly affect low-cycle fatigue life; they must be carefully controlled when optimum properties must be achieved.

Toughness

Toughness of titanium alloys, especially fracture toughness (Ref 6.7), plays an increasingly important part in design engineering for ballistic, hydrospace, and aerospace applications. The impact test, when conducted at low temperatures, combines the three primary promoters of brittle behavior: notches, high strain rate, and low temperature.

The temperature relation of Charpy impact energy of titanium alloys in different ranges of yield

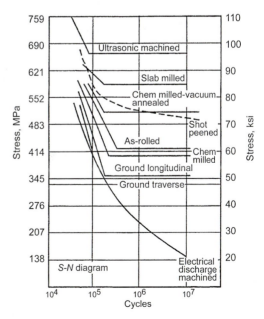

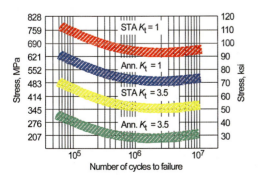

Fig. 6.16 Typical S-N curves for annealed (Ann.) and solution-treated and aged (STA) Ti-6Al-4V bar. K_t, theoretical stress-concentration factor

Fig. 6.17 Effect of various machining processes on reversed-bending fatigue strength of Ti-5Al-25Sn alloy

strength, as determined by the U.S. Naval Research Laboratory, is shown in Fig. 6.19. Impact energy decreases with decreasing temperature and increasing strength. Impact values for individual alloys do not undergo a sharp transition over a temperature range, as observed in some steels. Studies of impact fracture surfaces over a wide range of temperatures and alloy conditions show that the mode of fast fracture is dimpled rupture, a ductile mode of fracture.

Although the impact test is useful to compare alloys and measure the influence of thermomechanical processing on toughness, it is not as discriminating as some newer techniques. For medium-strength, large-section-size alloys, the U.S. Navy's dynamic tear test measures fracture toughness properties of the full section. This test requires a bar 457 by 127 by 25 mm (18 by 5 by 1 in.) that contains a brittle weld (to start a brittle crack) on the tension edge. When the specimen is struck on the edge opposite the weld, a crack starts and grows through the weld and into the material. The amount of energy required to break the specimen is a measure of crack propagation energy (fracture toughness).

The test apparatus is basically an impact testing machine of ~7 MJ (5000 ft · lbf) capacity. This test relates well to the prototype explosion tear test illustrated in Fig. 6.20. The test plate is 635 by 558 by 25 mm (25 by 22 by 1 in.) and contains a 51 mm (2 in.) flaw through the thickness. Two slots in the plate ensure uniaxial loading. The plate rests on a die with a rectangular opening. It is loaded by detonating an explosive charge from a predetermined distance. The blast forces material between the cut slots containing the flaw into the die opening. The amount of strain in extending a crack from the 51 mm flaw to the cut slot indicates resistance to crack growth.

The results of these two tests are correlated to form a fracture toughness index diagram for tita-

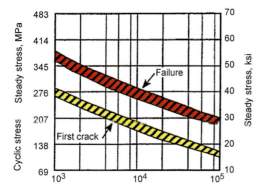

Fig. 6.18 Notched S-N curve for Ti-6Al-4V alloy

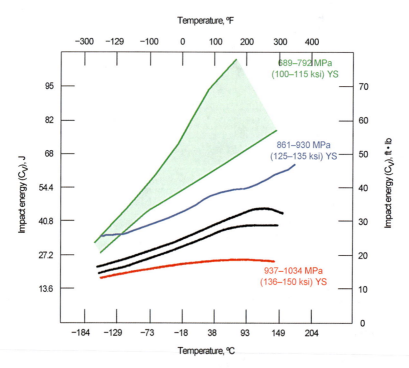

Fig. 6.19 Effect of temperature on Charpy V-notch energy for different ranges of yield strength (YS) in high-strength titanium alloy plate

nium alloys. This chart, shown in Fig. 6.21, indicates the relative resistance to fracture growth for specific energy levels. The cross-hatched bands indicate the transition from elastic to plastic strains required for extension to fracture. The upper line, or ceiling, is designated as the optimum materials trend line. This depicts the apparent upper limits for fracture toughness. In applications such as pressure hulls, an alloy must develop at least several percent of plastic strain to withstand sudden overloading. A dynamic tear energy of 2035 to 2712 J (1500 to 2000 ft · lbf) guarantees this level of fracture toughness.

As the strength level of metal increases, the ability for plastic flow diminishes in the presence of a crack. At this point, a crack propagates suddenly from a flaw and completely through the specimen without much plastic flow. The conditions under which this crack extends were first analyzed in terms of linear elastic stress. Through this analysis, the principal stresses associated with the crack were related to a stress-intensity factor, K. Fracture instability is assumed to occur when K reaches a critical value, which is characteristic of a material. This critical value of K is designated the fracture toughness and is a material constant.

The stress-intensity factor, K, depends on the crack length, geometry of the body containing the crack, and the manner in which external loads are applied. Generally, the stress-intensity factor in units of MPa $\sqrt{m}$ (ksi $\sqrt{in.}$) is expressed as:

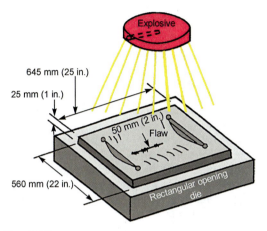

Fig. 6.20 Schematic diagram of explosion tear test

$$K = fg \sqrt{a}(Y) \qquad \text{(Eq 6.3)}$$

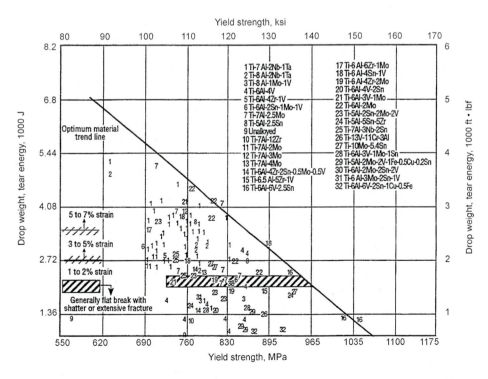

Fig. 6.21 Fracture toughness index diagram for titanium alloys. Most titanium alloys fail in a ductile manner and have reasonable toughness, a necessary property for pressure vessels.

where *fg* is gross stress in MPa (ksi), *a* is flaw size in mm (in.), and *Y* is a compliance factor relating component geometry and flaw size (nondimensional).

For structural shapes that fracture as a result of low ductility (and high strength), the lower limiting value of *K* is the plane-strain fracture toughness, K_{Ic}. This value corresponds to the stress intensity when unstable propagation of the crack occurs. In thin material, such as sheet, or in low-strength alloys, much plastic flow attends the cracking process. Thus, plane-strain conditions do not exist. Testing under these circumstances (plane-strain) yields stress intensities termed K_c.

Many specimen geometries and testing methods are available to determine fracture toughness. All specimens are preflawed, usually using a fatigue crack. The fracture toughness, *K*, can be calculated from fracture analysis of such a specimen. Figure 6.22 shows the range of K_{Ic} values for titanium alloys with 690 to 1240 MPa (100 to 180 ksi) yield strengths. Although often reported as true K_{Ic} values, the data at strength levels less than 895 MPa (130 ksi) could represent K_c, or plane-stress conditions, or a combination of both. Regardless, fracture toughness decreases as the strength level increases. This means that at a constant operating stress, the higher the strength of the alloy, the smaller the flaw size that can be tolerated. Using this fracture mechanics approach, a designer can calculate the maximum flaw size that a structure can maintain without brittle failure occurring.

The range of *K*-values for each strength level results from variations in testing technique, alloy composition, and thermomechanical processing. Figure 6.23 illustrates the influence of heat treatment on the fracture toughness of Ti-6Al-6V-2Sn. A solution-treating and aging sequence (line *b*) compared with annealing (line *a*) gives lower toughness but a better combination of strength and toughness.

The high-temperature solution treatment called beta annealing produces an acicular, or martensitic alpha, microstructure, which improves fracture toughness. This improvement is also achieved by fabrication, as illustrated in Table 6.7 for Ti-6Al-4V. The acicular structure is developed by beta forging. While beta forging or annealing increases fracture toughness, some loss of strength and tensile ductility occurs. (See the earlier discussion on the effect of alpha morphology on mechanical properties.) Many materials, including several titanium alloys, are susceptible to lower fracture toughness in special environments, such as saltwater. (For more details, see Chapter 14, "Corrosion," in this book.)

The technique used to evaluate this phenomenon is similar to those used in the fracture mechanics approach to fracture toughness. It uses a precracked specimen, and calculations for fracture toughness are made assuming conditions of plane strain. Susceptibility to cracking is measured by comparing the stress-intensity factor, *K*, obtained in a normal air test with the *K*-value obtained in liquid media.

In environments in which stress-corrosion cracking is a factor, the fatigue crack extends slowly (due to stress-corrosion cracking), and then it finally propagates to produce the typical rapid fracture. Figure 6.24 illustrates the influ-

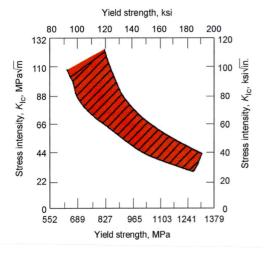

Fig. 6.22 Relationship between yield strength and reported plane-strain fracture toughness for titanium alloys

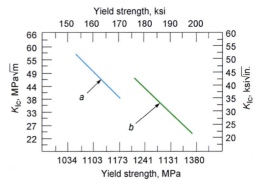

Fig. 6.23 Effect of heat treatment of the strength-toughness relationship of Ti-6Al-6V-2Sn alloy. Line *a* = annealed; line *b* = solution treated and aged

Table 6.7 Properties of alpha-beta and beta-processed Ti-6Al-4V forgings

Forging process	Heat treatment(a)	Tensile strength		Yield strength		Elongation, %	Reduction in area, %	K_{Ic}	
		MPa	ksi	MPa	ksi			MPa = $\sqrt{m}$	ksi = $\sqrt{in.}$
Alpha + beta	940 °C (1725 °F) 1 h, WQ 540 °C (1000 °F) 4 h, AC	1152	167	1063	154	13	48	49	45
Beta	940 °C (1725 °F) 1 h, WQ 540 °C (1000 °F) 4 h, AC	1159	168	1021	148	10	29	77	70
Alpha + beta	705 °C (1300 °F) 2 h, AC	1001	145	932	135	15	41	62	56
Beta	705 °C (1300 °F) 2 h, AC	980	142	890	129	13	35	100	91

(a) WQ, water quenched; AC, air cooled. Courtesy of RMI, Inc.

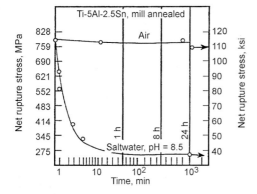

Fig. 6.24 Stress-rupture curves in air and 3.5% saltwater for center-fatigue-cracked specimens

Table 6.8 Sensitivity of titanium alloys to stress-corrosion cracking (SCC) in saltwater

Sensitive to SCC in the condition tested	Insensitive to SCC in the condition tested
Ti-2.5Al-1Mo-11Sn-5Zr-0.2Si	Ti-2Al-4Mo-4Zr
Ti-6Al-4V-1Sn	Ti-4Al-3Mo-1V(a)
Ti-6Al-4V-2Co	Ti-5Al-2Sn-2Mo-2V
Ti-13V-11Cr-3Al(a)	Ti-6Al-2Mo
Ti-7Al-2Nb-1Ta	Ti-6Al-2Sn-1Mo-1V
Ti-7Al-3Nb-2Sn	Ti-6Al-2Sn-1Mo-3V
Ti-8Al-1Mo-1V(a)	Ti-6Al-2Nb-1Ta-1Mo(a)
Ti-8Al-3Nb-2Sn	Ti-6.5Al-5Zr-1V
Ti-4Al-4Mn	Ti-6Al-4V (Extralow interstitial grade)(a)
Ti-5Al-2.5Sn(a)	
Ti-6Al-2.5Sn	
Ti-6Al-4V(a)	
Ti-6Al-3Nb-2Sn	
Ti-6Al-6V-2.5Sn(a)	
Ti-7Al-3Nb	
Ti-7Al-3Mo(a)	

(a) Commercial alloys. Other listed were not in commercial use at the time of writing. Courtesy of RMI, Inc.

ence of environment and time under stress on the fracture stress of a precracked specimen of Ti-5Al-2.5Sn. In this example, the saltwater environment surrounding the specimen lowered the fracture stress from 689 MPa (100 ksi) in air to 276 MPa (40 ksi) in saltwater. This diagram illustrates that a threshold type of behavior occurs in the liquid environment. The fracture stress decreases rapidly during the first 10 min and then levels off to form the threshold.

The susceptibility of titanium alloys to accelerated crack propagation in seawater is influenced by alloy content. The data indicate that the susceptibility occurs in alloys containing high aluminum, aluminum-tin, and high oxygen contents. Table 6.8 lists alloys that were screened for sensitivity to this phenomenon. The presence of the beta-isomorphous stabilizers molybdenum, vanadium, and niobium reduces the sensitivity of titanium alloys. Titanium alloys containing 5 to 7% Al plus 1 to 4% V or Mo (or both) are less sensitive.

The degree of susceptibility of some titanium alloys to the stress-corrosion cracking in saltwater can be altered by heat treatment or processing, or both. In general, heat treatment and pro-

cessing near the beta transus improves resistance, while aging at temperatures in the range of 480 to 705 °C (900 to 1300 °F) is detrimental.

Figure 6.25 shows the influence of beta fabrication on strength and fracture toughness in saltwater. The improvement shown over fracture toughness in air is related to the acicular alpha microstructure.

Oxygen content influences the fracture toughness; low-oxygen Ti-6Al-4V exhibits higher fracture toughness than high-oxygen material (Fig. 6.26) (Ref 6.2).

Fatigue Crack Growth Rate

Determination of fatigue crack growth rates (FCGRs) in aerospace materials, such as titanium and its alloys, is needed to use the so-called defect-tolerant (or fracture mechanics) approach to lifing. In this approach, it is assumed that defects and cracks exist in components. Safe operation is

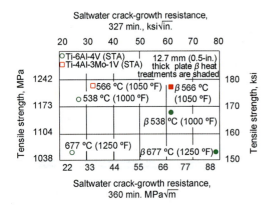

Fig. 6.25 Strength and saltwater crack-growth resistance of titanium alloys. Open symbols represent tensile strength-saltwater toughness for material processed below the beta transus and subsequently solution treated and aged (STA) at the indicated temperature. Solid symbols indicate beta annealing prior to STA treatment. Saltwater crack-growth resistance improves as a result of beta annealing.

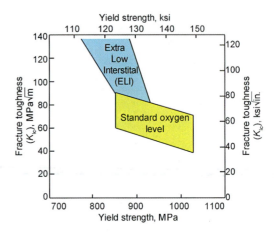

Fig. 6.26 Effect of oxygen content on fracture toughness. Extralow interstitial material contains less than 0.13 wt% O.

then obtained by regular inspections for defects and not allowing any crack to attain the critical size in the interval between inspections. The FCGR of titanium alloys in air is slower (better) than aluminum alloys but faster (worse) than steel. Seawater has little effect on the FCGR of titanium but a significant adverse effect on steel; the FCGR is influenced by factors such as the dwell time at maximum load. Fatigue crack growth rate scatter band data comparing Ti-6Al-4V cast and cast plus hot isostatic pressed material (see Chapter 8, "Melting, Casting, and Powder Metallurgy," in this book) with beta-annealed ingot metallurgy material is shown in Fig. 6.27.

The fracture mechanics approach to lifing is shown schematically in Fig. 6.28, which shows fatigue cracks propagating until they reach the unstable length corresponding to the fracture toughness at which failure occurs.

High-Temperature Near-Alpha Alloys

A list of conventional high-temperature near-alpha alloys is shown in Table 6.9 (Ref 6.1). The effect of annealing temperature is shown in Fig. 6.29. A fully transformed beta structure (consisting totally of lenticular alpha phase) favors creep, fracture toughness, and slow fatigue crack growth, while the presence of equiaxed primary alpha optimizes strength, low-cycle fatigue, and ductility (Fig. 6.29).

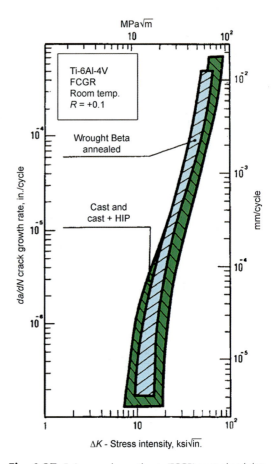

Fig. 6.27 Fatigue crack growth rate (FCGR) scatter band data comparing Ti-6Al-4V cast and cast plus hot isostatic pressed (HIP) material with beta-annealed ingot metallurgy material

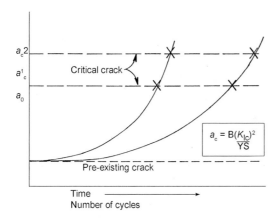

$$a_c = \frac{B(K_{Ic})^2}{YS}$$

Fig. 6.28 Fracture mechanics approach to life. The symbol "*a*" represents crack length, and the critical crack length (a_c) is calculated from the fracture toughness (K_{Ic}) and yield strength (YS).

Table 6.9 Conventional high-temperature near-alpha alloys

		Alloy			
		IM 829	IMI 834	Ti-6242	Ti-1100
Element, wt%	C	0.02	0.06	0.02	0.02
	Fe	0.03	0.02	0.08	0.02
	O	0.10	0.10	0.09	0.07
	Si	0.35	0.35	0.09	0.45
	Nb	1.0	0.7	...	...
	Mo	0.25	0.50	2.0	0.40
	Zr	3.0	3.5	4.0	4.0
	Sn	3.5	3.8	2.0	2.7
	Al	5.6	5.6	6.0	6.0

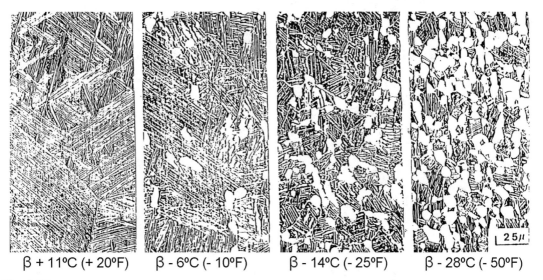

β + 11°C (+ 20°F) β - 6°C (- 10°F) β - 14°C (- 25°F) β - 28°C (- 50°F)

Fig. 6.29 Effect of annealing temperature on the microstructure of elevated-temperature near-alpha titanium alloys

Alpha-Beta Alloys

Alpha-beta titanium alloys contain both the alpha and beta phases. Aluminum is the principal alpha stabilizer and strengthens the alpha phase. Beta stabilizers allow the beta phase to be retained below the beta transus temperature and locked in by quenching. Beta stabilizers such as vanadium also provide strengthening and allow alpha-beta alloys to be hardened by solution heat treating and aging (STA). Alpha-beta alloys have a good combination of mechanical properties, rather wide processing windows, and can be used

at temperatures to 320 to 400 °C (600 to 750 °F). Alloy Ti-6Al-4V is used as a standard reference alpha-beta titanium alloy; it is often called the "workhorse" of the aerospace industry, because historically it accounted for approximately 60% by weight of the titanium used in aerospace industry and up to 80 to 90% by weight of that used in airframes.

Two newer alpha-beta titanium alloys poised to challenge Ti-6Al-4V are the fracture-resistant Ti-5553 (Ti-5Al-5V-5Mo-3Cr) (Ref 6.10), used on the Boeing 787, and a high-strength, cold-workable ATI 425 (Ti-4Al-1.5Fe-2.5V-0.25O$_2$)

titanium alloy developed and produced by Allegheny Technologies Incorporated (Ref 6.11).

The Ti-5553 exhibits excellent hardenability strength, high fracture toughness, and high-cycle fatigue behavior relative to Ti-6Al-4V. Because of these characteristics, it is used for a number of key high-load components in the Boeing 787, including the flap tracks, pylons, side of body chords, and landing gear.

Like all beta alloys, Ti-5553 (Ti-5Al-5V-5Mo-3Cr) must be heat treated to achieve its desired properties. On the Boeing 787, Ti-5553 is given an STA treatment to obtain these values:

- 1.24 GPa (180 ksi) ultimate tensile strength (UTS), 1.17 GPa (170 ksi) yield strength (YS), and 6% reduction in area (landing gear)
- 1.1 GPa (160 ksi) UTS, 0.96 GPa (140 ksi) YS, 6% elongation, and 71.3 MPa $\sqrt{m}$ (65 ksi $\sqrt{in.}$) K_{Ic} (nacelle structure)

The ATI 425 alloy is a high-strength, high-ductility titanium alloy available in a variety of product forms, including cold-rolled coil and sheet. Originally developed for use in ballistic armor applications, it is characterized for use in aerospace and industrial applications.

It is an alpha-beta titanium alloy that uses iron and vanadium as beta stabilizers, as well as aluminum as an alpha stabilizer. Lower aluminum and vanadium contents and higher oxygen and iron contents give the alloy a unique combination of ductility and tensile strength, which makes it useful for use in titanium applications that require cold forming, such as roll forming and bending, while still providing superior strength compared with low-alloy grades of titanium.

ATI 425 is a versatile alloy with higher strength than Ti-3Al-2.5V (nearly equivalent to Ti-6Al-4V) and similar corrosion-resistance properties. The availability of ATI 425 as a cold-rolled product make the material unique. The ability to obtain a fine, cold-rolled surface finish and a thin-gage strip product is desirable for some applications. It has significant corrosion resistance for a high-strength titanium alloy (Table 6.10).

The alloy flows much easier in forging and rolling than Ti-6Al-4V, is less prone to surface cracking, and requires a far lesser degree of surface conditioning for subsequent working. All forms of the alloy exhibit mechanical properties similar to Ti-6Al-4V but are slightly easier to form.

The alloy also demonstrates good bend ductility. Cold-rolled sheet up to 2.54 mm (0.100 in.) can be bent to a radius of 2.5 times the thickness (T), with the bend axis in either the longitudinal or transverse direction. Light-gage, hot-rolled plate (4.75 to 6.35 mm, or 0.1875 to 0.250 in.) can be bent to 3.5 T radius. As with all titanium alloys, the material must be totally free of alpha case and any surface anomalies, which could become stress raisers for cracking.

Fracture toughness values for plate vary with processing and anneal cycles. Duplex annealed ballistic plate, with a relatively coarse alpha-beta worked microstructure, has a plane-strain fracture toughness (K_{Ic}, average of four values) of 60.0 MPa $\sqrt{m}$ (54.6 ksi $\sqrt{in.}$). The fracture toughness obtainable through a duplex anneal rivals those values obtainable only via an extralow interstitial (ELI) formulation of Ti-6Al-4V. However, Ti-6Al-4V ELI formulations do not have the strength levels of conventional Ti-6Al-4V or ATI 425 titanium wrought products, such as cold-rolled sheet or small bar.

Results from fatigue tests demonstrate that fine-grained material, such as bar with a high percentage of alpha-beta reduction, has a much higher fatigue

Table 6.10 ATI 425 corrosion test results

| Media | Corrosion rate, mils/yr | | | Test duration, days |
	Ti-6Al-4V, published values	Ti-6Al-4V, Wah Chang Lab test	ATI 425	
25% nitric acid at BP 98 °C (208 °F)	26.4	24	3	7
5% HCl + 0.1% FeCl₃ at BP 101 °C (214 °F)	0.6(a)	166(a)	225	7
50% formic acid at BP 101 °C (214 °F)	315	259	293	1
1% HCl at BP 101 °C (214 °F)	99(a)	191(a)	399	1
8% HCl at BP 103 °C (217 °F)	1891	1830	2991	1
10% ferric chloride at BP 105 °C (221 °F)	Nil	Weight gain	Weight gain	7
Seawater at 21 °C (70 °F), crevice attachment	0.01	0	0	21
25% brine at 90 °C (194 °F), crevice attachment	Nil	Weight gain	Weight gain	21

Note: Results from test on hot-rolled sheet. BP, boiling point. (a) The published value cited by the Wah Chang Corrosion Laboratory lacked any documentation of test circumstances that could explain the difference between the published value and the Wah Chang obtained results for the Ti-6Al-4V control sample. For all practical purposes, given the experimental error for limited coupons, the values for Ti-6Al-4V and ATI 425 could be considered to be similar.

life than the coarse-grained ballistic plate with lesser amount of alpha-beta work (approximately 50%). To optimize one particular property, such as fatigue life, other characteristics may have to be sacrificed. The higher the stress level, especially as it nears the yield strength, the lower the fatigue life.

Product forms include hot-rolled plate, cold-rolled sheet, and forgings. Applications include fracture-critical airframe components, chemical process industry, components of high reciprocating mass, ballistic performance in both armor and containment vessels where mechanical integrity is critical, plate/frame heat exchangers made through superplastic forming, screen material, honeycomb packing support, and pressure vessels.

Beta Alloys

Beta alloys are sufficiently rich in beta stabilizers and lean in alpha stabilizers, so the beta phase can be completely retained with appropriate cooling rates (Ref 6.12). Beta alloys are metastable, and precipitation of alpha phase in the metastable beta is a method used to strengthen the alloys. Beta alloys contain small amounts of alpha-stabilizing elements as strengthening agents. As a class, beta and near-beta alloys offer increased fracture toughness over alpha-beta alloys at a given strength level. Beta alloys also exhibit better room-temperature forming and shaping characteristics than alpha-beta alloys, higher strength than alpha-beta alloys at temperatures where yield strength instead of creep strength is the requirement, and better response to STA in heavier

sections than the alpha-beta alloys. They are limited to service at approximately 370 °C (700 °F) due to creep. The improved (lower) flow stress of the beta alloys compared with Ti-6Al-4V is shown in Fig. 6.30.

The Ti-10V-2Fe-3Al features ease of processing (Fig. 6.30) and better fracture-toughness-to-yield-strength combinations (Fig. 6.31) than traditional alloys. Because they offer hot and cold formability advantages and metastable beta alloys are heat treatable, they are being used in an increasing number of applications (see Chapter 15, "Applications of Titanium," in this book).

Titanium Aluminides

Development of intermetallic titanium aluminides began in 1970. The three major compounds are gamma TiAl, alpha-2 (α_2) Ti_3Al, and $TiAl_3$ (Ref 6.3, 6.4). Despite a lack of fracture resistance (low ductility, fracture toughness, and FCGR), the titanium aluminides have potential for enhanced performance, which has attracted attention in the aerospace and automotive industries. Of the three compositions, TiAl is most significant commercially, followed by Ti_3Al and $TiAl_3$. $TiAl_3$ has little commercial interest. Interest in these materials is due to their low density and excellent mechanical properties at elevated

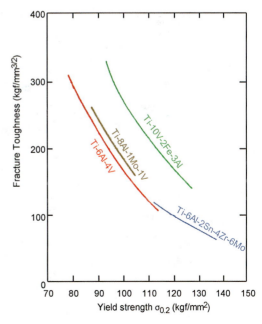

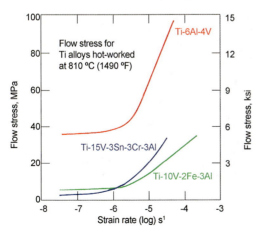

Fig. 6.30 Flow stress for two hot-worked beta-titanium alloys versus Ti-6Al-4V. Lower flow stress required relative to Ti-6Al-4V makes the beta alloys easier to form.

Fig. 6.31 The Ti-10V-2Fe-3Al alloy exhibits better fracture toughness to yield strength than the traditional alloys.

temperatures (over 600 °C, or 1110 °F), making them an attractive alternative to the nearly twice-as-dense, nickel-base, high-temperature superalloys used in aircraft turbines and automotive engines. Large-scale commercial use of titanium aluminides has only recently begun, with General Electric using gamma TiAl in the low-pressure (higher-temperature) stages 6 and 7 turbine blades of its GEnx (General Electric Next-Generation) engine, which powers the Boeing 787 and 747-8. The use of TiAl in place of nickel-base superalloys in these stages has contributed to increased thrust-to-weight ratios for this new class of engines. The automotive industry is attracted to titanium aluminides for use in internal combustion engines, high-temperature engine valves, and other components that would benefit from its high-temperature mechanical properties. The properties of titanium aluminides are shown relative to competing nickel- and iron-base superalloys in Table 6.11; note the density of each material. The titanium-aluminum phase diagram is shown in Fig. 6.32.

The enhanced creep performance of the titanium aluminides, particularly TiAl, compared with terminal titanium alloys is shown in Fig. 6.33.

Table 6.11 Properties of nickel, iron, and titanium aluminides

Alloyγ	Crystal structure(a)	Critical ordering temperature °C	°F	Melting point °C	°F	Material density g/cm³	lb/in.³	Young's modulus GPa	10⁶ psi
Ni₃Al	L1₂ (ordered fcc)	1390	2535	1390	2535	7.50	0.271	179	25.9
NiAl	B2 (ordered bcc)	1640	2985	1640	2985	5.86	0.212	294	42.7
Fe₃Al	D0₃ (ordered bcc)	540	1000	1540	2805	6.72	0.243	141	20.4
	B2 (ordered bcc)	760	1400	1540	2805	...	...	...	...
Ti₃Al	D0₁₉ (ordered hcp)	1100	2010	1600	2910	4.2	0.15	145	21.0
TiAl	L1₀ (ordered tetragonal)	1460	2660	1460	2660	3.91	0.141	176	25.5
TiAl₃	D0₂₂ (ordered tetragonal)	1350	2460	1350	2460	3.4	0.123	...	...

(a) fcc, face-centered cubic; bcc, body-centered cubic; hcp, hexagonal close-packed. Source: Ref 6.13

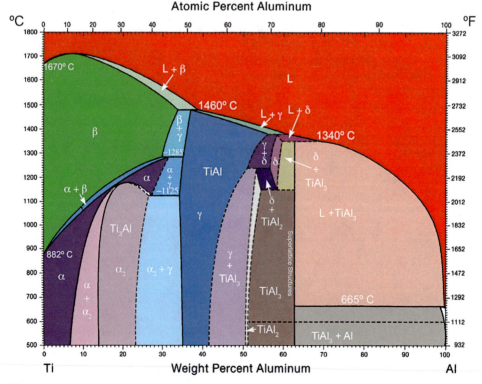

Fig. 6.32 Titanium-aluminum phase diagram showing the temperature-composition in which TiAl, Ti₃Al, and TiAl₃ are stable.

The difficulty in dislocation motion in the intermetallics (particularly TiAl) contributes to their good high-temperature creep behavior, but, at the same time, they exhibit low room-temperature ductility. This lack of ductility requires a change in philosophy by designers who generally use a fracture mechanics approach in structural design, assuming propagation of a pre-existing crack and a minimum level of ductility (4 to 5% elongation) as a further "comfort" factor.

Intermetallics exhibit good oxidation resistance; however, additions are needed for good high-temperature oxidation resistance. The oxida-

tion resistance of gamma alloys depends on niobium and niobium + tungsten contents; oxidation resistance is close to that of Inconel 718 with reasonable amounts of niobium. This means that some gamma alloys can be used at 870 °C (1600 °F) for 1000 h without requiring a coating. Their oxidation resistance is better (often significantly) than those of any so-called advanced alloys, including NbSi-X, MoSi-X, and high-entropy alloys up to 900 °C (1650 °F), where the gamma oxidation resistance has been measured.

The Ti3Al alloys exhibit a room-temperature yield stress of 952 MPa (138 ksi) in a 17 at.% Nb alloy (an "O" alloy) in combination with a fracture toughness of 28.3 MPa $\sqrt{m}$ (25.8 ksi $\sqrt{in.}$) and an elongation of 6%. The Ti-24Al-11Nb (at.%) has a yield strength of 831 MPa (120.5 ksi) and an elongation of 4.8%. However, the TiAl base intermetallic is considerably more attractive and has received much more attention.

Figure 6.34 shows the four basic microstructures developed in the TiAl intermetallic (Ref 6.4), and Table 6.12 lists the mechanical properties associated with these four microstructures. The fully lamellar (FL) microstructure is generated when cooled from an alpha treatment at a formation temperature (well below the alpha transus), which is a function of cooling rate and alloy

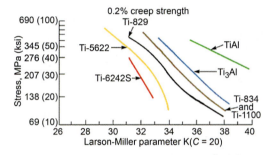

Fig. 6.33 Creep behavior of a number of terminal high-temperature titanium alloys and the intermetallic compounds Ti₃Al and TiAl, showing the enhanced creep behavior of the intermetallics, particularly equiatomic TiAl

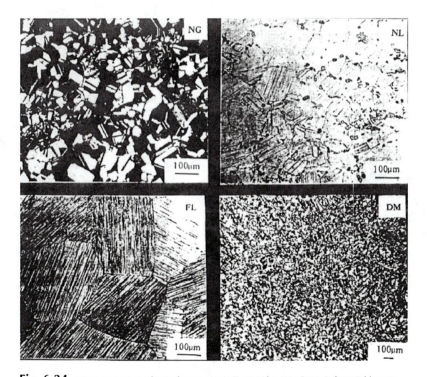

Fig. 6.34 Microstructure mechanical properties in Ti-47.5Al-2.5V-1Cr (γ). Refer to Table 6.12.

composition. Typical cooling rates range from 10 to 50 °C (20 to 90 °F) per minute. The near-lamellar (NL) microstructure is formed after annealing followed by cooling at 10 to 100 °C (20 to 180 °F) per minute from alpha transus −10 °C (−18 °F) to approximately alpha transus −25 °C (−45 °F). The duplex (DM) microstructure is formed upon annealing at and cooling from an alpha + gamma two-phase field temperature where the alpha:gamma volume ratio is approximately 1. The near-gamma (NG) microstructure is generated upon annealing near the eutectic temperature of 1125 °C (2057 °F), where the gamma/alpha or α_2 volume ratio is the greatest. This microstructure is dominated with medium-sized (10 to 50 µm) gamma grains, which appear to be single phase (Ref 6.4).

The GEnx engine specifies the gamma titanium aluminide Ti-48Al-2Nb-2Cr (at.%), with a density of 4.0 g/cm³ (0.144 lb/in.³) (compared to terminal titanium alloys with a density of 4.5 g/cm³, or 0.162 lb/in.³), in the form of a casting on the GEnx engine for use on the Boeing 787 and 747-8 airframes. The engine boasts a 20% reduction in fuel consumption compared to the CF6 engine, in part due to the use of this alloy in the sixth- and seventh-stage blades of the low-pressure turbine. Other engine manufacturers such as Safran-Snecma and Rolls-Royce (working on the Trent engine, which will power future Airbus models) are working with companies such as PCC Airfoils on cast gamma for use at temperatures above 700 °C (1290 °F). For further information, see Chapter 15, "Applications of Titanium," in this book.

Metal-Matrix Composites

There are two basic types of metal-matrix composites: continuous- and discontinuous-fiber composites (Ref 6.5). The characteristics of continuous silicon carbide (SiC) fibers produced by chemical vapor deposition are shown in Fig. 6.35.

Table 6.12 Microstructure and mechanical properties in Ti-46.5Al-2.5V-1Cr TiAl type

Microstructure(a)	Yield strength		Ultimate tensile strength		Elongation, %	K_{Ic}	
	MPa	ksi	MPa	ksi		MPa√m	ksi = √in.
FL	360	52	400	58	0.5	21	19
NL	430	62	480	70	2.3	17	15.5
DM	440–450	64–65	505–538	73–78	3.3–4.8	12	11
NG	387	56	468	68	1.7	17	15.5

(a) FL, fully lamellar; NL, near lamellar; DM, duplex; NG, near gamma

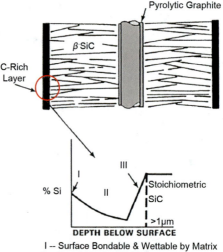

I -- Surface Bondable & Wettable by Matrix
II -- Broad Forgivability Zone
III -- Inner Gradient - Necessary for Maintaining Filament Strength

Fig. 6.35 The manufacture of continuous silicon carbide (SiC) fibers via chemical vapor deposition

Figure 6.36 shows tensile behavior of Ti_3Al reinforced by continuous SiC fibers compared with monolithic Ti_3Al and superalloys. Longitudinal strength is considerably better than the transverse strength (Table 6.13), so the composite must be oriented in a component with the fibers in the maximum stressed direction to make use of unidirectional reinforcement. The potential advantages of using continuous reinforced titanium composites are shown in Fig. 6.37.

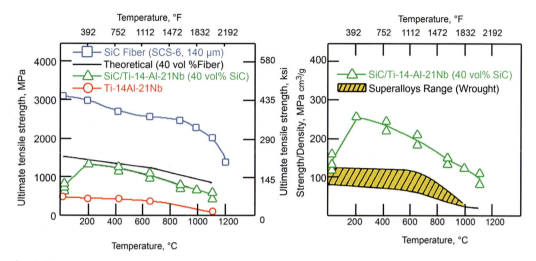

Fig. 6.36 Tensile behavior of Ti_3Al reinforced by continuous SiC fibers, compared with monolithic Ti_3Al and superalloys

Table 6.13 Characteristics of Ti-6Al-4V unidirectional composites

Material	Longitudinal ultimate tensile strength		Transverse ultimate tensile strength		Longitudinal modulus	
	MPa	ksi	MPa	ksi	GPa	ksi
Ti-6Al-4V	890	129	890	129	120	17.4
B_4C-B/Ti-6Al-4V	1055	153	310	45	205	29.7
SCS-6/Ti-6Al-4V	1455	211	340	49.3	240	34.8

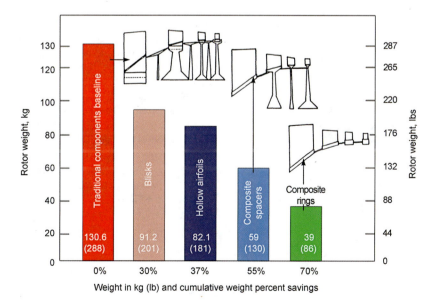

Fig. 6.37 The potential advantage of using continuous reinforced titanium composites is demonstrated for engine blades. Data courtesy of Allison

The microstructure of Ti-6Al-4V reinforced with TiC particles and components fabricated using a powder metallurgy approach are shown in Fig. 6.38. (For more details, see Chapter 8, "Melting, Casting, and Powder Metallurgy," in this book.) Table 6.14 lists the tensile properties of the composite material compared with monolithic Ti-6Al-4V.

Shape Memory Alloys

A shape memory alloy (SMA) is an alloy that "remembers" its original, cold-worked shape, returning to the predeformed shape when heated (Ref 6.14). The material is a lightweight, solid-state alternative to conventional actuators, such as hydraulic, pneumatic, and motor-based systems. Shape memory alloys have applications in industries, including medicine and aerospace. Shape memory alloys containing titanium include Ni-Ti (~55% Ni), Ni-Ti-Nb, and Ti-Pd. These materials are used in medicine, for example, as fixation devices for osteotomies in orthopedic surgery, and in dental braces to exert constant tooth-moving forces on the teeth.

Typical mechanical properties of nickel-titanium shape memory alloys include:

- Ultimate tensile strength: 754 to 960 MPa (110 to 140 ksi)
- Elongation to fracture: 15.5%
- Yield strength (high temperature): 560 MPa (80 ksi)
- Yield strength (low temperature): ~100 MPa (15 ksi)
- Elastic modulus (high temperature): ~75 GPa (11×10^6 psi)
- Elastic modulus (low temperature): ~28 GPa (4×10^6 psi)
- Poisson's ratio: ~0.3

Nitinol (nickel-titanium) was commercially introduced in the late 1980s as an enabling technology in a number of minimally invasive endovascular medical applications. The properties of Nitinol alloys manufactured to body temperature response provided an attractive alternative to balloon-expandable devices in stent grafts, where it gives the ability to adapt to the shape of certain blood vessels when exposed to body temperature. On average, 50% of all peripheral vascular stents currently available on the worldwide market are made of Nitinol.

Eyeglass frames made from titanium-containing SMAs are marketed under the trademarks Flexon (Marchon Eyewear Inc.) and TITANflex (Eschenbach Optik). These frames are usually made of SMAs that have their transition temperature set below the expected room temperature. This allows the frames to undergo large deformation under stress, yet regain their intended shape

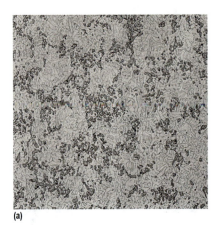

(a)

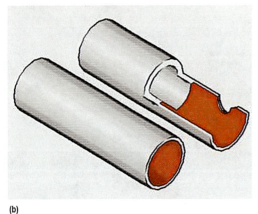

(b)

Fig. 6.38 (a) Microstructure of CermeTi material, TiC reinforcement. (b) Parts fabricated from this material. Courtesy of Dynamet Technology Inc.

Table 6.14 Typical properties of CermeTi versus Ti-6Al-4V

	Ultimate tensile strength		Yield strength		Elongation,	Elastic modulus		Hardness,
	MPa	ksi	MPa	ksi	%	GPa	msi	HRC
Ti-6Al-4V powder metallurgy	965	140	896	130	14	110	16.0	36
CermeTi-C metal-matrix composite (Ti-64 + TiC)	1034	150	965	140	3	130	18.9	42

once the metal is unloaded again. The very large, apparently elastic strains are due to the stress-induced martensitic effect, where the crystal structure can transform under loading, allowing the shape to change temporarily under load. This means that eyeglasses made of SMAs are more robust against being accidentally damaged.

Shape memory alloys are also used in orthopedic surgery as a fixation-compression device for osteotomies, typically for lower-extremity procedures. The device, usually in the form of a large staple, is stored in a refrigerator in its malleable form and is implanted into predrilled holes in the bone across an osteotomy. As the staple warms, it returns to its nonmalleable state and compresses the bony surfaces together to promote bone union.

Summary

The physical metallurgy and heat treatment of titanium alloys determines the microstructure and, in turn, their mechanical properties. The morphology (shape) of the alpha phase has a strong influence on the mechanical properties; a lenticular morphology favors fracture toughness, FCGR, and creep behavior, while an equiaxed shape favors ductility, low-cycle fatigue, superplastic forming/diffusion bonding, and moderately increased strength. The testing of titanium alloys is performed using the established methods used for other metallic alloys, and this chapter shows the mechanical properties of alpha, alpha-beta, and beta alloys under static and dynamic test conditions. Also presented in this chapter is the effect of temperature, section size, heat treatment, and alloying elements on mechanical properties. Titanium alloys have good-to-excellent mechanical strength, ductility, and elevated-temperature creep strength, S-N fatigue, fracture toughness, and FCGR (slow). The microstructures and mechanical properties of the intermetallics Ti_3Al and TiAl alloys are presented along with metal-matrix composites (both continuously and discontinuously reinforced) and shape memory alloys. The excellent mechanical properties of titanium-base materials make titanium alloys very useful materials in the aerospace, aircraft, marine, and pressure vessel industries.

ACKNOWLEDGMENT

The assistance of Dr. Y.-W. Kim of Gamteck in compiling the section on gamma titanium aluminides is greatly appreciated.

REFERENCES

6.1 F.H. Froes, D. Eylon, and H.B. Bomberger, Ed., *Titanium Technology: Present Status and Future Trends,* Titanium Development Association, Dayton, OH, 1985

6.2 R. Boyer, E.W. Collings, and G. Welsch, Ed., *Materials Properties Handbook: Titanium Alloys,* ASM International, 1994

6.3 H.A. Lipsitt et al., The Deformation and Fracture of Ti_3Al at Elevated Temperatures, *Metall. Trans. A,* Vol 11, 1980, p 1369–1375

6.4 Y.-K. Kim, *Acta Metall. Mater.,* Vol 40 (No. 6), 1992, p 1121–1134

6.5 F.H. Froes and J. Storer, Ed., *Proceedings of Conference on Recent Advances in Titanium Metal Matrix Composites,* TMS, Warrendale, PA, 1995

6.6 F.H. Froes, private communication, 2013

6.7 W.D. Callister, Jr. and D.G. Rethwisch, *Materials Science and Engineering: An Introduction,* John Wiley and Sons, Hoboken, NJ, 2010

6.8 H. Kuhn and D. Medlin, *Mechanical Testing and Evaluation,* ASM International, 2000

6.9 J.R. Davis, *Tensile Testing,* 2nd ed., ASM International, 2004

6.10 R.R. Boyer, Boeing Corporation, private communication, 2013

6.11 ATI Wah Chang, private communication, 2013

6.12 J.L. Murray, *Phase Diagrams of Binary Titanium Alloys,* ASM International, Metals Park, OH, 1987

6.13 *Properties and Selection: Nonferrous Alloys and Special-Purpose Materials,* Vol 2, *Metals Handbook,* 10th ed., ASM International, 1990

6.14 H.R. Chen, Ed., *Shape Memory Alloys: Manufacture, Properties and Applications,* Nova Science Publishers Inc., 2010

SELECTED REFERENCES

- K.G. Budinski and M.K. Budinski, *Engineering Materials: Properties and Selection,* 7th ed., Prentice Hall, 2002
- P. Lacombe, R. Tricot, and G. Beranger, Ed., *Proceedings of the Sixth International Conference on Titanium* (Nice, France), Metallurgical Society of AIME, Warrendale, PA, 1988
- G. Lutjering and J.C. Williams, *Titanium,* Springer, 2003

- G. Lutjering, U. Zwicker, and W. Bunk, *Proceedings of the Fifth International Conference on Titanium,* Metallurgical Society of AIME, Warrendale, PA, 1984
- *Proceedings of the Seventh International Conference on Titanium* (San Diego, CA), Metallurgical Society of AIME, Warrendale, PA, 1992
- *Proceedings of the Eighth International Conference on Titanium* (San Birmingham, U.K.), Metallurgical Society of AIME, Warrendale, PA, 1995
- *Proceedings of the Ninth International Conference on Titanium,* Metallurgical Society of AIME, Warrendale, PA, 2000
- *Proceedings of the Tenth International Conference on Titanium,* Metallurgical Society of AIME, Warrendale, PA, 2004
- *Proceedings of the Eleventh International Conference on Titanium,* Metallurgical Society of AIME, Warrendale, PA, 2008
- *Proceedings of the Twelfth International Conference on Titanium,* Metallurgical Society of AIME, Warrendale, PA, 2012

CHAPTER 7

Metallography of Titanium and Its Alloys*

METALLOGRAPHY, an important division of physical metallurgy, concerns itself with the internal structure and constitution of metals. Henry Sorby, an English metallurgist, developed the first successful metallographic procedures. His first micrographs, published in 1885, were regarded as attractive curiosities, and it took some 20 years for the importance of this work to be recognized. By 1920, metallography had become a necessary part of any metallurgical program.

A number of sophisticated tools aid metallurgists in the study of metal structures. These include the electron microscope, electron and x-ray diffraction, and the electron-probe microanalyzer. However, the instruments still most widely used to study metal structures are the light microscope and, to a lesser degree, the electron microscope. This chapter deals with titanium and titanium alloy structures as they are revealed by the light microscope and scanning electron microscope. Terms used in interpreting titanium microstructures are defined in the glossary at the end of this chapter.

Alloy content, method of fabrication, and heat treatment after fabrication influence the microstructures of titanium. Alloy content determines the type of phases present in an alloy (Ref 7.1–7.4). The method of fabrication affects grain size and grain orientation. Heat treatment influences the type and amount of phases present and their morphology.

Metallography plays an important part in all considerations in the thermomechanical processes of titanium. Proper sample-preparation procedures, coupled with accurate interpretation of microstructures, can aid in solving numerous metallurgical problems in titanium development and production.

Review of Physical Metallurgy—Alpha and Beta

The proper interpretation of microstructures of any alloy system requires understanding of the physical metallurgy of the system being studied. Three principal alloy types—alpha, alpha-beta, and beta—are briefly reviewed here.

Crystal Structure. Titanium can exist in two allotropic forms: alpha (a hexagonal close-packed crystal structure) and beta (a body-centered cubic structure) (Ref 7.1–7.4). In pure titanium, the alpha (α) phase is stable up to 880 °C (1620 °F), at which point it transforms to the beta (β) phase; the beta phase is stable from 880 °C (1620 °F) to the melting point. At room temperature, pure titanium consists of the alpha phase. However, the alloys can contain alpha, mixtures of alpha and beta, or beta phases, depending on the alloy content and conditions. Thus, the alloys are classified into these structural types: alpha (α), alpha-beta (α-β), and beta (β).

Alloy Classification. Alloy additions—except tin and zirconium, as subsequently noted—stabilize either the alpha or the beta phase. Elements that preferentially dissolve in the alpha phase tend to stabilize this phase and are known as alpha stabilizers. The elements more soluble in the beta phase tend to stabilize this phase and are termed beta stabilizers. Alloys that contain mostly alpha-stabilizing elements and consist

*Adapted and revised from Louis J. Bartlo and David J. McNeish, *Titanium and Its Alloys*, ASM International.

predominantly of the alpha phase (at room temperature) are classified as alpha alloys. Alloys that contain alloying elements that result in mixtures of alpha and beta phases are classified as alpha-beta alloys. Alloys that consist largely of the beta phase on air cooling from solution annealing temperatures are classified as beta alloys. Most commercial grades of unalloyed titanium and certain alpha alloys contain small amounts of beta-stabilizing elements, while the beta alloys contain small amounts of alpha-stabilizing elements as strengthening agents.

The Alpha System. Figure 7.1 represents a typical binary constitution diagram of alloying elements that stabilize the alpha phase, because they are more soluble in alpha than in beta. Adding such elements to titanium in increasing amounts stabilizes the alpha phase to higher temperatures (Ref 7.1–7.4). The alpha-beta and beta-transformation temperatures are higher, making it possible, and at times necessary, to process such alloys at higher temperatures.

Alloying elements of the substitutional type that belong to this system are aluminum, gallium, and germanium. The interstitial elements oxygen, nitrogen, and carbon are also of the alpha-stabilizing type and are included in this system.

The Beta System. Alloys containing beta-stabilizing elements are of two types (Ref 7.1–7.4). Those that are completely miscible with the beta phase are known as beta isomorphous, while elements that transform the beta phase in a eutectoid reaction are termed eutectoid. Figure 7.2 shows a typical binary constitution diagram for the beta isomorphous type of alloy addition. The addition ele-

ment is completely soluble in the beta phase, and transformation of beta to eutectoid products does not occur even under equilibrium conditions.

This system differs entirely from the alpha-stabilized system in that alloying additions lower the beta-transformation temperature rather than increase it as in the alpha system. This means that processing and annealing temperatures are somewhat lower than they are for alpha-type alloys. Alloying elements of the beta isomorphous type are vanadium, molybdenum, tantalum, and niobium.

The beta eutectoid system is represented by the constitution diagram in Fig. 7.3 (Ref 7.1–7.4). This system differs from the beta isomorphous system in that the alloying elements, under certain conditions, decompose the beta phase to alpha and an intermetallic phase. The beta eutectoid elements are of two types: the active eutectoid former and the sluggish eutectoid former.

Active eutectoid elements such as silicon and copper rapidly transform beta to an intermetallic phase or compound. Until recently, such elements were not used extensively in commercial alloys. However, the addition of such elements in small amounts has been beneficial to some elevated-temperature properties.

The other eutectoid formers—chromium, cobalt, nickel, iron, and manganese—are more sluggish in their eutectoid reactions and are not generally used in sufficient quantity to form compounds in most commercial alloys. The beta eutectoid elements, arranged in order of increasing activity, are shown in Table 7.1.

Two other alloying elements not shown in the table are zirconium and tin. Neither of them pro-

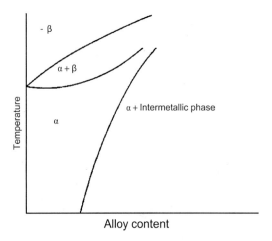

Fig. 7.1 Partial phase diagram of the alpha-stabilized system. The alpha-stabilizing elements are aluminum, germanium, gallium, carbon, oxygen, and nitrogen.

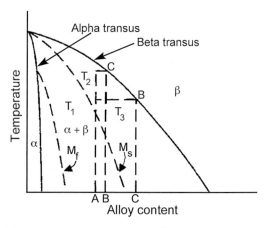

Fig. 7.2 Partial phase diagram of the beta isomorphous system. Alloying elements of the beta isomorphous type are vanadium, molybdenum, tantalum, and niobium.

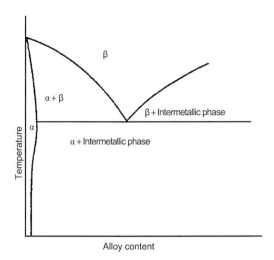

Fig. 7.3 Partial phase diagram of the beta eutectoid system. Alloying elements of the beta eutectoid type are manganese, iron, chromium, cobalt, nickel, copper, and silicon.

Table 7.1 Beta eutectoid elements arranged in order of increasing activity

Element	Eutectoid composition, wt%	Eutectoid temperature °C	°F	Composition for beta retention on quenching, wt%
Manganese	20	550	1022	0.65
Iron	15	600	1112	4.0
Chromium	15	675	1247	8.0
Cobalt	9	685	1265	7.0
Nickel	7	770	1418	8.0
Copper	7	790	1454	13.0
Silicon	0.9	860	1580	...

motes phase stability; that is, they do not strongly stabilize either the alpha or the beta phase. Both elements are extensively soluble in the alpha and beta phases and are important strengthening agents. Although they are more closely related to the alpha-stabilizing systems (and are sometimes used in large proportions in alpha or near-alpha alloys), they are useful additions to both alpha and beta alloys.

Terminology Used to Describe Titanium Alloys Structures

For proper interpretation of titanium structures, it is necessary to know the prior history of the material being examined, including the alloy content, working temperatures, method of fabrication, and thermal condition of the material. Equiaxed structures (polygonal-shaped grains) are usually produced by cold working and annealing above the recrystallization temperature. However, in certain instances, equiaxed alpha structures can be confused with equiaxed beta structures if the alloy content is unknown. Usually, the microstructures that result from beta annealing (or processing) are also easily recognized under the light microscope.

Cooling rates govern the grain morphology resulting from beta treatments. Acicular martensitic structures are produced by rapidly cooling beta-lean alloys from above their beta transus temperature. Such structures are recognized by the nee-

dle-shaped appearance of the grains. Platelike structures form on slow cooling alpha or alpha-beta alloys from temperatures in the beta field, and they are characterized by a wide, elongated grain shape. The slower the cooling rate, the coarser the platelike structure.

An explanation of the terms used in describing titanium microstructures, along with illustrative micrographs, is presented.

Elongated structures of the alpha or the beta phase are produced by unidirectional working. Figures 7.4 and 7.5 illustrate the fibrous shape of the elongated structure. Such structures appear in longitudinal sections prior to annealing or in materials that have not been fully recrystallized on annealing.

Equiaxed structures are recognized by the polygonal shape of the grains. Such structures appear in fully annealed, single-phase or nearly single-phase materials. Both the alpha and the beta phases can exist as equiaxed structures. Equiaxed alpha structures are produced by cold working and annealing above the recrystallization temperature but below the beta transus. Figure 7.6 shows equiaxed alpha in unalloyed titanium. The structure also shows particles of spheroidal beta stabilized by small amounts of iron in the material.

Equiaxed beta structures are produced in highly beta-stabilized alloys by annealing above the beta transus and cooling rapidly to retain an all-beta structure. Figure 7.7 shows an equiaxed beta structure in the Ti-13V-11Cr-3Al alloy.

Primary alpha should not be confused with alpha prime, which is discussed later. Primary alpha is alpha phase that remains untransformed as titanium is heated to temperatures in the alpha-beta field. The amount and size of the primary alpha grains diminish as the beta transus temperature is approached. Once the beta transus temperature has been exceeded for a sufficiently long time, no further evidence of primary alpha will be seen in the microstructure. Figure 7.8 shows primary alpha in the Ti-6Al-4V alloy. The alloy was heated to 955 °C (1750 °F), which is slightly below the beta transus of 995 °C (1825 °F). In Fig. 7.9, the alloy was heated to 1010 °C (1850 °F), which is above the beta transus, and the primary alpha grains are no longer evident.

Transformed beta is another term that is somewhat confusing. This is a general term that de-

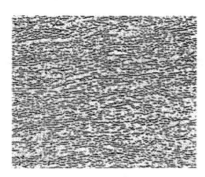

Fig. 7.4 Pure titanium sheet. Structure shows as-hot-rolled elongated alpha. Etchant: 10%HF-5%HNO$_3$. Original magnification: 250×

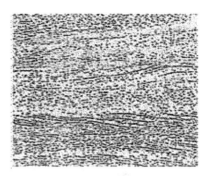

Fig. 7.5 Ti-13V-11Cr-3Al sheet. Structure shows as-hot-rolled elongated alpha-beta. Etchant: 2%HF-4%HNO$_3$. Original magnification: 250×

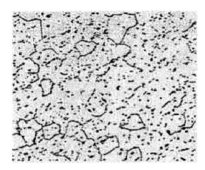

Fig. 7.6 Pure titanium sheet. Equiaxed alpha and beta spheroids resulted from heating specimen from Fig. 7.4 at 705 °C (1300 °F) for 1 h and air cooling. Etchant: 10%HF-5%HNO$_3$. Original magnification: 500×

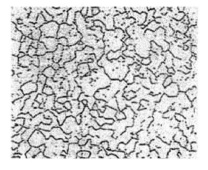

Fig. 7.7 Ti-13V-11Cr-3Al. Equiaxed beta (metastable) grains resulted from heating specimen from Fig. 7.5 at 790 °C (1450 °F) for ½ h and water quenching. Etchant: 2%HF-4%HNO$_3$. Original magnification: 250×

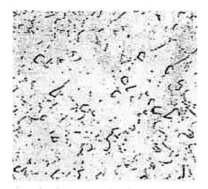

Fig. 7.8 Ti-6Al-4V bar. Primary alpha (untransformed alpha) and α (needlelike structure) after heating the bar at 955 °C (1750 °F), which is below beta transus, and water quenching. Etchant: 10%HF-5%HNO$_3$. Original magnification: 500×

Fig. 7.9 Ti-6Al-4V bar. Acicular α and prior-beta grain boundaries resulted from heating the bar at 1010 °C (1850 °F), which is above the beta transus, for 1 h and water quenching. Etchant: 10%HF-5%HNO$_3$. Original magnification: 500×

scribes the alpha phase as it forms directly from the beta phase. However, the transformation of beta to alpha can occur by one of two mechanisms, and the morphology of transformed beta structure varies considerably. Given sufficient time, transformation will occur through a nucleation-and-growth process, and it involves diffusion of atoms. The second mechanism is a diffusionless process, and it occurs through martensitic shear, which does not involve redistribution or mass movement of atoms in the structure. Thus, the term *transformed beta* gives little information regarding the morphology or the mechanism involved in the transformation process. Use of the term should be limited, and it should only be applied to describe structures in which transformation kinetics are unknown or when grain morphology cannot be resolved.

The terms *serrated*, *acicular*, *platelike*, *Widmänstatten*, and *alpha prime* are used to describe transformed beta structures in more detail.

Serrated alpha structures are produced by rapidly cooling high-purity titanium and certain alpha-stabilized alloys from above their beta transus temperature. The serrated structure is characterized by an irregular grain shape and jagged grain boundaries. Figure 7.10 shows a serrated alpha structure produced by water quenching unalloyed titanium from above the beta transus. Figure 7.11 illustrates a similar structure in an alpha-stabilized alloy rapidly cooled from above its beta transus. The alloy Ti-5Al-2.5Sn extralow interstitial was water quenched from 1065 °C (1950 °F).

The addition of a beta-stabilizing element to unalloyed titanium or to an alpha alloy changes the morphology of the transformed grains from a serrated to an acicular form. This occurs even when a small amount of beta stabilizer is used. Because most commercial unalloyed grades of titanium contain from 0.2 to 0.3% Fe (a beta stabilizer), these grades transform to the acicular rather than the serrated form. This is illustrated in Fig. 7.12, which shows unalloyed titanium (containing 0.3% Fe) water quenched from 1010 °C (1850 °F).

Acicular Alpha. The terms *acicular alpha* and *Widmanstätten* are generally interchangeable. Both terms describe alpha formed by nucleation and growth transformation from the beta phase. However, acicular by definition refers primarily to alpha phase with a fine needlelike appearance, whereas the Widmanstätten structure can exist as fine acicular or coarse platelike alpha phase.

Figure 7.13 shows acicular alpha in the Ti-6Al-4V alloy. The structure also shows an outline of the prior-beta grain boundaries that were formed by heating the alloy into the beta field (995 °C, or 1820 °F). On cooling from the beta field, alpha forms first at the prior-beta grain boundaries (given sufficient time), leaving a definite outline

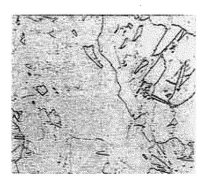

Fig. 7.11 Ti-5Al-2.5Sn extralow interstitial. Serrated alpha structure similar to that in Fig. 7.10 produced by water quenching the bar after heating it at 1065 °C (1950 °F) for ½ h (above its transus). Etchant: 10%HF-5%HNO₃. Original magnification: 100×

Fig. 7.10 Pure titanium bar. Serrated alpha structure produced by water quenching after heating the bar at 1010 °C (1850 °F), which is above beta transus, for ½ h. Etchant: 10%HF-5%HNO₃. Original magnification: 500×

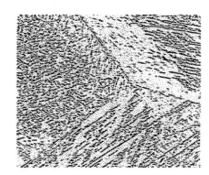

Fig. 7.12 Pure titanium bar. Acicular alpha structure produced by water quenching after heating at 1010 °C (1850 °F) for ½ h. Etchant: 10%HF-5%HNO₃. Original magnification: 250×

of the prior-beta grain size. Transformation of beta to alpha continues by nucleation growth of the alpha phase. The rate of cooling governs the size of the alpha transformation product.

Widmanstätten Structure. The Widmanstätten structure also results from nucleation and growth transformation of beta to alpha. The structure has a "basketweave" appearance, forming as alpha nucleates and growing on certain preferred crystallographic planes of the parent beta phase. The Widmanstätten structure can be produced in beta-lean alloys by using intermediate cooling rates from the beta field. In alpha-beta alloys, where the beta phase is rich in solute (beta-stabilizing elements), some beta is retained at room temperature between the alpha grains.

Figures 7.14 and 7.15 show Widmanstätten structures in the Ti-6Al-4V alloy. The microstructure in Fig. 7.14 was produced by working the alloy entirely in the beta field. Figure 7.15 shows a partial breakup of the Widmanstätten structure,

which resulted from finish-working the alloy slightly below the beta transus after prior heating above the beta transus.

Platelike alpha structure develops in alpha and alpha-beta alloys by slow cooling from the beta field. It has wide, elongated alpha phase. Figure 7.16 shows a platelike alpha structure in the Ti-5Al-2.5Sn alloy that was slow-cooled from the beta field. This type of microstructure also develops in the central portion of large billets that have been forged at temperatures in the beta field or high in the alpha-beta field. Figure 7.17 illustrates this condition in a 205 mm (8 in.) diameter Ti-6Al-4V billet.

Metastable Phases

Metastable, or nonequilibrium, phases that occur in titanium alloys are discussed in more detail in Chapter 4, "Principles of Beta Transforma-

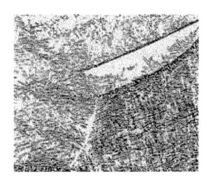

Fig. 7.13 Ti-6Al-4V. Acicular alpha structure and prior-beta grain boundaries formed on heating at 1060 °C (1940 °F) for ½ h and air cooling. Etchant: 10%HF-5%HNO₃. Original magnification: 100×

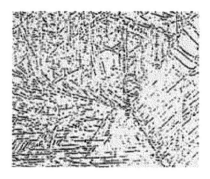

Fig. 7.14 Ti-6Al-4V plate. Widmanstätten alpha structure in plate rolled at 1040 °C (1900 °F) in the beta field. Etchant: 10%HF-5%HNO₃. Original magnification: 500×

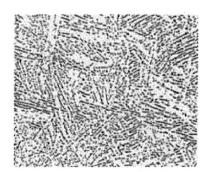

Fig. 7.15 Ti-6Al-4V plate. Partially broken-up Widmanstätten structure in plate rolled from 1020 °C (1870 °F) and annealed at 830 °C (1525 °F) for 1 h and furnace cooled to 595 °C (1100 °F), followed by air cooling. Etchant: 10%HF-5%HNO₃. Original magnification: 500×

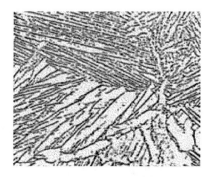

Fig. 7.16 Ti-5Al-2.5Sn billet. Course, platelike alpha in billet annealed in the beta field at 1040 °C (1900 °F) for 1 h and furnace cooled. Etchant: 10%HF-5%HNO₃. Original magnification: 100×

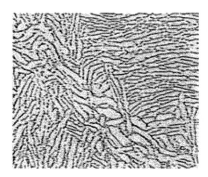

Fig. 7.17 Ti-6Al-4V billet. Platelike alpha and intergranular beta in central position of 205 mm (8 in.) diameter billet heated at 705 °C (1300 °F) for 2 h and air cooled. Etchant: 10%HF-5%HNO₃. Original magnification: 250×

tion and Heat Treatment of Titanium Alloys," in this book. These phases are reviewed briefly here with respect to their metallographic appearance. Of the three metastable phases (alpha prime, metastable beta, and omega), omega is the only one that cannot be detected using the light microscope.

Alpha prime, sometimes referred to as martensitic alpha, exists in two crystallographic forms. One form (alpha prime, or α′) has the hexagonal close-packed lattice while the other (alpha double-prime, or α″) is body-centered tetragonal. Little information is available on the nature of the α″ variant, but it is believed to be strain-induced martensite, which is produced in metastable structures by applying heat or stress.

The α′ phase is supersaturated alpha formed by a diffusionless shearlike transformation of the beta phase at or below the martensite start (M$_s$) temperature by rapid cooling beta-lean alloys. Figure 7.9 shows α′ in the Ti-6Al-4V alloy formed by water quenching the alloy from above the beta transus. Note the fine, needlelike structure, which is characteristic of the phase.

Because α′ is a nonequilibrium phase, aging at 480 to 590 °C (900 to 1100 °F) forms equilibrium alpha and beta that is enriched in solute. Tensile strength increases substantially with a slight loss in ductility. Aged α′ usually cannot be distinguished from unaged α′ with the light microscope.

Metastable beta is also a nonequilibrium phase. The phase occurs in alloys that contain sufficient amounts of beta-stabilizing elements to retain the beta phase following rapid cooling from the alpha-beta or beta fields. The presence of metastable beta is accompanied by low tensile strength and high bend ductility. Applying heat through aging treatments transforms the metastable beta to equilibrium alpha and beta that is

somewhat enriched in solute. (See Fig. 4.24 in Chapter 4, "Principles of Beta Transformation and Heat Treatment of Titanium Alloys," in this book.) When beta eutectoid elements are present in the alloy, aging of metastable beta transforms its equilibrium alpha and eutectoid products. The eutectoid products usually form after long aging times. These aging treatments are accompanied by high tensile strength and medium-to-low ductility.

To produce metastable beta, the alloy must contain a sufficient amount of beta-stabilizing element to depress the M$_s$ temperature below room temperature. If the amount of beta stabilizers in the alloy is low so that the M$_s$ temperature is above room temperature, or if the alloy is heated to temperatures high in the alpha-beta field so that the beta phase is lean in solute, α′ forms on quenching rather than metastable beta. This can be seen in Fig. 7.2.

If an alloy of composition A is heated to temperature T_1, the amount of beta stabilizer in the beta phase is composition B; on quenching, the M$_s$ is not encountered, and metastable beta is retained at room temperature. If the same alloy A is heated to temperature T_2, the alloy contains more beta and less alpha on a volume basis, but the beta is leaner in alloy content (composition C).

On quenching, the M$_s$ occurs at temperature T_3, and a mixture of alpha prime and metastable beta forms. Figure 7.18 shows a series of micrographs that illustrate this. The alloy is Ti-6Al-2Sn-4Zr-6Mo, an age-hardenable alpha-beta alloy. Figure 7.18(a) shows the alloy structure water-quenched from 870 °C (1600 °F). The microstructure consists of primary alpha in a metastable beta matrix.

The composition of the beta phase at d on the partial phase diagram in Fig. 7.18(d) is such that the M$_s$ does not occur on cooling, and metastable beta is retained. As the alloy is heated to 900 °C (1650 °F), the amount of primary alpha diminishes, while the volume percent of beta increases (Fig. 7.18b). As the alloy is heated to 925 °C (1700 °F), the amount of alpha continues to diminish, while the volume of beta increases. However, the composition of the beta phase (f) is somewhat leaner in alloy, and on quenching, the M$_s$ occurs at g and forms α′ (Fig. 7.18c).

As previously mentioned, aging metastable beta transforms it to equilibrium alpha or eutectoid products, or both. Figure 7.19 shows the Ti-6Al-4V alloy in the solution-treated and aged condition. The beta phase in the center of the figure contains a fine, needlelike structure, which is due to the precipitation of the fine equilibrium

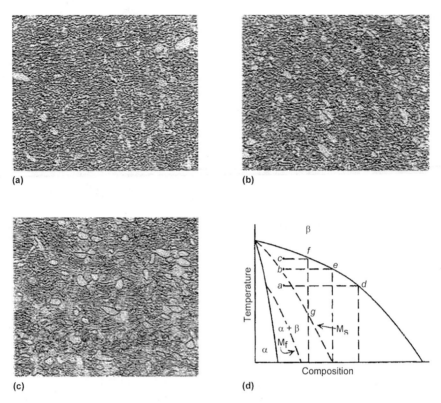

Fig. 7.18 Ti-6Al-2Sn-4Zr-6Mo billet. (a) Microstructure contains primary alpha in metastable beta matrix after heating the billet at 870 °C (1600 °F) for ½ h and water quenching. (b) Microstructure contains less primary alpha and more metastable beta in the matrix than in (a) after heating at 900 °C (1650 °F) for ½ h and water quenching. (c) Microstructure contains still less alpha than in (b) and more beta after heating at 925 °C (1700 °F) for 15 min, but the beta is now diluted and, after water quenching, alpha prime forms. (d) Partial phase diagram. Etchant: 10%HF-5%HNO₃. Original magnification: 500×

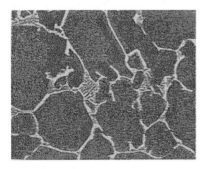

Fig. 7.19 Ti-6Al-4V billet. SEM micrograph shows the precipitation of fine, needlelike alpha phase in the metastable beta phase. Material was heated at 845 °C (1550 °F) for 1 h and water quenched, followed by aging at 540 °C (1000 °F) for 8 h and air cooled. Etchant: 10%HF-5%HNO₃. Original magnification: 2000×

alpha phase during the aging treatment. Eutectoid products are not present because vanadium is not a eutectoid former.

Many new all-beta or metastable beta alloys have been developed. Most of these alloys are available commercially. These alloys contain large amounts of beta-stabilizing elements, enabling them to retain an all-beta structure at room temperature after cooling from above the beta transus. The alloy contains so much beta stabilizer that an all-beta structure (metastable) can be retained by both air cooling and water quenching. Thus, the terms *solution treating* and *annealing* are synonymous for this alloy.

The metastable condition of the beta phase makes it possible to age the Ti-3Al-8V-6Cr-4Mo-4Zr alloy to high strength levels. Strengthening occurs through transformation of the beta phase to equilibrium alpha, beta, and (with sufficient aging time) TiCr₂. Figure 7.20(a) shows the metastable beta condition, and Fig. 7.20(b) the aged condition. It is unlikely that TiCr₂ is present, due to the short aging treatment. Aging for 90 to 100 h is necessary to form TiCr₂.

Omega Phase. The omega phase, sometimes referred to as beta prime, is submicroscopic (cannot be detected using the light microscope). The phase is associated with high hardness and brittle-

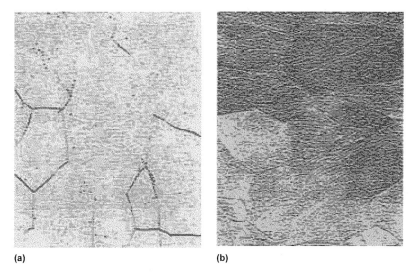

Fig. 7.20 Beta-C (Ti-3Al-8V-6Cr-4Mo-4Zr) sheet. (a) Structure shows metastable beta after heating at 980 °C (1800 °F) for 5 min and water quenching. (b) The sample in (a) was aged at 525 °C (975 °F) for 8 h and air cooled, resulting in a structure consisting of equilibrium alpha and beta. Etchant: 10%HF-5%HNO₃. Original magnification: 100×

ness (Ref 7.3). The reaction occurs during aging of certain metastable beta alloys at relatively low temperatures. During the initial stages of aging, metastable beta transforms to omega, leaving the remaining beta slightly enriched in solute.

As aging proceeds, the omega transforms to equilibrium alpha, with the remaining beta highly enriched in beta-stabilizing elements. In eutectoid-forming alloys, the reaction proceeds to alpha plus an intermetallic compound. When aluminum and tin are present in an alpha-beta alloy, the omega reaction is greatly suppressed. These elements prevent omega formation by increasing the sluggishness of the reaction.

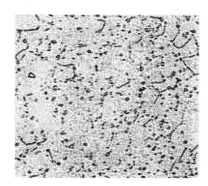

Fig. 7.21 Ti-5Al-2.5Sn sheet. Equiaxed alpha grains and beta spheroids produced by heating at 845 °C (1550 °F) for ½ h and cooling in air. Etchant: 10%HF-5%HNO₃. Original magnification: 250×

Related Terms

Spheroidal Beta. Small amounts of beta phase are common in unalloyed titanium and some alpha alloys. The phase appears spheroidal, and it is stabilized by small amounts of iron present in the material. Because iron is a strong beta stabilizer, small amounts of the element will retain beta. Figure 7.6 shows the phase in unalloyed titanium containing 0.3% Fe. Figure 7.21 shows the phase in the Ti-5Al-2.5Sn alpha alloy containing 0.35% Fe.

The beta phase in these grades maintains a fine alpha grain size by inhibiting grain-boundary migration. The presence of some beta phase also improves the formability and increases the hydrogen solubility limit of the alloy. However, iron is somewhat detrimental to corrosion resistance in certain environments.

Intergranular beta occurs between the alpha grains in alpha-beta alloys in which the alpha phase forms the matrix or continuous phase. Figure 7.22 shows an annealed Ti-8Al-1Mo-1V alloy in which alpha is the matrix and the beta phase is situated between the alpha grains.

Alpha Case. Because of the high diffusion rate of gases in titanium, the metal is readily contaminated by oxygen when heated in air or oxidizing atmospheres. The term *alpha case* describes the oxygen-enriched, alpha-stabilized surface that results from air contamination at elevated temperatures. Figure 7.23 shows the Ti-6Al-4V alloy with an alpha case 0.3 mm (0.012 in.) deep formed by heating in air at 1095 °C (2000 °F) for 2 h.

Oxygen contamination may be more difficult to detect metallographically in unalloyed titanium or alpha alloys. Oxygen is an alpha stabilizer. Because of its high solubility in the alpha phase, the contamination cannot always be observed metallographically. In such instances, microindentation hardness and bend-test methods are used. However, alpha case is readily observed in alpha, alpha + beta, and beta alloys processed in the alpha-beta and beta phase fields.

Because the rate of diffusion of oxygen in titanium is a function of temperature and time, it is desirable that minimum hot working temperatures be used and excessive heating time be avoided to minimize oxygen contamination.

Intermetallic Phases (Compounds). In addition to alpha and beta phases, intermetallic phases (sometimes called compounds) are occasionally present in terminal titanium alloys. Intermetallic phases form when solid solubility limits are exceeded and when transformation to eutectoid products occurs (Ref 7.1–7.4). Figure 7.24 illustrates intermetallic phases in a Ti-5Ni alloy. Ti_2Ni (dark phase) was identified by x-ray diffraction analysis. Figure 7.25 shows a fine dispersion of titanium silicide (Ti_5Si_3) in the Ti-679 alloy. The intermetallic phase, evident both within the alpha grains and at the grain boundaries, enhances the strength of the alloy at elevated temperatures.

Figures 7.26 and 7.27 illustrate titanium hydride in unalloyed titanium and the Ti-5Al-2.5Sn alloy. The titanium hydride appears as very fine needles.

Matrix. In titanium alloys, either the alpha or the beta phase can be the matrix, depending on alloy composition and thermal treatment. The matrix is the continuous phase, and, in most instances, it is the principal constituent in the struc-

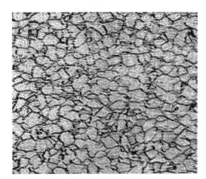

Fig. 7.22 Ti-8Al-1Mo-1V bar. Intergranular beta appears between alpha grains after heating at 1010 °C (1850 °F) for 1 h and furnace cooling to 595 °C (1100 °F), followed by air cooling. Etchant: 2%HF-4%HNO₃. Original magnification: 250×

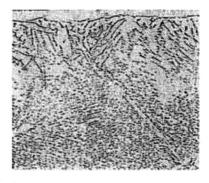

Fig. 7.23 Ti-6Al-4V plate. Alpha case 0.30 mm (0.012 in.) deep formed by heating at 1095 °C (2000 °F) for 2 h and air cooling. Etchant: 10% HF. Original magnification: 75×

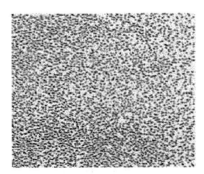

Fig. 7.24 Ti-5Ni sheet. Alpha and Ti2Ni (dark phase) are present after heating at 730 °C (1350 °F) for ½ h and air cooling. Etchant: 2%HF-4%HNO₃. Original magnification: 500×

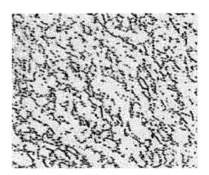

Fig. 7.25 Ti-11Sn-5Zr-2.5Al-1Mo-0.2Si bar. Primary alpha, transformed beta, and fine dispersions of titanium silicide after heating at 900 °C (1650 °F) for 1 h and air cooling, followed by reheating at 500 °C (930 °F) for 24 h and air cooling. Etchant: 10%HF-5%HNO₃. Original magnification: 1000×

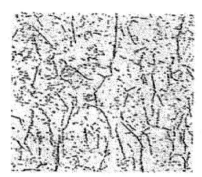

Fig. 7.26 Pure titanium tubing. Microstructure consists of alpha grains and titanium hydride (525 ppm). Etchant: 2%HF-4%HNO$_3$. Original magnification: 250×

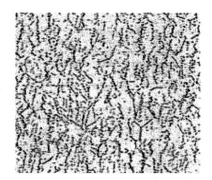

Fig. 7.27 Ti-5Al-2.5Sn alloy. Alpha grains and titanium hydride (655 ppm) appear after heating at 815 °C (1500 °F) for ½ h and air cooling, followed by hydrogenating in a sodium hydride bath at 400 °C (750 °F) for 4 h. Etchant: 10%HF-25%HNO$_3$, 45%glycerine-20%H$_2$O. Original magnification: 500×

ture. However, there are instances when the matrix phase is the minor constituent. The prior history of the material being examined must be known so the matrix phase can be established. Typically, alpha is the matrix in alpha alloys and in weakly beta-stabilized systems. In alloys containing large amounts of beta stabilizers, beta is usually the matrix. Figure 7.22 illustrates an example in which alpha is the matrix. In Fig. 7.28, beta is the matrix.

Ordered Intermetallic Compounds

Typical substitutional solid solutions have solute atoms randomly distributed at the lattice sites of the solvent. In certain alloy systems, this is true only at elevated temperatures. As these systems cool to some critical temperature, the solute atoms take up an orderly periodic arrangement (basically, the different atoms show a preference for their next-door neighbors) on the lattice sites of the solvent (Ref 7.3, 7.5, 7.6). This arrangement is not continuous initially and does not occur throughout the entire structure. This condition is termed short-range order. As the alloy cools further, the ordered arrangement progresses until it extends throughout the entire structure, a condition known as long-range order.

Because ordered structures result in a definite ratio of solvent to solute atoms, their composition can be expressed in a simple chemical formula (Ti$_3$Al, TiAl). However, it should be noted that the ordered structure is still a solid solution rather than an intermetallic phase. The basic difference is that intermetallic phases generally melt at a constant temperature. An ordered structure, on

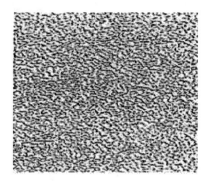

Fig. 7.28 Ti-8Mn sheet. Alpha appears in a beta matrix after heating at 650 °C (1200 °F) for 1 h, furnace cooling to 480 °C (900 °F), and air cooling. Etchant: 2%HF-4%HNO$_3$. Original magnification: 500×

heating to a critical temperature, becomes disordered and melts as an alloy over a range of temperatures.

Alloys based on the α$_2$ (Ti$_3$Al) and γ (TiAl) ordered structure are termed titanium aluminides (Ref 7.5, 7.6). Because of the attractive high-temperature properties of these ordered structures, the α$_2$ and γ alloy systems are receiving a large amount of interest. In these systems, it is normal to refer to the amount of the elements present in atomic percentage (at.%) rather than the convention for normal terminal alloys, where the amount of each element is given in weight percentage (wt%).

Metallographic examination of the α$_2$ alloy system indicates that microstructures obtained after heat treating can be predicted based on a knowledge of the more conventional titanium alloy systems. Typical microstructures of the Ti-14Al-21Nb alloy are shown in Fig. 7.29 (Ref

7.6). The Ti-14Al-21Nb (α_2 aluminide) is capable of producing an equiaxed Ti_3Al phase and an acicular Ti_3Al + beta phase when processed below, but near, its beta transus. This structure resembles that of Ti-6Al-4V when processed in a similar manner.

Figure 7.30 shows the four basic microstructures developed in the TiAl (γ) intermetallic (Ref 7.5). The fully lamellar (FL) microstructure is generated when cooled from an alpha treatment at a formation temperature (well below the alpha transus), which is a function of cooling rate and alloy composition. Typical cooling rates range 10 to 50 °C (18 to 90 °F) per minute. The near-lamellar (NL) microstructure is formed upon annealing at a temperature just below the alpha transus (15 to 30 °C, or 27 to 54 °F), followed by cooling at 10 to 100 °C (18 to 180 °F) per minute. The duplex (DM) microstructure is formed upon annealing at and cooling from an alpha + gamma two-phase field temperature, where the alpha:gamma volume ratio is 0.4 to 0.6. The near-gamma (NG) microstructure is generated upon annealing near the eutectic temperature (1125 °C, or 2057 °F), where gamma/alpha or α_2 volume ratio is the greatest. This microstructure is dominant, with medium-sized (10 to 50 μm) gamma grains, which appear to be single phase (Ref 7.5).

Effect of Fabrication and Thermal Treatment on Microstructure

Microstructural changes due to fabrication and subsequent heat treatments are numerous and varied, depending on alloy content and prior working history. If working is initiated and completed at temperatures in the beta field, the resulting micro-

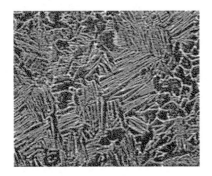

Fig. 7.29 Ti-14Al-21Nb bar. SEM micrograph shows equiaxed Ti_3Al phase and an acicular Ti_3Al + beta phase after processing near, but below, the beta transus. Etchant: 10%HF-5%HNO$_3$. Original magnification: 1000×

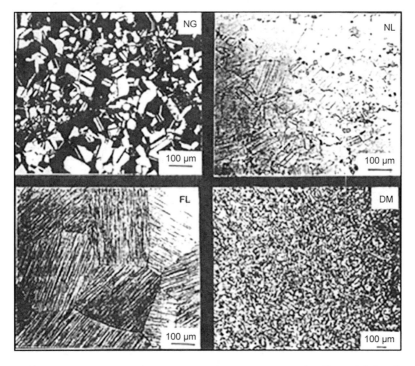

Fig. 7.30 Variety of microstructures in Ti-47.5Al-2.5V-1Cr (γ). NG, near gamma; NL, near lamellar; FL, fully lamellar; DM, duplex. See text for details on microstructure development.

structure is entirely transformed. The transformation product consists of acicular or platelike alpha, depending on section size and cooling rate. The structure also shows evidence of coarse, equiaxed prior-beta grains, as shown in Fig. 7.31(a), which is further evidence that fabrication was completed in the beta field. Structures developed in this manner are not changed significantly through the use of subsequent treatments. Fine acicular structures can be coarsened somewhat by heating in the alpha-beta field, but the coarse prior-beta structure can only be altered by further mechanical working at temperatures below the beta transus.

If working is initiated in the beta field and completed at some temperature in the alpha-beta field, the resulting structure is predominantly transformed beta. Prior-beta grain boundaries are distorted and partially broken up due to the lower finishing temperature, as shown in Fig. 7.31(a).

Work carried out entirely in the alpha-beta field or in the alpha field (as in an alpha alloy) yields a fine-grained structure with little or no transformation product. Figures 7.31 through 7.33 show the effect of forging temperature and reduction on the microstructure of several titanium alloys.

Figure 7.31 shows the effect of a 50% forging reduction on the Ti-6Al-4V alloy from three different temperatures. Figure 7.32 shows the effect of a 75% forging reduction on the same alloy from the same temperatures. In comparing Fig. 7.31(a) and 7.32(a), note the 25% greater reduction resulted in a partial breakup of the prior-beta grains. The greater reduction from the lower temperature also produced a finer-grained structure, as shown in Fig. 7.32(b) and (c). An equivalent reduction at a still-lower temperature (900 °C, or 1650 °F) yields an even finer alpha-beta structure. Figure 7.33 shows similar results on the Ti-8Al-1Mo-1V alloy.

The microstructural changes brought about through heat treatment are also numerous. Figure 7.34 shows the effect of heating Ti-6Al-4V to four different temperatures and cooling at three

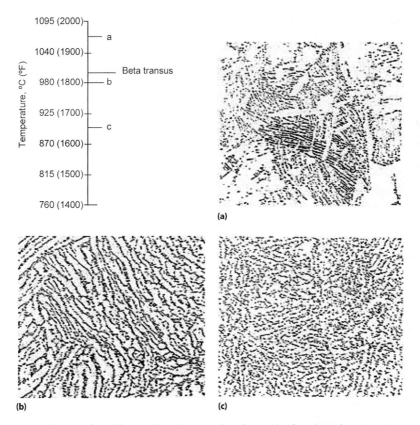

(a)

(b)

(c)

Fig. 7.31 (a) Acicular alpha (transformed beta) and prior-beta grain boundaries in bar forged 50% from 1065 °C (1950 °F), reheated at 730 °C (1350 °F) for 2 h, and air cooled. (b) Platelike and equiaxed alpha with some beta present in bar forged 50% from 980 °C (1800 °F), reheated at 730 °C (1350 °F) for 2 h, and air cooled. (c) Platelike alpha-beta structure in bar forged 50% from 900 °C (1650 °F), reheated at 730 °C (1350 °F) for 2 h, and air cooled.

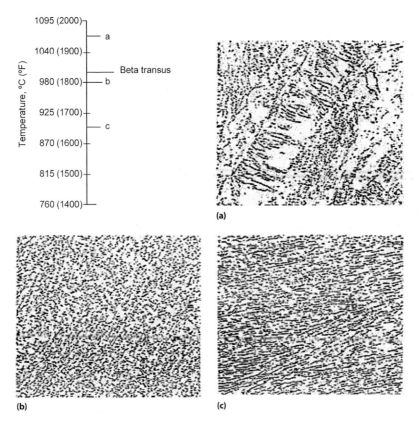

Fig. 7.32 Ti-6Al-4V bar. (a) Acicular alpha (transformed beta) and prior-beta grain boundaries in bar forged 75% from 1065 °C (1950 °F), reheated at 730 °C (1350 °F) for 2 h, and air cooled. (b) Platelike and equiaxed alpha with some beta present in bar forged 75% from 980 °C (1800 °F), reheated at 730 °C (1350 °F) for 2 h, and air cooled. (c) Platelike alpha + beta structure in bar forged 75% from 900 °C (1650 °F), reheated at 730 °C (1350 °F) for 2 h, and air cooled. Etchant: 10%HF-4%HNO₃. Original magnification: 250×

different rates from each temperature. The figure includes tensile properties for each annealing treatment and properties following a 4 h aging treatment at 540 °C (1000 °F).

Properties after aging provide some information about the phases present. Little, if any, change in properties can be expected when phases are in a nearly equilibrium condition prior to aging. Nonequilibrium phases, such as alpha prime and metastable beta, substantially increase ultimate tensile and yield strength properties after aging. Figure 7.34 shows that no response to aging occurs on furnace cooling from solution temperatures.

There is only a slight response on air cooling, while the greatest response is experienced with a water quench from the solution temperature. Good response to aging even occurs on water quenching from the beta field, but ductility values are quite low. The best combination of properties is produced by solution treating at temperatures relatively high in the alpha-beta field.

Metallographic Specimen Preparation

Procedures for metallographic preparation of titanium and titanium alloys vary somewhat from laboratory to laboratory. The preference of one procedure over another can usually be left to the discretion of the metallographer, but care must be exercised to avoid conditions that can lead to misinterpretation. The appendix in this chapter contains a brief outline of metallographic procedures, including a list of etchants.

Summary

This chapter reviewed the crystal structures of titanium found in the three principal terminal alloy types and titanium aluminides. The study of titanium microstructures is carried out primarily using the light microscope. Examples of microstructures resulting from fabrication and various

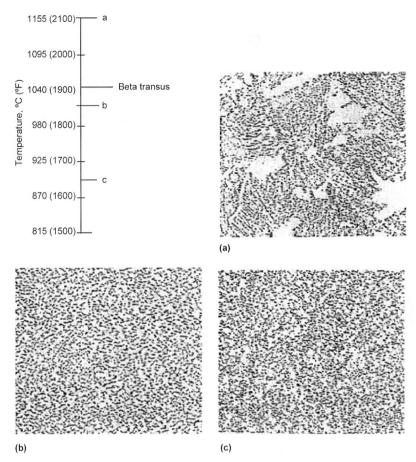

Fig. 7.33 Ti-8Al-1Mo-1V bar. (a) Acicular alpha and prior-beta grain boundaries in bar forged 70% from 1150 °C (2100 °F), reheated at 790 °C (1450 °F) for 8 h, and furnace cooled. (b) Equiaxed alpha and intergranular beta in bar forged 70% from 1010 °C (1850 °F), reheated at 790 °C (1450 °F) for 8 h, and furnace cooled. (c) Fine alpha-beta structure in bar forged 70% from 900 °C (1650 °F), reheated at 790 °C (1450 °F) for 8 h, and furnace cooled. Etchant: 10%HF-5%HNO₃. Original magnification: 200×

thermal treatments are presented in micrographs accompanying the text. These micrographs will help in identifying most typical titanium and titanium alloy microstructures. Sample-preparation techniques critical to the successful study of titanium microstructures are described in the appendix.

Glossary

acicular alpha. A fine, needlelike transformation product brought about through nucleation and growth.

alpha. The low-temperature allotrope of titanium with a hexagonal close-packed crystal structure.

alpha-beta structure. A microstructure that contains both alpha and beta as the principal phases.

alpha case. The oxygen-enriched alpha-stabilized surface that results from air contamination at elevated temperatures.

alpha prime. Also known as martensitic alpha. A supersaturated, nonequilibrium phase formed by a diffusionless transformation of the beta phase, which is lean in solute.

alpha stabilizer. An alloying element that dissolves preferentially in the alpha phase and raises the alpha-beta transformation temperature.

alpha transus. The temperature that designates the phase boundary between the alpha and alpha plus beta fields.

alpha 2. Intermetallic Ti₃Al ordered phase.

beta. The high-temperature allotrope of titanium with a body-centered cubic crystal structure.

beta eutectoid. Beta-stabilizer alloying elements resulting in the decomposition of beta to eutectoid products such as alpha and intermetallic compounds.

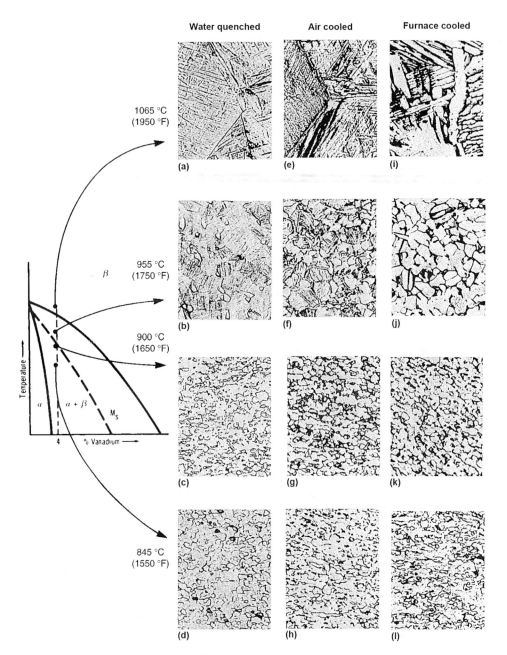

Fig. 7.34 Ti-6Al-4V bar. (a) α′ + β; prior beta grain boundaries. (b) Primary α and α′ + β. (c) Primary α and α′ + β. (d) Primary α and metastable β. (e) Acicular α + β; prior beta grain boundaries. (f) Primary α and acicular α + β. (g) Primary α and acicular α + β. (h) Primary α and β. (i) Plate-like α + β; prior grain boundaries. (j) Equiaxed α and intergranular β. (k) Equiaxed α and intergranular β. (l) Equiaxed α and intergranular β. Etchant: 10 HF, 5HNO₃, 85H₂O. 250×

beta isomorphous. Beta-stabilizer alloying elements, which are completely miscible in the beta phase.

beta stabilizer. An alloying element that dissolves preferentially in the beta phase and lowers the beta transformation temperature. Such elements promote the retention of beta at room temperature.

beta transus. The temperature designating the phase boundary between the alpha-beta and beta fields.

elongated alpha. A fibrouslike structure brought about by unidirectional fabrication.

equiaxed structure. A polygonal structure having approximately equal dimensions in all directions.

gamma. Intermetallic TiAl ordered phase.

hydride phase. The phase TiH_2 formed in titanium when the hydrogen content exceeds the solubility limit.

intergranular beta. Beta situated between alpha grains. Intermetallic compounds. An intermediate phase in an alloy system that has a narrow solubility range.

interstitial element. An element with a relatively small atomic size that can assume position in the interstices of the titanium lattice.

M_f. Martensite finish temperature. The temperature at which the martensite reaction is 100% complete.

M_s. Martensite start temperature. The maximum temperature at which alpha prime begins to form on cooling of the beta phase.

matrix. The constituent forming the continuous phase of a two-phase microstructure.

metastable beta. A nonequilibrium phase that can be transformed to alpha and eutectoid products by heat and stress.

morphology. A study of form and structure.

omega phase. Also referred to as beta prime. A nonequilibrium, submicroscopic phase that forms during the nucleation and growth transformation of beta to alpha. This phase occurs at lower temperatures than the alpha phase.

ordered structure. The orderly, or periodic, arrangement of solute atoms on the lattice sites of the solvent.

phase. A physically homogeneous and distinct portion of a material system.

platelike alpha. Alpha phase forming along preferred planes of beta during transformation of beta to alpha. Platelike alpha is characterized by relatively long, wide grains.

primary alpha. Alpha that remains untransformed as titanium is heated into the alpha-beta field.

prior-beta grain size. The grain size of the beta phase prior to partial transformation back to alpha.

serrated grain. Alpha phase characterized by an irregular phase size and jagged grain boundaries.

spheroidal structure. Phases having a circular or globular appearance.

substitutional elements. Alloying elements with a similar atom size that can replace or substitute for the titanium atoms in the titanium lattice.

transformed beta. Alpha phase formed by transformation from the beta phase.

transition elements. Elements containing valence electrons in more than one electron shell.

Widmanstätten structure. A structure brought about by the formation of a new phase along preferred crystallographic planes of the prior phase. The Widmanstätten structure is a transformation product of the beta phase.

Appendix—Metallographic Preparation

Step 1: Sectioning. Care should be exercised during sectioning and cutting of metallographic samples so the specimen is not overheated. Abrasive cutting can be used if adequate coolant is provided during the cutting operation. Hacksawing is recommended for small samples where heating effects can be a problem.

Step 2: Mounting. Titanium can be mounted in several molding materials. The most common are Bakelite, transoptic, and diallyl phthalate. Diallyl phthalate is recommended when edge preservation is important. The temperatures encountered in using these materials generally are not a problem. If heating is of concern, samples can be mounted in a room-temperature self-curing plastic material, or polyester resin. Specimens are usually left unmounted if they will be electropolished. However, if mounting is necessary, conductive mounting materials such as Olsen resin No. 2 are available. When using these resins, a small amount of Bakelite can be added to the mold first, because the electrolyte attacks the Olsen resin during electropolishing.

Step 3: Grinding. Preliminary grinding of titanium is similar to that of other metals. Either silicon carbide or emery papers can be used. Specimens are ground on successive grades of paper starting with No. 3 emery or 180-grit silicon carbide through 3/0 emery or 600-grit silicon carbide. The specimen should be rotated 45° to the previous grinding lines when proceeding to the next-finer-grit abrasive paper. Wet grinding is preferred. However, dry grinding can be used if care is exercised to avoid overheating of the specimen.

Step 4: Polishing. Polishing can be accomplished using electrolytic, mechanical, and vibratory methods.

Electropolishing is carried out directly from the final grinding stage and is considerably faster than mechanical or vibratory processes. The electrolyte typically consists of 590 mL methyl alcohol, 350 mL ethylene glycol, and 60 mL perchloric acid. Polishing time is usually 15 to 25 s at a current density of 1 to 112 A/cm. Polishing time and amperage vary somewhat with specimen size and polishing area. The electrolyte with a low concentration of perchloric acid is nonexplosive and can be stored safely for several weeks. However, *care should be exercised in handling perchloric acid, because it can react explosively with organic materials.*

Mechanical Polishing. Satisfactory results can also be achieved using mechanical polishing techniques. The mechanical polishing process is a two-step operation and is carried out as follows:

1. Preliminary polishing on red felt or billiard cloth using AB polishing alumina No. 1 (5.0 μm levigated alumina) or 6 μm diamond paste on a wheel
2. Final polishing on microcloth using AB gamma polishing alumina No. 3 (0.05 μm levigated alumina). Final polishing should consist of several iterations of polishing and etching to ensure removal of disturbed metal. Adding 0.5% HF to the polishing lubricant is sometimes helpful in the final polishing stage.

Vibratory polishing, although a slower process than electrolytic or mechanical polishing, produces good results. The vibratory process is also a two-step operation and is carried out as follows:

1. Preliminary polishing for 2 to 4 h on canvas cloth using a 5.0 μm slurry of levigated alumina
2. Final polishing for 2 to 4 h on microcloth with a 0.05 μm slurry of levigated alumina.

Unalloyed titanium has a tendency to smear in the vibratory process, and several short polishing and etching iterations should be used to remove disturbed metal.

Step 5: Etching. Etching can be accomplished by swabbing or immersing. Swabbing is preferred when mechanical or vibratory polishing is used. Etching times are usually short, 3 to 20 s, depending on the etchant and the alloy being etched. Table 7.2 lists a number of etchants for titanium and titanium alloys. Nearly all etchants contain some hydrofluoric acid and an oxidizing agent. The 10%HF-5%HNO$_3$ etchant is most widely used to reveal general structure.

REFERENCES

7.1. J.L. Murray, *Phase Diagrams of Binary Titanium Alloys,* ASM International, Metals Park, OH, 1987
7.2. *Titanium Technology: Present Status and Future Trends,* Titanium Development Association, Dayton, OH, 1985
7.3. G. Lutjering and J.C. Williams, *Titanium,* Springer, 2003
7.4. R. Boyer, E.W. Collings, and G. Welsch, Ed., *Materials Properties Handbook: Titanium Alloys,* ASM International, 1994

Table 7.2 Etchants for titanium and titanium alloys

Alloy	Etchant	Function
Unalloyed Ti and most Ti alloys	10 mL HF 5 mL HNO$_3$ 85 mL H$_2$O	Reveals general structure
Unalloyed Ti and most Ti alloys	10 mL HF 30 mL HNO$_3$ 50 mL H$_2$O	Reveals general structure
Ti-8Mn	2 mL HF	Reveals general structure
Ti-13V-11Cr-3Al (aged) and other beta alloys	4 mL HNO$_3$ 94 mL H$_2$O	Reveals general structure, removes stain
All Ti and Ti alloys	1 mL HF 2 mL HNO$_3$ 50 mL H$_2$O$_2$ 20 mL H$_2$O	...
Ti-6Al-6V-2Sn	10 mL 40% KOH 5 mL H$_2$O$_2$ 20 mL H$_2$O	Stains alpha and transformed beta; retained beta remains white
Ti-7Al-12Zr	18.5 g benzalkonium chloride 35 mL ethanol 40 mL glycerine 25 mL HF	Reveals general structure
Most alloys	2 mL HF 98 mL H$_2$O	Reveals surface contamination (oxygen)
Ti-5Al-2.5Sn	10 mL HF 25 mL HNO$_3$ 45 mL glycerine	Reveals hydrides

7.5. Y.-W. Kim, *Acta Metall. Mater.,* Vol 40 (No. 6), 1992, p 1121–1134

7.6. H.A. Lipsitt et al., The Deformation and Fracture of Ti$_3$Al at Elevated Temperatures, *Metall. Trans. A,* Vol 11, 1980, p 1369–1375

SELECTED REFERENCES

- S. Abkowitz, The Emergence of the Titanium Industry and the Development of the Ti-6Al-4V Alloy, *JOM Monograph Series,* Vol 1, TMS, Warrendale, PA, 1999
- R. Boyer, E.W. Collings, and G. Welsch, Ed., *Materials Properties Handbook: Titanium Alloys,* ASM International, 1994
- M.J. Donachi, *Titanium: A Technical Guide,* 2nd ed., ASM International, 2000
- *ISTFA 2013: Proceedings from the 39th International Symposium for Testing and Failure Analysis,* ASM International, 2013
- E.H. Kraft, "Opportunities for Low Cost Titanium in Reduced Fuel Consumption, Improved Emissions, and Enhanced Durability Heavy-Duty Vehicles," Oak Ridge National Laboratory, July 2002
- E.H. Kraft, "Summary of Emerging Titanium Cost Reduction Technologies," Oak Ridge National Laboratory, Dec 2003
- P. Lacombe, R. Tricot, and G. Beranger, Ed., *Proceedings of the Sixth International Conference on Titanium* (Nice, France), Metallurgical Society of AIME, Warrendale, PA, 1988
- G. Lutjering and J.C. Williams, *Titanium,* Springer, 2003
- G. Lutjering, U. Zwicker, and W. Bunk, *Proceedings of the Fifth International Conference on Titanium,* Metallurgical Society of AIME, Pittsburgh, PA, 1984
- *Mechanical Testing and Evaluation,* Vol 8, *ASM Handbook,* ASM International, 2000
- *Proceedings of the Seventh International Conference on Titanium* (San Diego, CA), Metallurgical Society of AIME, Warrendale, PA, 1992
- *Proceedings of the Eighth International Conference on Titanium* (San Birmingham, U.K.), Metallurgical Society of AIME, Warrendale, PA, 1995
- *Proceedings of the Ninth International Conference on Titanium,* Metallurgical Society of AIME, Warrendale, PA, 2000
- *Proceedings of the Tenth International Conference on Titanium,* Metallurgical Society of AIME, Warrendale, PA, 2004
- *Proceedings of the Eleventh International Conference on Titanium,* Metallurgical Society of AIME, Warrendale, PA, 2008
- *Proceedings of the Twelfth International Conference on Titanium,* Metallurgical Society of AIME, Warrendale, PA, 2012

CHAPTER 8

Melting, Casting, and Powder Metallurgy*

THIS CHAPTER DISCUSSES techniques for melting and casting of titanium and its alloys and describes applications of powder metallurgy (PM) to these engineering metals (Ref 8.1–8.3). The decrease in machining required for near-net shapes such as castings and PM components (and consequent reduced-cost components) is discussed in Chapter 1, "History and Extractive Metallurgy," in this book.

Melting

The consolidation of titanium sponge to a solid article presented very difficult engineering challenges during the early development of this industry in the late 1940s (Ref 8.1–8.3). There were no facilities available at that time for melting the reactive metals without severe contamination. Liquid titanium reacts rapidly with or dissolves all solids, liquids, and gases except the inert gases, including argon and helium.

The first serious efforts to consolidate titanium were made by William J. Kroll, the inventor of the magnesium-reduction process bearing his name. Two basic methods were studied by Kroll, the U.S. Bureau of Mines, and private industries. One method used PM techniques for cold compaction and sintering. These steps were followed, in some cases, by hot or cold working to produce small amounts of metal products suitable for inspection and testing. However, such products had serious shortcomings that could not be corrected at that time, despite major efforts. Residual magnesium chloride ($MgCl_2$) salts entrapped within the metal particles impaired the mechanical properties and weldability. High hydrogen levels also lowered property values.

The other method explored by Kroll was an electric arc furnace invented by Werner von Bolton for melting tantalum. This small furnace had three particularly important features: a water-cooled copper hearth, a consumable electrode, and a vacuum system. Kroll modified the furnace to use a nonconsumable thoriated-tungsten electrode. He also replaced the hearth with a small, water-cooled crucible. These initial efforts, with numerous improvements by others, eventually led to the consumable electrode vacuum arc remelt (VAR) electric furnaces.

Large furnaces of this type serve almost exclusively for producing high-quality titanium, zirconium, and other active and refractory metals, some superalloys, and a few steels.

Conventional Vacuum Arc Remelting

Furnace Designs. The vacuum system used in the early furnace was essential to avoid air contamination. It also had another benefit by removing troublesome hydrogen and chloride salts. Water-cooled copper hearths and crucibles provided a convenient and practical container for the aggressive molten metal. As long as the copper is cooled properly, molten metal freezes quickly on its surface and forms a layer of solid titanium or "skull," which contains the molten metal. Such crucibles are not damaged, and titanium and other metals are not contaminated significantly with copper.

Other changes soon followed. To avoid contamination, nonconsumables were replaced with consumable electrodes having the desired ingot

---

*Adapted and revised from F.H. Froes, Daniel Eylon, and Howard B. Bomberger, *Titanium and Its Alloys*, ASM International.

composition. The need for better alloy homogeneity and metal degassing led to remelting ingots one or two more times using consumable electrodes. This is done by using correspondingly larger crucibles to provide clearances of 50 to 80 mm (2 to 3 in.) or more. However, remelt electrodes and furnaces are first cleaned carefully to remove foreign materials, including heavily oxidized metal and salts. Salts in the furnace and on the electrodes are hygroscopic and interfere with pumpdowns. These materials can add unwanted oxygen to the melt.

Figure 8.1 illustrates the basic requirements of a modern VAR, or consumable electrode vacuum arc furnace. The main features include a vacuum system, a consumable electrode supported by a drive mechanism, a copper crucible placed within a water-cooling jacket, a direct-current power supply, and both automatic and manual controls (not shown). With the negative polarity on the electrode, a greater amount of the energy (approximately 65%) heats the pool, providing good ingot and surface quality.

An electromagnetic coil (not shown) usually installed around the water jacket to control the motion of the arc and the molten pool serves three important functions: minimize arc damage to the crucible, mildly stir or mix the pool, and provide the desired uniform conditions for solidification.

The last two effects are important for good product quality. Furnaces are also designed to avoid extraneous magnetic fields from power leads, steel structures, and other electric furnaces and equipment in the area. Such equipment tends to distort the natural magnetic field from the melting current and can seriously interfere with arc and pool control. Adjoining furnaces "talk" to each other unless they are shielded. Such behavior can be observed on furnace-viewing systems.

Good arc control is also important for safety. At the high power levels used, a crucible can rupture quickly if the arc dwells on its surface. Subsequent steam and hydrogen explosions can cause extensive damage. Additional safety equipment includes furnace-viewing systems, automatic shutdown equipment, barricades, and explosion ports. The sodium-potassium alloy eutectic is used in the United Kingdom as a coolant instead of water, in part because of safety concerns.

Major advancements in automatic electronic controls provide near-optimum, reproducible conditions throughout the melt cycle. Direct benefits include better surfaces and internal quality of ingots, improved metal yields, more effective hot topping, and more efficient melting. Sensitive load cells continuously monitor electrode weight and provide useful information on melt rate and hot topping. Better arc-gap control results in improved surfaces, internal quality, and safety.

Furnaces as described are used routinely to produce titanium and titanium alloy ingots 700 mm to 1 m (30 to 40 in.) in diameter and weighing 4,500 to 14,000 kg (10,000 to 30,000 lb). A few furnaces are used to melt 1.2 m (48 in.) diameter ingots weighing 20,000 kg (40,000 lb), and a few 1.5 m (60 in.) ingots have been melted weighing 27,000 kg (60,000 lb).

Melting furnaces used in titanium casting shops have the same essential features. Most, but not all, use consumable electrodes and relatively small crucibles. Melts range from approximately 100 to less than 1500 kg (a few hundred to less than 3000 lb) of molten metal. A tilting mechanism allows the molten metal to be poured into casting molds.

Melt Shop Practices. Although titanium melt practices are well established, differences exist in the equipment used, in the methods used for handling materials, in operating parameters, in ingot sizes, in number of remelts, in alloys produced, and in product forms. The most common steps in a titanium melt process involve selection and blending of raw materials, pressing the mixture into blocks, welding the blocks and scrap into electrodes, and melting.

Melt Charges. Production of good-quality products requires the use of clean, uniform, and well-characterized titanium sponge and other raw

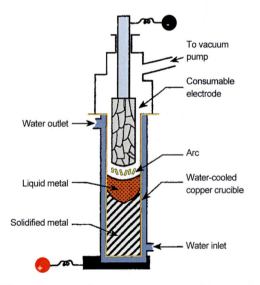

Fig. 8.1 Diagram for a vacuum arc consumable electrode furnace

materials free from harmful inclusions. Alloying elements that melt and dissolve readily (e.g., aluminum, chromium, copper, iron, manganese, tin, and zirconium) can be added in elemental forms, but silicon and the more refractory metals (e.g., molybdenum, niobium, tantalum, and vanadium) are conveniently and economically added as aluminum-reduced master alloys. The aluminum content of such binary and ternary alloys is generally 6% (by weight) or less. Oxygen is often added as well-dispersed TiO_2 powder. Such melt charges are most commonly blended, with or without granular scrap, and are then compacted into blocks by pressing. Block sizes, shapes, and densities vary with local preferences and pressing facilities. Shapes include quarter-round rectangles and partially and fully round cylinders. Common weights range from approximately 20 to 60 kg (50 to 150 lb). Electrodes are assembled from the pressed blocks, usually with appreciable amounts of clean bulk scrap. However, if large amounts of scrap are used, some adjustment may be required in block composition to obtain an acceptable ingot composition. For example, if the oxygen content of the scrap is undesirably high, it can usually be diluted to an acceptable level by increasing the amount of lower-oxygen sponge. Generally the electrodes, based on block and bulk scrap shapes and densities, are designed and assembled to achieve a maximum fill ratio. Poor fill ratios can result in undesirably long electrodes or small ingots.

A minimum clearance of 50 to 75 mm (2 to 3 in.) between the electrode surface and the crucible should be provided. Such spacing is needed for gas flow for good degassing and for clearance to avoid shorting and arcing to the side wall. Templates are useful for checking the spacing, but the spacing should be reconfirmed after the electrodes are suspended in the furnace.

The blocks and massive scrap are assembled into an electrode and welded together by consumable arc or plasma electrode methods. All welds must be well shielded or made in an inert gas chamber to avoid contamination and formation of refractory oxides and nitrides. Tungsten-tipped electrodes can contribute refractory inclusions and must not be used.

Block density and welds must provide both adequate electrical conductivity and strength. During a melt, the electrode is under nonstatic conditions where power surges and strong magnetic fields can cause the electrodes to sway and vibrate. Straps (referred to as tie bars) of scrap metal, generally made of commercially pure titanium, can be used to reinforce the electrode.

Use of Revert. Revert (sometimes referred to as scrap) is less expensive than virgin sponge, so its use, when feasible, results in a lower-cost ingot (or casting; see the section "Casting" in this chapter). Revert includes turnings and other scrapped items such as mismachined rejected parts. This material must be rigorously cleaned (degreased), and magnetic and high- and low-density particles must be removed. Removal of flame-cut materials is mandatory, because they have regions of nitrogen, oxygen, and carbon enrichment.

Turnings are inspected by means of x-ray to check for broken tungsten carbide (WC) cutting tools and other high-density inclusions, which could end up in the ingot. However, as noted previously, if large amounts of scrap are used, some adjustment may be required in block composition to obtain an acceptable ingot composition. For example, if the oxygen content of the scrap is undesirably high, it can usually be diluted to an acceptable level by increasing the amount of lower-oxygen sponge.

Titanium for rotating (engine) use is generally restricted to lower-revert use than that intended for airframe applications, and commercially pure grades have no restriction on the use of revert. The incorporation of revert in cold hearth melting (see the following) is considerably easier and more economical than for VAR, because it can be fed directly into the melting process without prior consolidation.

Melting. Complete electrodes are suspended in vacuum furnaces, as shown in Fig. 8.1 (Ref 8.1–8.3). After sealing, the furnace is evacuated and the various requirements are checked, including leak rate, coolant flow, controls, and electrode clearance. Melting is initiated at a low power setting by lowering the electrode to contact a small bottom charge placed earlier in the crucible. Operating voltages usually range between 35 and 50 V, depending on a number of variables including the gas content of the electrode, current requirements, arc gap, electrode resistance, and ingot size.

After the arc is initiated, power is increased slowly to the level desired. Maximum current levels range between 500 and 1000 A per 25 mm (1 in.) of crucible diameter. However, it is important that the level be selected carefully, because it directly determines the melt rate, the depth of the molten pool, and the surface quality. Excessive power can be undesirable by producing very deep pools that require extralong hot topping and that can result in surface and internal quality problems. Insufficient power results in low melt rates and poor ingot surfaces. Typical melt rates are approximately 1 kg/kWh (2 lb/kWh) in vacuum and approximately 0.8 kg/kWh (1.7 lb/kWh) under argon pressure.

Arc length is important and commonly ranges between 25 and 50 mm (1 and 2 in.) or more. They can be maintained using automatic equipment based on voltage control. Very short arc gaps can result in frequent shorts and degraded ingot quality. It is important that arc length does not greatly exceed the minimum clearance between the electrode and the crucible. Longer gaps encourage the arc to move to the crucible wall, where it can cause damage. This tendency is especially great if the arc is not controlled by a magnetic field and if the electrodes are not centered properly.

Furnace pressure and gas content of the charge can affect arc behavior and product quality. At certain low pressures, the arc can become very unstable, wander erratically, and have a tendency to move to the crucible wall. To minimize this problem, furnace pressure is often maintained in a range of approximately 100 to 700 µm on second and third melts, using selected pumping rates and valve settings. High arc voltage, which infers long arc, should also be avoided.

Relatively high pressures can be encountered on the first melt because of residual hydrogen gas. This is especially true if the electrode contains a high percentage of leached sponge. Gases from other volatile constituents, including magnesium, sodium, and their salts also influence arc behavior. Pressures of these condensable gases are difficult to measure and may have little effect on observed furnace pressures. Generally, as the gas pressure in the arc zone is decreased, the arc becomes more diffused, power surges and fluctuations are decreased, and there is less damage to equipment. Rapid evaporation of constituents can cause extensive metal spatter, poor ingot walls, ingot porosity, and impaired visibility in the furnace.

After approximately 90% of the electrode for the final melt is consumed, the power is decreased to allow the pool (up to 65% of the ingot length) to solidify without the development of harmful defects. A few hours may be required for hot topping larger ingots, and sufficient electrode metal must be available. It is especially important that the top of the ingot not be allowed to freeze over prematurely. The top should be the last portion to solidify to avoid large shrinkage and the development of gas cavities.

After melting is completed, some cooling is allowed in the furnaces before the ingots are removed. The time required for cooling depends on preferred practices and the equipment used. If the ingot is to be remelted, heavy oxidation of the surface should be avoided. Some furnaces permit almost immediate transfer, without oxidation, to the next melt chamber. Final ingots of alloys sensitive to thermal cracking (for example, the intermetallic titanium aluminides) should be removed from furnaces while hot and cooled slowly (e.g., in vermiculite) to avoid surface cracking and poor workability.

Safety precautions should be taken when handling hot ingots and on opening hot furnaces. Fine deposits of magnesium, sodium, and titanium in the furnaces and on ingots can ignite and be hazardous. Handling ingots with large molten interiors is also dangerous.

As noted, a solenoid around the water jacket can provide control of the arc and the molten pool, but they are not very useful for overcoming extraneous magnetic effects from poor furnace designs and locations. Well-designed equipment and coils encourage arcs to sweep smoothly and slowly around the electrode without dwelling on the crucible wall; such behavior can give the desired uniformly good walls on final ingots. Under such conditions, the molten pool will also rotate slowly and gently. Rapid stirring can produce sloshing, high skulls, poor walls, and internal ingot defects. Poor rotation and no rotation result in variable wall quality. However, good wall and internal quality on final melts also require that other conditions be met as well, including optimum power levels, electrode-to-wall spacing, arc lengths, and low gas content.

Continuous operation of a coil in one direction promotes growth of grains in a preferential direction as the metal solidifies. Such grains influence the primary forging characteristics of the metal and can result in a distinct twist best seen in non-round billet products. This problem can be avoided by automatically reversing the magnetic field at regular short intervals.

Product Quality. Vacuum arc melting is used routinely to produce products to the highest-quality standards. However, the reactive nature of titanium and the special conditions required present many opportunities for the formation of defects unless the operations are well developed and well controlled. Inclusions and porosity are especially objectionable, because they can serve as stress raisers and nuclei for fatigue failures. Some quality problems with titanium products are listed as follows.

Low-Density Hard Inclusions. Alpha-stabilized particles containing relatively large amounts of nitrogen and/or oxygen are hard and brittle. They are difficult to detect, but the larger ones, when associated with voids from mechanical working, are normally found by ultrasonic inspection. These defects, called type I, have been attributed to air contamination during sponge production. An example is shown in Fig. 8.2. However, research shows that any opportunity for extensive localized oxidation, including burning and furnace leaks, is a source for such defects. Poor weld shielding, foreign materials, poorly torch-cut scrap, and unblended additions of TiO_2 are also considered to be sources of this type of defect. Triple melting was introduced to correct this problem, but it is not a complete solution for the larger and more refractory particles, or when the melting practice itself is a source.

High-Density Inclusions. The two types of high-density inclusions (HDIs) include the refractory metals (tungsten, molybdenum, tantalum, and niobium) and the refractory metal carbides. The main source of such materials is poorly cut torch scrap. Machine turnings cut with carbide tools are the main source of carbide particles. Tungsten particles have often been traced to nonconsumable welding electrodes. These materials have relatively high melting points and densities. They resist melting and sink to the skull at the bottom of the molten pool. Improvements in recycling scrap, which remove tungsten carbide particles prior to VAR melting, eliminated HDIs as a problem in turnings scrap.

Soft Alpha Segregates. Large alpha-stabilized regions, called type II defects, are occasionally found in alloys containing aluminum and tin. Such defects are unique to titanium alloys and form in shrinkage pipes. They appear as light and dark etching lines in the macrostructure of product made from ingot pipe locations, as shown in Fig. 8.3. These lines represent original pipe surfaces slightly depleted or enriched by evaporation and redeposition of aluminum, tin, and other relatively volatile constituents. Good hot-topping practices can prevent such defects. These flaws can impair transverse mechanical properties.

Ingot Porosity. Vacuum remelting essentially eliminated ingot porosity, except in pipe sections. Porosity is associated with residual chloride salts from sponge. Salts in sponge have essentially no solubility in the solid metal but form vapor bubbles in the liquid phase. They result in unhealed pipe in billet, some weld porosity, and porosity in blended-elemental powder compacts. Porosity in the pipe sections, resulting from metal shrinkage and volatile constituents, can be largely avoided by good hot-topping practices.

Segregation of alloying elements during solidification is affected by several factors, including the alloy partitioning coefficient, rate of solidification, natural and forced movement of the liquid phase, diffusion, and grain size. Of these, the partitioning coefficient is especially relevant.

The partitioning coefficient, K, is the ratio of the concentration of the solute in the solid to the concentration of the solute in the liquid, or $K = C_S:C_L$, as suggested by Fig. 8.4. The values in Table 8.1 are determined from binary phase diagrams at solute concentrations of approximately 5%. Variations in K can be explained by differences in phase diagrams and how they are used. Nevertheless, such information is quite useful. Deviations from the ideal K-value of unity indicate a natural ten-

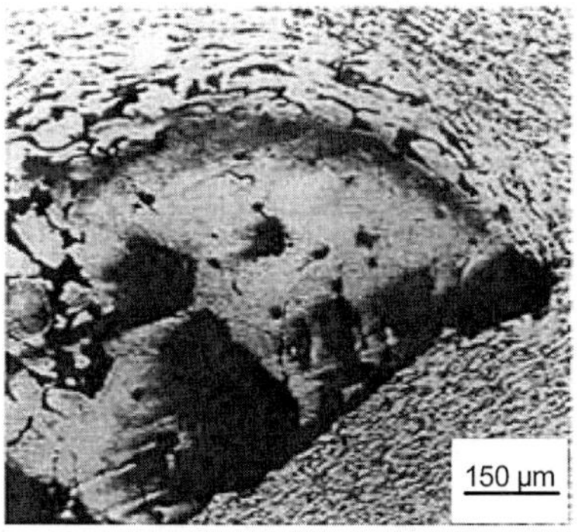

Fig. 8.2 Hard, low-density, alpha-stabilized inclusion in Ti-6Al-4V known as a type 1 defect. Original magnification: 100×

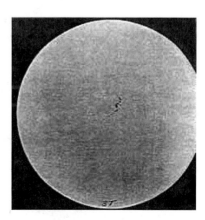

Fig. 8.3 Soft, alpha-stabilized type II defect that formed in the pipe section of a Ti-6Al-4V ingot, which was then forged to a 200 mm (8 in.) diameter billet

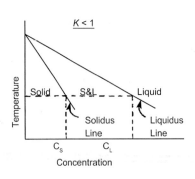 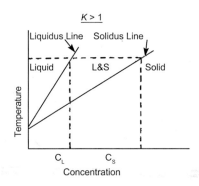

Fig. 8.4 Partial binary phase diagrams. Note that the compositions of the liquid and solid phases of an alloy are not the same at a given temperature but are given by C_l and C_s, respectively, where K, the partition ratio, is given by $K = C_s/C_l$.

dency for an alloying element to segregate on a micro- or macroscale. Specific examples of segregation related to unfavorable K-values include beta flecks and tree rings (Ref 8.3).

Beta flecks are a form of macrosegregation having relatively large, localized concentrations of beta-stabilizing elements. By heating an alpha-beta alloy close to its beta transus temperature and cooling rapidly, beta-rich regions (i.e., flecks) are retained in the macrostructure, as shown in Fig. 8.5. Similar defects, known as "freckles," are found in some specialty steels and superalloys.

Elements having a small K-value, such as chromium, copper, iron, and manganese, are especially prone to produce such defects in titanium products. They partition to the liquid phase and lower its melting point. Some of this enriched liquid flows interdendritically as solidification and shrinkage occur. It is most noticeable where solidification rates are slow and grain sizes are large, as found, for example, in large ingots and on hot topping.

Beta flecks can have a significant effect on heat treated properties but less on annealed properties. Annealed Ti-6Al-6V-2Sn shows that flecks have little effect, but a loss in low-cycle fatigue is observed with the Ti-5Al-2Sn-2Zr-4Cr-4Mo (Ti-17) alloy. Corrective actions favor the use of more rapid solidification, smaller ingots, short hot topping, and ingot soaking.

Tree rings are concentric patterns in the macrostructures. They outline liquid/solid interfaces of variable composition that exist during melting. The rate at which the interface moves affects the extent of alloy partitioning and therefore local alloy concentrations. Thus, variations in the solidification rate result in small local variations in composition, and a macro pattern is established. Variations in heat transfer, erratic arc

Table 8.1 Partitioning coefficients for binary titanium alloys

Solute	Coefficient(a), K
Aluminum	0.9
Cobalt	0.5
Chromium	0.4
Copper	0.3
Iron	0.3
Manganese	...
Molybdenum	3.5
Nickel	0.4
Niobium	4.0
Silicon	0.35
Tin	...
Vanadium	0.5
Zirconium	...

(a) Two data sources

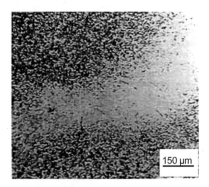

Fig. 8.5 Typical beta fleck found in Ti-5Al-2Sn-2Zr-4Cr-4Mo alloy after it was solution treated within 25 °C (45 °F) of its beta transus. Original magnification: 100×

behavior, and stirring can influence the rate at which the interface advances and therefore its composition. Such patterns can be prevented by avoiding abrupt or erratic changes in an operation during the final melt, including power input, stirring, and shorting. An example of this defect is shown in Fig. 8.6.

Variations in oxygen and nitrogen levels also occur, but these appear to be associated primarily with small furnace leaks, electrode cleanliness, and the moisture content of sponge and less to partitioning.

Ingot Surfaces. The quality of ingot surfaces affects metal yields and conditioning costs and is related to a number of melting variables. Large gas content results in much spatter, vapor deposition on the crucible, and a high skull, which are especially noticeable on first-melt ingots. However, if the electrode is brushed to remove condensed salts, adequate power is selected, and electrical conditions permit the arc to sweep smoothly around the electrode tip, then good ingot surface can be obtained. Poor electrode spacing and centering along with noncoaxial electrical systems and improper solenoid operation result in nonuniform ingot surfaces. Some areas may have good quality and others may have numerous cold shuts and other defects.

Trends in Titanium Melting

Product Quality. As discussed previously, the technology for controlling most quality problems is known. An exception consists of the hard, low-density particles sometimes referred to as type I defects. These inclusions consist mainly of the alpha phase stabilized by greater-than-normal amounts of nitrogen and/or oxygen; they have little or no ductility and can serve as nuclei for premature failure of critical parts.

Although the nature, cause, and a variety of potential sources of these defects are generally known, complete control and detection are very difficult to achieve. All raw materials and melting are suspect. However, because of the potential consequences of undetected type I inclusions, changes in melting technology are being developed and evaluated.

One approach is to cold hearth melt the initial charge using electron beam (EB) (Fig. 8.7) or plasma heat sources, and, in some cases, remelt the cold-hearth-melted ingot in a conventional VAR furnace. Both types of cold hearth melting have now achieved commercial status. Studies show that on cold hearth melting, dense inclusions (e.g., dense refractory metals and carbides) settle into and remain in the hearth skull and that low-density type I inclusions tend to dissolve or melt in the hearth. However, removal of such defects in the hearth (due to the relatively long time that they are held in molten titanium) and their avoidance in later remelting can require additional work, including damming arrangements, to demonstrate the level of effectiveness.

Another approach is to develop plasma arc hearth cold melting for removing low- and high-density inclusions.

Melting Facilities. Although consumable electrode melting is used almost exclusively in the production of titanium ingots and cast parts, other melting methods, such as EB and plasma arc hearth cold melting, offer a practical means for improving product quality, as noted previously. Plasma melting does not require the very low vacuum levels required by EB. In addition,

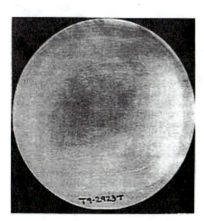

Fig. 8.6 Typical tree ring segregation found in the transverse section of a Ti-6Al-4V billet caused by periodic melting irregularity. Much of the billet forging was performed with a square shape and accounts for the noncircular pattern. Original magnification: 1×

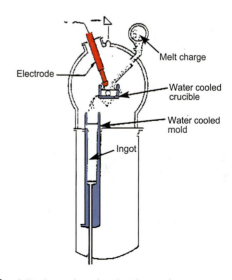

Fig. 8.7 Electron beam hearth melting and ingot casting

plasma melting does not result, as does EB, in an evaporation loss of each element in the alloy, with a significantly disproportionate loss of aluminum. This loss varies with operating practices but is reasonably reproducible. Thus, adjustments to the input chemistry can be made to allow the correct final ingot chemistry to be achieved.

Success with these developments has resulted in a higher order of confidence in final ingot quality and a more uniform product. There is also evidence that a single cold hearth melt is sufficient for some titanium products.

Although less advanced than EB and plasma technology, induction melting is another interesting approach to melting. It is possible to induction melt small amounts (approximately 16 kg, or 35 lb) of titanium in segmented, water-cooled crucibles.

Nonconsumable water-cooled copper electrodes have limited use. Arc damage to the electrode and melt contamination are avoided either by rotating the arc with a magnetic field or by mechanically rotating the electrode.

Casting

Casting is the oldest metals net-shape technology. For thousands of years it provided inexpensive, complex-shaped products of iron and copper alloys and, more recently, also of aluminum-, magnesium-, and nickel-base alloys. Unfortunately, basic cast microstructures, coupled with inherent porosity, reduce the mechanical properties of these base materials well below those of ingot metallurgy (IM) products, making those alloy castings often unsuitable for demanding applications such as required in the aerospace industry.

In comparison, titanium alloy castings have recently demonstrated a combination of lower cost and mechanical property levels equal or close to IM products (Ref 8.4, 8.5), especially after densification by hot isostatic pressing (HIP) (Ref 8.6). The good mechanical properties of titanium alloy castings are a result of the high cleanliness of titanium castings, the ability to close and completely heal pores by the HIP operation (essentially, diffusion bond surfaces together to give a metallurgically sound region), and the control of microstructure, which is possible in titanium alloys (discussed later in this chapter).

The beta-to-alpha phase transformation that takes place during solid-state cooling from the solidification temperature results in a fine lenticular alpha structure (Fig. 8.8) with desirable fracture- and creep-resistance properties. Methods have been developed to modify this microstructure to also give higher strength with very acceptable ductility and crack-initiation behavior.

The relatively low cost of titanium castings (the result of the near-net-shape capability) and

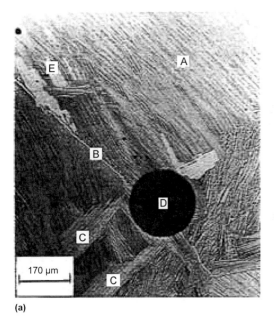

(a)

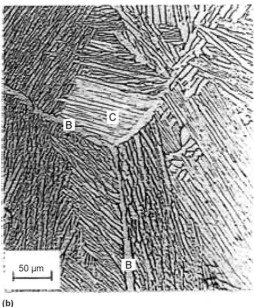

(b)

Fig. 8.8 Lenticular alpha microstructure in (a) as-cast Ti-6Al-4V and (b) cast + hot isostatic pressed Ti-6Al-4V. A, prior-beta grain; B, grain-boundary alpha; C, alpha plate colony; D, gas pore; E, shrinkage pore

their good mechanical properties have contributed to the introduction of increasing amounts of titanium into airframes, gas turbine engines, marine systems, and corrosion-resistant industrial systems. Casting is the most cost-effective and widespread titanium net-shape technology.

The high reactivity of molten titanium presents a great challenge in melting and mold-making technology. Melting of net-shape castings is generally done by vacuum arc, basically the same method used for producing titanium alloy ingots. As a result, most alloys that have been successfully produced as ingots can also be produced in a net-shape cast form. A wide range of titanium alloys have been used in casting, from the rich, high-strength beta and near-beta alloys, to high-temperature near-alpha alloys, to corrosion-resistant lean alloys, to commercially pure titanium, and the relatively brittle titanium aluminides.

Most cast components are made of commercially pure (CP) titanium and Ti-6Al-4V. The Ti-6Al-4V alloy is emphasized because it is the most commonly used titanium alloy, with a wide database available from ingot, cast, and PM products.

This casting section covers the major aspects of alloy melting and shape-making methods, with special emphasis on rammed graphite and investment casting net-shape-making methods.

Casting Technology

Melting for Net-Shape Casting. The most common process for melting titanium cast alloys is consumable electrode arc melting. A schematic of the casting equipment is shown in Fig. 8.9. A more detailed description of the process was presented previously in the "Melting" section of this chapter.

The oxygen level of cast parts is primarily controlled by the melt stock; low-oxygen ingot material results in low-oxygen castings. The use of properly recycled material (revert) typically increases the oxygen level but lowers the product cost. For chemical homogeneity reasons, it could be preferable to produce an ingot from revert prior to casting, although an additional remelting stage increases the oxygen level and the cost. The preferred practice by some casting houses is to carefully control the blend of revert and virgin material to the desired chemical composition and to directly cast.

Rammed Graphite Mold. This method is commonly used to produce relatively large components for industrial applications (Fig. 8.10a), although large aerospace components such as heli-

copter rotor hubs (Fig. 8.10b) have been manufactured. Rammed graphite was the earliest commercial titanium casting mold-making technique practiced in the United States. Traditionally, a mixture of graphite powder, pitch, corn syrup, starch, and water is rammed against a wooden or fiberglass pattern to form a mold section. Corn syrup and starch give the mold some green strength after the rammed mold has been dried in air for 24 h or for shorter periods at 200 °C (400 °F) in a drying furnace. The mold segments are fired for 24 h at 1025 °C (1875 °F) under a suitable shield, causing all the constituents to carburize and harden. In some cases, water-soluble binders are used in the mixture, which eliminate the need for a high firing temperature. By using this method, the minimum practical cast part wall thickness is 3.8 mm (0.15 in.), and a typical surface finish is 250 rms (root mean square). A rammed graphite mold section is shown in Fig. 8.11.

Molds for large and complicated shapes are assembled from many segments. The graphite mold is so hard that it is necessary to chisel it off the cast parts, which is a labor-intensive operation. Castings are typically cleaned in an acid bath, chemically milled, weld repaired if necessary, and sand blasted for good surface appearance. Care is required to prevent hydrogen pickup during the acid-cleaning operation. In large mold segments, it is sometimes difficult to control the precise shape of the mold during the drying and

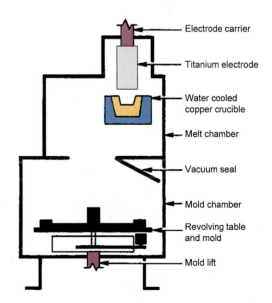

Fig. 8.9 Consumable electrode titanium vacuum arc melting furnace with centrifugal casting table. Courtesy of Howmet

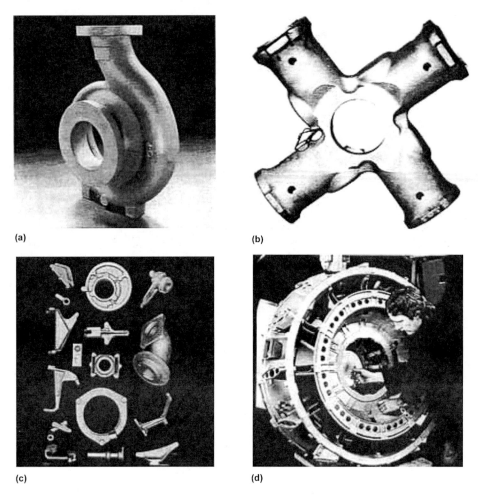

Fig. 8.10 Cast titanium components produced by (a) rammed graphite method, 300 mm (12 in.) long chemically pure titanium pump housing for the chemical process industry; (b) rammed graphite method, Ti-6Al-4V helicopter hub; (c) investment casting airframe components; and (d) investment casting, 135 kg (300 lb) Ti-6Al-4V fan frame for the General Electric CF6-80C gas turbine engine case. Courtesy of (a) OREMET, (b) TiTech, (c) Howmet, and (d) PCC

Fig. 8.11 Rammed graphite mold section. Courtesy of TiLine

Fig. 8.12 Ceramic mold. Courtesy of TiLine

firing stages, which limits the dimensional accuracy of the final product.

Typical thickness tolerances are ±0.75 up to 150 mm (±0.030 up to 6 in.). Dimensional tolerances of large components can be improved by using shell cores, which are light, hard, and accurate, and by using ceramic molds. The ceramic mold segments (Fig. 8.12) are produced from wooden patterns in a proprietary process that maintains good mold accuracy and reproducibility. This mold has a higher cost than the rammed graphite and is also more difficult to separate from the cast parts.

Sand Casting. Although sand casting is the lowest-cost and most widely used practice throughout the general casting industry, it has not been used for titanium because of reactivity problems with conventional mold sands. The U.S. Bureau of Mines developed a sand casting method for titanium alloys using sand molds coated with zirconium oxide. Cast parts have a very limited alpha-case depth, but the surface finish is inferior to that obtained by the rammed graphite method. Sand casting has the potential of producing lower-cost components and is primarily aimed at marine and chemical applications.

Investment Casting. In this method, a wax pattern is produced using an injection molding technique. The wax injection tooling cavity is produced slightly oversized to account for shrinkage of the wax, of the ceramic shell, and of the alloy, as well as dimension reduction during the chemical milling stage needed to remove the alpha case. A wax pattern gating system is added, and the assembly is dipped in ceramic slurries and stuccoes and dried several times to build a shell with enough strength to sustain the molten metal pressure after being hardened by firing. The wax pattern is removed in a steam autoclave, followed by firing, which leaves the mold cavity ready for casting. The inner shell face coat is designed to minimize the alpha-case reaction zone, the result of a reaction between the molten metal and the mold material. A typical shell assembly is shown in Fig. 8.13.

To improve productivity, many duplicate components can be cast in clusters, and robots are used to perform the shell-dipping operation, which also improves product reproducibility. The ceramic shells are placed inside the mold chamber of a vacuum arc furnace. Casting can be done on a centrifugal table to assist the metal flow (shown in Fig. 8.9) or, more simply, by gravity pouring, which requires a higher preheat temperature of the shells to improve molten metal fluidity and mold fill. The ceramic shell and the gating system are removed after casting, followed by chemical milling to remove the alpha case.

Investment casting (Ref 8.4) provides very good dimensional control and is suitable for production of both small and large high-quality aerospace airframe and engine components (Fig. 8.10c, d). Typical thickness tolerance is ±0.25 up to 25 mm (±0.010 up to 1 in.). Currently, components up to 300 kg (650 lb) can be produced using this method. The minimum practical cast part wall thickness is 1.3 mm (0.05 in.), and the typical surface finish is 125 rms.

Casting Alloys. Commercial titanium castings are predominantly made of Ti-6Al-4V and various grades of CP titanium. However, a number of other alloys have recently been cast (Table 8.2). In almost all cases, these alloys are simply cast versions of conventional IM alloys. The major

Fig. 8.13 Investment casting shell assembly. Courtesy of PCC

Table 8.2   Titanium alloys used for producing cast parts

Alloy designation	Alloy class	Remarks
Commercially pure titanium	Oxygen levels and minor alloy additions constitute different grades.	Corrosion resistance, chemical and energy applications
Ti-6Al-4V	Alpha + beta alloy	The most commonly used alloy for aerospace and industrial applications
Ti-6Al-6V-2Sn	Medium-strength alpha + beta alloy	Strength higher than Ti-6Al-4V
Ti-6Al-2Sn-4Zr-2Mo	Near-alpha, high-temperature alloy	Wrought material contains silicon for higher creep resistance.
Ti-6Al-2Sn-4Zr-6Mo	Alpha + beta high-strength alloy	For engine rotating components requiring higher strength than Ti-6242
Ti-6Al-2Sn-2Zr-2Mo-2Cr-0.25Si	Deep-hardenable alpha + beta alloy	Applications requiring some deep hardenability
Ti-5Al-2.5Sn	Alpha alloy	Good cryogenic properties, highly weldable
Transage 175 (Ti-2.5Al-13V-7Sn-2Zr)	Deep-hardenable, martensitic alloy	Strain transformable, high strength
Ti-10V-2Fe-3Al	Near-beta forgeable alloy	High strength, good fracture resistance
Ti-15V-3Al-3Sn-3Cr	Near-beta alloy	High-strength, cold-formable alloy
Beta C (Ti-3Al-8V-6Cr-4Zr-4Mo)	Metastable beta alloy	High strength and fatigue resistance
Beta III (Ti-11.5Mo-6Zr-4.5Sn)	Metastable beta alloy	High strength, good ductility, deep hardenability

difference between wrought products and cast parts stems from the subsequent hot working and heat treatment of the ingot material, which allows microstructural manipulations not possible in cast parts, particularly the development of an equiaxed recrystallized alpha structure.

Hot Isostatic Pressing. A heated, argon-filled pressure vessel (HIP autoclave, Ref 8.6) is used to densify titanium alloy castings. Hot isostatic pressing eliminates casting solidification shrinkage voids and gas porosity (Fig. 8.8). However, surface-connected porosity cannot be healed and must be weld repaired. Hot isostatic pressing of Ti-6Al-4V castings is typically done in the temperature range of 890 to 955 °C (1650 to 1750 °F) under pressures ranging from 70 to 105 MPa (10 to 15 ksi) for 2 to 4 h. Similar conditions are used for titanium prealloyed powder HIP consolidation.

Titanium castings are HIPed without a can or mold, making it a less expensive operation than powder HIP consolidation, which requires an evacuated mold for powder containment. Hot isostatic pressing enhances critical properties such as fatigue resistance while maintaining properties such as fracture toughness, fatigue crack growth rate (FCGR), and tensile strength at acceptable levels. Therefore, fatigue-critical cast parts are always HIPed, regardless of whether they are for airframe components, engine parts, or orthopedic surgical body implants. Some casting houses do not HIP all products unless specified by the user, because for many marine or chemical process applications, mechanical property levels are not as critical as the general corrosion resistance. It is also possible to reduce the level of porosity in a cast part by increasing the number of strategically placed gates and risers, thereby causing shrinkage porosity to be relegated to locations that are later removed from the cast part. However, some casting producers find the additional expense associated with increased gates and risers to be higher than the HIP operation, and they consequently HIP all parts.

Weld repair is a common practice for filling surface porosity, post-HIP surface depressions, and cold shuts (the result of premature solidification). Inert gas tungsten arc welding is typically used with alloy filler rods of regular or extralow interstitial material, followed by a stress-relief heat treatment. Weld repair does not have an adverse effect on tensile properties, high- or low-cycle fatigue, FCGR, fracture toughness, creep rate, or creep-rupture strength of castings. This is because the weld zone develops a cast structure that is very similar to the overall cast part microstructure.

Heat Treatment. Two types of heat treatments are used with titanium alloy castings. The first is a stress-relief treatment primarily intended to relieve the residual stresses induced during casting and HIP. The second type is aimed at changing the cast microstructure for mechanical property improvement. This latter type is carried out at higher temperatures than the stress-relieving treatment, typically close to or above the beta transus temperature, and is discussed in the section on microstructure.

Stress-relief heat treatments are generally carried out between 700 and 850 °C (1300 and 1550 °F) for 2 h in a vacuum or inert gas environment to protect the cast part surface from oxidation. This type of heat treatment does not change the as-cast microstructure (Fig. 8.8a) on an optical microscopy level and is equivalent to mill annealing of IM alloys. Because of relatively slow cooling rates, especially in large components produced by the rammed graphite method, some castings are used without a stress-relief treatment. In some other cases, the HIP cycle can be considered as a stress-relief treatment due to the relatively slow cooling rates associated with this operation. However, because HIP is carried out at temperatures relatively close to the beta transus temperature, a more significant microstructural change could occur.

Microstructure and Mechanical Properties. The mechanical properties of titanium alloy castings result from both the basic cast structure and various casting features (Ref 8.4). Because the as-cast microstructure of alpha-beta alloys is very similar to IM beta-annealed and relatively slowly cooled structures, the mechanical properties of the latter can be used to interpret those of cast parts.

Microstructure of Alpha + Beta Alloy Castings. A typical cast Ti-6Al-4V microstructure is shown in Fig. 8.8a and 8.8b. The major constituents of this microstructure are discussed as follows.

Beta Grain Size ("A" in Fig. 8.8a). The beta grain size, typically ranging from 0.5 to 5 mm (0.02 to 0.2 in.), develops during solid-state cooling through the beta phase field, with slower cooling rates resulting in larger beta grains. As the beta grain size increases, properties such as FCGR improve, but properties such as high- and low-cycle fatigue strength may deteriorate.

Grain-Boundary Alpha ("B" in Fig. 8.8a). These alpha plates develop along the boundaries of the beta grains during cooling through the alpha-beta phase region. They reduce fatigue life both at room and elevated temperatures.

Alpha Plate Colonies ("C" in Fig. 8.8a). Alpha plates form within the beta grains on cooling below the beta transus temperature. When cooling is slow, as in thick-wall castings, plates are arranged in colonies or packets that are similarly aligned and have a common crystallographic orientation. These colonies lead to early fatigue crack initiation by a mechanism of intense shear across the colony plates. Large colonies can improve FCGR. Alpha plate thickness ranges from 1 to 5 μm, and colony size typically ranges from 50 to 500 μm. As a general rule, slow cooling through the alpha-beta phase region leads to larger colonies and thicker alpha plates. For this reason, slowly solidified thicker casting sections exhibit larger beta grains, thicker alpha plates, and larger alpha plate colonies.

Casting Porosity. Porosity is considered as part of the overall as-cast structure because it can strongly influence properties such as fatigue strength. Two types of pores are found. One is the result of trapped gas and has a spherical shape ("D" in Fig. 8.8a). The other is shrinkage porosity, a result of volume reduction upon solidification, and can be as small as a few micrometers ("E" in Fig. 8.8a) and as large as a few millimeters (0.1 in.) in thick sections. Large pores are detrimental to fatigue life if not healed by HIP or weld repair. As previously mentioned, it is possible to locate most of the porosity in regions that are later removed from the cast part, such as gates and risers.

Microstructure Modifications. In practice, three methods used to modify the basic cast structure are HIP (Ref 8.6), heat treatment, and thermohydrogen processing (THP) by temporary alloying with hydrogen (Ref 8.7, 8.8).

Hot isostatic pressing is primarily done to close casting porosity and shrinkage. The thermal cycle associated with an alpha-beta phase field HIP operation thickens the alpha plates, even if performed at relatively low temperatures. Higher subtransus HIP temperatures increase the tensile strength by 10% when followed by a 955 °C (1750 °F) heat treatment with 85 °C (150 °F)/min cooling rate. At the same time, coarsening of the alpha plates leads to lower fatigue strength but higher fracture toughness. Optimization of HIP parameters (temperature in particular) and the use of faster cooling rates could lead to further property enhancement.

Because the cast structure does not contain residual work, it is very stable in the alpha + beta

temperature range. Attempts to alter the cast morphology must be done close to or above the beta transus temperature. The largest change in microstructure can be accomplished by a beta solution treatment followed by intermediate and high cooling rates. This results in a refinement of the alpha plates (often precipitated during a subsequent aging treatment) with a smaller colony size or even a basketweave structure, the latter having better fatigue resistance than the original cast lamellar colony structure.

Beta solution treatment followed by 815 °C (1500 °F) aging for 24 h yields a broken-up structure (BUS) of alpha plate (Fig. 8.14). Subtransus solution treatment (close to the transformation temperature) produces a duplex structure, which retains some vestige of the cast structure in a matrix of transformed beta. After adequate aging, this microstructure can result in higher strength levels than the as-cast structure, but the remnants of the grain-boundary alpha phase are sometimes detrimental to fatigue strength. It also appears that cooling rates associated with both types of solution treatments are important for achieving good fatigue properties. Intermediate cooling rates achieved by circulating argon (known as gas fan cooled) produce improved fatigue strength in Ti-6Al-4V castings, which is related to the finer alpha plate structure produced.

It is also possible to modify the microstructure with THP using hydrogen as a temporary alloying element. In the method, up to 2 wt% hydrogen is added to the cast parts, allowing an alloy such as Ti-6Al-4V to behave as an alloy with an 800 °C (1475 °F) eutectoid temperature. By implement-

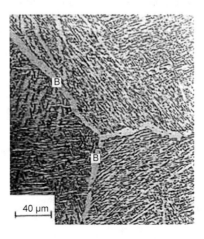

Fig. 8.14 Cast Ti-6Al-4V broken-up structure, the result of a 1025 °C (1880 °F) solution treatment followed by an 815 °C (1500 °F) aging for 24 h. B, grain-boundary alpha

ing various heat treatments and dehydrogenating practices, it is possible to obtain a range of fine microstructures that are totally different from the as-cast structures. A commercial process, termed constitutional solution treatment, was based on this concept and was used for a short time (but subsequently discontinued) by a major casting house in the United States.

Grain Refinement. Further control of the microstructure in titanium cast parts can be accomplished by way of refinement of the beta grains using inoculants in a manner similar to those already used for aluminum and superalloys. For certain casting conditions, the beta grain size of Ti-6Al-4V can be reduced from 1 to 0.3 mm (0.04 to 0.01 in.) by adding Ti-5wt%C inoculants. A similar beta grain size reduction was reported in cast Transage-175 alloy using inoculants.

Mechanical Properties. The tensile properties and fracture toughness of cast parts in various conditions are summarized in Table 8.3. Some of the data is taken from cast test coupons. The properties of actual cast parts, especially large components, can be lower due to a coarser grain structure. The tensile and fracture toughness properties of all conditions are very similar to those exhibited by beta-annealed IM material. The FCGR behavior of cast Ti-6Al-4V is compared to beta-annealed material in Fig. 8.15 (Ref 8.4, 8.5). No significant differences were found between the FCGR of wrought beta-annealed, cast, and cast + HIP material.

Figure 8.16 compares the scatterbands of room-temperature smooth axial fatigue properties of cast and cast + HIP Ti-6Al-4V with those of annealed wrought material (Ref 8.4, 8.5).

Hot isostatic pressing of cast parts can significantly increase fatigue strength into the lower IM regime, mainly because of pore closure. However, it is still generally below that of wrought material results, because the basic cast structure has lower resistance to fatigue crack initiation due to the alpha colony structure and the grain-boundary alpha (Fig. 8.8b). Higher fatigue strength can be achieved by innovative heat treatments, as shown in Fig. 8.17 for 16 mm (0.63 in.) diameter investment-cast test coupons that were heat treated (after HIP) to the microstructure in Fig. 8.14. The fatigue results of cast test coupons with the BUS are well within the wrought material scatterband shown in Fig. 8.16.

Another approach to improvement of fatigue strength by modification of the Ti-6Al-4V cast microstructure is by the THP treatment discussed previously (Ref 8.7, 8.8). The fatigue curve of a sample treated by hydrogen using a process termed high-temperature hydrogenation (HTH) is shown in Fig. 8.18. The microstructure of material given this treatment is similar to that shown in Fig. 8.14. The fatigue strength of the HTH-treated cast coupons is higher than that for the mill-annealed IM material shown in Fig. 8.16. This type of treatment opens the gate for titanium cast parts to be used in fatigue-critical applications.

Trends in Casting Technology

Trends in titanium casting technologies focus on development of technology in shape making, melting technology, alloy design, and postcasting treatments.

Shape Making. The trend in net-shape making in casting is in investment casting technology (Ref 8.4). More casting houses are converting exclusively to this method. Improvements in melting techniques to produce superheated melts enable casting of more complex shapes and thinner sections. Improvements in mold materials and mold coats improve surface quality, reduce surface contamination, and minimize the need for chemical milling. The use of robotics in mold dipping reduces cost and increases product repro-

Table 8.3 Typical mechanical properties of Ti-6Al-4V castings at room temperature

Material condition	Yield strength		Ultimate tensile strength		Elongation, %	Reduction in area, %	Fracture toughness	
	MPa	ksi	MPa	ksi			MPa√m	ksi = √in.
As-cast	895	130	1000	145	8	16	71–77	65–70
Cast + HIP (HIP at 890 °C, or 1650 °F)	870	126	960	139	10	18	91	83
Cast + HIP + near-beta heat treatment(a)	870	126	965	140	8	17	...	...
Cast + HIP + beta heat treatment(b)	870	126	970	141	8	19	...	...
Wrought beta anneal	860	125	955	139	9	21	84	76

HIP, hot isostatic pressing. (a) 990 °C (1810 °F) for 1 h, gas fan cool (GFC) + 540 °C (1000 °F) for 8 h, GFC. (b) 1015 °C (1860 °F) for 1 h, GFC + 540 °C (1000 °F) for 8 h, GFC

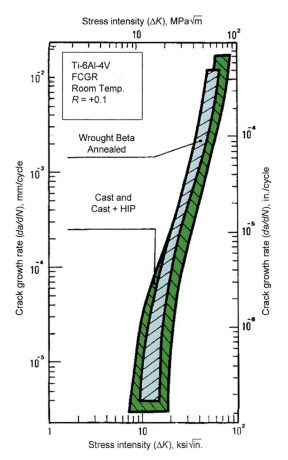

Fig. 8.15 Fatigue crack growth rate (FCGR) scatter band data comparing Ti-6Al-4V cast and cast + hot isostatic pressed (HIP) material with beta-annealed ingot metallurgy material

ducibility. The use of computer-aided mold design and solidification computer models enables true net shapes to be produced with fewer design iterations, less porosity, and more uniform microstructures. The use of advanced core-making technologies enables the production of thin-web hollow parts required for the fan, compressor, and turbine section parts of gas turbine engines.

Melting Technology. Advanced and emerging melting methods discussed in the "Melting" section of this chapter enable the melting of cleaner alloys for improved cast part integrity and superheated melting for better shape definition and thinner sections. Superheated melts do not require centrifugal force mechanisms for improved mold fill.

Alloy Design. Currently, all casting alloys are basically IM alloys. However, with the continuing increase in the volume of the titanium casting industry, the development of specific casting alloys could be considered. These can be lower-melting-temperature alloys for easier melting and reduced mold reactions and alloys that yield finer grain structure in cast products and therefore improved properties. Alloys could be developed that are more amenable to postcast processes such as HIP, heat treating, thermohydrogen treating (Ref 8.6, 8.8), and weld repair.

Postcasting Treatments. These treatments can improve cast properties above IM levels. Improvements in HIP cycles, such as controlled cooling rates after the pressure cycle, could eliminate the need for subsequent heat treatments and could improve properties.

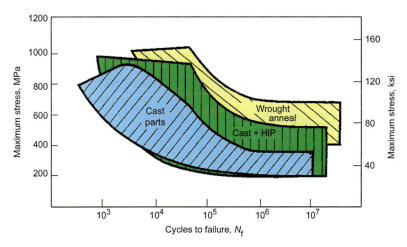

Fig. 8.16 Room-temperature smooth axial fatigue data comparing Ti-6Al-4V cast and cast + hot isostatic pressed (HIP) material with annealed ingot metallurgy material

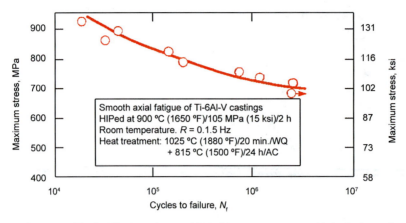

Fig. 8.17 Fatigue (S-N) curve of Ti-6Al-4V hot isostatic pressed (HIPed) investment-cast parts with the "broken-up" structure. WQ, water quenched; AC, air cooled

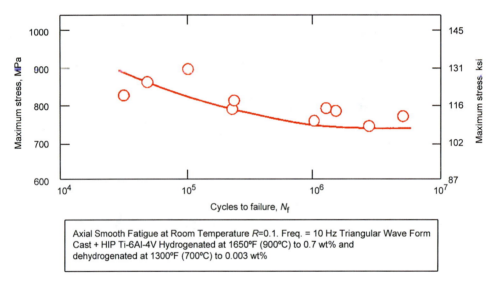

Fig. 8.18 Fatigue (S-N) curve of Ti-6Al-4V hot isostatic pressed (HIPed) investment-cast parts treated by temporary alloying with hydrogen in the high-temperature hydrogenation process

Titanium Powder Metallurgy

Various categories of titanium PM, their major characteristics, and their current status are summarized in Table 8.4 (Ref 8.9–8.15). Figure 8.19 shows where PM, in general, and powder injection molding, in particular, fit in with other fabrication processes.

Powder metallurgy is the production, processing, and consolidation of fine particles to make a solid metal. A primary advantage of the PM approach over other methods is the more efficient use of material (Ref 8.16–8.20). Other advantages include greater shape flexibility and reduced processing steps. ADMA Products has recently used hydrogenated titanium sponge to produce high mechanical property levels (Ref 8.12, 8.14).

A large, well-established PM industry involving materials such as iron-, copper-, and nickel-base alloys has been in existence for many years. The main driver for PM is cost reduction, where mechanical properties can play a secondary role. Powder metallurgy of titanium alloys was established during the 1980s. Cost and material savings are major goals using two PM approaches: the prealloyed (PA) technique, which involves HIP compaction, and the blended-elemental (BE) method, which involves a cold press-and-sinter operation, mainly on Ti-6Al-4V, the "workhorse" alloy of the titanium industry.

The PA approach is typically used to produce demanding aerospace components in which mechanical property levels (particularly fatigue behavior) must be equivalent to cast and wrought IM levels. The industry met these requirements by establishing clean powder production and handling procedures.

In contrast, titanium compacts produced using the BE method until recently have not achieved fatigue levels required for critical aerospace components because of inherent salt and porosity. These products are aimed at use in applications that do not require high dynamic (fatigue) properties. In the late 1980s, low-chloride starting stock became available, which led to improved density and enhanced fatigue behavior. The BE technique was enhanced to the point where mechanical properties compare well with cast and wrought product. Production costs for the BE technique are lower than those for the PA technique, although larger, more complex near-net shapes are possible using the PA + HIP approach.

Other titanium PM methods include additive manufacturing (where a component is gradually built up by melting successive layers), powder injection molding, spray deposition, and porous materials.

Rapid solidification (RS) offers some of the same advantages for titanium alloys as those already demonstrated for other systems, such as aluminum. In particular, RS allows extended solid solubility formation of metastable phases, structure refinement, and freedom from segregation/workability problems.

Categories of Titanium Powder Metallurgy

Blended-elemental processing includes the following considerations.

Powder Production. In the BE approach, sponge fines, the small (100-mesh), irregular grains of titanium powder normally rejected during the conversion of ore to ingot, have been used as starting stock (Fig. 8.20) (Ref 8.12, 8.14). Alloying elements are blended with these fines, normally in the form of a powdered master alloy, so the desired bulk chemistry is achieved after sintering. The sponge fines used are almost exclusively from either conventional Hunter or Kroll sponge, which inherently contains 1500 ppm salt (sodium or magnesium base, respectively), or electrolytic sponge with an order of magnitude lower salt content. Fines produced from ingot stock, using a hydride-dehydride comminution process, have been produced with very low chloride content.

Table 8.4   Categories of titanium powder metallurgy

Category	Features	Status
Additive manufacturing	Powder feed melted with a laser or other heat source	Pilot production
Powder injection molding	Use of a binder to produce complex, small parts	Production
Spraying	Solid or potentially liquid	Research base
Near-net shapes	Prealloyed and blended elemental	Commercial
Far-from-equilibrium processes	Rapid solidification, mechanical alloying, and vapor deposition	Research base
Thermohydrogen processing	Use of hydrogen as a temporary alloying element	Research base
Porous materials	Less than 100% dense materials	Research base

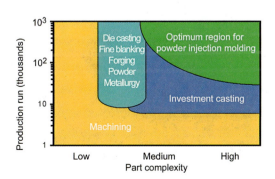

Fig. 8.19 Diagram showing where powder metallurgy, in general, and powder injection molding, in particular fit in with other fabrication processes. Courtesy of Krebsöge, Radevormwald

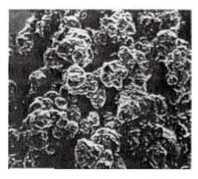

Fig. 8.20 Scanning electron micrograph of porous titanium sponge fines used as starting stock in blended-elemental powder metallurgy

Several innovative extraction processes were investigated during the 2000s and early 2010s, which, in many cases, produced powder that can be used in the BE approach to component fabrication. (For process summaries, see Table 1.14 in Chapter 1, "History and Extractive Metallurgy," in this book.)

Consolidation. The powder blend is cold compacted under pressures up to 400 MPa (60 ksi) to a green density of 85 to 90% of theoretical density, either isostatically or with a relatively simple mechanical press. The "green" compact is then sintered at a temperature of approximately 1260 °C (2300 °F) to produce 95 to 99.5% of theoretical density and to homogenize the composition. A further increase in density can be achieved by postsinter HIP, which generally improves mechanical properties (Table 8.5). Postsinter densification can also be achieved with a "soft" tooling method using a granular pressing medium around a sintered compact and conventional hot forging equipment.

Direct Processing. Powder metallurgy component production generally involves powder production and consolidation into a net shape as separate steps. However, additional cost reduction is possible by converting the powder directly to a mill product. Foil, sheet, and plate stock can be produced using the BE approach. Plate is produced by rolling a compacted billet. Foil and sheet are fabricated directly from powder, as shown in Fig. 8.21.

Preforms. Forging of BE preforms offers inexpensive forging stock and production of more equiaxed and fatigue-resistant microstructures. Compacts isothermally forged up to 80% using both regular and low-chloride (electrolytic) sponge exhibit a microstructure having a significantly reduced alpha aspect ratio, resulting in only limited fatigue enhancement due to residual porosity.

Alloy Types Studied. Most studies involve Ti-6Al-4V alloy. However, the technique is equally applicable to alloys such as Ti-6Al-6V-

2Sn, Ti-6Al-2Sn-4Zr-2Mo, Ti-6Al-2Sn-4Zr-6Mo, Ti-10V-2Fe-3Al, and other more recent alloys. The use of other alloys enables production of blended-elemental PM components having higher strength and higher elevated-temperature properties.

Microstructure. Typical Ti-6Al-4V BE compact microstructures produced by mechanical pressing and isostatic compaction and exhibiting residual porosity are shown in Fig. 8.22(a) and (b), respectively. The mechanically pressed-and-sintered compact displays a low-aspect-ratio alpha structure (Fig. 8.22a). This is a consequence of the very small beta grain size resulting from the particular process used and the residual porosity (which pins grain-boundary motion). The alpha plate structure of the isostatically compacted material (Fig. 8.22b) is more typical of material sintered above the beta transus temperature, where a large beta grain size develops, and subsequently cooled slowly to form a coarse alpha plate colony structure. These basic microstructures can be modified, along with a concurrent improvement in fatigue behavior, by use of an innovative heat treatment to produce a BUS, shown in Fig. 8.22(c), or by using temporary alloying with hydrogen (THP) (Ref 8.8, 8.12, 8.14), again to refine the structure (Fig. 8.22d). The material shown in Fig. 8.22(d) is produced from chlorine-free powder, which was subsequently HIPed to 100% density, followed by hydrogen treatment.

Mechanical Properties. Typical tensile properties of BE compacts produced using several routes are compared with IM properties in Table 8.5. In critical components, such as rotating parts, fatigue behavior is very important. Fatigue results for BE and PA powders are compared with IM data in Fig. 8.23, where it is observed that the remnant salt from the sponge-making process and associated voids cause early crack initiation in the BE material. Lower chloride content and control of the microstructure by subsequent processing heat treatment or THP (Fig. 8.22c, d) can enhance fatigue behavior. One method of producing pow-

Table 8.5   Typical tensile properties of blended-elemental Ti-6Al-4V compacts compared to mill-annealed wrought products

Material(a)	0.2% yield strength		Ultimate tensile strength		Elongation, %	Reduction in area, %
	MPa	ksi	MPa	ksi		
Cold isostatic: press and HIP (CHIP)	827	120	917	133	13	26
Press and sinter (no HIP)	868	126	945	137	15	25
Wrought mill anneal	923	134	978	142	16	44
Typical minimum properties (MIL-T-9047)	827	120	896	130	10	25

(a) HIP, hot isostatic pressing; CHIP, cold and hot isostatic pressing. Data courtesy of Dynamet Technology, Inc.

der with very low chloride content is by comminution of titanium ingot stock.

Other mechanical properties, such as fracture toughness and FCGR, appear to be at the same levels as those expected from IM material of the same chemistry and microstructure when the density is at least 98% of full density.

The effect of density on tensile and fatigue behavior is shown in Fig. 8.24 and 8.25. Tensile and yield strengths increase linearly with sinter density, while fatigue endurance limit increases at a higher rate as density approaches 100% of full density. When 99.8% density is achieved by post-sinter HIP of low-chloride BE material, fatigue strength is improved up to the lower region of the wrought alloy scatterband (Fig. 8.23).

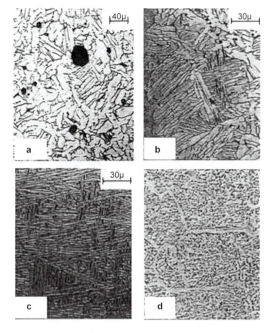

Fig. 8.22 Typical blended-elemental Ti-6Al-4V compact microstructures after cold pressing and sintering with some residual porosity. (a) Mechanically pressed. (b) Isostatically pressed. (c) After the "broken-up" structure heat treatment. (d) After thermohydrogen processing

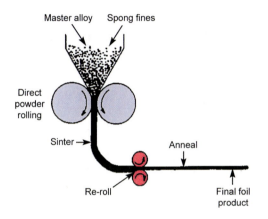

Fig. 8.21 Schematic processing sequence for production of Ti-6Al-4V foil from blended-elemental starting stock

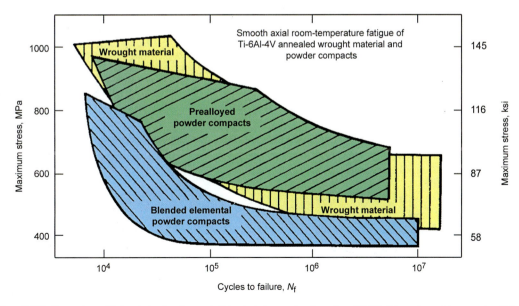

Fig. 8.23 Comparison of fatigue behavior of annealed blended-elemental and prealloyed Ti-6Al-4V powder metal compacts with ingot metallurgy material

Shape-Making Capabilities. Very complex shapes, such as the impeller shown in Fig. 8.26, can be produced by cold isostatic pressing (CIP) using elastomeric molds. However, part size currently is limited to a 600 mm (2 ft) diameter by the availability of CIP equipment. Dimensional tolerance is approximately ±0.5 mm (±0.02 in.).The press consolidation technique can produce parts up to 30 m (~100 ft), but the shape-making capability is typically not as good as in CIP. The connector link arm shown in Fig. 8.27 is an example of a part made by the press consolidation method.

Larger BE components are difficult to fabricate by welding smaller parts, because the material does not weld well due to inherent porosity and chloride content. However, weldability is improved at chlorine levels below 150 ppm.

The economics of the BE approach are attractive; the price of sponge fines blended with master alloy additions currently is approximately 20% that of PA powder.

Cold and Hot Isostatic Pressing Process. The Dynamet Technology Inc., now part of RTI, cold and hot isostatic pressing (CHIP) process (Fig. 8.28) is used to produce near-net-shape parts for

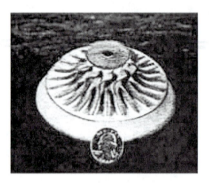

Fig. 8.26 Cold isostatically pressed blended-elemental Ti-6Al-4V impeller produced using an elastomeric mold. Courtesy of Dynamet Technology Inc.

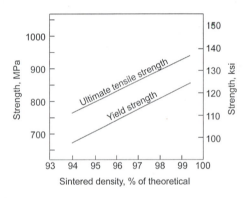

Fig. 8.24 Effect of density on yield and tensile strengths of pressed-and-sintered Ti-6Al-4V

Fig. 8.27 Pressed-and-sintered blended-elemental Ti-6Al-4V connector link arm for the Pratt & Whitney F100 engine. Courtesy of Imperial Clevite Inc.

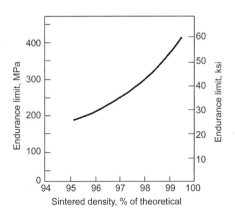

Fig. 8.25 Effect of density on the fatigue endurance limit of pressed-and-sintered Ti-6Al-4V

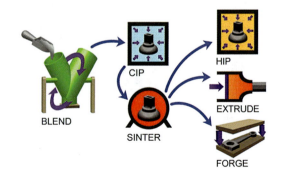

Fig. 8.28 Schematic of Dynamet Technology Inc.'s CHIP (cold and hot isostatic pressing) process.

finish machining to high-tolerance configurations (Ref 8.12, 8.14). The process can also be used to make forging preforms and mill product shapes for subsequent processing, such as billet for casting, extrusion, and hot rolling. In the case of as-sintered material, full density is achieved during subsequent processing.

The CHIP process, a green manufacturing technology, is an acceptable process for producing military, industrial, and medical components. The advanced PM process uses titanium powder (typically Kroll-process hydride-dehydride powder) blended with master alloy powder such as aluminum-vanadium. The blended powder is compacted to shape by CIP in elastomeric tooling. With proper selection of powders, well-designed CIP tooling, and appropriate pressing conditions, a shaped powder compact can be produced and readily extracted from the PM tooling with sufficient green strength for handling. It must also have sufficient uniformity and intimate contact of the powder particles for densification and homogeneous alloying in the subsequent sintering process.

A wide range of shapes have been produced, with size limited only by the capacity of the equipment (Fig. 8.29). The size of the CIP unit is usually the limiting factor, because vacuum furnaces and HIP units are available in larger sizes than CIP units. Size capability also depends on the powder fill characteristics, product configuration, and tooling parameters. Product weights range from a few grams to hundreds of kilograms (ounces to several hundred pounds).

Major cost benefits of this clean, energy-efficient manufacturing process include the use of relatively low-cost raw materials, avoidance of costly melt processes, and relatively little material lost during processing. The capability to produce to a near-net shape conserves raw material and also reduces costs for machining to finished parts. Cold-pressed compacts are sintered in vacuum to high or nearly full density. Alloying titanium with other elements is accomplished by solid-state diffusion during the sintering process. By selecting the proper powders and sintering parameters, a homogeneous alloyed material with sufficiently high density, free of interconnected porosity, is achieved.

The sintering process was established to reach a minimum density level at which the material had no interconnected porosity. At this density threshold, the material could be HIPed without the processing expense of HIP encapsulation, making the HIP process economically viable. Process developments have enabled the achievement of greater than 98% sintered density. This results in as-sintered tensile properties (Tables 8.5, 8.6) equivalent to those of wrought properties and superior to those of castings. This reduces the

Fig. 8.29 Examples of near-net titanium shapes produced by Dynamet Technology, Inc. OD, outside diameter

need for the HIP operation and further strengthens the economic advantage of this PM cold isostatic pressing/sinter manufacturing technology. Qualification of the process by Boeing Co. for use on its commercial aircraft was a major advance for the technology.

The BE technique has also been used for the fabrication of metal-matrix composites using particulate and a CHIP combination, or forging, extrusion, and rolling of the CHIP preform (Ref 8.14). The CermeTi (Dynamet Technology, Inc.) family of titanium alloy matrix composites incorporates TiC and TiB$_2$ particulate ceramic (Fig. 8.30a) or intermetallic (TiAl) as a reinforcement with minimal particle/matrix interaction (Ref 8.14). A shot sleeve liner for use in the aluminum die-casting industry is shown in Fig. 8.30(b). The liners are now standard equipment in the industry. Mechanical properties of CermeTi material are shown in comparison with PM Ti-6Al-4V in Table 8.7. Seven-layer armor and dual-hardness gears have been made of CermeTi material.

ADMA Products' Hydrogenated Titanium Process. The use of titanium hydride powder instead of titanium sponge fines led to achieving essentially 100% density in complex parts using a simple cost-effective press-and-sinter technique (Ref 8.12, 8.14). In this work, hydrogenated non-Kroll powder was used along with 60:40 aluminum:vanadium master alloy to produce components made of Ti-6Al-4V alloy. The non-Kroll powder is produced by Advance Materials

Table 8.6   Ti-6Al-4V alloy: ASTM E 8 tensile properties

Material(a)	% of theoretical density	Ultimate tensile strength		Yield strength		Elongation, %
		MPa	ksi	MPa	psi	
AMS 4928 (min)	...	896	130	827	120	10
Typical wrought	...	965	140	896	130	14
Typical PM CIP/sinter	98	951	138	841	122	15
Typical PM CHIP	100	965	140	854	124	16

(a) PM, powder metallurgy; CIP, cold isostatic pressed; CHIP, cold and hot isostatic pressed. Courtesy of Dynamet Technology, Inc.

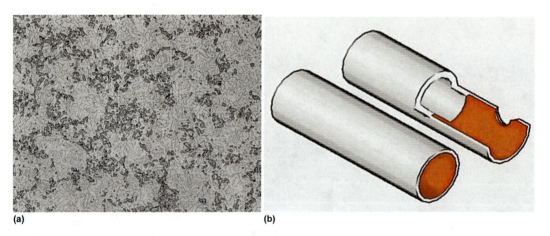

(a)   (b)

Fig. 8.30   (a) Microstructure of CermeTi material with TiC reinforcement. (b) Parts fabricated from this material. Courtesy of Dynamet Technology, Inc.

Table 8.7   Typical properties of CermeTi versus Ti-6Al-4V

Material(a)	Ultimate tensile strength		Yield strength		Elongation, %	Elastic modulus		Hardness, HRC
	MPa	ksi	MPa	ksi		GPa	10^6 psi	
Ti-6Al-4V PM	965	140	896	130	14	110	16.0	36
CermeTi-C MMC (Ti-64 + TiC)	1034	150	965	140	3	130	19.3	42

(a) PM, powder metallurgy; MMC, metal-matrix composite. Data courtesy of Dynamet Technology, Inc.

(ADMA) Products Inc. by cooling the sponge produced in a Kroll process with hydrogen rather than the conventional inert gas to yield a lower-cost titanium hydride powder. The press-and-sinter densities achieved using this novel fabrication technique are shown in Fig. 8.31. The associated microstructure and typical mechanical properties are shown in Fig. 8.32 and Table 8.8, respectively, after cold pressing, sintering, forging, and annealing. The mechanical properties compare well with those exhibited by cast and wrought products. The low cost of this process, in combination with the attractive mechanical properties, makes this approach well suited to applications in the automotive industry (Fig. 8.33).

In the Kroll process, the removal of titanium sponge from the retort and its subsequent crushing is time and energy intensive. In comparison, ADMA's process produces TiH_2 that, unlike titanium sponge, is very friable (Fig. 8.34) and easily removed from the retort, with no need for an expensive sizing operation. ADMA's vacuum distillation processing time is also at least 80% less than in the Kroll process, because phase transformations/lattice parameter changes of the hydride sponge in the presence of hydrogen accelerate the distillation removal of $MgCl_2$. Atomic hydrogen is released during sintering-dehydrating of the TiH_2 powder and serves as a scavenger for impurities (e.g., oxygen, chlorine, magnesium, etc.),

Table 8.8 Room-temperature tensile properties of a hydrogenated titanium compact (after dehydrogenation)

	Material	
Properties	Ti-6Al-4V (PM) 3.5 cm (1.376 in.) thick	ASTM
Ultimate tensile strength, MPa (ksi)	994–1028 (144–149)	897 (130)
Yield strength, MPa (ksi)	911–938 (132–136)	828 (120)
Elongation, %	14.0–15.5	10
Reduction in area, %	34–38	25

PM, powder metallurgy

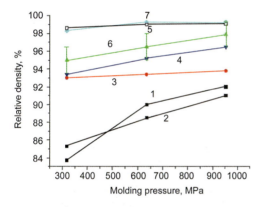

Fig. 8.31 Density of Ti-6Al-4V compacts after sintering. Conditions 5 and 7 used hydrided powder and show the highest, most uniform densities.

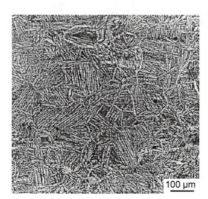

Fig. 8.32 Microstructure of sintered Ti-6Al-4V material

Fig. 8.33 Ti-6Al-4V parts produced using a pressed-and-sintered approach and titanium hydride. 1) Connecting rod with big end cap. 2) Saddles of inlet and exhaust valves. 3) Plate of valve spring. 4) Driving pulley of distributing shaft. 5) Roller of strap tension gear. 6) Screw nut. 7) Embedding filter, fuel pump. 8) Embedding filter. Courtesy of Ukrainian Academy of Sciences

resulting in titanium alloys having low interstitials, which at least meet properties of IM alloys. ADMA Products' hydrogenated titanium powder manufacturing capacities are 113,400 kg (250,000 lb) per year (Fig. 8.35), with an upgraded pilot capacity of 1.8 M kg (4 M lb) per year scheduled for 2014. Major aircraft companies and the U.S. Departments of Energy and Defense agencies tested this material and reported that the properties of the PM titanium alloys meet Aerospace Material Specification requirements and meet or exceed titanium wrought alloys made by conventional IM approaches.

Blended-Elemental Components. Figures 8.36 to 8.38 illustrate a few examples of BE titanium products manufactured by PM.

Prealloy Approach. The PA approach (Ref 8.9, 8.10, 8.12–8.14) involves the use of PA powder, generally spherical in shape, produced through melting by using a technique such as plasma rotating electrode processing (PREP), shown in Fig. 8.39; gas atomization, shown in Fig. 8.40; and by conversion of angular powder to a spherical morphology by plasma processing (the Tekna technique), shown in Fig. 8.41. Table 8.9 lists various commercially available spherical powders. Spherical powders generally are hot consolidated by HIP. Initial complex shape-making research was carried out using ceramic molds. However, his process was discontinued due to concerns that ceramic or other particle contamination could result in degradation of mechanical properties (particularly S-N fatigue, Fig. 8.42). Additional discussion is given in Ref 8.21. Prealloyed parts produced using a shaped metal can and removable mild steel inserts (removed by chemical dissolution) are shown in Fig. 8.43 and 8.44 and are commercially available.

Powder Processing. A number of powder-cleaning processes have been evaluated, but they have not seen commercial implementation. Production of clean powder and maintaining cleanliness through a carefully controlled handling procedure is still the only viable method to guarantee contamination-free compacts.

Because cost considerations strongly favor production of near-net shapes, the use of HIP and

Fig. 8.34 TiH$_2$ powder. Courtesy of ADMA Products Inc.

Fig. 8.35 Pilot-scale unit at ADMA Products for manufacturing hydrogenated titanium powder. Annual capacity is 113,400 kg (250,000 lb). Courtesy of ADMA Products Inc.

Fig. 8.36 The Toyota Altezza, 1998 Japanese car of the year, was the first family automobile in the world to feature titanium valves. Inset: Ti-6Al-4V intake valve (left) and TiB/Ti-Al-Zr-Sn-Nb-Mo-Si exhaust valve (right). Courtesy of Toyota Central R&D Labs, Inc.

Fig. 8.37 Titanium metal-matrix composite golf club head (reinforced with TiB). Courtesy of Toyota Central R&D Labs, Inc.

Fig. 8.38 Susan Abkowitz of Dynamet Technology, Inc. holds a softball bat with a powder metallurgy titanium alloy outer shell. Courtesy of Dynamet Technology, Inc.

vacuum hot pressing generally does not permit significant working of the powder during consolidation. Thus, the normal method of controlling the microstructure in titanium alloys is not available. The strain-energizing process (SEP), which involves working the powder particles by deformation in a rolling mill, enables modifying the alpha morphology to an equiaxed morphology for fatigue strength enhancement. Additionally, THP using hydrogen as a temporary alloying element (whereby hydrogen is introduced, the material is transformed, and the hydrogen is removed) enables adjustment of the microstructure (Fig. 8.45).

It has also been demonstrated that alpha + beta forging of powder compacts followed by an alpha-beta solution treatment can significantly improve the low-cycle fatigue strength of Ti-6Al-4V by changing the as-HIPed microstructures to a duplex structure of low-aspect-ratio primary alpha surrounded by a finer alpha in a beta matrix.

Consolidation. By far, the most often used compaction process is HIP. Other consolidation methods include fluid die compaction (FDC), vacuum hot pressing (VHP), and the Ceracon (ceramic consolidation) process. The FDC process involves production of shaped cavities or dies that are filled with powder. A refinement of the process is rapid omnidirectional compaction (ROC). In the FDC process, the dies are typically made of mild steel and behave as a viscous liquid

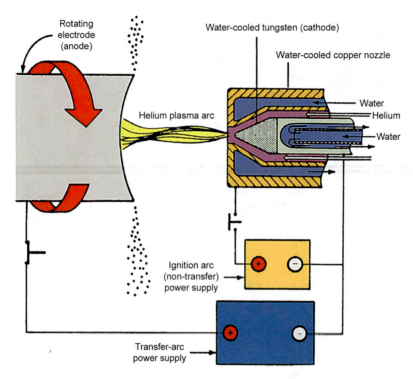

Fig. 8.39 Schematic of plasma rotating electrode process for producing prealloyed titanium powder

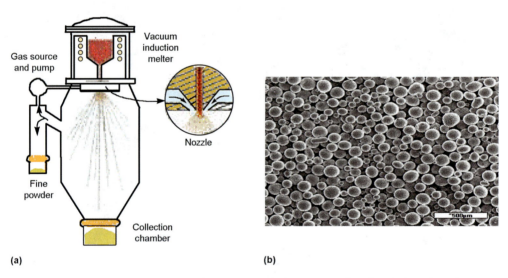

(a)   (b)

Fig. 8.40 (a) Gas atomization setup. (b) Scanning electron micrograph of a gas-atomized prealloyed spherical Ti-6Al-4V. Courtesy of Affinity International

under pressure at temperature. In this case, the pressure can be applied either in an autoclave or in a forge press. This latter approach allows lower-temperature, shorter-time compaction cycles, which can be a major advantage in mini-mizing diffusion-controlled reactions such as precipitate reactions/coarsening and reaction between dissimilar materials in hybrid concepts. In addition, modification of the microstructure with a fatigue improvement has also been reported. The

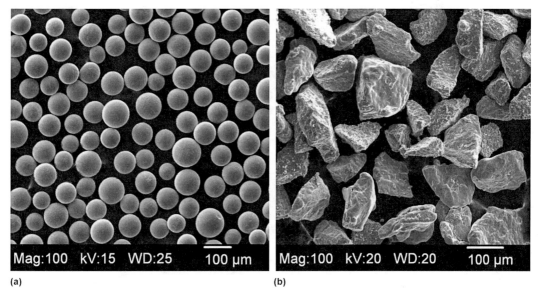

Mag:100   kV:15   WD:25   100 µm   Mag:100   kV:20   WD:20   100 µm

(a)   (b)

Fig. 8.41 Scanning electron micrographs of (a) spherical powder produced by processing (b) angular hydride-dehydride titanium to a spherical morphology using the Tekna technique

Table 8.9   Spherical titanium powders

Type	Producer	Product	Cost (Ti-6Al-4V), U.S. dollars/lb
Gas atomized	ATI Powder (formerly Crucible)	−150 to +45 µm	130
Plasma rotating electrode process (PREP)	Advanced Specialty Metals	−150 to +45 µm	189
Gas atomized (plasma)	Raymor/Pyrogenesis	−30 to +250 µm, 0.09 wt% O_2	40
PREP	Baoji Orchid Titanium	−210 to +45 µm, 0.13 wt% O_2	84
Gas atomized (electrode induction)	ALD Vacuum Technologies	...	...
Gas atomized	Sumitomo Sitex	0.08–0.13 wt% O_2	91
Gas atomized	TLS Technik	53–150 µm, 0.13 wt% O_2	73
Induction plasma	Tekna	Converts irregular, −150 to +37 µm	73
Gas atomized	Iowa State University/Iowa Powder Atomization Technologies	<34 µm	...
PREP	Phelly Materials	−180 to +45 µm, 0.13 wt% O_2 max	72
PREP and gas atomized	Affinity International	177 µm	40

VHP method involves hot compaction of powder in a forge press (fitted with a vacuum system), using permanent and reusable dies to produce fully dense shapes. The Ceracon process involves hot compaction in a granular sand pressing medium. Attributes include a rapid compaction cycle.

Direct processing to mill products (e.g., Ti-6Al-4V plate) can be cost-effective using PA powder, as it is for the BE method. A PM preform was fabricated to final plate dimensions. The mechanical properties produced were very attractive.

Titanium powders are also consolidated by extrusion in an evacuated can. In the case of Ti-6Al-6V-2Sn, tensile and fracture toughness properties equivalent to wrought levels were obtained. An advantage of extrusion is lower-temperature compaction than by HIP. The FDC or VHP methods have an increased importance for RS alloys as they minimize microstructural coarsening.

Preforms. Production cost analysis of isothermally forged components indicates that a large portion of the cost is involved in fabrication of the preform, which is isothermally forged. The use of Ti-10V-2Fe-3Al powder compacts as preforms appears to be a viable approach to produce high-integrity, low-cost isothermal forgings having properties comparable to or exceeding wrought products.

Alloy Types Studied. In addition to extensive studies on Ti-6Al-4V, work has also been carried

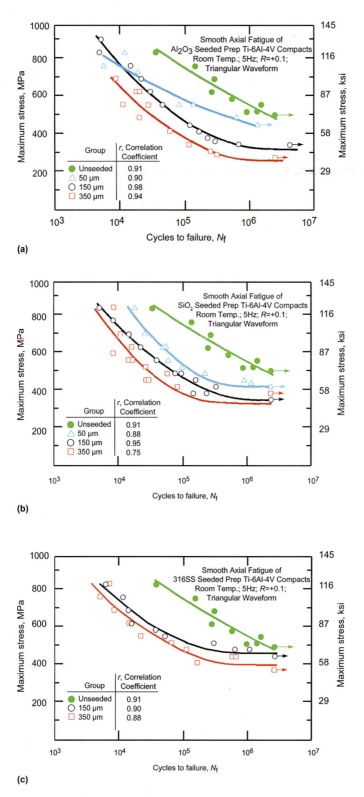

Fig. 8.42 Data points and computer-generated *S-N* curves for all fatigue specimens initiated at seeded (a) Al₂O₃, (b) SiO₂, and (c) 316 stainless steel contaminants, compared with contaminant-free unseeded baseline

Fig. 8.43 Near-net shape Ti-6Al-4V engine casing fabricated using the prealloyed metal can method. Courtesy of Synertech PM Inc./P&W Rocketdyne

Fig. 8.44 Selectively net shape extralow interstitial Ti-6Al-4V impeller for a rocket engine turbopump. Fabricated using the prealloyed metal can method. Courtesy of Synertech PM Inc./P&W Rocketdyne

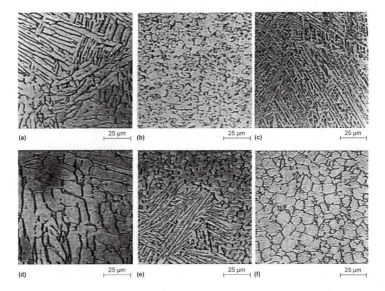

Fig. 8.45 Microstructures of prealloyed Ti-6Al-4V compacts. (a) Hot isostatic pressed (HIP) at 900 °C (1650 °F), 105 MPa (15 ksi), for 2 h. (b) Thermohydrogen processed (THP) using HIP at 900 °C (1650 °F), 105 MPa (15 ksi), for 2 h. (c) THP at 900 °C (1650 °F), 105 MPa (15 ksi), for 12 h to result in a broken-up structure. (d) HIP at 925 °C (1700 °F), 105 MPa (15 ksi), for 4 h in a ceramic mold. (e) Strain-energizing processed (SEP) powder using HIP at 870 °C (1600 °F), 105 MPa (15 ksi), for 2 h. (f) SEP powder fluid die compacted at 925 °C (1700 °F)

out on Ti-6Al-6V-2Sn, Ti-6Al-2Sn-4Zr-6Mo, Corona 5 (Ti-4.5Al-5Mo-1.5Cr), Beta III (Ti-11.5Mo-6Zr-4.5Sn), Ti-10V-2Fe-3Al, and the high-temperature alloys IMI-685 (Ti-6Al-5Zr-0.5Mo-0.25Si) and IMI-829 (Ti-5.5Al-3.5Sn-3Zr-1Nb-0.3Si). In all cases, use of clean powder allows the achievement of mechanical properties equivalent to IM levels.

Microstructures. Hot isostatic processing of Ti-6Al-4V alloy below the beta transus temperature results in a microstructure consisting of alpha plates in a beta matrix. The aspect ratio of these plates is predominantly determined by the amount of work that the alpha (or precursor martensite) receives, which is not large in the case of HIP. Fatigue initiation resistance is improved with a small, equiaxed alpha morphology rather than a coarser/lenticular shape. Modification of the HIP cycle results in a more equiaxed alpha morphology (Fig. 8.45d). To further decrease the alpha aspect ratio, it is necessary to increase the amount of strain in the alpha particles to promote relaxation of the alpha. This can be achieved by either deforming the powder prior to compaction (e.g., SEP) by a high-strain-rate compaction method such as FDC and ROC, or by compacting the powder at relatively low temperatures and high pressures. After SEP, the as-compacted microstructure (Fig. 8.45e) shows areas retaining the original particle shape that were not recrystallized. This is a result of small powder particles that attach to larger particles (i.e., satelliting) and are not cold rolled sufficiently due to their smaller size (Fig. 8.45e). These unrecrystallized areas result in marginal fatigue strength improvement of the SEP compacts over rotating electrode process compacts with the same level of contaminants. However, a fully recrystallized structure was obtained in SEP powder compacted by the FDC process (Fig. 8.45f) with good tolerance to contaminants.

The hot press consolidation method enables the achievement of a more uniform recrystallized alpha structure either after VHP or by use of FDC. The VHP microstructures can also tolerate some level of porosity without significant loss of fatigue strength. This tolerance to defects and porosity is an important consideration in process selection. The VHP process is not yet optimized, and it is expected that the use of ultraclean PREP powders will enable the achievement of fatigue results superior to those of wrought material.

The microstructure of PA compacts can also be modified by the BUS treatment (Fig. 8.22c) previously discussed in relationship to BE material, by thermomechanical processing, and by THP using temporary alloying with hydrogen (Fig. 8.45b,c) (Ref 8.8).

Mechanical Properties. For both IM and PM titanium products, mechanical properties are strongly dependent on alloy microstructure and foreign particles and contaminants. These contaminants have little effect on static properties such as tensile behavior or cyclic propagation, as in fatigue crack growth, but can significantly degrade initiation-related properties such as fatigue (Fig. 8.42). For IM material, the chance of a foreign particle or contaminant being present is

extremely small but finite, and the very small chance of a failure in a component due to this reason is accepted on a statistical basis. In the case of PM, improvements in powder-making and handling techniques also led to low contaminant levels, so the powder technique should be accepted in a similar manner to the IM route. In addition, work is now in progress to improve nondestructive testing techniques for identifying the smallest possible inclusions in titanium alloy compacts. The mechanical properties of powder products, both static (Tables 8.10, 8.11) and, most importantly, critical-fatigue behavior (Fig. 8.23, 8.46),

Table 8.10   Properties of Ti-6Al-4V prealloyed powder compacts

0.2% yield strength		Ultimate tensile strength		Elongation, %	Reduction in area, %	Fracture toughness	
MPa	ksi	MPa	ksi			MPa√m	ksi√in.
930	135	992	144	15	33	77	70

Table 8.11   Hot isostatic pressed powder metal Ti-6Al-4V fracture toughness at 20 °C (68 °F)

Specimen	Fracture toughness (K_{Ic})	
	MPa√m	ksi√in.
1	94.0	85.5
2	96.5	87.8
3	92.5	84.2
Forged Ti-6Al-4V	55	50

Note: Compact tension specimens. BS 7448-1 (1991). Type 1 load vs. displacement. Plane-strain criteria confirmed. Good consistency. K_{Ic} values relatively high. Courtesy of Dr. Wayne Voice, Rolls-Royce, U.K.

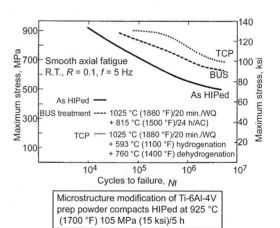

Fig. 8.46 Fatigue data of as-compacted plasma rotating electrode process powder and compacts with a modified microstructure. BUS, broken-up structure; TCP, thermochemical process; WQ, water quenched; AC, air cooled

are at least equivalent to IM levels (including welded material).

Mechanical properties can be enhanced using the BUS and THP (Ref 8.8) processes. These processes result in the high-cycle fatigue strength of clean Ti-6Al-4V PREP compacts being raised to levels at least equal to levels obtained in IM product (Fig. 8.46). A duplex structure of low-aspect-ratio primary alpha surrounded by a finer alpha in a beta matrix can be produced by alpha-beta forging and solution treatment, resulting in significantly improved low-cycle fatigue strength. Additionally, use of higher-strength alloys such as Ti-10V-2Fe-3Al allows the achievement of tensile strengths and fatigue performance exceeding IM Ti-6Al-4V material.

Parts Produced Using Metal Cans. The commercial practice of producing a part in a shaped metal can is carried out by pouring titanium pow-

ders under clean-room conditions into a metal can (with the metal inserts), followed by compaction (Ref 8.12, 8.14). Despite the 30 to 35% volume shrinkage (typical for HIP of PA spherical powders), advanced process modeling allows net surfaces to be achieved and minimal machining stock on the near-net surfaces. Also, these near-net-shape titanium parts are made up to the size of existing HIP furnaces, that is, up to 3 m (10 ft), which is larger than the capabilities of the other technologies discussed in this chapter.

Parts produced in metal cans exhibit static mechanical properties superior to those of conventional cast and wrought (IM) components (Fig. 8.47, 8.48 and Tables 8.10, 8.11) because of the refined microstructure and lack of directionality. Figure 8.47 shows actual tensile strength levels obtained in cast and wrought product compared with data from PM product. However, the

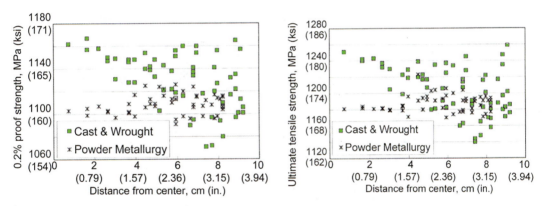

Fig. 8.47 Comparison of ingot and powder metallurgy tensile properties. Courtesy of Prof. Igor Polkin, VILS, Russia

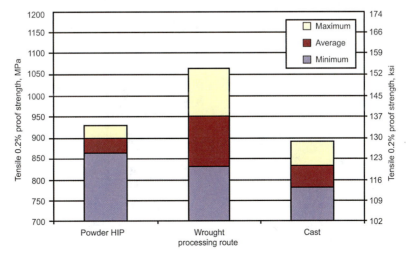

Fig. 8.48 Comparison of powder hot isostatic pressed (HIP) wrought and cast tensile properties

minimum values (which are used in design) for the PM material is above that for the conventionally fabricated material (Fig. 8.48). Fracture toughness of the PM product is superior to cast and wrought material (Table 8.10), as is the *S-N* fatigue strength (Fig. 8.23, 8.46).

Additive manufacturing (AM) uses the principle of slicing a solid model in multiple layers, storing the data in a computer, and building up the part layer by layer with powder following the sliced model data (Ref 8.22). Two basic approaches to AM (Ref 8.22) are direct-energy deposition (DED), shown in Fig. 8.49, and powder-bed fusion (PBF), shown in Fig. 8.50. The DED approach has the advantages of large build envelopes, higher deposition rates, the capability of multiple material depositions in a single build, and the ability to add material to existing parts (including repair of parts). The PBF technique allows buildup of complex features, hollow cooling passages, and production of high-precision parts.

The strength of AM Ti-6Al-4V is 1104 MPa (160 ksi) with 5 to 6% elongation as-formed, and after a HIP operation it is 965 MPa (140 ksi) with 15% elongation, equivalent to cast and wrought levels (data courtesy of B. Dutta, the POM Group). Tensile strength, yield strength, and elongation of Ti-6Al-4V alloy built using various AM processes are shown in Fig. 8.51.

The *S-N* fatigue performance of additive-layer manufacturing (ALM) is similar to that of conventional material levels (Fig. 8.52). However, the main issue influencing growth in deployment

of ALM for titanium alloys is related to raw material supply. With material cost being typically 40 to 50% of total manufacturing cost for AM titanium, material cost is a major issue. Material supply chain is an issue for both powder and wire; sustainable sources are not always available, and supplies of certain alloys have limited availability.

The direct manufacturing capability of AM technologies also helps reduce manufacturing cost in the case of high buy-to-fly ratio parts. Researchers at the U.S. Department of Energy's Oak Ridge National Laboratory built a Ti-6Al-4V bleed air leak-detect (BALD) bracket for the Joint Strike Fighter engine using electron beam melting technology (Fig. 8.53) (Ref 8.22, 8.23). Traditional manufacturing from wrought Ti-6Al-4V

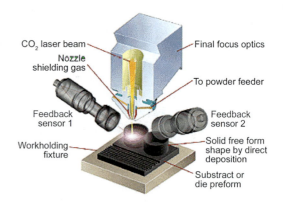

Fig. 8.49 Schematic showing direct-metal-deposition technology. Courtesy of DM3D Technology

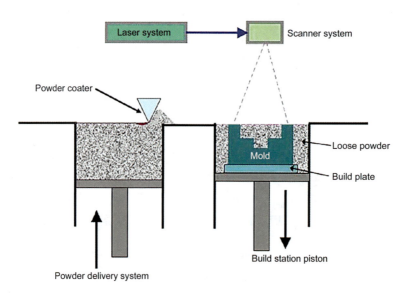

Fig. 8.50 Schematic showing powder-bed fusion technology. Courtesy of Jim Sears

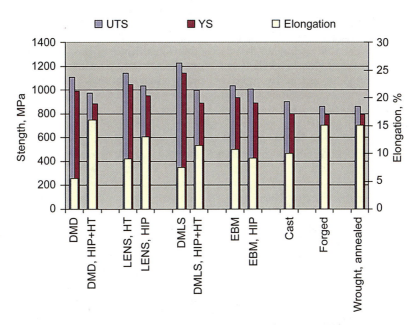

Fig. 8.51 Ultimate tensile strength (UTS), yield strength (YS), and elongation of Ti-6Al-4V alloy produced using various additive manufacturing processes. DMD, direct-metal deposition; HIP, hot isostatic pressing; HT, heat treatment; LENS, laser-engineered net shaping (Ref 8.16); DMLS, direct-metal laser sintering; EBM, electron beam melting

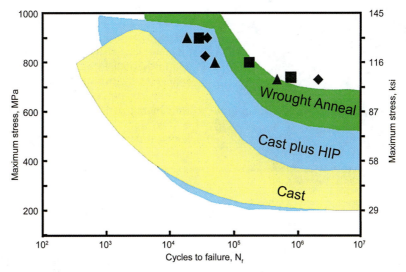

Fig. 8.52 Comparison of room-temperature fatigue properties of Ti-6Al-4V processed by additive manufacturing and a comparison with conventionally fabricated Ti-6Al-4V. ■, ♦, and ▲ represent properties in the three orthogonal directions: x, y, and z, respectively. HIP, hot isostatic pressing. Courtesy of EADS/Jim Sears

plate costs almost $1000/lb due to a high (33:1) buy-to-fly ratio as opposed to just over a 1:1 ratio for the AM-built part. Estimated savings through AM is approximately 50%. The characteristics/advantages of fabricating the BALD component are listed in Table 8.12. The advantage of attaching features is demonstrated in Fig. 8.54 and 8.55,

and the capability of using ALM in repair of a part is illustrated in Fig. 8.56.

Powder Injection Molding. Metal powder injection molding (PIM) is based on the injection molding of plastics (Ref 8.24–8.26). The process was developed for long production runs of small (normally below 400 g), complex-shaped titanium parts in a cost-effective manner. The technique involves melting and pelletization of a mixture of titanium powder and a binder that is injected into a die, where the binder is removed chemically/thermally, and the part is sintered. By increasing the metal (or ceramic) particle content, the process evolved into a process for production of high-density metal, intermetallic, and ceramic components (Fig. 8.57). The method enables the fabrication of components having good mechanical properties, provided the chemistry (particularly oxygen) is controlled. Typical shapes are shown in Fig. 8.58. By incorporating a porous layer on the surface of body-implant parts, Praxis Technology Inc. is able to cause bone ingrowth and an improved bonding between the implant and bone.

Fig. 8.53 Bleed air leak-detect bracket for Joint Strike Fighter built using electron beam melting technology. Courtesy of Oak Ridge National Laboratory

The majority of early work on developing a viable titanium PIM process was plagued by the unavailability of suitable powder, inadequate protection of the titanium during elevated-temperature processing, and less-than-optimum binders for a material as reactive as titanium. However, PIM practitioners learned that titanium is the universal solvent and must be treated accordingly.

Titanium has a high capacity to readily form interstitial solutions with a wide range of commonly encountered elements, including carbon, oxygen, and nitrogen, which has presented several challenges for titanium PIM development efforts. These interstitial solutions are undesirable because they significantly degrade the ductility of sintered titanium PIM parts. Therefore, it is advantageous to use a binder, which can be completely removed from the green PIM part without leaving these detrimental impurities behind. This is particularly true when fabricating structural aerospace and medical implant parts, which can require oxygen impurity levels below 300 ppm to meet ASTM F 167 standards.

Unfortunately, unlike conventional ceramic and ferrous alloy PIM processing, there is a significantly narrower processing window between the debinding cycle and the temperature where impurity diffusion becomes significant within titanium. In general, this requires that the titanium PIM binder be essentially removed from the green part at temperatures typically below 260 °C (500 °F) to prevent introducing impurities into the sintered parts. Additionally, the binder must exhibit high chemical stability and not undergo catalytic decomposition in the presence of titanium metal powder surfaces during molding operations, even when held for long isothermal holds within the injection molding machine.

Attempts to adapt conventional ceramic and metal PIM binder systems for titanium processing

Table 8.12 Ti-6Al-4V bleed air leak-detector bracket fabricated using additive manufacturing

	Traditional manufacturing	Additive manufacturing
Buy:fly ratio	33:1	Nearly 1:1
Machining	Extensive—97% of raw material is removed by machining.	Slightly oversized part produced using Arcam AB electron beam melting requires machining of 1.27 mm (0.05 in.) on all surfaces.
Average mechanical properties after hot isostatic pressing		
Yield stress, MPa (ksi)	...	876 (127)
Ultimate tensile strength, MPa (ksi)	...	945 (137)
Elongation, %	...	14.4
Chemistry	Ti-6Al-4V	Ti-6.06Al-4.08/ 4.10V-0.17/0.14O_2
Cost (2014)	~$1000	<$500

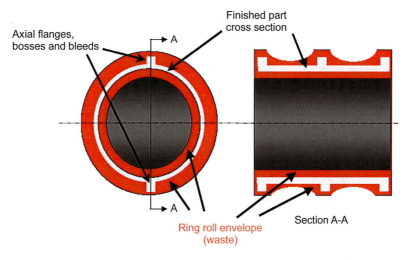

Roll forged case profile—forging waste

Finished part cross section

Axial flanges, bosses and bleeds

A

A

Ring roll envelope (waste)

Section A-A

Discrete features such as axial flanges and bosses produce a disproportionate increase in forging size and weight.

Fig. 8.54 Material waste in machining features on a forged preform in conventional manufacturing. Material shown in red is removed.

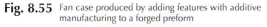

Fig. 8.55 Fan case produced by adding features with additive manufacturing to a forged preform

Fig. 8.56 Additive manufacturing repair of gas turbine components

have met with limited success. This is due to the fact that these systems often contain significant amounts of thermoplastic polymer in their formulations. Unfortunately, even some of the more common polymer binders known for their ability to readily thermally "unzip" to their starting monomers (e.g., polymethyl methacrylate, polypropylene carbonate, poly-a- methylstyrene) still

tend to introduce impurities into the sintered titanium PIM bodies, because their depolymerization occurs close to those temperatures where impurity uptake initiates. Alternative binder systems based on catalytic decomposition of polyacetals are promising but require expensive capital equipment to handle the acid vapor catalyst, as well as a suitable means of eliminating the formaldehyde

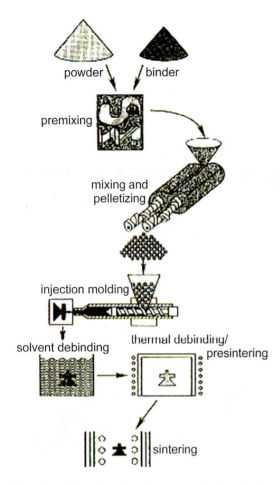

Fig. 8.57 Schematic of the steps involved in powder injection molding, in which a polymer binder and metal powder are mixed to form the feedstock, which is molded, debound, and sintered

oligomers that form as polymer decomposition by-products. However, there are a number of binder systems that appear to have the necessary characteristics to be compatible with titanium.

Currently, titanium PIM parts are made up to 30 cm (~12 in.) long; parts over 10 cm (4 in.) and weighing 50 g (~2 oz) are not common (Fig. 8.58). The limiting factors at this time are dimensional reproducibility and chemistry. Due to the shrinkage, large parts become dimensionally more difficult to make due to loss of shape during shrinkage. If the parts have flat surfaces to rest on the setter, they come out fairly consistently. However, parts with multiple surfaces that require setters in complex shapes become less practical as the size increases. Further, large overhanging areas become difficult to control dimensionally due to gravity. With experience, the packing density of titanium powder mixes will be increased, especially as new binders become available and shrinkage can be decreased, making the dimensional problems less of a factor.

The current estimate is that the worldwide titanium PIM part production is currently at approximately the 3 to 5 ton/month level. This market is poised for expansion. What is needed is low cost (less than $20/lb, or $44/kg), generally spherical (for good flowability) powder of the right size (less than approximately 40 μm), and good purity (which is maintained throughout the fabrication process). For nonaerospace applications, the purity level of the Ti-6Al-4V alloy can be less stringent; for example, the oxygen level can be up to 0.3 wt% while still exhibiting acceptable ductility levels. (Aerospace requires a maximum

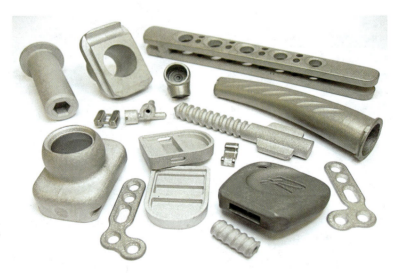

Fig. 8.58 Titanium powder injection-molded components. Courtesy of Praxis Technology Inc.

oxygen level of 0.2 wt%.) For the commercially pure (CP) grades, oxygen levels can be even higher, up to at least 0.4 wt%. (Grade 4 CP titanium has a specified limit of 0.4 wt%.) In fact, grade 4 CP titanium (550 MPa, or 80 ksi, ultimate tensile strength), while lower in strength than regular Ti-6Al-4V (930 MPa, or 135 ksi, ultimate tensile strength), may well be a better choice for many potential PIM parts where cost is of great concern. Grade 4 offers the use of lower-cost starting stock and higher oxygen content in the final part. Further in the future, beta alloys, with their inherent good ductility (body-centered cubic structure), and the intermetallics, with attractive elevated-temperature capability, are potential candidates for fabrication by way of PIM. The science, technology, and cost are now in place for the titanium PIM market to show significant growth.

Typical Ti-6Al-4V PIM characteristics are:

- Oxygen content: 0.14 to 0.32 wt%
- Relative density: 96 to 98%
- Ultimate tensile strength: 828 to 931 MPa (120 to 135 ksi)
- Elongation: 9 to 14%

Factors affecting the mechanical properties of PIM titanium include:

- Oxygen level increase leads to higher strength and lower ductility.
- Relative density increase leads to higher strength and higher ductility.
- Beta grain size decrease leads to higher strength and higher ductility.
- Smaller powder size increases the oxygen level and decreases the beta grain size.

A further list of the Ti-6Al-4V PIM mechanical properties is given in Table 8.13. Figure 8.59 shows the effect of oxygen content on the strength and ductility of sintered CP titanium powder.

Spraying. Spray forming involves molten metal and solid powder (Ref 8.12, 8.14). Because of its very high reactivity, the challenges associated with molten metal spraying of titanium are considerable. However, spray forming in both an inert environment and under reactive conditions has been achieved with appropriately designed equipment.

Recently there has been increased interest in cold spray forming that involves solid powder particles. Cold spray < 500 °C (930 °F) can produce both monolithic "chunky" shapes and coated components. In this process, solid powder is introduced into a de Laval-type nozzle and expanded to achieve supersonic flow. Powders are in the range of 1 to 50 μm, at relative low temperatures (<500 °C, or 930 °F), and with a velocity in the range of 300 to 1200 m/s (984 to 3937 ft/s). Both monolithic chunky shapes and coated components can be produced (Fig. 8.60); coatings can even be applied to the inside of tubular components. The density of the sprayed region is less than full density, but this can be increased to 100% density by a subsequent HIP operation. This technique is also very useful in bonding metals such as titanium and steel (Fig. 8.61).

Rapid solidification (RS) includes the following considerations.

Powder Production. Because of its extreme reactivity in the molten state, titanium powder is difficult to produce with techniques used routinely for aluminum and less-reactive metals (Ref 8.27, 8.28). However, there are a number of techniques available that have been used to rapidly solidify titanium, including local melting, melt extraction, layer glazing, and gas atomization. In the gas atomization technique, either a titanium skull must be formed or some reaction with guide tubes, orifices, and atomization nozzles must be accepted.

Processing and Consolidation. The microstructure of rapidly solidified material must be controlled through processing and consolidation steps so that the benefits of RS can be fully used to achieve attractive mechanical property levels

Table 8.13 Attractive Ti-6Al-4V powder injection molding mechanical properties

Oxygen content, wt%	Relative density, %	Ultimate tensile strength		Elongation, %	Binder system
		MPa	ksi		
0.24	96.0	971	140.8	12.0	Polypropylene-ethyl vinyl acetate, paraffin wax, carnauba wax, dioctyl phthalate
0.25–0.28	95.5	840	121.9	14.0	Polypropylene, polymethyl methacrylate, paraffin, steric acid
0.25	95.5	836	121.2	13.4	Specially developed on polymer base

in the final article. This is challenging, particularly with alloying additions such as the rare earths (lanthanides), where there is minimal solid-state solubility. That is, once a second-phase dispersion coarsens, it cannot be refined without returning to the liquid state. Consequently, the general guideline for processing of RS titanium alloys is to avoid high temperature and extended time excursions. Thus, while conventional processing, as discussed for PA powder, may be acceptable in some instances, schemes such as

high-pressure, low-temperature HIP, ROC (perhaps done in an isothermal mode; that is, iso-ROC), and high-strain-rate techniques (explosive or gun compaction) have been evaluated.

Alloy Types. A comprehensive study done on RS conventional titanium alloys has used the Ti-6Al-4V alloy. However, some data have been presented on a similar alpha-beta Ti-Mo-Al alloy (Table 8.14) in which both strength and ductility were increased as the cooling rate was increased (Ref 8.28). The work on the Ti-6Al-4V used cooling rates varying from 10^2 to 10^7 °C/s (180 to 1.8×10^7 °F/s) (splat). From this work, two significant effects were noted: the beta grain size decreased as the cooling rate was increased (Fig. 8.62), and at the higher cooling rate studied, an equiaxed alpha was produced on subsequent high-temperature annealing (Fig. 8.63). This latter effect is not observed on annealing either conventional powder or unworked IM material and probably results from the fine beta grain size. As discussed in the section on PA powders, this fine, equiaxed alpha is likely to have attractive fatigue-initiation properties.

Conventional alloys are those that can be produced using the conventional IM method. However, there are other titanium alloys that can only be produced by RS methods. They require extremely rapid cooling to maintain solute in solution or to avoid the segregation and workability problems that occur in IM material. An example of an alloy class that generally cannot be processed to give a fine dispersion of precipitates by conventional means and therefore requires the RS

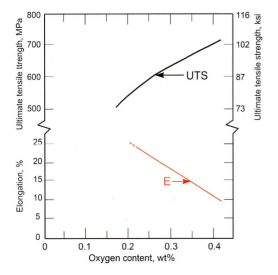

Fig. 8.59 Effect of oxygen content on the strength and ductility of sintered commercially pure titanium powder. UTS, ultimate tensile strength; E, elongation. Courtesy of Daido Steel Co.

Fig. 8.60 Components produced at the Commonwealth Scientific and Industrial Research Organization, by cold spraying commercially pure powder

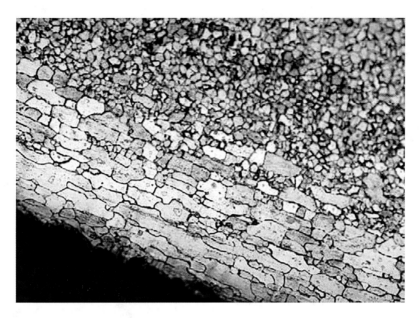

Fig. 8.61 Optical micrograph showing the excellent bonding of titanium (bottom) to steel (top). Courtesy of Ktech

Table 8.14 Effect of cooling rate on tensile properties of a Ti-Mo-Al alloy

Production method	Ultimate tensile strength		Reduction in area, %
	MPa	ksi	
Casting	896	130	12
Cast + hot work	1068	155	12
10^3 to 10^4 °C/s (1,800 to 18,000 °F/s) cooling + hot isostatic pressing	1068	155	22
Flake 10^6 to 10^7 °C/s (1.8×10^6 to 10^7 °F/s) cooling + hot isostatic pressing	1137	165	30

method is titanium combined with rare earth additions such as yttrium, erbium, and neodymium. The increase in fineness of the microstructure is shown in Fig. 8.64. An indication of the possible strength improvement is given in Fig. 8.65, and an order of magnitude creep-rate and stress-rupture improvement was found in the same program. However, cooling rate is very critical, and rates of 10^2 to 10^5 °C/s are necessary to retain 0.7 at.% Er in pure titanium using techniques such as laser surface melting. A fine dispersion (50 to 250 Å) is then obtained on aging at 700 °C (1290 °F). If alloys can be developed that can increase the service temperature of titanium by 165 °C (300 °F), substantial weight savings can be achieved in advanced engines by replacement of more-dense superalloy components.

Other research is evaluating eutectoid-former additions (nickel, copper, silicon, iron, tungsten, etc.), which have been studied in detail as IM prod-

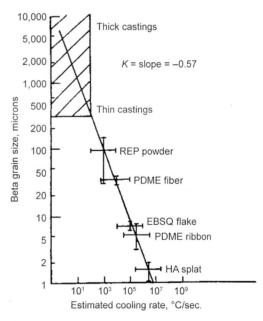

Fig. 8.62 Effect of cooling rate on beta grain size of Ti-6Al-4V produced using various methods. REP, rotating electrode process; PDME, pendent drop melt extraction; EBSQ, electron beam splat quenching; HA, hammer and anvil

uct. However, these additions are prone to segregation, and PM may be the only way to successfully process such alloys. These alloys have the potential to achieve extremely high strength with good ductility, although here the high cooling rate may not be required, only a nonsegregated product.

Fig. 8.63 Equiaxed alpha produced in rapidly solidified Ti-6Al-4V after annealing at 965 °C (1765 °F)

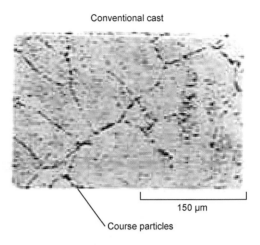

Fig. 8.64 Comparison of structures produced by conventional casting (top) and rapid solidification (bottom) of titanium rare earth alloys

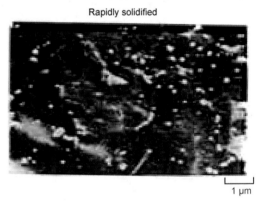

Fig. 8.65 Strength increase with rare earth additions to commercially pure titanium

Thermohydrogen Processing. This process is still under development. The titanium-hydrogen phase diagram is shown in Fig. 8.66. By intentionally adding hydrogen to a titanium alloy such as Ti-6Al-4V with a typical PM microstructure, the microstructure can be refined in the dehydrogenated condition (Fig. 8.66) with an enhancement in mechanical properties (Ref 8.8).

Porous Structures. A novel type of porous, low-density titanium alloy can be produced by HIP consolidation of alloy powder in the presence of an inert gas such as argon (Fig. 8.67) (Ref 8.12, 8.14). Tensile strength decreases in a linear manner as the porosity level increases, following the "rule of mixtures" relationship at least up to the 30% porosity level. This material exhibits excellent damping characteristics, suggesting a generic area of application. There may also be applications in body implants, with the foam integrated in various locations to facilitate growth of bone/flesh into the porous regions, promoting a stronger joint. Tensile testing of the bond between foam and dense material indicated a bond strength in excess of 85 MPa (12.3 ksi), well above the Food and Drug Administration (FDA) requirement of 22 MPa (3.2 ksi) for porous coatings on orthopedic implants.

Porous structures with potential use in honeycomb structures, in sound attenuating, and in firewall applications are now possible with very precisely controlled porosity levels and architecture using a novel blended metal-plastic approach (Fig. 8.68).

Safety

Safety should always be foremost in the mind of the titanium practitioner. There have been many fires/explosions since the beginning of the titanium industry in the late 1940s, with a number of fatalities. It should be recognized that titanium is a very reactive metal that readily combines

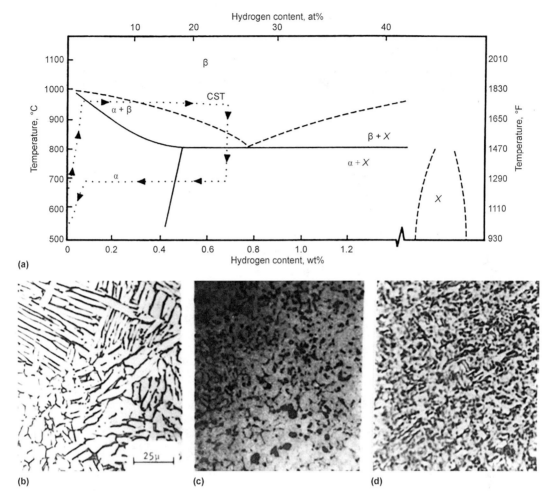

(a)

Fig. 8.66 (a) Pseudobinary phase diagram for Ti-6Al-4V. X represents the hydride phase. CST, constitutional solution treatment. Refinement of the microstructure of Ti-6Al-4V powder compact using the thermohydrogen processing technique is shown in (b) as hot isostatic pressed coarse alpha laths, (c) hydrogenated then compacted, and (d) hydrogenated in the compacted state. The conditions in (c) and (d) are after dehydrogenation, both showing a refined alpha microstructure: (c) equiaxed grains and (d) fine alpha laths.

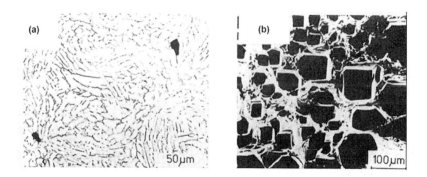

Fig. 8.67 (a) Optical micrograph of pores in hot isostatic pressed Ti-6Al-4V containing argon after annealing at 700 °C (1290 °F). (b) Scanning electron micrograph of sample containing up to 40% porosity after annealing at a temperature higher than 1000 °C (1830 °F)

with oxygen to form its oxide bulk. Titanium has a thin surface layer of oxide, which provides its excellent corrosion resistance (see Chapter 14, "Corrosion," in this book). There are a number of situations in which a titanium fire/explosion can occur. All manufacturers' data and product sheets

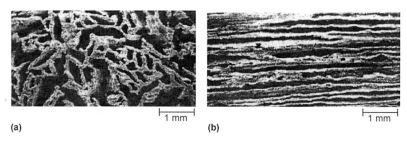

Fig. 8.68 Micrographs of foams produced by die compaction and extrusion. (a) Cross section is perpendicular to the compaction direction. (b) Cross section is parallel to the extrusion direction.

should be consulted, as well as all Occupational Safety and Health Administration (OSHA), Environmental Protection Agency (EPA), and state and federal regulations. The International Titanium Association, from which the following text is taken, has additional resources that may be of help in titanium safety (Ref 8.29).

"The first rule of thumb for housekeeping in fire prevention is material segregation: safely storing titanium scrap, chips, and "swarf" in metal containers. *Swarf* is a term used to collectively identify extremely fine unoxidized titanium scrap normally created from wet sanding operations. It represents a major fire hazard because small-particle scrap is more susceptible to ignition and burning. Small scrap always should be stored in quantities easily moved with the materials handling equipment available. Any facility processing titanium scrap should have a "hot work" program that prevents torch cutting, grinding, and welding in areas where titanium and other combustibles are stored.

Material segregation also involves separating small titanium scrap from other metal scrap, especially steel dust. The mixing of rusting steel and titanium scrap along with water could create an exothermic reaction or heat, which leads to burning materials. Titanium, a reactive metal, wants to return to its oxide state, such as titanium dioxide, when it burns.

Beyond material segregation, fire safety should include preventing the buildup of metal dust in a titanium facility. Develop a plant-maintenance regime that involves regular inspection and cleaning of a facility's equipment, air ducts, and dust collectors. Titanium powder and unoxidized dust can flare and burn when suspended in the air.

Other housekeeping points involve the periodic inspection of a plant's lighting and electrical systems. Static-charge sparks can be an ignition source for titanium. To avoid static buildup, use grounding and bonding wires on all equipment used to process titanium chip, dust, and powder. Do not forget to look at building-support structures or suspended ceilings for dust accumulation. Many facilities forget to look in areas that are not easily accessible.

Companies are urged to put together a comprehensive fire safety program to educate employees. Companies should consult with an industry fire-prevention expert or contact state, county, or municipal fire inspectors. (The International Titanium Association is a resource to obtain this type of information.) Of course, any common-sense fire-safety program involves designating safe, restricted areas for those employees who smoke.

What happens if there is an accident and a fire breaks out at a titanium facility? A typical titanium fire burns hot and steady; rarely is there a sudden explosion or a violent burst of flame. Users and processors should be aware of areas where titanium dust may build up, and they should take action to prevent accumulations in tight areas.

If a fire does occur, *do not use water* to douse the blaze. The best way to extinguish a titanium fire is to smother it with salt (regular table salt), sand, and dirt; 50 lb bags or 55 gal drums of sand or salt can be purchased at industrial-supply outlets. These bags and drums should be stored on the factory floor in specially designated areas where they can be easily reached."

For further details, including a discussion on safe melting practices, the reader is referred to Ref 8.29.

Future Developments in Titanium Powder Metallurgy

Advances in the BE area stem from the use of low-chloride sponge in combination with densification operations such as HIP and forging (either conventional or isothermal) and innovative treatments such as the BUS treatment or THP. The qualification of Dynamet Technology BE powder metallurgy material by Boeing is a major step forward, as is the use of hydrogenated sponge by ADMA Products.

Prealloyed technology using metal-can technology is at the point of acceptance as a cost-effective manufacturing method by the extremely demanding aerospace industry. The availability of a lower-cost spherical powder would greatly help this technology become commercially viable. The PA approach should also offer inherent advantages for difficult-to-work and -machine alloys such as intermetallic titanium aluminides (Ti_3Al and TiAl).

The AM technique has tremendous potential for the cost-effective fabrication of complex titanium parts and for repair of damaged components. Metal injection molding is an attractive method for the manufacture of small (½ kg, or 1 lb) titanium parts. Sprayed components and porous titanium could have niche markets.

Rapid solidification offers many of the same advantages in the titanium system as in other systems, such as aluminum. This technique offers advantages in systems such as the titanium-lanthanides because it enables circumventing constitutional constraints placed on the system by conventional IM. Fine, stable dispersions of rare earth compounds are being studied for potential use in high-temperature applications. Additions such as the eutectoid formers (chromium, nickel, tungsten) and metalloids (carbon, boron) should lead to new higher-strength alloys. Rapid solidification may help significantly in optimizing the mechanical properties of the titanium aluminides. A fine grain size combined with oxygen "gettering" by rare earth additions could enhance the ductility of this normally brittle class of alloys. As an integral part of the overall exploration of the potential for RS titanium alloys, controlled processing must be practiced. Generally, the requirements here will be for reduced thermal exposure during compaction, so that the microstructural and constitutional advantages offered by RS can be maintained in the final product, giving enhanced mechanical property behavior.

Summary

Melting, casting, and PM of titanium alloys were reviewed in this chapter. The traditional melting technique for titanium, consumable arc melting, has been re-evaluated because of concerns over ingot quality in demanding applications. Cold hearth melting to reduce/eliminate defect content, followed in some cases by a subsequent consumable arc melting, is now a commercial entity.

Casting is the lowest-cost net-shape processing route. This technique has shown a steady, healthy growth, a result of improving shape-making capabilities and enhanced mechanical properties from microstructural control.

Powder metallurgy can be subdivided into blended elemental (BE), prealloyed (PA), additive manufacturing (AM), powder injection molding (PIM), spray forming, and rapid solidification (RS). The BE approach offers low cost and mechanical behavior at or approaching ingot levels. In contrast, PA is higher in cost but is capable of producing large, complex shapes. The AM approach is an attractive processing alternative and should expand significantly. Powder injection molding is already a viable fabrication option for small, complex parts (less than approximately ½ kg, or 1 lb, in weight). The RS technique is high cost but also offers the potential for higher strength, increased temperature capability, and lower-density alloys than any other processing method. However, it remains far from commercialization. Thermohydrogen processing and porous materials were also briefly discussed.

Glossary of Acronyms

BE	blended elemental.
BUS	broken-up structure.
CHIP	cold and hot isostatic pressing.
CIP	cold isostatic pressing.
CP	commercially pure (titanium).
CSC	continuous shotcasting.
CST	constitutional solution treatment.
EB	electron beam.
EBRD	electron beam rotating disc.
ELI	extralow interstitial (content).
FCGR	fatigue crack growth rate.
FDC	fluid die compaction.
GA	gas atomization.
GFC	gas fan cooled.
HCF	high-cycle fatigue.
HDH	hydride-dehydride.
HDI	high-density inclusion.
HIP	hot isostatic pressing.
HPLT	high pressure, low temperature.
HTH	high-temperature hydrogenation.
IM	ingot metallurgy.
iso-ROC	isothermal rapid omnidirectional compaction.

LCF   low-cycle fatigue.
PA   prealloyed.
PM   powder metallurgy.
PREP   plasma rotating electrode process.
REP   rotating electrode process.
rms   root mean square.
ROC   rapid omnidirectional compaction.
RS   rapid solidification.
SEP   strain-energizing process.
TCP   thermochemical process (temporary alloying with hydrogen).
TMP   thermomechanical processing.
VAR   vacuum arc remelt.
VHP   vacuum hot press.

REFERENCES

8.1   F.H. Froes, *Kirk-Othmer Encyclopedia of Chemical Technology,* 5th ed., Nov 2006, p 24, 838

8.2   M.A. Imam, F.H. Froes, and K.L. Housley, Titanium and Titanium Alloys, *Kirk-Othmer Encyclopedia,* March 2010, p 1–41 (online only)

8.3   F.H. Froes, D. Eylon, and H.B. Bomberger, Ed., *Titanium Technology: Present Status and Future Trends,* Titanium Development Association, Dayton, OH, 1985

8.4   D. Eylon, J.R. Newman, and J.K. Thorne, Titanium and Titanium Alloy Castings, *Properties and Selection: Nonferrous Alloys and Special-Purpose Materials,* Vol 2, *ASM Handbook,* ASM International, 1990

8.5   F.C. Teifke, N.H. Marshall, D. Eylon, and F.H. Froes, Effect of Processing on Fatigue Life of Ti-6Al-4V Castings, *Advanced Processing Methods for Titanium,* D. Hasson, Ed., The Metallurgical Society, 1982, p 147–159

8.6   H.D. Hanes, D.A. Seifert, and C.R. Watts, *Hot Isostatic Processing,* Battelle Press, 1979, p 55

8.7   R.R. Wright, J.K. Thorne, and R.J. Smickley, Technical Bulletin TB 1660, Ti-Cast Division, Howmet Turbine Components Division, Howmet Corp., 1982

8.8   F.H. Froes, O.N. Senkov, and J.I. Qazi, Hydrogen as a Temporary Alloying Element in Titanium Alloys: Thermohydrogen Processing, *Int. Mater. Rev.,* Vol 49 (No. 3–4), 2004, p 227–245

8.9   F.H. Froes et al., Developments in Titanium Powder Metallurgy, *J. Met.,* Vol 32 (No. 2), 1980, p 47–54

8.10   F.H. Froes and J.E. Smugeresky, Ed., *Powder Metallurgy of Titanium Alloys,* TMS/AIME, Warrendale, PA, 1980

8.11   D. Eylon, P.R. Smith, W. Schwenker, and F.H. Froes, Status of Titanium Powder Metallurgy, *Industrial Applications of Titanium and Zirconium: Third Conference,* STP 830, R.T. Webster and C.S. Young, Ed., ASTM, 1984, p 48–65

8.12   F.H. Froes, Powder Metallurgy of Titanium Alloys, *Advances in Powder Metallurgy,* I. Chang and Y. Zhao, Ed., Woodhead Publishing, 2013, p 202

8.13   M. Qian, Ed., *Powder Metallurgy of Titanium,* Trans Tech Publications, Durnten-Zurich, Switzerland, 2012

8.14   F.H. Froes, Titanium Powder Metallurgy: Developments and Opportunities in a Sector Poised for Growth, *Powder Metall. Rev.,* Vol 2 (No. 4), Winter 2013, p 29–43

8.15   J.H. Moll, *J. Mater.,* Vol 52 (No. 5), May 2003, p 32

8.16   F.H. Froes, M.A. Imam, and D. Fray, Ed., *Cost-Affordable Titanium,* TMS, Warrendale, PA, 2004

8.17   M.N. Gungor, M.A. Imam, and F.H. Froes, Ed., *Innovations in Titanium Technology,* TMS, Warrendale, PA, 2007

8.18   M.A. Imam, F.H. Froes, and K.F. Dring, Ed., *Cost-Affordable Titanium III,* Trans Tech Publications, Zurich-Durnten, Switzerland, 2010

8.19   M.A. Imam, F.H. Froes, and R.G. Reddy, Ed., *Cost-Affordable Titanium IV,* Trans Tech Publications, Zurich-Durnten, Switzerland, 2013

8.20   M.J. Donachie, Jr., *Titanium: A Technical Guide,* ASM International, 2000

8.21   S.W. Schwenker, D. Eylon, and F.H. Froes, Influence of Foreign Particles on Fatigue Behavior of Ti-6Al-4V Prealloyed Powder Compacts, *Metall. Trans. A,* Vol 17, Feb 1986, p 271

8.22   B. Dutta and F.H. Froes, The Additive Manufacture (AM) of Titanium Alloys, *Adv. Mater. Process.,* Feb 2014, p 18–23

8.23   R. Dehoff, C. Duty, W. Peter, Y. Yamamoto, W. Chen, C. Blue, and C. Tallman, Case Study: Additive Manufacturing of Aerospace Brackets, *Adv. Mater. Process.,* Vol 171 (No. 3), March 2013, p 19–22

8.24   R.M. German, *Powder Metallurgy Science,* 2nd ed., MPIF, Princeton, NJ, 1994, p 192 8.25. F.H. Froes and R.M. German, Titanium Powder Injection Molding (PIM),

Met. Powder Rep., Vol 55 (No. 6), 2000, p 12

8.26 R.M. German, *Powder Inject. Mold. Int.,* Vol 3 (No. 4), Dec 2009, p 21–37

8.27 H. Jones, *Rapid Solidification of Metals and Alloys,* Monograph No. 8, The Institution of Metallurgists, London, 1982

8.28 F.H. Froes and D. Eylon, Ed., *Titanium: Rapid Solidification Technology,* TMS, Warrendale, PA, 1986

8.29 "Safety Guidelines for Handling Titanium," International Titanium Association, www.titanium.org/?page=SafetyManual

CHAPTER 9

Primary Working*

TITANIUM AND TITANIUM ALLOYS are fabricated using many methods to produce mill products, including billet, bar, rod, wire, plate, sheet, strip, extrusions, pipe, and tubing (Ref 9.1–9.13). Most integrated titanium mills have primary working equipment designed for titanium, but, in some cases, conversion work is done on equipment designed for stainless steel and other specialty steels. This presents difficulties because of the behavior and properties of titanium. Hot working temperatures must be more closely controlled, generally using the lowest temperatures practical. Titanium readily seizes and galls and requires thoroughly conditioned surfaces to remove any cracks or contamination that could propagate into the base metal on subsequent processing. In addition, anisotropy (the characteristic of exhibiting different values of a property in different directions) must be controlled.

The requirements of higher working pressures, special heating and furnace atmosphere controls, and removal of contaminated surface material created the need for equipment (Ref 9.14) and procedures specifically designed for titanium (Ref 9.15–9.20). Mill processing operations have been established and equipment designed and fabricated to meet the special needs of this reactive metal. Special operations and equipment include:

- Furnaces with better temperature uniformity
- Heavier forging presses and rigid rolling mills
- Vacuum annealing furnaces
- Improved grinding and pickling facilities

By using properly designed processes, titanium alloys can be worked as readily as many nickel-base and other specialty steels.

Crystal Structure

Two basic crystalline phases of titanium in its commonly used form are alpha (α) and beta (β) (Ref 9.1, 9.11). The crystal structures of alpha and beta are hexagonal close-packed and body-centered cubic, respectively. The alpha structure exists in pure titanium to a temperature of approximately 880 °C (1620 °F). Above this temperature, allotropic transformation occurs, and the beta structure is present to the melting point. The transformation temperature can be raised or lowered and the amounts of each phase can be altered by adding alloying elements. These effects are discussed in more detail in Chapters 2 to 4 in this book.

Alloying with aluminum stabilizes the alpha phase and produces all-alpha alloys, which possess good elevated-temperature properties and can be welded easily but are not heat treatable. Tin and zirconium additions also strengthen the alpha phase, but they are not classified as alpha stabilizers, rather being classified as neutral additions. Additions of molybdenum, vanadium, iron, chromium, and most other metals stabilize the beta phase. In sufficient amounts, they can produce an all-beta alloy.

Beta alloys can be quite ductile, and although they can be heat treated to high strength levels, they have no exceptional properties or stability at elevated temperatures. Lesser additions of beta stabilizers, with or without alpha stabilizers, produce two-phase alpha-beta (α-β) alloys, also referred to as alpha-beta alloys. These alloys, which combine the best properties of the two crystal structures, are the most widely used.

Influence on Deformation. The type of alloy influences the deformation behavior of titanium. Hot working in the beta phase field (above the alpha-beta-to-beta transformation

---

*Adapted and revised from James S. Myers, *Titanium and Its Alloys*, ASM International.

temperature, or beta transus) is done readily in nearly all the alloys (see Chapters 5 and 6 in this book). At lower, subtransus temperatures, differences exist in working characteristics, depending on alloy type. For example, all-alpha or near-alpha alloys, which contain little or no beta phase at lower temperatures, are considerably more difficult to work without cracking at reduced temperatures than alpha-beta alloys. The latter have at least some of the more ductile beta phase at lower temperatures.

All-beta alloys may require higher working pressures for hot deformation, but they are not as susceptible to rupturing as the temperature is lowered. Although most titanium alloys can be cold worked (at least to a limited extent), unalloyed and all-beta alloys are most amenable to cold deformation.

Forging

Titanium reacts readily at elevated temperatures with air (interstitial contamination) and forms a heavy oxide and nitride scale at temperatures above 650 °C (1200 °F). Oxygen and nitrogen dissolve readily in titanium, and their presence is detrimental to the properties of the fabricated product. Oxygen is more reactive than nitrogen and diffuses into the metal more rapidly. Therefore, oxygen contamination is the more serious problem in processing titanium. Once formed, the oxide scale is brittle and spalls off during working.

Certain precautions must be taken in heating titanium in hot working operations. Air-atmosphere furnaces are generally used. Although electric furnaces are preferred, conventional oil- and gas-fired furnaces can be used if a slightly oxidizing atmosphere is maintained to minimize hydrogen pickup. Because direct impingement of the flame on the metal must be prevented, muffle-type furnaces with baffles are preferred.

Also, care must be taken during heating to prevent contact of titanium with steel scale, which can result in an exothermic thermite-type reaction in which iron oxide is reduced by titanium. The reaction can damage furnace hearths, forging dies, and the workpiece. Titanium can be preheated at a temperature between 700 and 760 °C (1300 and 1400 °F), then heated to the forging temperature as rapidly as possible. Preheating avoids thermal-stress cracking, which can result from excessively rapid heating directly to the higher forging temperature. Minimizing the time

at the forging temperature lowers surface oxidation and contamination and restricts grain growth.

Ingot Breakdown

Commercial vacuum-arc-remelted titanium ingots range from 508 to 1067 mm (20 to 42 in.) in diameter and weigh 1,800 to 18,000 kg (4,000 to 40,000 lb). Ingots may require conditioning after melting, depending on the cast surface quality to remove surface pinholes and spongy areas, which could contribute to tearing during the ingot breakdown stage. These melt-related surface imperfections are removed by spot grinding localized areas or by lathe turning the ingot surface. Although all types of forging presses and hammers are available, ingot breakdown is normally done on a hydraulic press with open flat dies. The deformation rate can be controlled more closely using a hydraulic press, and the as-cast structure can be worked with less danger of tearing. The primary forging practice has a pronounced effect on the structure and mechanical properties of titanium alloys.

Hydraulic presses used for titanium range in capacity from 900 to 10,800 metric tons (1,000 to 12,000 tons). Ingots can be straight-forged, with the cross-sectional area reduced and the axial dimension lengthened, to obtain billets and forged bars. Alternatively, they may be upset 20 to 50% initially and then forged to lengthen the material in the axial direction. The latter combination of upsetting and forging permits greater deformation to produce a given size, with a corresponding refinement in the structure. This is especially useful if large-section billets are to be produced.

An upsetting operation may also be used if slabs ranging from 1.2 to 1.5 m (47 to 60 in.) wide are required. This method produces greater widths than could be expected from straight forging of ingots ranging from 711 to 813 mm (28 to 32 in.) in diameter. An ingot at the start of forging is shown in Fig. 9.1.

Protective coatings are used on ingots and billets to lower surface contamination levels of hydrogen and oxygen and to help reduce the galling effect of the titanium on the forging die. Coatings are often proprietary, but commercial coatings containing a silicon base to form a seal and serve as a lubricant are available that meet the requirement for titanium.

Coatings have been developed for use at specific temperature ranges and requirements. Therefore, several different coatings can be used, depending on temperature, alloy, and type of forging operation. Such coatings are generally mixtures of oxides and glasses, which are applied by spraying, dipping, and brushing. The ingot can be preheated to 95 to 150 °C (200 to 300 °F) before applying the protective coating.

Titanium Processing. Ingot-breakdown forging is normally conducted from temperatures above the beta transus because the body-centered cubic beta phase is more ductile, and working pressures are generally lower (Ref 9.15–9.20). Forging temperature ranges for titanium alloys are listed in Table 9.1. Breakdown forging is typically done between 925 and 1175 °C (1700 and 2150 °F).

After the as-cast ingot structure is deformed by forging to some smaller cross-sectional size, the temperature is reduced. Intermediate forging temperature ranges from 845 to 1095 °C (1550 to 2000 °F). A predetermined amount of deforma-

Fig. 9.1 Titanium ingot at the start of press forging. Ingot breakdown is done in a press with open flat dies so that the deformation rate can be controlled more closely and the as-cast ingot structure can be worked with less danger of cracking.

tion in the alpha-beta field is necessary prior to the phase-change transformation or recrystallization anneal (of the beta phase). During recrystallization, the material goes through recovery and recrystallization and into the grain-growth phase. The recrystallization temperature is usually approximately 40 to 85 °C (75 to 150 °F) above the beta transus. It should be noted that control of the time at temperature is necessary at this point. Excessive grain growth must be avoided, but sufficient time to drive the phase-change transformation is necessary. In many cases, recrystallization is followed by a water quench. On completion of recrystallization, the material is forged in the alpha-beta range to final size. The extent of microstructural refinement is related to the percentage reduction of area in this range.

There is no set pattern on reheating. At the first signs of shallow surface ruptures or an undue increase in working pressure, the material should be reheated. Reheat time should be limited to the minimum necessary to bring the material to the desired temperature. All-alpha or near-alpha alloys, such as Ti-5Al-2.5Sn and Ti-8Al-1Mo-1V, require more frequent reheating because surface rupturing occurs as the temperature drops into the single-phase alpha field. If appreciable surface rupturing occurs during forging, the billet should be conditioned by grinding before further forging.

As shown in Fig. 9.2, forging pressures for the all-beta alloy Ti-13V-11Cr-3Al are considerably higher than for the alpha-beta alloy Ti-6Al-4V. Additional reheating of the all-beta material may be necessary due to limitations in available working pressures. Also, Ti-13V-11Cr-3Al is more sensitive to oxygen contamination than most alpha and alpha-beta alloys.

Table 9.1 Forging temperatures for selected titanium alloys

Alloy	Forging temperature, °C (°F)			
	Beta transus	Ingot breakdown	Intermediate	Finish
Commercially pure	890–955 (1650–1750)	955–980 (1750–1800)	890–925 (1650–1700)	815–890 (1500–1650)
Alpha or near-alpha				
Ti-5Al-2.5Sn	1030 (1890)	1120–1175 (2050–2150)	1065–1095 (1950–2000)	1010–1040 (1850–1900)
Ti-6Al-2Sn-4Zr-2Mo	995 (1820)	1095–1150 (2000–2100)	1010–1065 (1850–1950)	955–980 (1750–1800)
Ti-8Al-1Mo-1V	1040 (1900)	1120–1175 (2050–2150)	1065–1095 (1950–2000)	1010–1040 (1850–1900)
Alpha-beta				
Ti-8Mn	800 (1475)	925–980 (1700–1800)	845–900 (1550–1650)	815–900 (1500–1550)
Ti-4Al-3Mo-1V	960 (1760)	1095–1150 (2000–2100)	980–1035 (1800–1900)	900–955 (1650–1750)
Ti-6Al-4V	995 (1820)	1095–1150 (2000–2100)	980–1035 (1800–1900)	925–980 (1700–1800)
Ti-6Al-6V-2Sn	945 (1735)	1035–1095 (1900–2000)	955–1010 (1750–1850)	870–940 (1600–1725)
Ti-7Al-4Mo	1000 (1840)	1120–1175 (2050–2150)	1010–1065 (1850–1950)	955–980 (1750–1800)
Beta				
Ti-13V-11Cr-3Al	720 (1325)	1120–1175 (2050–2150)	1010–1065 (1850–1950)	925–980 (1700–1800)

Intermediate Conditioning. Titanium and its alloys are conditioned using all methods known to the steel industry (Fig. 9.3). The most common method used is grinding. As a general rule, titanium can be ground at the same rate as hardened high-speed steels and die steels.

Precautions. Titanium can crack if it is ground under the conditions normally used for steel. Several precautions should be taken to minimize the potential for cracking, including:

- Selection of the proper type of wheel
- Use of low wheel speed and feed with high wheel pressure
- Selection of high-quality, heavy-duty automatic or semiautomatic grinders

If the proper wheel is not selected, loading can occur, whether the wheel is sharp or dull. As loading progresses, the grinding action decreases until burnishing occurs. This burnishing action results in increased temperature, causing high residual tensile stress. If these residual tensile stresses exceed the elastic limit, deformation occurs, which can result in surface cracks.

A pickling solution of hydrofluoric and nitric acid is used to remove "smearing" of the ground titanium surface, followed by spot conditioning to remove cracks that may have been smeared. Pickling solutions that are out of balance can cause cracks if the nitric acid content is too low and high tensile stresses exist at the material surface.

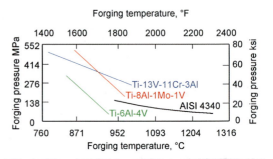

Fig. 9.2 Effect of forging temperature on forging pressure. Lower forging pressures are required in hot working titanium alloys at higher temperatures. A curve for AISI 4340 low-alloy steel shows that much higher forging pressures are required for working titanium alloys within a lower forging range.

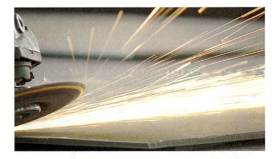

Fig. 9.3 Two conditioning processes are intermediate grinding, which is performed on semifinished billets to minimize yield loss, and final conditioning, where billets are completely ground to remove scale. Any surface defects remaining are later removed by spot grinding. This ensures that a superior product is provided when machining, centerless grinding, and bar peeling are not specified.

Forged Billets and Bars

In producing forged billets and bars, working temperatures are lower than those used for ingot breakdown and intermediate sizes to develop a fully wrought structure and to refine grain size. Table 9.1 lists the lower finishing temperatures for various alloys. A refined structure is desirable in billets for forgeability during subsequent hot working operations. These usually involve low-temperature finishing, but beta forging is sometimes used to improve the elevated-temperature properties and fracture toughness of certain alloys.

Grain Refinement. Macrostructural and microstructural refinement is necessary for satisfactory mechanical properties, because forged bar products are often fabricated into parts without additional deformation. Both the beta grains and the alpha grains require refinement. The beta grains are refined during the early processing procedures by beta working, followed by recrystallization at a higher temperature in the beta phase field. The alpha grains are refined during subsequent alpha-beta processing (see Chapter 6, "Mechanical Properties and Testing of Titanium Alloys," in this book). Adequate grain refinement is also necessary to provide structures that lend themselves to stringent ultrasonic testing of billets for use in critical applications (Fig. 9.4, 9.5). Optimum forging temperatures are determined by performing upset, forged wedge, reforged, and torsion tests on samples of the materials at various temperatures to evaluate microstructure, forgeability, and special properties required.

Starting ingot size generally limits the size of the final forged billet to a maximum diameter or section thickness of approximately 610 mm (24 in.). Billets are produced in round, square, rectangular, and octagonal sections, depending on the processing and end use of the material. Forged bars are usually finished as small-diameter rounds and rect-

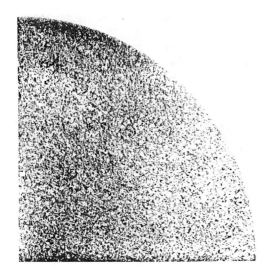

Fig. 9.4 Satisfactory macro/microstructure consisting of 100% uniform, fine-grained, recrystallized beta and alpha phases

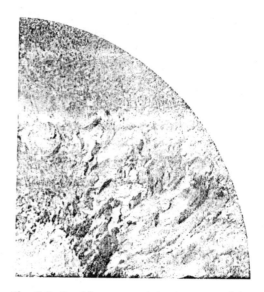

Fig. 9.5 Unsatisfactory macro/microstructure consisting of nonuniform, large, unrecrystallized beta grains resulting from an insufficient amount of deformation below/above the beta transus temperature, followed by a beta anneal substantially above the beta transus temperature, so that the beta grains did not recrystallize to the fine, uniform grain size shown in Fig. 9.4

angular sections that range in thickness or diameter from approximately 50 to 150 mm (2 to 6 in.).

Forging presses are used in billet production, while both presses and hammers are used in processing smaller forged bars. Variations in strain rate have little influence on forgeability of alpha or alpha-beta alloys, but the beta alloys show a marked increase in strength at higher strain rates. Thus, at 790 °C (1450 °F), beta alloys require 50% more energy at a typical hammer velocity of 5.1 m/s (200 in./s) than at a typical press velocity of 38 mm/s (1.5 in./s).

For equal reductions, the energy required for forging the beta alloys from 790 °C (1450 °F) at hammer velocities is nearly 4 times that for AISI 4340 steel at 1260 °C (2300 °F). The difference in forging pressure between steel and titanium alloys is shown in Fig. 9.2. It illustrates why much higher working forces are needed for titanium compared with low-alloy steel, particularly as the workpiece cools, or when lower temperatures are intentionally used during finish-forging operations.

Finishing Billets and Bars. Forged billets and bars are usually annealed, straightened, and finished to particular specification requirements. Finishing is necessary to remove surface defects and oxygen contamination. It serves as a descaling operation as well. Round material can be peeled or lathe-turned, while squares, rectangles, and octagons are surface ground.

Testing. The surface finish of the mill product is determined by the ultrasonic specification. The ultrasonic test is used to identify internal defects

in the material. In addition to ultrasonic inspection, tests for mechanical property and metallographic conformance to specifications are performed to qualify the lot. Typical tests include:

* Etched specimens for structural examination
* Notched stress-rupture tensile test
* Room-temperature tensile test
* Elevated-temperature tensile test
* Creep test
* Fracture toughness test

These tests may be taken from an upset or forge sample of the billet to demonstrate the capability of the material after further hot working.

Rolling

Equipment ordinarily used for rolling stainless steel is satisfactory for rolling titanium, as long as the important differences between steel and titanium processing procedures are recognized. In forging titanium, higher working pressures, closer control of temperatures, and more frequent surface conditionings are recognized and used. In rolling, these characteristics of titanium led to installing rolling equipment specifically designed for titanium. This special rolling equipment includes four-high hot mills, cluster mills, and continuous vacuum annealing furnaces for close-tolerance sheet and strip, and improved heating furnaces and surface-finishing facilities for various types of rolled products.

Cylindrical rolls produce flat products, such as sheet, plate, and strip, while grooved rolls produce rounds, squares, rectangles, and structural shapes. Figure 9.6 shows the most commonly used mill designs for rolling. Two-high (reversing and nonreversing) and three-high mills are normally used for both breakdown and finishing of flat products and shapes. Reversing two-high mills are often used for processing heavy pieces and long lengths of slabs, blooms, plates, and rounds.

Two-high and three-high rolling mills are to produce narrow, flat-rolled material where thickness control is not critical and thinner gages are not required. For wider material, such as wide plate, four-high mills provide better roll rigidity and closer thickness control. Four-high rolling mills are also used to produce both hot and cold rolled plate, sheet, and strip.

Cluster mills, like those manufactured by Sendzimir and Schloemann, are used for cold rolling very thin strip and foil where very close control of thickness must be maintained. Configurations for the four-high and cluster mills (Sendzimir and Schloemann) are shown in Fig. 9.7.

Hot rolling temperatures are typically lower than those used for forging, as shown in Table 9.2. In addition, heating times for rolling are generally shorter because of the smaller starting section sizes. Bar-rolling temperatures often are higher than indicated in Table 9.2 for sheet and plate rolling, because reheating at an intermediate stage is usually not possible during rolling of long lengths of bar.

Higher temperatures are similarly used in some hot strip-rolling operations in which hot bands and hot rolled coils are produced in tandem mills following single heating of the starting slab. Most rolled bar, hand-mill sheet, and alloy-grade plate are hot finished, although sheets are sometimes cold finished to improve surface condition and gage tolerance. By contrast, continuously processed strip and foil receive substantial amounts of cold rolling after hot rolling to intermediate size.

Radial Precision Forging Machines

Starting material for bar depends on the size of the machine available. Currently, the largest unit used on titanium can accept input material up to 610 mm (24 in.) in diameter (Fig. 9.8). German Fabrication Machines' (GFM's) radial precision forging machines (American GFM Corp.) are constructed to forge bars from stock from 25 to 610 mm (1 to 24 in.) in diameter and up to 400 mm (16 in.) square.

The speed of the dies varies with machine size, ranging from 1800 to 2400 strokes/min with force available on each die ranging from 50 to 800 metric tons (55 to 90 tons), respectively. Feed-in rate is infinitely variable between 0 and 7.6 m/min (0 and 25 ft/min) in larger machines. The GFM's machine has four connecting rods with a constant, very short stroke compared with conventional crank and eccentric presses. This allows high stroke rates, whereby the deformation speed on the titanium bar remains within the range of conventional presses. This type of operation gives GFM's machines a high production potential.

Temperature ranges for GFM forging are listed in Table 9.1 for large rounds, bars, squares, and rectangles. Shapes, mandrel forging, and contouring of titanium are discussed in Chapter 10, "Secondary Working of Bar and Billet," in this book.

Rolled Rod and Bar

Starting material for rod and bar rolling is typically a forged square billet. Starting billets are generally either 200 to 250 mm (8 to 10 in.) or

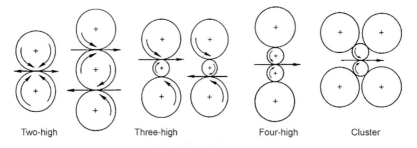

Two-high   Three-high   Four-high   Cluster

Fig. 9.6 Typical rolling mill designs used in rolling plate, sheet, strip, and foil

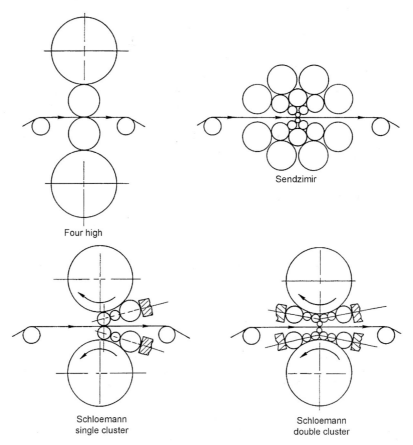

Fig. 9.7 Four-high and cluster mills manufactured by Sendzimir and Schloemann. These rolling mills lend themselves to cold rolling thin strip and foil to very close tolerances.

Table 9.2 Rolling temperatures for selected titanium alloys

Alloy	Rolling temperature, °C (°F)		
	Bar	Plate	Sheet
Commercially pure	760–815 (1400–1500)	760–790 (1400–1450)	705–760 (1300–1400)
Alpha or near-alpha			
Ti-5Al-2.5Sn	1010–1065 (1850–1950)	980–1035 (1800–1900)	980–1010 (1800–1850)
Ti-6Al-2Sn-4Zr-2Mo	955–1010 (1750–1850)	955–980 (1750–1800)	925–980 (1700–1800)
Ti-8Al-1Mo-1V	1010–1065 (1850–1950)	980–1035 (1800–1900)	980–1035 (1800–1900)
Alpha-beta			
Ti-8Mn	705–760 (1300–1400)	705–760 (1300–1400)	...
Ti-4Al-3Mo-1V	925–955 (1700–1750)	900–925 (1650–1700)	900–925 (1650–1700)
Ti-6Al-4V	955–1010 (1750–1850)	925–980 (1700–1800)	900–925 (1650–1700)
Ti-6Al-6V-2Sn	900–955 (1650–1750)	870–925 (1600–1700)	870–900 (1600–1650)
Ti-7Al-4Mo	955–1010 (1750–1850)	925–955 (1700–1750)	925–955 (1700–1750)
Beta			
Ti-13V-11Cr-3Al	955–1065 (1750–1950)	980–1035 (1800–1900)	730–890 (1350–1650)

100 to 125 mm (4 to 5 in.) square, and they are thoroughly conditioned prior to rolling. Rolled rods and bars are produced as rounds, squares, rectangles, and hexagons ranging in size from approximately 3 to 50 mm (0.1 to 2 in.) in diameter or thickness. Smaller sizes are usually rolled as coils, while larger sizes are processed as straight lengths up to 9 m (30 ft) long.

Rolling temperatures for several alloys are listed in Table 9.2. Heating times are kept to a minimum, because excessive scale results in surface cracks and defects. Starting bar-rolling temperatures are usually on the high side, because reheating ordinarily is not possible. Rolling must be completed before the material temperature drops below the desired finishing range. In most instances, this results in substantial final reductions below the beta transus. This is desirable to achieve optimum microstructural refinement and the best combination of strength and ductility.

After rolling, smaller rod sizes (6.4 to 9.5 mm, or 0.25 to 0.38 in.) are chemically descaled in hot caustic baths and pickled in an aqueous mixture of nitric and hydrofluoric acids (10 to 35% HNO_3 and 1.5 to 5% HF). If straight lengths are required, the material is annealed either in air or vacuum, straightened, and cut. If further processing is required to produce fastener stock or wire, the rod is annealed, thoroughly cleaned, and left in coil form.

After rolling, larger bar sizes are annealed, straightened, descaled, and surface conditioned. Air cooling from the lower annealing temperatures (700 to 730 °C, or 1300 to 1350 °F) is common practice. Square and rectangular bars can be furnished with a descaled, pickled finish or surface ground to remove all defects and contamination.

Round stock for further fabrication, such as blade forging, is usually centerless-ground, although it can also be machined in bar-turning equipment. Final inspection consists of ultrasonic examination for internal defects. This is followed by tensile testing in the annealed or heat treated condition. Additional tests include structural examination and measurement of mechanical properties at room and elevated temperatures.

Plate, Sheet, Coil, and Foil Rolling

The distinction between sheet and plate is based on dimensions. Plate is flat rolled to 4.76 mm (0.1875 in.) or more in thickness and over 254 mm (10 in.) wide.

Dimensional differences between sheet and strip are not as readily defined. For this discussion, sheet is defined as a relatively small, individually cross-rolled piece, while strip implies the material has been processed unidirectionally as a long, continuous coil. Often, strip is defined as less than 610 mm (24 in.) wide and sheet as wider than 610 mm. A coil can be cut into individual flat lengths, and the separate pieces are designated as sheets.

Starting materials for sheet and plate products are forged and bloomed slabs conditioned by grinding. Unalloyed and alloyed plates up to 3.8 m (150 in.) wide and up to 15.2 m (600 in.) long are produced with the limitation on thickness and length being the size of the starting ingot or billet. Most plate sizes are considerably smaller.

Hand-mill sheet sizes are less than 1.2 m (48 in.) wide by 3.7 m (145 in.) long, with gages typically not less than 0.50 mm (0.020 in.). For these, individual sheet bars are cut from forged and bloomed slabs. The exact size of the sheet bar depends on the size of the final sheet product and expected processing losses. Sheet rolling on a two-high mill is shown in Fig. 9.9. Rolling is often performed on four-high reversing mills.

Sheet rolling is usually done in two steps. The first consists of cross rolling the sheet bar to an intermediate thickness (or breakdown size) that is approximately 3 to 4 times the final rolled thickness. The second involves finish rolling (often in a pack or sandwich assembly) to the desired hot rolled size. Cross rolling is used to equalize the longitudinal and transverse properties. Rolling temperatures are shown in Table 9.2.

Both rolling steps are carried out from the same temperature, although for some of the more difficult-to-work alloys, a higher temperature is used for initial rolling. Except for special process-

Fig. 9.8 German Fabrication Machines' radial precision forging machine. Courtesy of Timet

Fig. 9.9 Two-high rolling mill. This type of mill lends itself to rolling titanium sheet and plate. Courtesy of Timet

ing and for beta alloys, sheet rolling temperatures are typically 40 to 85 °C (70 to 150 °F) below the beta transus.

Plate rolling is usually done by cross rolling the starting slab section to final plate size in a one-step operation, unless intermediate conditioning is necessary. After cross rolling to the intermediate size, the material is descaled and conditioned to remove surface defects. If the final sheet thickness is to be greater than 2.5 mm (0.1 in.), assembly into a pack is not necessary. For thinner gages, intermediate rolled pieces are often assembled in sandwich fashion into packs with one to six pieces. Steel cover plates and side bars are all welded together to protect the titanium from contamination.

Parting agents, such as lime and aluminum oxide, are used between the individual sheets to prevent sticking during rolling. The number of titanium pieces assembled in a pack decreases with increasing final sheet thickness, so overall pack thickness prior to finish rolling is fairly constant. Pack rolling of this type provides improved sheet surfaces and better control of sheet thickness, and it enables rolling thinner, wider sheets satisfactorily.

After finish hot rolling, the sheets are sheared out of the packs, trimmed, chemically descaled in a hot caustic bath, and annealed. Sheet and plate flatness after hot rolling usually is not sufficient to meet the stringent requirements typical for titanium. A combination of annealing and creep flattening is used by placing weight plates on the sheets in the furnace.

Temperatures for the annealing/creep-flattening process range from 680 to 790 °C (1250 to 1450 °F) and provide good mechanical properties. For certain applications, this is followed by finish annealing at a higher temperature, or by solution treatment with or without subsequent aging. Other flattening procedures, such as roller and stretcher leveling, are used on titanium, but the lower elastic modulus and small spread between yield and ultimate tensile strengths limit the use of these leveling operations.

Surface finishing of annealed sheets and plates is done by pickling in a mixture of nitric and hydrofluoric acids (10 to 35% HNO_3 and 1.5 to 5% HF) if surface defects and smoothness are such that other finishing is unnecessary. However, most sheets are surface ground with successively finer-grit belts to remove defects, irregularities, and contamination and to improve gage uniformity. Acid pickling follows surface grinding to smooth out the residual grind lines. Cold rolling

is also used to improve the surface condition of sheets. Both tensile and bend tests are conducted on the finish material.

Strip and Foil. Continuously rolled coils are produced in widths to 1.2 m (48 in.) and gages as thin as 0.25 mm (0.010 in.) in the commercially pure grades. Even thinner material is available in narrower widths of 610 mm (24 in.). Several alloy grades, including Ti-6Al-4V, Ti-3Al-2.5V, Ti-5Al-2.5Sn, Ti-13V-11Cr-3Al, and Ti-15V-3Cr-3Sn-3Al, are processed as strip, although in heavier gages and narrower widths than unalloyed material. Alloy development and processing improvements have enabled production of thinner, wider alloy strip. In addition to Ti-15V-3Cr-3Sn-3Al (Ti-15-3), Ti-15Mo-3Nb-3Al-0.2Si (Beta 21S) was developed as an alloy from which to produce cold rolled strip.

Hot rolling converts slab material into a single longer piece that can be coiled at an intermediate-thickness (between 3.2 and 3.8 mm, or 0.125 and 0.150 in.) hot band (Fig. 9.10). This is processed to thinner finish gages by cold rolling and annealing. Hot bands are produced in widths slightly greater than the final strip width. The extra width permits shearing to remove cracks and thinner edge material.

Starting stock for hot rolling is either forged or bloomed slabs between 76 and 152 mm (3 and 6 in.) thick, which have been thoroughly conditioned. Because of the higher temperatures and generally longer heating times used for slabs, a protective coating is sometimes applied on alloy material to reduce oxidation and subsequent surface defects. Induction heating is often used.

Although hot rolling equipment and procedures for strip vary, typical processing of material ranging from 0.9 to 1.2 m (36 to 48 in.) wide consists of rolling the slab in a universal mill to approximately 19 mm (0.75 in.) thick, then rolling through several stands to a final band thickness

Fig. 9.10 Hot rolling converts slab material into a single longer piece that can be coiled at an intermediate-thickness hot band. Courtesy of Timet

between 3.18 and 3.8 mm (0.125 and 0.150 in.). This is carried out without reheating from slab temperatures above the beta transus ranging from 925 to 1230 °C (1700 to 2250 °F), depending on the grade. Material from the last rolling mill stand is coiled hot as the final operation.

The hot band is descaled by uncoiling from one end of the line and running it through a grit blast or other mechanical scale-removing unit. This is followed by line annealing, if required, and a light rolling mill pass for flattening plus annealing and descaling. At this point, the hot rolled strip surface is carefully examined for slivers and other hot rolling defects. If necessary, the coil is surface ground, acid pickled, and edge trimmed.

Cold Rolling. Titanium and its alloys are cold rolled on four-high mills to between 0.50 and 0.81 mm (0.020 and 0.032 in.) gages for standard tolerances on strip less than 915 mm (36 in.) wide. For close gage control on thinner and wider material, particularly in the alloy grades, cluster mills, such as Sendzimir or Schloemann mills, are mandatory. The Sendzimir cluster mill is used to produce both narrow and wide strip.

The cold workability of commercially pure titanium allows reductions up to 50% before annealing is required. Appreciably lower reductions are required in processing the alpha and alpha-beta alloys. The excellent cold rolling behavior of beta alloys allows reductions as high as, or higher than, achieved in commercially pure strip. Line annealing for the heavier gages is done in air-atmosphere furnaces. Continuous annealing in a vacuum furnace virtually eliminates contamination, with no loss in thickness and no need for pickling. Generally, the material is at temperature only 5 to 7 min. Therefore, higher annealing temperatures in the range of 760 to 955 °C (1400 to 1750 °F) are used.

The strip is then passed through a hot caustic bath for descaling and a nitric-hydrofluoric acid bath for pickling. As processing continues, this combination of cold rolling, annealing, descaling, and pickling is repeated several times. Strip in some alloy grades, particularly alpha-beta alloys such as Ti-6Al-4V, is annealed at higher temperatures. In coil form, it is batch vacuum annealed for several hours to achieve substantial recrystallization and to reduce directionality.

Because of surface contamination during line annealing in air and of accompanying descaling and pickling problems, it is impractical to use air annealing and pickling on thicknesses less than approximately 0.38 to 0.58 mm (0.015 to 0.020 in.). Vacuum annealing facilities were established for this reason and to eliminate descaling-pickling time and attendant metal loss (Fig. 9.11).

The alternatives are batch annealing in an inert atmosphere or vacuum furnace. Batch annealing has disadvantages, including:

- Welding of adjacent wraps occurs at temperatures exceeding 760 to 790 °C (1400 to 1450 °F).
- "Coil set" is produced if the annealing treatment is the finish-annealing cycle.
- A longer heating and cooling cycle is required.

After final annealing, the strip is slit to desired width, cut into individual lengths, and inspected. Samples are sheared for tensile and bend tests plus chemical analyses.

Foil processing to gages thinner than 0.25 mm (0.010 in.) is a continuation of the cold rolling and annealing cycles described previously. Handling problems are magnified due to the thin gage. Widths are usually less than those for thicker material. Contamination-free annealing is essential due to the high surface-to-volume ratio. The foil surface must be thoroughly cleaned of any residual rolling oil and other foreign material before it enters the annealing chamber. At these thin gages, the number of rolling-annealing cycles becomes relatively large in reducing the thickness.

Extrusion

In extrusion, a billet is forced under compression to flow through a die opening to form a product of smaller, uniform cross section. The procedures used in extruding titanium are similar to practices used in extruding steel. The process produces rounds, shapes, tubes, and hollow shapes.

Fig. 9.11 Continuous vacuum annealing is used to process thin strip free from surface contamination. Courtesy of Timet

Cold extrusion of titanium including hydrostatic extrusion has been performed, but it is not a commercial process. Hot extrusion is used to produce long sections. Shapes can also be made by rolling, but the process is practical only if sufficient quantities are produced to justify the high cost of tooling.

Horizontal and vertical presses and high-energy-rate machines are used for extrusion, but accumulator-driven hydraulic horizontal presses are most often used for hot extrusion operations. They are available with capacities up to 10,900 metric tons (12,000 tons), but most presses fall in the range of 1,800 to 3,600 metric tons (2,000 to 4,000 tons).

Extruded titanium shapes are supplied in numerous configurations, including basic angle, tee, hollow, and channel shapes. Section thicknesses vary from 3.2 to 32 mm (0.125 to 1.250 in.) within circumscribing circles ranging from 38 to 280 mm (1.5 to 11 in.). Extruded sections less than 2.5 mm (0.1 in.) thick have been produced.

Grades produced include commercially pure titanium, Ti-6Al-4V, Ti-6Al-6V-2Sn, and Ti-5Al-2.5Sn. Various sizes of pipe and tube hollows are also produced, primarily from commercially pure titanium and Ti-3Al-2.5V. Much of this is used in further processing to supply seamless tubing. Pipe up to 510 mm (20 in.) in diameter and 2.5 to 31.8 mm (0.1 to 1.25 in.) wall thickness is extruded on an 11,000 metric ton (12,000 ton) press.

Lubrication. Tool surfaces for hot extruding titanium must be lubricated to reduce friction between the billet and tools. Lubrication is particularly important to reduce galling. Two basic types of lubricants used for extruding titanium are greases and glasses. Metallic copper coatings are also used with grease lubricants, particularly in extruding the commercially pure grades. Grease mixtures offer little or no thermal protection for the die; therefore, die wear limits their use.

Sejournet Glass Process. Most extruders of titanium use the Sejournet glass process (a process invented in 1950 by Ugine Séjournet of France that uses glass as a lubricant for extruding steel). Because of the high reactivity of titanium, it is important that proper glass compositions be selected to:

- Prevent reaction between the glass and titanium
- Achieve suitable viscosity
- Provide chemical stability
- Protect billet surfaces from contamination

Glass also serves as an insulator to protect the tools from direct contact with the hot billet during extrusion, which prevents overheating of the tools, improving tool life.

Figure 9.12 shows a schematic illustration of the Sejournet glass-lubrication extrusion process. Billets are transferred from the heating furnace to the extrusion-press charging table, where they are rolled over a glass-fiber sheet or through a layer of glass powder that fuses to the billet surface. A pad of glass or glass fiber is also placed in front of the die to serve as a reservoir of glass at the die face during extrusion.

Two types of metal flow can occur during extrusion, one desirable and the other to be avoided. The desirable type is parallel metal flow, in which the surface of the billet becomes the surface of the extrusion. By comparison, shear metal flow, in which the billet surface penetrates inward and creates a stagnant metal zone at the die shoulder, is undesirable because it prevents effective die lubrication and can cause interior and surface defects in the extrusion. To achieve parallel metal flow, the die is flat faced with a contoured die opening. To provide parallel metal flow and a reservoir of glass on the die face, the entry to the die opening is conical in shape.

The billet exterior becomes the extrusion surface. Therefore, billet surface preparation and billet heating contribute to the quality of the extrusion surface. The billet should be machined overall and the front outside edge should be rounded to obtain better die lubrication and lower breakthrough pressures. It is also necessary to maintain (as near as possible) a scale-free surface during heating. This not only reduces the adverse effect of scale on the extrusion surface finish, but it also avoids any reaction between the oxide scale and the molten glass lubricant, which may

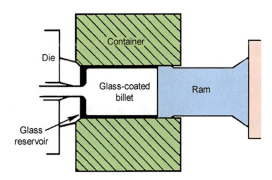

Fig. 9.12 Sejournet glass-lubricated extrusion process. Billet surface preparation and heating are important in contributing to extruded surface quality.

alter the thermal properties and viscosity of the glass. For this reason, heating the billet in a muffle furnace under an argon atmosphere or by low-frequency induction in an inert atmosphere is preferred.

High speeds are used for extruding titanium, regardless of whether grease or glass lubricants are used. Grease offers little protection to the die at high extrusion temperatures. With glass lubrication, high speeds are necessary to ensure a continuous flow of molten glass without depleting the reservoir at the die face. Actual ram speed during extrusion varies with alloy composition, extrusion temperature, and extrusion ratio. It is usually in the range of 5 to 8 m/min (~200 to 315 in./min). The actual temperature of the extrusion often increases sufficiently at these speeds to heat it into the beta field, with a resulting microstructure of heavily deformed, transformed beta. This is more pronounced in the alloy grades at the higher extrusion ratios, which require higher billet temperatures to achieve satisfactory metal flow. Extrusion temperatures between 760 and 1180 °C (1400 and 2150 °F) are used for most titanium alloys. Lubricants are used at the lower temperatures for Ti-3Al-2.5V.

Extruded Shapes. Production of most extruded shapes is confined to the alloy grades of titanium. Billets up to 711 mm (28 in.) in diameter and 1.7 m (67 in.) long can be extruded in the largest presses, but the more common billet sizes are between 100 and 230 mm (4 and 9 in.) in diameter and 610 to 760 mm (24 to 30 in.) long. Billet containers and liners are usually made of, at least in part, hot work tool steels such as H12. They are typically heated using gas-fired and electrical-resistance units to temperatures between 205 and 260 °C (400 and 500 °F) and preferably between 425 and 480 °C (800 and 900 °F), to reduce chilling of the billet surface and glass lubricant.

Die materials with which grease lubricants are used include carbides, hot work tool steels, high-speed tool steels, and high-carbon, high-chromium cast steels. Somewhat less stringent high-temperature properties are needed for dies used with glass lubrication; therefore, common hot work tool steels, including cast tool steels, are satisfactory. Die dimensions often cannot be maintained to provide a reasonably economic die life, particularly with some of the thinner, more complex shapes. In many instances, the die opening and land areas are coated with a ceramic such as zirconium oxide. Coated dies of this type result in closer-tolerance extrusions and have much better die life.

Processing methods after extrusion vary with shapes produced using glass and those produced using grease lubrication, but all require cleaning after extrusion. Those made using the glass method are quenched, and the glass is removed either mechanically by shot blasting or chemically in hot caustic followed by acid pickling.

Straightening and detwisting operations are required for structural shapes. The operations are usually conducted in hydraulic torsional-stretching machines, although roll or punch straightening is also used. Commercially pure titanium is straightened either cold or hot, but alloy shapes require hot straightening at a temperature between 425 and 540 °C (800 and 1000 °F) due to their high yield strengths and tendency for springback.

Resistance heating is used, and short-time annealing and stress relieving is often combined with straightening to minimize any deleterious effects of residual stresses. Annealing treatments for various times at temperatures between 650 and 790 °C (1200 and 1450 °F) are used, and small quantities of heat treatable alloys have been produced in the higher-strength, solution-treated and aged condition. The problems of distortion and surface contamination during heat treatment limit the production of the latter.

Extrusion finishing consists of descaling in hot caustic and pickling in a nitric-hydrofluoric acid mixture to remove the contaminated surface layer. Other finishing operations involve grinding, chemical milling, roll forming and sizing, and hydrostatic re-extrusion for sizing and surface improvement. Such operations would be performed to produce shapes that can be used directly without additional machining. One such finishing procedure used successfully is warm drawing. T-sections of Ti-6Al-4V and other alloys are drawn at temperatures ranging from 480 to 540 °C (900 to 1000 °F) to section thicknesses ranging from 1.1 to 2.0 mm (0.043 to 0.080 in.), with improved surface finish and dimensional tolerances at the desired straightness level.

Integrally stiffened Ti-6Al-4V and Ti-6Al-6V-2Sn panels are satisfactorily extruded on a 11,000 metric ton (12,000 ton) press to produce four ribbed panels 406 mm (16 in.) wide with stiffeners 10 mm (0.4 in.) thick.

Extruded Pipe and Hollows. The same considerations apply to the extrusion of pipe, tube shells, and hollows as those for extruded shapes. The nature of the product requires consideration of factors such as eccentricity, mandrel lubrication, and billet preparation. Hollow billets are prepared by machining and piercing. Because of the unfavorable stresses that would be imposed by piercing, holes smaller than 50 mm (2 in.) in diameter are machined. A larger-sized hole can be formed by piercing prior to extrusion, either in the extrusion press or in an auxiliary vertical press. Billet length is limited to 7 times the diameter of the pierced hole. Glass is often used to lubricate the piercing mandrel. After piercing, the billet is machined before extrusion.

Mandrels are made of hot work tool steel, such as H11, due to the severe conditions imposed on it during hot extrusion. For glass lubrication, either a fibrous glass sock is placed over the mandrel, or powdered glass is sprinkled on the inside surface of the hollow billet. Grease, glass, and copper coatings are used as lubricants in the extrusion of tubes. The glass extrusion process is used widely for tubes because of the greater lubricating and insulating qualities of glass, which provide more protection for the mandrel and enable production of longer tubes with better overall surfaces.

Postextrusion processing of tubes is the same as that for extruded shapes, although extruded tubing typically is not stretcher straightened unless it is severely distorted. Most tubes are straightened during subsequent tube-making operations. Extruded tubes require cleaning. Those made using the glass process are quenched, and the glass is removed by shot blasting and by immersion in hot caustic followed by acid pickling to remove surface contamination. If annealing is required, the latter pickling operation is done in a postannealing cycle.

Wire and Tube Processing

Wire Products. Titanium wire products include weld filler wire and fastener stock in commercially pure titanium and many titanium alloy grades. Fastener materials typically are restricted to a few of the higher-strength titanium alloys, such as Ti-6Al-4V and Ti-6Al-6V-2Sn. The most common method of wire production is conventional die drawing of hot rolled rod to final wire sizes, although roll drawing also is used. Tungsten carbide and diamond dies with included angles of 25° are usually used for die drawing, with draw benches or bull blocks supplying the drawing force.

Commercially pure titanium wire is drawn at room temperature, while the alloy grades are often drawn warm at temperatures between 540 and 650 °C (1000 and 1200 °F). In either case,

adequate lubrication is required to minimize galling and seizing of titanium to the die surface. Intermediate annealing and cleaning are done as required to restore ductility for further drawing passes. Both air annealing and vacuum annealing are used. For smaller-diameter wire, the latter is preferred to eliminate surface contamination.

Tube Products. Unalloyed titanium accounts for most tubular products. Welded and seamless tube sections are available. Welded tubing is processed by roll forming and seam welding, while seamless tube processing consists of tube reducing or die drawing, or both, of extruded tube shells to thinner-wall tubing.

Welded tube sections and seamless tube products are made of commercially pure titanium, Ti-6Al-4V, Ti-3Al-2.5V, Ti-0.2Pd, Ti-Code 12, and Ti-15V-3Cr-3Al-3Sn. As with wire drawing, lubrication must be provided for the outside and inside surfaces in tube reducing. Intermediate annealing is done as needed to restore ductility for further tube-reducing passes. Equipment used in reducing tubing is either the Pilger mill or drawing stand.

Wire Processing. Wire typically is produced in diameters in the range of 2.5 to 8 mm (0.010 to 0.312 in.) and 90 to 150 m (~300 to 490 ft) long coils. Commercially pure titanium wire is made in more sizes and smaller diameters than most alloy grades. Longer continuous lengths can be made in finer sizes, and, for some applications, relatively short, straight lengths can be produced in the medium-to-large diameters. The starting size for commercially pure and alloy grades is coil ranging in wire diameter from 6.4 to 12.7 mm (0.250 to 0.500 in.), which is annealed and cleaned.

Commercially pure titanium wire is almost always cold drawn. Lubricants and coatings of graphite, copper powder, lead oxide, soaps, and various proprietary materials are used in drawing. Pretreating in a fluoride-phosphate bath to develop a chemical conversion coating is recommended prior to applying the lubricant. Such coatings are also used with conventional lubricants such as graphite-containing grease. Reductions of 20 to 50% are achieved between intermediate annealing treatments.

Prior to annealing, all residual lubricants should be removed by pickling to minimize contamination from foreign material. Although larger wire sizes are annealed in air, vacuum annealing is often used for both large- and small-diameter wire to prevent oxygen contamination and to remove hydrogen.

After annealing, wire must be recoated and lubricant applied before additional drawing passes can be made. Alloy titanium wire is commonly die-drawn warm at a temperature between 540 and 650 °C (1000 and 1200 °F), at least in the larger diameters. Many of these lubricants are used satisfactorily, including lime and other oxides or oxide mixtures. Induction heating is often used to heat the wire just prior to entering the die. After the final drawing passes, the wire is cleaned, finish annealed (welding wire is not finish annealed), and, where necessary, acid pickled to remove any oxygen-rich surface layer. Wire larger than 2.54 mm (0.100 in.) in diameter is also centerless-ground as a surface-finishing procedure. Samples are cut for tensile testing and chemical analyses, and the wire product is inspected and packaged for shipment. Generally, the major problems encountered in the production of wire are surface defects and contamination.

Tubing Processing. Welded tubing is produced in sizes ranging from 12 to 60 mm (0.5 to 2.5 in.) in diameter with wall thicknesses from 0.5 to 2.3 mm (0.020 to 0.090 in.) up to 30 m (100 ft) long. This size range was extended in some instances by forming and welding large-diameter ducting in fixtures. The larger the diameter, the thicker the wall must be to achieve satisfactory tube forming and welding. Nearly all welded tubing is made of commercially pure titanium and Ti-0.2Pd, although Ti-6Al-4V, Ti-3Al-2.5V, Ti-Code 12, and Ti-15V-3Cr-3Al-3Sn are produced as welded tube sections.

Starting material for welded tubing is typically strip that is slit to the required width for a given tube diameter. It is formed into a circular cross section in a mill, which bends and shapes the strip in a series of contoured rolls into the desired cylindrical configuration. A gas tungsten arc welding unit adjacent to the final set of forming rolls welds the two strip edges at speeds up to 2.5 to 7.6 m/min (100 to 300 in./min). In the weld zone, both the outside and the inside of the tube are shielded with inert gas to prevent oxidation of the titanium in the weld and heat-affected zones.

After welding, the tubing passes through a series of sizing rolls to ensure that the required circular cross section is obtained in the final product. The tubing is stress-relief annealed and cut to desired lengths.

After stress relieving, tube samples are tensile, flare, and flatten tested. The finished tube is inspected using eddy current and ultrasonic nondestructive testing techniques, and leak tested using pneumatic pressure.

Welded tubing can be die drawn to deform and work the weld bead and to serve as a tube-sizing operation. With close control of roll forming and welding conditions, proper tube size and weld bead shape are achieved so subsequent die drawing for sizing is not necessary.

Seamless Pipe and Tubing. Seamless pipe sizes ranging from 100 to 150 mm (4 to 6 in.) in diameter are produced, primarily of commercially pure titanium and Ti-3Al-2.5Sn. These represent extruded pipe that has been conditioned or cold deformed sufficiently to provide the required surface finish and size. Commercially pure titanium seamless tubing is produced in sizes ranging from 6.4 to 63.5 mm (0.25 to 2.50 in.) in diameter and 0.3 to 6.4 mm (0.012 to 0.25 in.) in wall thickness. The most popular seamless tube sizes are in the range of 19 to 25 mm (0.75 to 1.00 in.) in diameter and 0.8 to 1.3 mm (0.030 to 0.050 in.) in wall thickness. Other alloy grades, such as Ti-6Al-4V, are also produced in limited quantities.

The starting material for seamless tube processing is an extruded tube hollow that has been cleaned and is free of serious surface defects. If the surface of the extruded hollow allows, reduction of wall thickness and diameter is done by die drawing over a plug or bar mandrel (Fig. 9.13).

Deformation processing is done in a tube reducer, or Pilger mill, as shown in Fig. 9.14. Lubrication is important in die and mandrel drawing, as it is for wire drawing. Adequate lubrication must also be supplied to the interface between the mandrel plug or bar and the inside surface of the tube. Pretreatments (such as conversion coatings to develop surface subcoats) are used for tube drawing.

As indicated in Fig. 9.14, a tube reducer decreases the diameter and wall thickness by a swaging action. The workpiece is compressed between two semicircular rolls with matching tapered grooves, and a stationary tapered mandrel is aligned at the centerline of the pass. The tube is rotated approximately 60° between each reciprocating motion of the rolls to ensure concentricity. Because of the compressive forces and incremental working, large reductions up to 85% in area per pass are possible, which is more than twice those achieved in die drawing. Tube reducing also eliminates some of the intermediate cutting, annealing, pickling, and pointing operations required in drawing. Drawing is more suitable for producing small lots, because dies are relatively inexpensive and easier to change. The tube-reducing process is used for most seamless titanium tubing.

Almost all deformation of commercially pure titanium and titanium alloy tubing is performed at

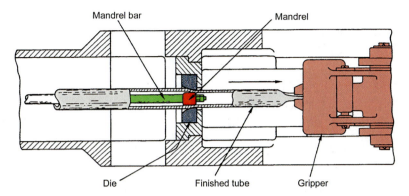

Fig. 9.13 Die drawing seamless tubing. Starting material is an extruded tube hollow free from surface defects. Wall thickness and diameter are reduced by die drawing over a plug or bar mandrel.

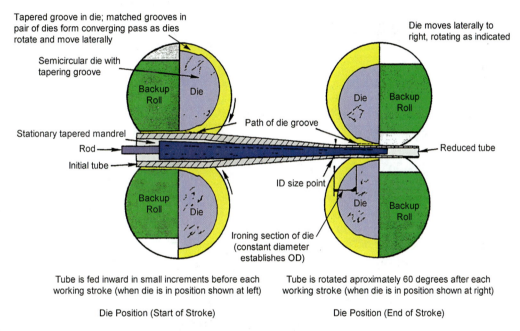

Die Position (Start of Stroke)   Die Position (End of Stroke)

Fig. 9.14 Schematic of a tube reduction. Vertical section through the tube reducer shows dies at start and end of stroke. Because of the compressive forces and incremental working, tube reducing provides much higher reductions than die drawing. ID, inside diameter; OD, outside diameter

room temperature. Some die drawing is performed warm to increase the amount of reduction per pass. Tubing requires intermediate cleaning, annealing, descaling, pickling, and lubricant application in several cycles to reduce the material to the required size. In mandrel bar drawing, a multiple-roll rotary straightener is used to "reel and spring" the tube so it can be removed from the mandrel. In such drawing, the wall is reduced to the desired thickness, and then the tube diameter is reduced to the required size in one or two additional free-sinking passes without a mandrel.

After final drawing or tube reducing to finished size, the material is cleaned, annealed, pickled, straightened, and ultrasonically inspected. Mechanical testing of the final product consists of tensile, flare, flatten, and sometimes pressure tests.

Summary

Titanium and titanium alloys are produced in nearly every form of mill product. Differences in the behavior of titanium and that of the specialty steels led to the use of mill processing procedures and equipment designed specifically for titanium. The characteristics and properties of titanium and titanium alloys that received special attention in primary working include:

- High-capacity rigid deformation equipment is required due to the high loads needed at the practical hot working temperatures typically used to obtain the optimum microstructure (refined beta and alpha grains), which provides the best combination of mechanical properties.

- Due to the reactivity of titanium above a temperature of approximately 650 °C (1200 °F), surface contamination is minimized by using the shortest possible heating times or by heating in vacuum or inert atmosphere. The latter is particularly important in products with a high surface-to-volume ratio.

- Due to surface oxidation and oxygen contamination, frequent conditioning operations are required during processing to remove scale, surface defects, and oxygen-rich surface metal.

- Hydrogen contamination, which in sufficient quantity causes embrittlement, can be avoided by using a slightly oxidizing atmosphere during heating and by keeping the ratio of nitric to hydrofluoric acids above 7:1 during pickling.

- The severe seizing and galling behavior of titanium in drawing operations is essentially eliminated by using proper lubricants and die design.

- Problems of directionality of two-phase alloys and anisotropy of the hexagonal close-packed alpha structure are controlled by cross rolling (in plate and sheet) and thermal treatment (in strip).

- Deformation characteristics of the different titanium alloys vary significantly, especially among the three common types: alpha, alpha-beta, and beta. As a result, processing procedures were established to accommodate these variables, including specifying hot working temperature, reduction per pass and intermediate reheating or annealing, annealing temperature and cooling rate, extent of oxygen and hydrogen contamination, and the amount and type of surface finishing.

REFERENCES

9.1 F.H. Froes et al., Ed., *Titanium Technology: Present Status and Future Trends,* Titanium Development Association, 1985

9.2 S. Abkowitz, The Emergence of the Titanium Industry and the Development of the Ti-6Al-4V Alloy, *JOM Monograph Series,* Vol 1, TMS, Warrendale, PA, 1999

9.3 G. Lutjering, U. Zwicker, and W. Bunk, *Proceedings of the Fifth International Conference on Titanium,* Metallurgical Society of AIME, Warrendale, PA, 1984

9.4 P. Lacombe, R. Tricot, and G. Beranger, Ed., *Proceedings of the Sixth International Conference on Titanium* (Nice, France), Metallurgical Society of AIME, Warrendale, PA, 1988

9.5 *Proceedings of the Seventh International Conference on Titanium* (San Diego, CA), Metallurgical Society of AIME, Warrendale, PA, 1992

9.6 *Proceedings of the Eighth International Conference on Titanium* (San Birmingham, U.K.), Metallurgical Society of AIME, Warrendale, PA, 1995

9.7 *Proceedings of the Ninth International Conference on Titanium,* Metallurgical Society of AIME, Warrendale, PA, 2000

9.8 *Proceedings of the Tenth International Conference on Titanium,* Metallurgical Society of AIME, Warrendale, PA, 2004

9.9 *Proceedings of the Eleventh International Conference on Titanium,* Metallurgical Society of AIME, Warrendale, PA, 2008

9.10 *Proceedings of the Twelfth International Conference on Titanium,* Metallurgical Society of AIME, Warrendale, PA, 2012

9.11 G. Lutjering and J.C. Williams, *Titanium,* Springer, 2003

9.12 R. Boyer, E.W. Collings, and G. Welsch, Ed., *Materials Properties Handbook: Titanium Alloys,* ASM International, 1994

9.13 M.J. Donachi, *Titanium: A Technical Guide,* 2nd ed., ASM International, 2000

9.14 T. Altan et al., *Forging Equipment, Material, and Practices,* Air Force Materials Laboratory, Wright-Patterson Air Force Base, Dayton, OH, 1973

9.15 R.G. Broadwell, R.B. Sparks, and J.E. Coyne, The Effect of Processing and Heat Treatment Variables on Some Critical Mechanical Properties of Ti-8Al-1Mo-1V and Ti-6Al-2Sn-4Zr-2Mo Compressor Wheel Forgings, *Met. Eng. Q.,* American Society for Metals, Aug 1968

9.16 J.E. Coyne, G.H. Heitman, J. McClain, and R.B. Sparks, The Effect of Beta Forging on Several Titanium Alloys, *Met. Eng. Q.,* American Society for Metals, Aug 1968

9.17 G.H. Heitman, J.E. Coyne, and R. Galipeau, The Effect of Alpha Beta and Beta Forging on the Fracture Toughness of Several High-Strength Titanium Alloys, *Met. Eng. Q.,* American Society for Metals, Aug 1968

9.18 *Forging Industry Handbook,* Forging Industry Association, Cleveland, OH, 1966

9.19 E. Bohanek, and H.D. Kessler, An Advanced Titanium Base Alloy for Service at Temperatures in Excess of 800 °F, *Reactive Metals,* Interscience Publishers, Inc., New York, 1959

9.20 L. Croan and F. Rizzitano, *The Influence of Forging Temperature on Mechanical Properties of Ti-6Al-4V Titanium Alloys,* Watertown Arsenal, Watertown, MA, 1957

CHAPTER 10

Secondary Working of Bar and Billet*

AFTER CONVERTING A TITANIUM alloy ingot into bar or billet, it can be further worked into useful shapes using essentially the same processes and equipment used to work other metals and alloys. The principal shaping processes are open-die and closed-die forging, hot die and isothermal forging, ring rolling, and extruding (Ref 10.1–10.15).

These metalworking processes are used to efficiently and economically produce shapes and to develop more desirable microstructural and mechanical properties than can be achieved in heavy bar and billet (Ref 10.1). Generally, the same tensile and other mechanical properties are achieved as those in small-diameter bar and sheet.

Physical Metallurgy

Pure titanium goes through an allotropic transformation at 885 °C (1625 °F) from a hexagonal close-packed crystal structure (α) below this temperature to a body-centered cubic (bcc) structure (β) above it. The bcc, or high-temperature, structure (β) deforms more easily because more slip systems are available for deformation.

Alloys that are essentially single-phase alpha at ambient temperatures (the commercially pure compositions and alloys that contain alpha stabilizers and only small amounts of beta stabilizers) are designated alpha (α) alloys. There is a temperature range in titanium alloys where the high-temperature phase and the low-temperature phase coexist. Titanium alloys in this range

are called alpha-beta (α-β) alloys and are usually hot worked to achieve optimum microstructure and mechanical properties, particularly tensile ductility. Other alloys, called beta (β) alloys, are usually hot worked above the beta transus because the beta structure and its properties are desired for final use. Alloying elements called alpha phase stabilizers, including aluminum (Al), carbon (C), oxygen, and nitrogen (in the forms O_2 and N_2), raise the transformation (to beta) temperature. Other alloying elements called beta stabilizers, including vanadium (V), chromium (Cr), iron (Fe), and molybdenum (Mo), lower it. Figure 10.1 shows phase diagrams illustrating the effect of alloying on the transus.

Titanium alloys are usually hot worked at a temperature approximately 30 °C (50 °F) below the beta transus temperature (Ref 10.16, 10.17), to develop the optimum mechanical properties associated with globular alpha in a transformed beta-matrix (α-β) microstructure. Therefore, it is critical to know the beta transus temperature (α + β/β) for each heat or lot of material.

Steels are worked at 80 to 90% of their melting temperature (T_m), many nickel-base alloys are hot worked at 85 to 95% of T_m, and titanium alloys are hot worked at 60 to 70% of T_m because of concern for the transus. This is the first departure from traditional metalworking practices, because the titanium alloys have higher solidus temperatures than the steel and nickel-base alloys.

Four major methods of plastically deforming titanium and its alloys are forging, ring rolling, extrusion, and bar or profile rolling (Ref 10.18).

---

*Adapted and revised from C.J. School and Donald E. Batzer, originally from James E. Coyne, *Titanium and Its Alloys,* ASM International.

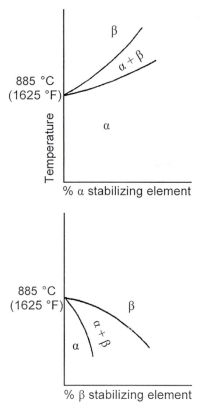

Fig. 10.1 Hypothetical phase diagrams. Curves originate at the transformation temperature 885 °C (1625 °F) and show the effects of alpha- and beta-stabilizing elements on the α and β transus.

Fig. 10.2 Fan blades, compressor discs, and many other engine components use forged titanium parts.

Fig. 10.3 Titanium hip joint

Fig. 10.4 Forged titanium golf club heads

Fig. 10.5 Forged titanium sporting equipment

Forging

Titanium and its alloys are forged into components for use in many aerospace applications, including jet engines, airframes, missiles, and spacecraft. Forged titanium alloys are also widely used in medical and performance sports equipment (Fig. 10.2–10.5). Forgings are produced using the same types of metalworking equipment used to forge aluminum alloys, steels, and nickel-base alloys, including hammers, mechanical and hydraulic presses, high-energy units, and hot die and isothermal hydraulic presses (Ref 10.18, 10.19).

Forgeability. Titanium alloys in general are readily forgeable. Forgeability and extrudability refers to the capacity of a material to be plastically deformed without serious rupturing and cracking. An alloy with good forgeability can be subjected to a large amount of plastic deformation over a broad temperature range before significant cracking occurs.

Factors that govern forgeability include:

- Amount of energy needed to deform the alloy
- Rate of strain hardening and strain-rate sensitivity of the alloy
- Temperature range over which the alloy can be deformed without chill cracking and rupturing

Figure 10.6 shows the yield strength at two strain rates versus temperature for Ti-6Al-4V and Ti-8Al-1Mo-1V. Yield strength increases rapidly with decreasing temperature, and these alloys are sensitive to strain rate. As with all metals and alloys, yield strength decreases with increasing temperature. As with most other alloys, titanium alloys are also sensitive to strain rate. Yield strength and flow stress increase as the rate of straining or deformation increases.

Most titanium alloys have a narrow forge temperature range and are sensitive to cracking at the lower end of the range. The upper end of the range is generally controlled by the beta transus for the individual alloy, not by a problem with hot shortness, as there is with alloys forged closer to their T_m. Rather, the problem lies in strain hardening and surface cracking at lower temperatures.

Alloy chemical composition plays an important part in forgeability. Highly alpha-stabilized alloys such as Ti-5Al-2Sn and Ti-8Al-1V-1Mo have a narrower forge temperature range than the highly beta-stabilized alloys such as Ti-10V-2Al-3Fe and Ti-6Al-2Sn-4Zr-6Mo. Also, a fine-grained, heavily wrought billet is less prone to rupturing than is a coarse-grained billet, because of the greater randomness of slip systems available for deformation.

As mentioned previously, forgeability of titanium alloys can vary considerably. Table 10.1 lists many commercial titanium alloys and their beta transi, forging temperature ranges and pressures, and resistances to cracking when forging.

Forge Temperatures. Titanium alloys are typically forged at temperatures below the beta transus. The beta transus for Ti-6Al-4V is nominally 995 °C (1825 °F) and 1040 °C (1900 °F) for Ti-8Al-1Mo-1V. Material is heated to approximately 30 °C (50 °F) below the beta transus temperature prior to forging to produce an optimum balance of mechanical properties and the most frequently required microstructure of α + β. Transformed beta (alpha phase in a beta phase matrix) structures are generally not acceptable except where superior fracture toughness is required.

The yield strength of Ti-6Al-4V at 970 °C (1775 °F), which is 30 °C (50 °F) below the beta transus, is 41.4 MPa (6 ksi) at a strain rate of 5 mm/mm/min (5 in./in./min); at 930 °C (1700 °F), yield strength is approximately doubled. The Ti-8Al-1Mo-1V alloy at 1010 °C (1850 °F), which is 30 °C (50 °F) below the beta transus, has the same yield strength as Ti-6Al-4V at 970 °C. By comparison, the yield strength of Ti-8Al-1Mo-1V is 125 MPa (20 ksi) at 970 °C, which is 70 °C (125 °F) below the beta transus.

The alloy Ti-8Al-1Mo-1V becomes crack-sensitive close to 970 °C or lower. Both alloys are sensitive to strain rate. Because strain rates experienced in a hammer (approximately 2000 mm/mm/min, or 2000 in./in./min) and in a hydraulic press (20 to 40 mm/mm/min, or 20 to 40 in./in. min) are greater than the 5 mm/mm/min (5 in./in. min) plotted in Fig. 10.6, Ti-8Al-1Mo-1V is more difficult to forge than Ti-6Al-4V.

Although titanium alloy forgings do not typically encounter service temperatures above 540 °C (1000 °F), the energy needed to plastically deform them is quite high. This is because most other alloy families are forged closer to their melting temperatures (>0.80 T_m) than the titanium alloys (0.6 to 0.7 T_m). The other alloy families referred to here include nickel-base superalloys, which are designed for high-temperature service (650 to 980 °C, or 1200 to 1800 °F).

Titanium alloys have a very narrow temperature range for forging, as indicated in Table 10.1. This means that selection of proper forging temperature and conservation of heat are important. Adequate deformation at the selected forge temperature could be a consideration, and geometry (part shape) influences the degree of sophistication that can be realized due to the rate of heat

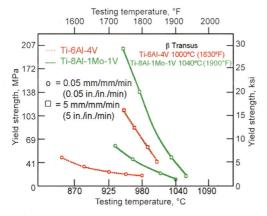

Fig. 10.6 Effect of strain rate and forging temperature on yield strength of Ti-6Al-4V and Ti-8Al-1Mo-1V

Table 10.1 Forging characteristics of representative titanium alloys

Alloy	Type	α transus(a) °C	°F	β transus(b) °C	°F	Die-forging range(b) °C	°F	Pressure to crack(c) MPa	ksi	Resistance
Commercially pure	α	905	1660	915	1675	815–900	1500–1650	450–515	65–75	Excellent
Ti-5Al-2.5Sn	α	945	1735	1040	1900	970–1010	1775–1850	520–585	75–85	Fair
Ti-8Al-1V-1Mo	Near α	930	1700	1040	1900	970–1010	1775–1850	520–585	75–85	Fair
Ti-3Al-2.5Sn	α-β	...	...	940	1725	815–915	1500–1675	450–515	65–75	Excellent
Ti-6Al-4V	α-β	...	...	1000	1830	900–970	1650–1775	515–585	75–85	Good/excellent
Ti-6Al-4V ELI(d)	α-β	...	...	970	1775	900–940	1650–1725	515–585	75–85	Good/excellent
Ti-6Al-6V-2Sn	α-β	...	...	945	1735	860–915	1575–1675	450–515	65–75	Excellent
Ti-6Al-2Sn-4Zr-2Mo	α-β	...	...	985	1810	900–970	1650–1775	515–585	75–85	Good/excellent
Ti-6Al-2Sn-4Zr-6Mo	α-β	...	...	950	1740	870–930	1600–1700	550–620	80–90	Excellent
Ti-10V-2Al-3Fe	β	...	...	795	1460	760–870	1400–1600	550–620	80–90	Excellent
Ti-3Al-8V-6Cr-4Mo-4Zr(e)	β	...	...	795	1460	815–1010	1500–1850	550–620	80–90	Excellent
Ti-13V-11Cr-3Al	β	...	...	720	1325	870–980	1600–1800	585–690	85–100	Excellent

(a) ±11 °C (20 °F). (b) ±14 °C (25 °F). (c) For forging in a hydraulic press 635 mm/min (25 in./min). Double these pressures for hammer forgings. (d) ELI, extra low interstitial. (e) Known as Beta-C

loss, because various shapes lose heat at different rates depending on their surface-to volume ratio.

A considerable amount of frictional heat buildup and metal temperature rise can occur during forging, particularly when using a fast deformation process such as a hammer, screw press, and high-energy impacter. Heat buildup is sometimes referred to as adiabatic heating. If the degree of adiabatic heating causes the titanium to exceed the beta transus, an undesirable "overheated" microstructure of transformed beta can be formed.

Forge temperatures should be selected to match the forge method strain rate, amount of deformation, and desired final microstructure (Ref 10.3, 10.4, 10.16). For most traditional requirements (globular alpha in a transformed beta matrix), the beta transus temperature of the alloy and "heat" of the material to be forged must be known and a forge temperature selected that is approximately 30 °C (50 °F) or more below the transus temperature to prevent overheating.

Classes of Forgings

Forgings are made in one or more heating and working operations, depending on the complexity of configuration. Forgings, whether made on a hammer or a press, are classified either as open die or closed die (Ref 10.18). Closed-die forgings produced in a hydraulic press can be further classified into conventional, hot die, and isothermal. Equipment and processes are similar to those used for other alloy systems.

Open-die forgings (also called flat-die, blacksmith, and hand forgings) are made on flat or simple dies by repeated strokes or blows, with the workpiece manipulated into a simple shape. An example of a titanium open-die forging (hand forging) is shown in Fig. 10.7, a shape that will be further processed into a closed-die geometry. Open-die forging is frequently used to shape bar and billet into a more useful distribution of material prior to closed-die forging. Useful distribution is defined as a distribution that will minimize the input weight required to fill the impression and the formation of forge defects (laps) by allocating the material over the impression according to the volume requirements of the impression.

An open-die forging may be the method of choice when the shape is simple, the quantity small, and the cost of dedicated closed-die tooling excessively high compared with the cost of input material and the additional machining time required to remove the extra input material. However, it must be possible to obtain the desired microstructure and mechanical properties from the forging to make it an acceptable method for shaping.

Closed-Die Forging. Most titanium forgings are produced as closed-die forgings (also called impression-die forgings), because it is relatively easy to justify the tooling and labor cost through input material cost-savings resulting from the refined product shape. Also, it is possible to produce more uniform, better-controlled microstructure and therefore mechanical properties due to the additional controlled thermomechanical working required to produce a closed-die part.

While the closed-die forge process appears simple—forcing heated metal into a die cavity machined to a predetermined shape—the process

Fig. 10.7 Example of titanium forged in open or flat dies as in blacksmithing (hand forging)

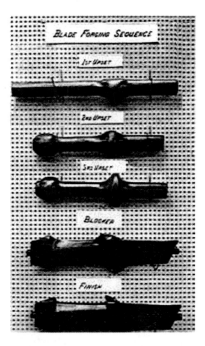

Fig. 10.8 Fan blade forging sequence. From top to bottom: first upset, second upset, third upset, blocker, and finish

can be very complex, depending on the required shape.

Factors that must be analyzed in the overall design of the closed-die forging process include:

* Shape complexity and volume distribution of the forging; secondary working of bar or billet
* Quantity, method, and configuration of preforms or blockers to distribute the volume in a defect-free progression
* Flash dimensions in the dies and the additional metal volume required for flash in the preforming and finish operations
* Forging load, energy, and the off-center load for each forging operation

Often, the final closed-die forging is preceded by open-die forging to allocate the metal to properly fill the impression. Figure 10.8 shows a typical sequence of operations used in making titanium blades.

Other factors to consider in designing forge operations are billet material flow stress and its variation with temperature drop and section thickness, lubrication required to overcome friction at the die-material interface, and die cavity geometry, all of which determine forging load and energy requirements. The hammer and press-head velocity under load and the contact times influence the flow stress and workability of titanium and thus influence die fill and energy and load requirements.

The forge process design and equipment selection is often based on past experience. However, to better understand the process and to minimize the time and expense of "cut-and-try" development, computer-aided design and computer-aided manufacturing and sophisticated computer modeling are used to be right the first time.

Hot Die Forging. To reduce the input material weight and subsequent machining costs, designers consider near-net shape, the close-tolerance titanium forgings produced on hydraulic presses using the hot die approach (Ref 10.18).

In open- and closed-die forging, dies are usually maintained at a temperature below 590 °C (1100 °F) maximum and below 480 °C (900 °F) maximum, respectively. For hot die forging, dies are heated and maintained at a temperature in the range of 590 to 930 °C (1100 to 1700 °F). The benefit of less metal chilling from die contact enables the use of less energy to produce a given shape, or the production of a more refined shape using the same energy. Precision forgings requiring no machining other than hole drilling, for example, are produced using this approach. Hot die forging is typically performed in air and is used primarily to produce small airframe structural shapes.

Isothermal Forging. Open- and closed-die titanium forgings are being produced using isothermal forge presses (Ref 10.1, 10.3, 10.18, 10.19). The process consists of preheating the material to be forged in a vacuum or protective atmosphere and forging in a vacuum or protective atmosphere with the dies heated and maintained at or near the metal forge temperature.

Benefits of isothermal processing include:

- Significantly lower cut weights
- Closer control of preheat and die temperature throughout the forge cycle
- More latitude and control of strain rates during the forge cycle, because isothermal units are controlled at a nominal 0.2 mm/mm/min (0.2 in./in./min), while most hydraulic presses operate at approximately 10 to 40 mm/mm/min (10 to 40 in./in./min)
- Lower forge energy requirements
- Ability to produce a near-net-shape geometry

Titanium alloys are isothermally forged using the same equipment and procedures used to produce other alloys, particularly consolidated powder nickel-base superalloys.

Precision forging produces a highly specialized forged titanium product. These forgings require little or no machining prior to assembly. Jet engine blades requiring machining only of the attachment detail have been precision forged since the 1950s. An example of the shape development of a precision blade is shown in Fig. 10.8. Because the air foil surface of the blade is not machined after forging, the grain flow is maintained. Improved fatigue, corrosion, and erosion properties are maintained over those of conventionally forged and machined blades. The use of precision-forged blades reduces machining to a minimum. As a result, savings are realized in bar stock input weight and in special machining operations.

Precision forging is relatively expensive, requiring sophisticated process design and close control of process variables and tool conditions. Thus, close control of furnace temperature, absolute prevention of scale by atmosphere control and by coating the stock, uniform and adequate lubrication, and strict dimensional and microstructural control of the forgings are critical. The design and manufacture of precision-forging dies requires particular attention to local shrinkage and elastic deformation.

Precision forging of small rib and web structural shapes up to 0.25 m² (~400 in.²) in planview area is also cost-effective. All machining except for the simplest of milling operations and drilling of holes is eliminated.

Generally, larger part-production volumes (hundreds to thousands) are required to cover the increased cost of forge process engineering, more costly forge tools, and dimensional inspection fixtures associated with precision forging.

Ring roll forging produces symmetrical and conical-shaped cylinder lengths. In the process of deformation, the thickness of the ring blank is decreased as the diameter and/or height is increased (Ref 10.3, 10.18, 10.19).

The typical sequence of operations begins with a hollow ring blank made by piercing a pancake into a doughnut, forging a dog-bone shape from which the center is machined out, and cutting lengths from an extrusion. Figures 10.9 to 10.11 illustrate these operations. The material to be rolled is heated as in conventional forging and is deformed by the force of an idler roll inside the blank that forces the blank against a rotating drive or pressure roll outside the blank. Figure 10.12 illustrates a typical ring roller setup.

Metallurgical factors pertinent to rolling of titanium rings are the same as for forging. Alpha and alpha-beta alloys are rolled at a temperature 30 °C (50 °F) below the beta transus. When the temperature of the ring blank drops to where the stock will no longer deform, the blank is reheated, and the rolling operation is continued. With proper control of temperature and sufficient deformation and a reduction in wall thickness of 2:1 or 3:1, the microstructure of the rolled ring is similar to that of a forged part. The main difference is grain flow, which, in a ring-rolled forging, is predominantly tangential. Figure 10.13 shows the macrostructure of a Ti-6Al-4V ring, and Fig. 10.14 shows the macrostructure of a closed-die

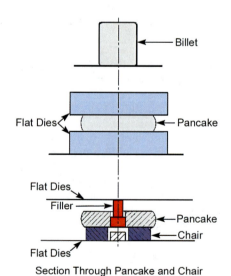

Fig. 10.9 Blank for processing in the ring mill is first pierced (blank piercing).

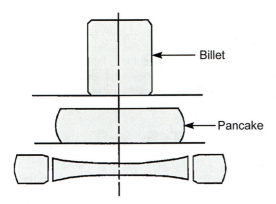

Fig. 10.10 Dog-boning operation for preparing a blank for the ring mill

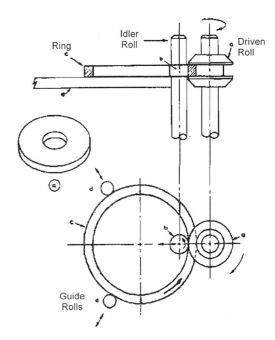

Fig. 10.12 Ring roller setup to produce symmetrical cylindrical or conical shapes. A preformed heavy-walled ring is heated for conventional forging and deformed between a driver and idler roll.

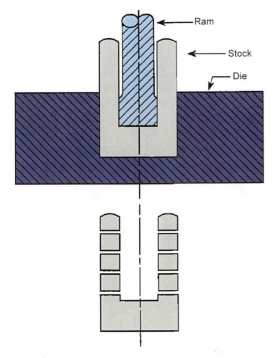

Fig. 10.11 Extruded ring blank. Ring blanks can be made by indirect, or back, extrusion.

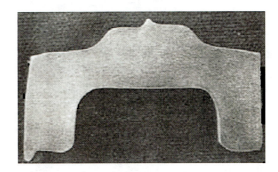

Fig. 10.13 Macrostructure of rolled Ti-6Al-4V ring illustrating the predominantly tangential grain flow

Fig. 10.14 The grain flow shown in this macrostructure of a closed-die radial section of a compressor wheel forging differs from that of the rolled ring in Fig. 10.13.

forging. Typical tensile properties of a Ti-6Al-4V rolled ring are listed in Table 10.2.

Extrusion

Another method of working titanium and its alloys is extrusion. In the extrusion operation, billet stock is heated to the required temperature, placed in a chamber, and compressed by a mov-ing punch. When sufficient pressure is achieved, the billet is forced through an orifice into the desired shape. Extrusions are classified as forward or back (reverse) extrusion, depending on the direction of metal movement relative to the punch. Figure 10.15 shows simple forward and backward extrusions. Both solid and hollow extruded shapes are possible. Extruded shapes are shown in Fig. 10.16.

Extrusions more than 18.3 m (60 ft) long and weighing more than 455 kg (1000 lb) are pro-

duced. Production titanium extrusions generally use high-speed horizontal presses capable of extruding at extremely high rates. Extrusion, like the forging process, requires lubricants that are stable at high temperatures to reduce friction and die erosion caused by hot metal as it passes through the die. Greases are effective at low temperatures. Molten glass and other ceramic lubricants are used at high temperatures (Fig. 10.17).

Titanium extrusions are produced from the same stock as forged and ring-rolled products. The titanium billet is cut into desired lengths, heated, and extruded. Heating is done in conventional gas-fired furnaces, electric furnaces, and by induction. Induction minimizes heating time and surface contamination. The extrusion process is an excellent means of making long, uniform shapes. It enhances maximum use of material and minimizes machining.

Table 10.2 Typical tensile properties of Ti-6Al-4V rolled ring

0.2% yield strength		Tensile strength		Elongation, %	Reduction of area, %
MPa	ksi	MPa	ksi		
970	140.6	1065	154.4	12.5	36.3
1009	146.4	1075.6	156.0	13.5	44.2
965	140.0	1054.9	153.0	12.5	34.1
974	141.2	1038.4	150.6	15.0	44.8
978	141.8	1060.4	153.8	12.5	41.1

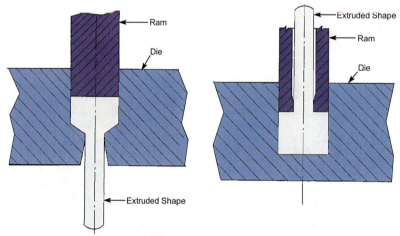

Fig. 10.15 Extrusion methods: forward extrusion (left) and backward extrusion (right)

(a)

(b)

Fig. 10.16 Representative extruded titanium shapes

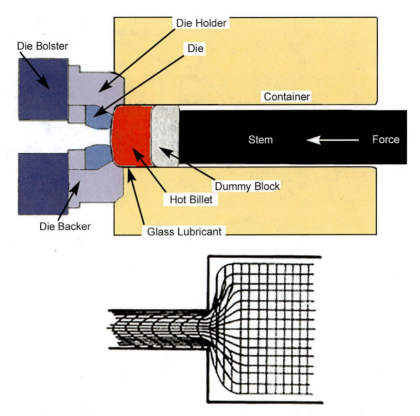

Fig. 10.17 Extrusion setup. Top diagram illustrates the operation of an extrusion press (note location of the glass lubricant). Bottom diagram illustrates the streamline flow of the titanium billet as it moves through the extrusion orifice.

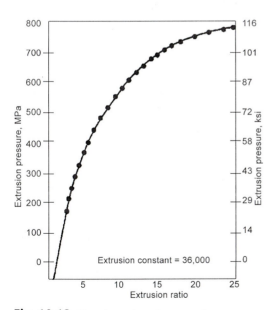

Fig. 10.18 Effect of extrusion ratio on extrusion pressure

High extrusion ratios are used to achieve maximum benefits of the process. The extrusion ratio describes the relationship between the cross-sectional area of the blank being extruded (Ab) to the area of the extruded shape (Ae), or Ab/Ae. The higher this ratio, the greater the energy needed to start metal flow. Figure 10.18 shows the effect of extrusion ratio for Ti-6Al-4V alloy on required extrusion pressure.

High extrusion ratios require that the billet be heated above the beta transus temperature. Ratios of 30:1 are common, and much higher ratios are possible. The only extrusion accomplished below the beta transus are those at ratios of less than 3:1. The structure of extruded beta titanium is acicular alpha (transformed beta).

Extrusion requires the same control as with forging in using beta-forging temperatures. If heating exceeds the beta transus, the transformed structure should be highly worked, or mechanical properties will be adversely affected. Fortunately, many thin-section parts cool rapidly, and the worked structure is retained. Experiments conducted with beta-forged Ti-6Al-4V at heavy reductions (up to 80%) and at temperatures ranging from 30 to 170 °C (50 to 300 °F) above the beta transus indicate that the worked structure reverts to recrystallized equiaxed grain structure in approximately 3 min at temperatures above the beta transus.

Microstructure and Mechanical Properties

Mechanical properties of titanium alloys are so dependent on microstructure that the two can be discussed together (Ref 10.1, 10.5–10.14). Also, most alloy specifications include microstructural requirements specifically to assure various mechanical property characteristics.

Alpha-Beta-Processed Alloys. Prior to the mid-1970s, titanium was processed using forging temperatures and increments of deformation that would ensure sufficient deformation (approximately 50% minimum) below the beta transus to produce a microstructure (Fig. 10.19) of equiaxed primary alpha in a beta matrix (see Chapter 6, "Mechanical Properties and Testing of Titanium Alloys," in this book).

Also, material exhibiting fully transformed structures of coarse Widmanstätten platelets in equiaxed beta grains heavily outlined with alpha was considered to be beta-embrittled due to its significantly lower tensile ductility. These structures generally were caused by inadvertent heating over the beta transus with little or no subsequent deformation.

If a workpiece is not solution treated after forging, its microstructure is governed almost entirely by the forging operation (temperature and work). Figure 10.20 shows the amount of primary alpha present in Ti-6Al-4V alloy as a function of temperature.

Table 10.2 lists typical room-temperature tensile properties. Metal heated above the beta transus, and apparently improperly worked, suffered a significant loss of ductility. Also, the percent of primary alpha decreases with increasing forging temperature (Fig. 10.20).

Alpha-stabilized alloys such as Ti-8Al-1Mo-1V and Ti-6Al-2Sn-4Zr-2Mo were formulated for use in high-temperature (430 to 510 °C, or 800 to 950 °F) applications, primarily in the mid-to-late compressor stages of aircraft jet engines. The creep strength of these alloys improves with processing that lowers the percentage of alpha present. The percent alpha can be reduced by either working or solution treating closer to, but below, the beta transus, as illustrated by the data given in Tables 10.3 and 10.4.

Beta-Processed Alloys. Prior to the mid-1970s, titanium forgings were produced by processing

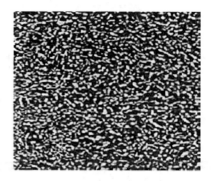

Fig. 10.19 Processed structure of alpha-beta titanium. A desirable equiaxed primary alpha forms with sufficient working (approximately 50%) below the beta transus temperature, followed by an alpha-beta anneal and slow cool. White particles are equiaxed primary alpha. Etchant: 2%HF-2%HNO$_3$-96%H$_2$O. Original magnification: 250x

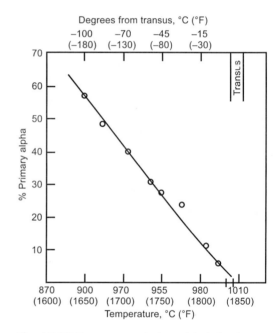

Fig. 10.20 The percentage of primary alpha in the microstructure decreases sharply as the forging or heat treating temperature approaches the transus (1000 °C, or 1830 °F).

below the beta transus to yield a structure of fine, equiaxed primary alpha in a beta matrix to achieve what were considered optimum properties (Ref 10.1), particularly tensile ductility. However, prior to that time, tensile ductility and the Charpy test were the accepted measures of toughness and resistance to crack propagation.

By the 1970s, experimental work indicated that properly conducted forging from above the beta transus can be highly beneficial to certain me-

Table 10.3   Tensile properties at 21 °C (70 °F) for the Ti-6Al-4V alloy(a)

Forging temperature			0.2% yield strength		Tensile strength			
°C	°F	Primary alpha, %	MPa	ksi	MPa	ksi	Elongation, %	Reduction of area, %
915	1675	57	979	142	1036	150	12.5	33.8
930	1700	44	986	143	1063	154	12.5	29.9
940	1725	36	982	142	1067	155	13.0	31.9
955	1750	26	965	140	1020	148	12.5	40.1
975	1785	18	958	139	1027	149	14.0	37.0
985	1805	11	1014	147	1057	153	13.0	33.2
995	1825	5	972	141	1036	150	12.0	30.5
1005	1840	0	951	138	1020	148	6.0	10.1
Heat treatment(a)								
Ti-8Al-1Mo-1V (with different microstructures)								
A(b)		0	799	116	916	133	13.0	22.0
A(b)		50	861	125	951	138	19.0	35.0
A(b)		70	868	126	951	138	15.0	35.0
Ti-6Al-2Sn-4Zr-2Mo (with different microstructures)								
B(c)		0	861	125	985	143	12.0	25.0
B(c)		40	875	127	985	143	16.0	46.0
B(c)		60	889	129	1006	146	15.0	40.0

(a) Beta transus 1000 °C (1830 °F); annealed 704 °C (1300 °F) 2 h, air cooled (AC). (b) A = 1010 °C (1850 °F) 1 h, AC + 595 °C (1100 °F) 8 h, AC; (c) B = 975 °C (1785 °F) 1 h, AC + 595 °C (1100 °F) 8 h, AC

Table 10.4   Creep properties at 21 °C (70 °F) of some titanium alloys

Heat treatment	Primary alpha, %	Temperature		Stress		Time to 0.1% plastic deformation, h
		°C	°F	MPa	ksi	
Ti-8Al-1Mo-1V (with different microstructures)						
A(a)	0	455	850	345	50	117
A(a)	50	455	850	345	50	44
A(a)	70	455	850	345	50	41
A(a)	0	510	950	207	30	114
A(a)	50	510	950	207	30	25
A(a)	70	510	950	207	30	14
Ti-6Al-2Sn-4Zr-2Mo (with different microstructures)						
B(b)	0	455	850	448	65	38
B(b)	40	455	850	448	65	41
B(b)	60	455	850	448	65	28
B(b)	0	510	950	310	45	26
B(b)	40	510	950	310	45	12
B(b)	60	510	950	310	45	19

(a) A = 1010 °C (1850 °F) 1 h, air cooled (AC) + 595 °C (1100 °F) 8 h, AC; (b) B = 975 °C (1785 °F) 1 h, AC + 595 °C (1100 °F) 8 h, AC

chanical properties. Also, progress in the study of fracture mechanics and the acceptance of the compact tension test as a mathematically valid measure of resistance to crack propagation (K_{Ic}) reinforced metallurgists' data. Fracture toughness tests showed that the torturous crack path between the laths of alpha and transformed beta phases in a transformed structure is more resistant to crack growth than the globular alpha in a transformed beta matrix (see Chapter 6, "Mechanical Properties and Testing of Titanium Alloys," in this book).

The combination of the good forgeability of the beta structure with the higher forge temperature possible for beta forging makes possible significant improvements in refinement of shape and part definition. Figure 10.21 shows the level of improvement possible when beta forging from 1150 °C (2100 °F) compared with conventional forging at 955 °C (1750 °F).

An excellent example of a complex shape achieved by beta forging is shown in Fig. 10.22. This is an 18 kg (40 lb) structural forging of aluminum design produced as beta-forged Ti-6Al-4V.

Beta-forged parts can exhibit significant improvement in strength, creep resistance, fatigue, and fracture toughness values. Figure 10.23 shows the effects on impact strength of forging from above the beta transus. Other examples of toughness improvement are the highly beta-stabilized alloys Ti-17 (Ti-5Al-2Sn-2Zr-4Mo-4Cr) and Ti-6Al-2Sn-4Zr-6Mo. When alpha-beta forged, solution heat treated, and aged, these alloys have fracture toughness values (K_{Ic}) of approximately 33 MPa$\sqrt{m}$(30 ksi$\sqrt{in}$). When beta forged, alpha-beta solution heat treated, and aged, the fracture toughness values exceed 55 MPa$\sqrt{m}$ (50 ksi$\sqrt{in}$).

Loss of tensile ductility is the most significant reduction of mechanical property values. Adequate tensile ductility can be achieved by properly controlled beta-forge practices. Typical tensile properties of alpha-beta forged and properly and improperly beta-forged Ti-6Al-4V are shown in Table 10.5.

The difference in properly versus improperly beta-forged material is best observed by viewing the structure at low magnifications. Figures 10.24 and 10.25 show properly and improperly forged material, respectively. Properly forged material has a distorted macro grain, while improperly forged material has an equiaxed macro grain. The equiaxed macro grain indicates insufficient reduction after heating above the beta transus. It is for this reason that strikeovers (i.e., reheating to the forge temperature above the beta temperature and producing little deformation) are generally prohibited. However, striking over is acceptable when working in the alpha-beta range.

Beta forging requires proper control of the temperature and deformation, as shown in Fig. 10.26, for example, where Ti-6Al-4V stock was heated and worked in the beta field. Tensile specimens were removed after each operation. The material heated in the beta field and highly deformed plastically has good tensile properties, but the same forged material heated in the beta field without adequate plastic deformation has significantly lower tensile ductility. The material must be worked sufficiently subsequent to each heating into the beta field when high tensile ductility is

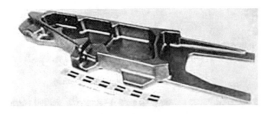

Fig. 10.21 Hammer forging. Left: workpiece forged from the beta field at a temperature of 1150 °C (2100 °F). Right: workpiece forged from the alpha-beta field at 955 °C (1750 °F). The piece on the left is filed. Dies were designed originally for forging aluminum.

Fig. 10.22 Flat-track forging. Originally of aluminum design, this 18 kg (40 lb) Ti-6Al-4V structural part was forged from the beta field.

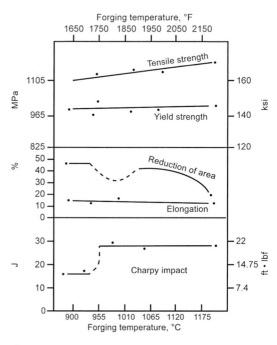

Fig. 10.23 Increased impact strength can be attained in Ti-6Al-4V by forging above the beta transus. Heat treatment: 940 °C (1725 °F) for 1 h, water quenched (565 °C, or 1050 °F) for 1.5 h; air cooled.

Table 10.5   Typical tensile properties of Ti-6Al-4V processed in various ways

How forged	0.2% strength		Yield strength		Tensile elongation, %	Reduction of area, %
	MPa	ksi	MPa	ksi		
Alpha-beta, 955 °C (1750 °F)	924	134	1000	145	14	36
Beta (properly)	889	129	958	139	12	30
Beta (improperly)	883	128	945	137	8	12

Fig. 10.24 Structure of properly forged material showing distorted beta grain. Etchant: 10%HF-10%HNO₃-80%H₂O. Original magnification: 8x

Fig. 10.25 Structure of improperly forged material showing equiaxed beta grain. Equiaxed grains indicate insufficient working after heating in the beta field. Etchant: 10%HF-10%HNO₃-80%H₂O. Original magnification: 8x

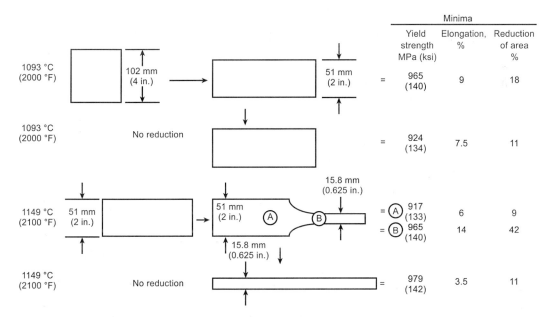

Fig. 10.26 Effect of deformation on tensile properties of heating Ti-6Al-4V in the beta field. The controlling factor is the amount of metal deformation after beta heating.

desired. Die design considerations must be such as to provide the proper temperature and deformation relationship on intricate shapes of varying cross sections. As can be seen in Fig. 10.26, the amount of metal deformation after beta heating is the controlling factor.

The qualitative effect of beta forging on mechanical properties is summarized in Table 10.6. With proper control, there is very little difference in most properties, and, in some instances, certain properties are improved, such as fracture toughness and creep strength.

The emergence of the hot die and isothermal press forge capability significantly improved and expanded the parameter controls by which beta forging can be performed. Metal and die temperatures and strain rates are now controlled to provide more optimum combinations of microstructure and mechanical properties. As an example, if all or most of the beta-processed properties were desired (with some sacrifice in tensile ductility), both the metal temperature and die temperature could be held above the beta transus. If a combination of most of the improved beta processing (creep, impact, fracture toughness) and improved tensile ductility was desired, metal temperature can be controlled above the beta transus and the die temperature below the beta transus.

Three alloys used primarily in the beta-forged condition are Ti-13V-11Cr-3Al for sheet, plate, and billet for structurals; Ti-17 (Ti-5Al-2Sn-2Zr-4Mo-4Cr) for jet engines; and Ti-10V-2Al-3V for airframe structural parts.

Ti-13V-11Cr-3Al is a metastable beta alloy, which is quite strain-rate sensitive and therefore better suited to press forging than hammer forging. The beta phase is easily retained by quenching and fast air cooling from the beta phase field. The alpha phase precipitates on aging, thereby developing 1105 to 1170 MPa (160 to 170 ksi) yield strength in heavy sections > 100 mm (4 in.). It is not as tough in the presence of a notch as some other titanium alloys (such as Ti-6Al-4V), but proper processing can improve its toughness.

Table 10.7 shows how mechanical properties are affected by the microstructure. It is important in the Ti-13V-11Cr-3Al alloy to create sufficient lattice distortion during forging to produce random nucleation of the aging precipitate in the postforge age cycle. This is achieved by working the material in the low end of the forging range (<980 °C, or 1800 °F). A properly processed Ti-13V-11Cr-3Al forging has a microstructure similar to that shown in the electron micrograph in Fig. 10.27, with alpha particles randomly distributed. An improperly processed Ti-13V-11Cr-3Al forging has the structure shown in Fig. 10.28, with the alpha particles essentially aligned in the same direction.

Table 10.6 Qualitative comparison of beta-forged versus alpha-beta-forged parts

Mechanical properties	Ti-6Al-4V	Ti-8Al-1Mo-1V	Ti-6Al-2Sn-4Zr-2Mo
Yield strength at 20 °C (70 °F)	Slightly lower	Slightly lower	Slightly lower
Tensile strength at 20 °C (70 °F)	Slightly lower	Slightly lower	Slightly lower
Elongation	Slightly reduced	Slightly reduced	Slightly reduced
Reduction of area	Reduced	Reduced	Reduced
Notched tensile strength ($K_t = 10$)	Improved	Improved	Improved
Notched time fracture ($K_t = 3.8$)	Improved	Improved	Improved
Fatigue	...	...	...
Creep strength	Improved	Improved	Improved
Creep stability	Same	Same	Same
Fracture toughness	Improved	Improved	Improved

Table 10.7 Properties of Ti-13V-11Cr-3Al forgings showing effect of process history

0.2% yield strength		Strength		Tensile elongation, %	Reduction of area, %	R/T(a)	Precracked Charpy impact	
MPa	ksi	MPa	ksi				J/cm²	in.· lb/in.²
Improperly processed								
...	...	1103	160	...	2.0	1.040	4.55–5.25	260–300
1102	160	1143	166	2.5	3.9	32		
1127	163	1185	172	2.5	2.4	36		
Properly processed								
1110	161	1176	171	5.7	7.8	24	6.65–7.18	380–410
1145	166	1207	175	5.7	5.5	28		
1117	162	1169	170	4.0	7.0	22		

(a) Radius (R) of bend over thickness (T) of a side-notched specimen bent slowly around a mandrel

The near-beta alloys Ti-17 (Ti-5Al-2Sn-2Zr-4Mo-4Cr) and Ti-10V-2Fe-3Al also retain the beta phase on cooling from above the transus. Subsequent solutionizing and aging precipitates finely dispersed alpha particles.

For a Ti-10V-2Fe-3Al alloy, this same procedure produces excellent toughness and strength but very poor tensile ductility (approximately 3%). To improve tensile ductility for this alloy, the processing is modified to a beta-temperature-blocking operation with as much deformation as possible, followed by a carefully designed 10 to 20% reduction, an alpha-beta finish forge operation, to promote some globular alpha in the microstructure. However, too much alpha-beta deformation has a significant negative impact on fracture toughness, because the alloy is very sensitive to the thermomechanical work cycle.

Considerable success has been achieved using hot die or isothermal forging to control metal and die temperature around the beta transus to ensure obtaining desired combinations of microstructure and mechanical properties.

Surface Effects of Heating

Titanium is a highly reactive element at the temperatures used for secondary forming. It readily reacts with hydrogen and oxygen. These reactions must be controlled and their effects removed before the titanium part can be used.

Oxygen Contamination. Titanium and its alloys react with the oxygen in air to form an oxide. At 650 °C (1200 °F), the oxide is light and powdery. As the exposure temperature increases to approximately 1040 °C (1900 °F), the oxide changes to a white, light, flaky layer that is easily scraped off. Above 1040 °C (1900 °F), the scale becomes thick and brown-colored and looks like steel scale. This scale can be removed by shot blasting using chilled steel shot and by a caustic salt bath at a temperature of 480 °C (900 °F), followed by an acid dip.

Beneath the scale is an oxygen-enriched surface layer 0.127 to 0.38 mm (0.005 to 0.015 in.) deep. Because oxygen is an alpha-stabilizing element and raises the beta transus, this layer is seen as a continuous alpha phase. The oxygen-enriched alpha layer is hard and brittle and is called alpha case. Severely reduced fatigue life and poor machinability result if the layer is not removed. Figure 10.29 shows an alpha case layer on an alpha-beta alloy. Alpha case is typically removed by acid pickling in an aggressive HNO_3-HF acid solution. It can also be removed by machining using carbide tools, but tool life is reduced. Most materials specifications require a surface measurement after all processing to ensure the removal of alpha case.

Hydrogen has a small atomic diameter and readily diffuses into titanium. If too much hydrogen is absorbed, it forms a brittle titanium hydride compound. All materials specifications have a maximum-allowed hydrogen level on the order of 0.0125 to 0.0150% (125 to 150 ppm) and generally require testing after all processing (including chemical milling to remove alpha case). Billet material typically contains between 40 and 80 ppm hydrogen. Heating for secondary working operations can add another 20 to 40 ppm. Chemi-

Fig. 10.27 Microstructure of properly processed Ti-13V-11Cr-3Al consisting of uniformly distributed alpha particles. This is a result of working the alloy in the low end of the forging range (<980 °C, or 1800 °F) and aging. This is the structure of the high-ductility material referred to in Table 10.7. Original magnification: 10,000x

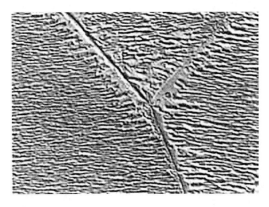

Fig. 10.28 Microstructure of improperly processed Ti-13V-11Cr-3Al consisting of alpha precipitate aligned in one direction, as contrasted with the uniform distribution shown in Fig. 10.27. This is the structure of the low-ductility material referred to in Table 10.7. Original magnification: 10,000x

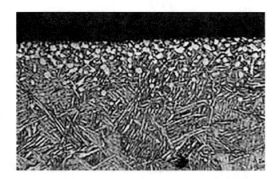

Fig. 10.29 Hard, brittle, oxygen-enriched alpha case forms on the surface of titanium when exposed to elevated temperatures in air. This undesirable layer must be removed because it cracks easily and reduces fatigue life. Etchant: 2%HF-2%HNO₃-96%H₂O. Original magnification: 100x

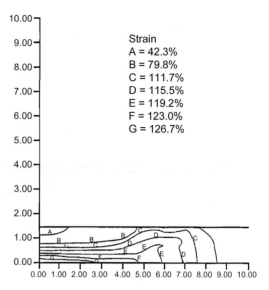

Fig. 10.30 Modeled strain distribution in commercially pure titanium after upset. (Only one-fourth of the pancake is modeled because of twofold symmetry.) The strain varies from 42.3% at "A" at the top center to 126.7% at "G" at the middle center of the pancake.

cal milling, if not properly controlled, can add a similar amount.

To minimize hydrogen pickup during heating, the billet receives a glass/ceramic coating, which serves as a lubricant and a diffusion barrier. Further, titanium should be heated in either electric or gas- and oil-fired furnaces where an oxidizing or "excess-air" fired atmosphere is used to ensure the complete combustion of the hydrogen in the fuel.

If the hydrogen level allowed by the specification is exceeded, it can be reduced by vacuum annealing. A cycle of 2 h at 760 to 790 °C (1400 to 1450 °F) at a good vacuum level of 0.0013 Pa (10^{-5} torr) should reduce the level to 30 or 40 ppm. Lower temperatures or softer vacuums take longer but are also effective. However, a vacuum anneal should be avoided if possible, because it is expensive.

Modeling

One of the more useful tools commercially available is a computer program or code that describes the shape development, strain, and thermal energy change imparted to a shape via the hot deformation process (Ref 10.18).

The programs are based on a mathematical system called finite-element measurement. The codes minimize the time and investment involved in "cut-and-try" methods in the shop. Using these codes, it is possible to design a deformation process that can produce the desired shape, defect free, and impart the desired amounts of strain in a particular area of a shape or a more uniform amount of strain throughout a shape. Similarly,

the temperature changes in a part can be predicted, although their control is more difficult. To achieve this control, the preliminary forge design, metal and die temperatures, and rate of deformation must be defined. Also, accurate physical and mechanical data on the alloy to be processed must be available.

Examples of models produced using a commercially available code are shown in Fig. 10.30 and 10.31. A billet of commercially pure titanium 190 mm round by 343 mm long (7.6 in. by 13.6 in.) was heated to 840 °C (1550 °F) and press forged on flat dies heated to 370 °C (700 °F) using a head speed of 380 mm/min (15 in./min) to upset the billet to 80 mm (3 in.) thick. The resulting strain diagram is shown in Fig. 10.30 and the resulting temperature diagram in Fig. 10.31.

Summary

Titanium bar and billet alloys are shaped using the principal hot working processes. Whatever the process used, successful deformation of titanium alloys into various shapes requires knowledge of the physical metallurgy of the alloy system.

The key to ease of forming and control of mechanical properties is the beta transus temperature of the alloy being worked. Selection of the work-

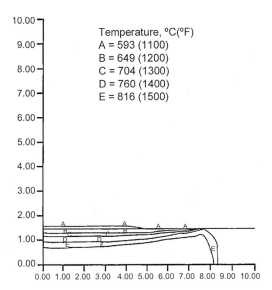

Fig. 10.31 Modeled temperature distribution in commercially pure titanium after upset. The temperature varies from 595 °C (1100 °F) at "A" at the top center to over 815 °C (1500 °F) at "E," representing most of the pancake. The isotherm represented by "A" for the die shows that its surface temperature increased to over 540 °C (1000 °F).

ing temperature and subsequent deformation determine the microstructure. Because titanium alloys are structure-sensitive, components are produced to controlled microstructure and tailored for specific mechanical properties.

Glossary

bender. An open die that simply bends input stock and draw/roll part. Contour of bender generally shapes stock to the outline of the die impression.

blocker. The closed-die impression that precedes finish operation. Shapes the forging 60 to 95% of final geometry. More gentle and flared compared with finish operation.

box. Prefix used with finish, blocker, preblocker, and so on to indicate a special flashline design used when control of material flow is particularly critical.

closed die. Dedicated tooling to achieve specific geometry as defined by die impression. Usually requires die-to-die contact, thus the term *closed die*.

coin. Closed-die forge operation usually performed after finish forge to accomplish special-geometry dimension conformance, sometimes in localized areas.

control flatten. A closed-die geometry consisting of an input stock locator, uniform part thickness, rounded corners, and center spike locator. Usually designed to produce minimum flash development.

draw. An open-die-shaping operation using specialized or stock die tooling. Reduces input stock cross section for entire length or localized areas while increasing the overall length of the input stock. Principally used for structural parts.

edger. A contoured open die that performs the function of offsetting material away from the basic centerline and usually follows the draw/roll operations of structural geometries. The contour of the edger usually follows the plan view of the forging, allowing proper positioning of shaped stock in the next die cavity.

finish. The closed-die impression that, in most cases, yields the final forge geometry.

flatten. An operation using stock tool flat dies with intent to reduce the overall height of the piece being forged.

open die. Use of tools (stock or special) to shape material in a hand forging manner. Tools usually do not come die-to-die, thus the term *open die*.

pot. A closed-die geometry used prior to finish- or blocker-die geometries for shafts and hubs and as the last closed-die impression prior to preparing the ring blank for the ring-roll operation. For shafts, sometimes called skirted-pot die due to flashline design.

preblock. The closed die that usually precedes the blocker operation. Shapes the forging more gently and more flared compared with the blocker. It is sometimes possible to go preblock to finish. Used mainly to accomplish proper material distribution on complex structural parts.

prefinish. An operation using the finish die on a part-way-down basis. Predominantly used on lower part-production volumes where additional die geometries are not cost-justified.

preform. A closed-die impression of gentle, slow transitions to provide for initial forming of flatten and/or control flatten, shaped, and so on.

roller. An open-die shape that provides for refining contour/gathering of drawn material to provide proper stock distribution and reduce metal-flow defects. Ensures consistency of length for next die geometry.

shape. An open-die operation (hand forging) using flat dies. Shapes round stock to square to fit into next die cavity and so on.

swedge. An open-die geometry that angularly shapes (tapers) one end of the input stock for shafter hubs prior to pot, blocking, and/or finish operation.

upset. An open-die geometry that provides material movement when stock is placed in vertical position. Used where radial grain flow or localized areas of increased volume from initial stock size is required.

REFERENCES

10.1 F.H. Froes et al., Ed., *Titanium Technology: Present Status and Future Trends,* Titanium Development Association, 1985

10.2 S. Abkowitz, The Emergence of the Titanium Industry and the Development of the Ti-6Al-4V Alloy, *JOM Monograph Series,* Vol 1, TMS, Warrendale, PA, 1999

10.3 *Forming and Forging,* Vol 14, *Metals Handbook,* 9th ed., ASM International, 1988

10.4 T. Byrer, Ed., *Forging Handbook,* Forging Industry Association, Cleveland, OH, and American Society for Metals, Metals Park, OH, 1985

10.5 G. Lutjering, U. Zwicker, and W. Bunk, *Proceedings of the Fifth International Conference on Titanium,* Metallurgical Society of AIME, Warrendale, PA, 1984

10.6 P. Lacombe, R. Tricot, and G. Beranger, Ed., *Proceedings of the Sixth International Conference on Titanium* (Nice, France), Metallurgical Society of AIME, Warrendale, PA, 1988

10.7 *Proceedings of the Seventh International Conference on Titanium* (San Diego, CA), Metallurgical Society of AIME, Warrendale, PA, 1992

10.8 *Proceedings of the Eighth International Conference on Titanium* (Birmingham, U.K.), Metallurgical Society of AIME, Warrendale, PA, 1995

10.9 *Proceedings of the Ninth International Conference on Titanium,* Metallurgical Society of AIME, Warrendale, PA, 2000

10.10 *Proceedings of the Tenth International Conference on Titanium,* Metallurgical Society of AIME, Warrendale, PA, 2004

10.11 *Proceedings of the Eleventh International Conference on Titanium,* Metallurgical Society of AIME, Warrendale, PA, 2008

10.12 *Proceedings of the Twelfth International Conference on Titanium,* Metallurgical Society of AIME, Warrendale, PA, 2012

10.13 G. Lutjering and J.C. Williams, *Titanium,* Springer, 2003

10.14 R. Boyer, E.W. Collings, and G. Welsch, Ed., *Materials Properties Handbook: Titanium Alloys,* ASM International, 1994

10.15 M.J. Donachi, *Titanium: A Technical Guide,* 2nd ed., ASM International, 2000

10.16 P. Dadras and J.F. Thomas, Jr., Characterization and Modeling for Forging Deformation of Ti-6Al-2Sn-4Zr-2Mo-0.1Si, *Metall. Trans. A,* AIME, Nov 1981

10.17 L.J. Bartlo, "The Metallurgy of Titanium and Titanium Alloys," RMI Research Report R483, RMI Titanium Co., Niles, OH, 1967

10.18 T. Altan et al., *Forging Equipment, Material, and Practices,* Air Force Materials Laboratory, Wright Patterson Air Force Base, Dayton, OH, 1973

10.19 *Forging Industry Handbook,* Forging Industry Association, Cleveland, OH, 1966

CHAPTER 11

Forming of Titanium Plate, Sheet, Strip, and Tubing*

THIS CHAPTER DISCUSSES the type of fabrication referred to as secondary working (Ref 11.1–11.14) as distinct from primary working processes described in Chapter 9, "Primary Working," in this book. Secondary working refers to manufacturing processes that transform mill products resulting from primary working into finished parts. For the most part, the equipment and methods used in forming metals such as stainless steel and aluminum are applicable to forming titanium. However, certain precautions must be observed because the metallurgical characteristics of titanium and especially titanium alloys place restrictions on the forming methods compared, for example, with stainless steels.

Forming Considerations

Forming Temperatures. Hot forming requires expensive tooling and handling procedures (Ref 11.1–11.14). Therefore, selection of forming temperatures involves consideration of several factors, including the alloy being formed, the type and dimension of the part, and the effect on material properties. Advantages and disadvantages of hot and cold forming are summarized in Table 11.1. Hot sizing is often preceded by one or more cold forming steps, while hot forming is usually a single-step operation. Generally, the decision of which process to use depends on the type of forming equipment available and the shapes to be produced. If the equipment can accommodate heated dies, hot forming is usually chosen.

Metallurgical factors related to forming were discussed in Chapter 10, "Secondary Working of Bar and Billet," in this book. In general, it is more difficult to form titanium sheet than the more familiar aluminum alloys and alloy steels. Despite the additional controls required, titanium is formed to the same tolerances as stainless steels and aluminum alloys for aerospace parts (Ref 11.1, 11.2).

Titanium-base alloys and commercially pure titanium behave like cold rolled stainless steel when formed at room temperature. At 650 °C (1200 °F), titanium alloys form like annealed stainless steel at room temperature. Various formability ratings can be applied to different materials. Table 11.2 lists factors governing several forming operations and causes for failure in each instance. Table 11.3 lists the alloys in order of decreasing formability for some common forming operations.

Titanium alloys resist sudden deformation, although impact forming is possible (Ref 11.15–11.17). Stretching and pressing operations with controlled rate of load application are recommended. Slow forming speeds improve formability. At elevated temperatures, some alloys such as Ti-6Al-4V have better formability at relatively high forming speeds, while others such as Ti-13V-11Cr-3Al have lower ductility at higher forming rates.

Elevated temperatures increase the ductility of metal (Ref 11.12, 11.14). This is a major factor in the improved formability in hot forming. Heating the workpiece to approximately 540 °C (1000 °F) lowers its yield strength and forming force requirements. Table 11.4 lists bend test

---

*Adapted and revised from Oren J. Huber and John R. Schley, *Titanium and Its Alloys,* ASM International.

Table 11.1 Advantages and disadvantages of hot forming versus cold forming/hot sizing

Hot forming	Cold forming/hot sizing
Advantages	
Fewer operations	Forming can be done on all available types of forming machines
Lower forming pressures	Reduced dwell time in cold forming press
Material is at elevated temperature for shorter time than hot sizing	Parts are stress relieved on hot sizing
Parts can be made in most alloys that could not be produced by cold forming	Can use lower-cost tooling materials in cold forming
Disadvantages	
Requires heat-resistant tool materials	Requires additional equipment (hot sizing presses)
Tools must be adapted for heating	Long dwell times in hot sizing press (30 min)
Requires use of slow press with some dwell time (5 min or more)	Requires two sets of dies (one set heat resistant)
Limited to forming operations on equipment that can use heated tools	Some parts cannot be made by cold forming procedures

Table 11.2 Types of failures in sheet forming processes and material parameters controlling deformation limits

The parameters can be determined in tensile and compressive tests.

Process	Cause of failure		Ductility parameter(a)	Buckling parameters(b)
	Splitting	Buckling		
Brake forming	x	...	6.3 mm (ε in 0.25 in.)(c)	...
Dimpling	x	...	51 mm (ε in 2.0 in.)(d)	...
Beading	...	...	...	...
Drop hammer	x	...	13 mm (ε in 0.5 in.)(c)	...
Rubber press	x	...	51 mm (ε in 2.0 in.) (S_u)	...
Sheet stretching	x	...	0.5 mm (ε in 0.02 in.)	...
Joggling	x	x	0.5 mm (ε in 0.02 in.)	E_c/S_{cy}
Linear stretching	x	x	51 mm (ε in 2.0 in.)(e)	E_t/S_{ty}
Trapped rubber, stretching	x	x	51 mm (ε in 2.0 in.)(f)	E_t/S_{ty}
Trapped rubber, shrinking	...	x	...	E_c/S_{cy} and $1/S_{cy}$
Roll forming	...	x	...	E_t/S_{ty}(g) and E_c/S_{cy}(h)
Spinning	...	x	...	E_c/S_{cy} and E_t/S_u
Deep drawing	...	x	...	E_c/S_{cy} and S_{ty}/S_{cy}

(a) ε indicates natural or logarithmic strain; the dimensions indicate the distance over which it should be measured. (b) E_c = modulus in compression; E_t = modulus in tension; S_{cy} = compressive yield strength; S_{ty} = tensile yield strength; S_u = ultimate tensile strength. (c) Corrected for lateral contraction. (d) For a standard 40° dimple. (e) The correlation varies with sheet thickness. (f) The correlation is independent of sheet thickness. (g) For roll forming heel-in sections. (h) For roll forming heel-out sections

Table 11.3 Relative formability of annealed titanium alloys for sheet forming operations (in descending order)

Brake forming minimum bend radius at room temperature	Drop hammer at 454–482 °C (850–900 °F) (max stretch), %	Hydropress (trapped rubber)	
		Maximum stretch at 316–371 °C (600–700 °F), %	Maximum shrink at 316–371 °C (600–700 °F), %
15V-3Cr-3Al-3Sn: 2.5T	15V-3Cr-3Al-3Sn: 16	15V-3Cr-3Al-3Sn: 10	15V-3Cr-3Al-3Sn: 5
13V-11Cr-3Al: 2.5T	13V-11Cr-3Al: 16	13V-11Cr-3Al: 10	13V-11Cr-3Al: 5
5Al-2.5V: 4T	6Al-4V: 13	6Al-4V: 5	6Al-4V: 4
6Al-6V-2Sn: 4.5T	5Al-2.5Sn: 13	5Al-2.5Sn: 5	5Al-2.5Sn: 3
6Al-4V: 4.5T	...	...	...
6Al-2Sn-4Zr-2Mo: 4.5T			
Joggle (runout/depth ratio)			
At room temperature	At 316–371 °C (600–700 °F)	Stretch wrap (max) at room temperature, %	Skin stretch (max) at 454–510 °C (850–950 °F), %
15V-3Cr-3Al-3Sn: 1.6	15V-3Cr-3Al-3Sn: 1.6(a)	5Al-2.5Sn: 8	15V-3Cr-3Al-3Sn: 15(a)
13V-11Cr-3Al: 1.6	13V-11Cr-3Al: 1.6(a)	15V-3Cr-3Al-3Sn: 5.5	13V-11 Cr-3Al: 15(a)
5Al-2.55Sn: 4	6Al-4V: 3	13V-11 Cr-3Al: 5.5	5Al-2.5Sn: 12.5
6Al-4V: 4.5	5Al-2.55Sn: 4.5	6Al-4V: 3.5	...

(a) Room temperature

Table 11.4 Typical minimum bend test data for various titanium alloy sheets

Temperature			Minimum bend radius (T)		
°C	°F	Bend axis(a)	Ti-15-3(b)	Ti-5Al-2.5Sn	Ti-6Al-4V
21	70	L	2	7	4½
		T	2	6	4½
204	400	L	...	6	3½
		T	...	4½	3½
316	600	L	...	6	3
		T	...	4½	3
427	800	L	...	5½	3½
		T	...	4	2½
538	1000	L	...	5	2
		T	...	3½	2
649	1200	L	...	3½	1½
		T	...	3	1½
760	1400	L	...	2½	1
		T	...	2½	1
816	1500	L	...	1½	1
		T	...	2	1

(a) L = bend axis perpendicular to rolling direction; T = bend axis parallel to rolling direction. (b) This alloy is formed at room temperature.

data for several titanium alloys at various temperatures and shows the increased formability with increasing temperature.

Handling and Cleaning. Blanks for forming must be clean and free from surface defects to avoid failure during forming, part failure in service, and to produce a blemish-free surface on the manufactured part. In addition, all blanks must be deburred prior to heating and forming. This precautionary measure minimizes the tendency for parts to edge crack during forming. Oxygen-enriched surfaces must be avoided or removed prior to forming, because they lead to notch sensitivity. Contaminated surfaces must also be removed after hot forming.

Greases, oils, and cleaning agents containing chlorides (or other halogens) should be avoided on parts to be hot formed, heat treated, and welded. Titanium alloys could become susceptible to stress corrosion after such exposures.

Scale Removal. Dark oxides that form on titanium above 540 °C (1000 °F) are removed by mechanical and chemical methods. Mechanical descaling methods include grit and vapor blasting, tumbling, and other more drastic methods. Wire brushing is possible, but stainless steel wire brushes rather than steel wire brushes should be used to avoid contaminating the titanium surface with iron, which can cause adverse corrosion effects.

Chemical descaling uses alkaline and acid baths and sequential treatments involving both. Acid baths usually are aqueous solutions with 10 to 50% nitric acid (HNO_3) and 1 to 3% hydrofluoric acid (HF). The ratio of HNO_3 to HF is usually 10 to 1, or greater. Molten salt baths containing sodium hydride or sodium hydroxide are used to precondition heavy scale so it can be subsequently removed by the HF-HNO_3 acid bath solution. Scale removal is discussed in detail in Chapter 9, "Primary Working," in this book.

Grease Removal. Grease must be removed before hot forming and prior to acid immersion. Heavy oils, greases, mill marking, shop soils, and fingerprints are removed using procedures similar to those used for stainless steels and high-temperature alloys. Vapor degreasing in trichloroethylene (now disallowed) or perchloroethylene is possible when the treatment is followed by acid pickling to ensure the removal of any residue that could lead to stress-corrosion cracking in subsequent processing. However, in some carefully controlled operations where there is no chance for entrapment of degreasing solvents, pickling sometimes is not done. Methyl ethyl ketone (MEK) and acetone are preferred solvents for surface greases, because they do not leave a harmful residue, which is characteristic of chlorinated organic solvents.

Preparation for Forming

High-quality blanks are required to produce consistently reliable formed parts. Quality control and inspection are required at all stages of blank preparation. The raw material must be inspected for thickness variations, surface condition, and edge quality. Flat mill products are marked to show gage, grade, heat number, rolling direction, and specification number. A sample from appropriate locations in the lot of material should be tested for mechanical properties, including tensile and yield strengths, and should undergo bend and other special formability tests.

Paper interleaving protects surfaces of titanium sheet in storage and in warehouse handling. Bare sheets should be prevented from sliding over each other or over work tables to avoid surface damage. Plastic protective coverings are also used in many instances. Properly designed storage facilities allow flat stock removal or resupply without stock damage and keep the material clean while in storage.

Blank Preparation. Many cutting processes are used to shape blanks from flat titanium mill stock. Mechanical operations such as shearing, sawing, and nibbling are preferred. Methods of blank cutting, shaping, and edge preparation are discussed in Chapter 13, "Machining and Chemical Shaping of Titanium," in this book.

Layout procedures and cutting sequences should be planned so that the raw stock is used efficiently. Shear cracks and other stress raisers should not be left in finished blanks. Surface damage must be prevented, and defects should be removed before forming operations begin.

Surface Preparation. Scratches and marks (as from grinding) that are coarser than those produced by 180-grit emery abrasive are removed by surface sanding. This is a hand operation, and it can be time-consuming and expensive. Solvent (MEK) and other appropriate degreasing is used on blanks that will be hot formed. Similar degreasing and surface cleaning precedes acid etching where surface scratches are removed in this man-

ner. "White glove" handling prevents fingerprints after this step in blank preparation. Fingerprints can leave sodium chloride deposits, which could cause stress-corrosion cracking when there is a subsequent heating operation.

Heating Methods

Different temperature ranges are used in hot forming of titanium and titanium alloys, including 200 to 315 °C (400 to 600 °F), 480 to 540 °C (900 to 1000 °F), and 650 to 815 °C (1200 to 1500 °F), for the more difficult-to-form alloys and more complex parts. Severity of deformation and mechanical properties of individual alloys determine the required temperatures. Stronger alloys, such as Ti-8Al-1Mo-1V, could require higher forming temperatures. Most alloys are easier to form at higher temperatures, but some show degradation of properties at elevated temperatures. Therefore, maximum and minimum temperatures may be limited. Table 11.5 shows typical forming temperatures for some alloys and various processes.

Numerous methods are available for heating blanks and tooling for forming. Because uniform heating is desirable, torch heating is not practical. Overheating and the resulting degradation of properties are hard to control with such heating. Time at temperature should be kept to a minimum; heating time should not be much longer than necessary to achieve heat penetration. For this reason, forming blanks are usually heated one at a time, regardless of the heating method used.

Table 11.5 Typical temperatures for forming operations

| Alloy | Forming temperature, °C (°F) | | Forming method(b) | | | | | | | | |
	Mild forming	Severe forming(a)	A	B	C	D	E	F	G	H	I
Unalloyed titanium	205–316 (400–600)	...	x	x	x	x	x	x	x	...	...
	...	482–538 (900–1000)	x	...	...	...	x	...	...	x	x
	...	538–705 (1000–1300)	...	...	...	x	...	...	...	...	...
Ti-6Al-4V	205–316 (400–600)	...	...	x	...	...	...	...	...	...	...
	...	482–538 (900–1000)	...	...	...	...	...	x	x	x	...
	...	538–649 (1000–1200)	...	...	x	...	...	...	...	...	x
	...	538–788 (1000–1450)	x	...	...	x	x	x	...	...	...
Ti-5Al-2.5Sn	205–316 (400–600)	...	...	x	...	...	...	...	...	...	...
	...	482–538 (900–1000)	...	...	...	...	...	x	x	...	...
	...	538–760 (1000–1400)	x	...	x	x	x	...	...	...	x
Ti-8Al-1Mo-1V	...	To 732 (1350)	x	x	x	...	...	...	...	...	...
	343 (650)	649–760 (1200–1400)	x	...	...	...	...	...	...	...	...
	...	788 (1450)	...	...	...	...	x	...	x	...	x

(a) Temperatures should be held to a minimum to reduce scaling. Time at temperature is important, because titanium is embrittled by oxygen above 538 °C (1000 °F) as a function of time and temperature. Generally, 2 h is maximum for 704 °C (1300 °F); 20 min is maximum for 871 °C (1600 °F). These are accumulated times to include heating times for single or multirange forming, intermediate stress relief, and final stress relief. (b) A = drop hammer; B = hydropress; C = brake; D = spin; E = draw; F = matched die; G = hydroform; H = finish die; I = creep form

Furnace Heating. Electric furnaces with air atmosphere are most suitable for heating titanium blanks. Gas-fired furnaces are also acceptable if the flame does not contact the surface of the part and the atmosphere is slightly oxidizing. Heat zones should be uniform in temperature. Portable furnaces reduce handling between furnace and forming equipment.

Insulating blankets reduce heat losses during transfer of heated blanks from the furnace to forming machines. Heat losses during transfer and during contact with the dies are appreciable due to the high surface-to-volume ratio of the blanks. Consequently, it is not practical to heat the thinner forming blanks in furnaces.

Some precautions are necessary to avoid contamination and undesirable property changes during furnace heating. Furnace atmospheres, hearths, and refractories must be clean. Purging for several hours is recommended for furnaces previously used with other atmospheres; gas-fired furnaces should be operated with excess oxygen. As emphasized in previous chapters, hydrogen contamination during heating must be avoided to prevent catastrophic degradation of properties. All scale and other extraneous matter should be removed from the furnace before heating forming blanks. Furnace temperature control should be automatic and the temperature maintained within 15 °C (25 °F) of the set point.

Resistance heating is practical for titanium because titanium has relatively high electrical resistivity (Ref 11.2, 11.15, 11.17). Irregular sections are troublesome because of the resulting variations in current density. Therefore, higher temperatures occur in portions of the blank where the cross section is smaller. Adjacent steel tooling may also present a problem in the form of a parallel electrical conduction path. However, resistance heating is effective for many types of forming.

Heat losses are kept low when it is possible to use resistance heating with the blank supported near the forming faces. Forming can then be done with little handling time and heat loss. The forming operation often begins within several seconds of the time the power is shut off, or forming may be done with the heating power on. Consequently, temperature control is relatively good compared with furnace heating.

Radiant Heating. Infrared heat sources such as quartz lamps are often used to heat blanks during forming when one side of the blank is exposed to radiation. Control is difficult when the heat pattern is not uniform. Distance from the part to the heat source and the angle of heat incidence at the surface of the workpiece determine the heat pattern.

Hot Die Heating. Forming blanks for simple operations (such as brake forming) are often adapted to hot die heating. The blank and preheated die are in contact until the blank reaches the forming temperature. Applied forces then shape the part. Only the blank area in contact with the die is heated. The heated blank is not transported, and therefore, handling is minimized.

Heated forming dies are frequently used, especially at the higher forming temperatures. Slow deformation processes require heated dies to reduce heat losses, which can occur when a heated blank contacts a relatively cold die. Electrical heating is commonly used, and flame heating should be avoided. Internal resistance heating or an external heat source is used. Special heaters that match the die contour are common, such as ceramic bodies with embedded resistance wires. In other instances, simple strip heaters or cartridge heaters are adapted to the process. Uniform temperature of the die surface is also essential.

Forming Lubricants

Lubricants used in titanium forming minimize friction, which reduces energy requirements; reduce galling and seizing between surfaces of the blank and tooling; and reduce heat transfer between blank and die in hot working. Nonchlorinated organic oils are effective lubricants for cold forming. Solid additives (graphite and molybdenum disulfide) are added to the oils for hot forming. Typical lubricants for some forming operations are listed in Table 11.6.

Tooling Materials

The economics of materials selection for tooling in titanium-forming operations is complicated. Tolerances, number of parts to be made, forming temperatures, ease of modification of design, and material cost are some factors that must be considered (Ref 11.2, 11.15, 11.17). Table 11.7 lists some typical tool materials for various forming operations.

Forming Processes

The various forming processes combine guided and free bending with relatively little or no change

Table 11.6 Typical lubricants for forming operations

Forming operation	Forming temperature	Lubricants used
Stretch forming skins	Cold	Grease-oil combinations
		Wax-type lubricant
	Hot	Colloidal graphite
Stretch forming sections	Cold	Wax-type lubricant plus flake graphite (10:1 by volume)
Stretch forming extrusions	Hot	Molybdenum disulfide
		Colloidal graphite
Brake forming	Cold	Colloidal graphite
Contour rolling sections	Cold	Colloidal graphite
Roll forming	Cold	Heavy oil (SAE 60)
	Hot	Colloidal graphite
Hot sizing	Hot	Colloidal graphite
Draw forming (matched dies)	Hot	Colloidal graphite plus alcohol, xylene
Hammer forming	Hot	Colloidal graphite
Hydropress forming	Hot	Colloidal graphite

Table 11.7 Typical tooling materials for forming operations

Forming operation	Type forming	Tool material
Stretch forming skins	Cold	Cast aluminum with epoxy face
	Hot	Cast ceramic
Stretch forming sections	Cold	Kirksite form block; Ampco Bronze No. 21 wiper die
	Hot	H11, H15 tool steels; high-silicon cast iron
Stretch forming extrusions	Cold	Mild steel; AISI 4130 steel
	Hot	AISI 4130 steel; type 310 stainless steel
Brake forming	Cold	AISI 4340 steel (36–40 HRC)
	Hot	H11, H13 tool steels; Incoloy 802
Contour rolling sections	Cold	O2 tool steel
Yoder roll forming	Cold	O2 tool steel
	Hot	H11, H13 tool steels
Hot sizing	Hot	Mild steel; high-silicon cast iron; high-silicon nodular cast iron; H13 tool steel; type 310 stainless steel; RA 330 stainless steel; Inconel X; Hastelloy X; Incoloy 802
Draw forming	Hot	High-silicon cast iron; Incoloy 802
Hammer forming and hydropress forming	Cold or hot	Kirksite dies with stainless steel caps; lead punches with stainless steel caps; high-silicon cast iron; RA 330 stainless steel; Inconel X; Incoloy 802

in thickness. Rigid dies are used singly and combined with a matching die and, in some instances, a hard rubber pad. Forces are applied through mechanical and hydraulic systems. Some typical setups are briefly described here.

Brake forming titanium is similar to brake forming cold-worked stainless steels, except springback in titanium alloys is much higher and difficult to predict. Parts having large bend radii present no problem, but smaller radii can require hot forming to avoid cracking.

A typical die for brake forming titanium alloys is shown in Fig. 11.1. The punch and channel die can be made of various tool steels. Hard rubber inserts in the channel die reduce surface scratches as the metal moves over the channel die face.

The punch surface is polished to be free of scratches and other defects. Zinc alloys and other inexpensive materials are used for the punch in short runs if a protective cover sheet prevents direct contact between the punch and the titanium

workpiece. Stainless steel is satisfactory for separation. Both blanks and tooling can be heated for hot brake forming. Table 11.8 lists temperatures satisfactory for brake forming titanium alloys.

Table 11.4 shows the effect of temperature on formability of several titanium alloys. As noted previously, formability improves with higher temperature. Heated punches and blanks are recommended for most production operations. Because the area of contact between the punch and blank is small, blanks are heated by means other than die heating.

Springback in brake forming is a function of the forming temperature and the shape of the part. Figure 11.2 shows the effect of bend-radius-to-material-thickness ratio on springback for different temperatures and thicknesses of Ti-6Al-4V. These typical curves illustrate that there is considerably less springback when material is formed at elevated temperature and when the radius-to-thickness ratio is low, which is behavior characteristic of titanium alloys and other materials.

Stretch forming produces single-curved contours and certain more-complex contours in sheet. The metal elongates around a form block or die in processing. Grips hold the blank securely as the work is pulled into the desired shape. These grips should apply a uniform stress over a wide area while not damaging the sheet by creating stress raisers, which could cause premature failure. Typical setups for forming a dome-shaped contour in sheet and for forming an angle section are shown in Figs. 11.3 and 11.4, respectively.

Stretch forming is often done cold to eliminate expensive die materials and handling procedures, and also to obtain optimum properties in the formed part. Cold forming leads to high stretching forces and slow stretching speeds, so that elevated temperatures may be necessary in many instances.

The sheet process uses a blank that can be flat as cut from the raw sheet, or brake formed and stress relieved. The sheet is positioned over the forming die, securely clamped in the grips, and then formed. Forming begins as the blank contacts the die block. Forming continues as the part is pulled around the die or as the die is pushed against the blank. In some instances, hand tapping assists forming in local areas. The stretch forming operation requires precise control.

In stretch forming, blanks are heated for approximately 5 min in place on the tooling, which is also heated. Resistance and radiant heating are commonly used; radiant heat is adaptable to continuous use during pulling. Resistance heating can be used effectively only when the blank is not connected electrically to the stretching equipment or when electrical insulation is possible.

Springback allowances are usually designed into the tooling, particularly for stretch forming at room temperatures. Variations between lots of a

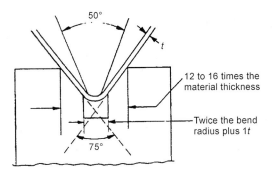

Fig. 11.1 Schematic of typical die for brake forming

Table 11.8 Brake forming temperatures for some titanium alloys

Titanium alloy	Temperature, °C (°F)	
	Blank	Punch/die
Commercially pure	204–316 (400–600)	260 (500)
Ti-6Al-4V	538–649 (1000–1200)	260 (500)
Ti-5Al-2.5Sn	538–704 (1000–1300)	260 (500)
Ti-8Al-1Mo-1V	593–649 (1100–1200)	621 (1150)

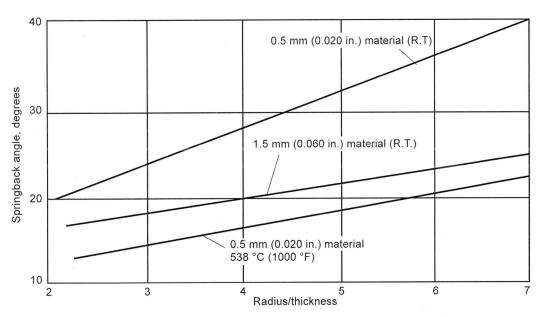

Fig. 11.2 Springback in Ti-6Al-4V alloy as a function of radius-to-material-thickness ratio. R.T., room temperature

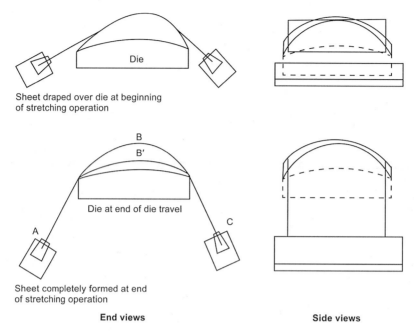

Fig. 11.3 Process for stretch forming sheet into a dome-shaped contour

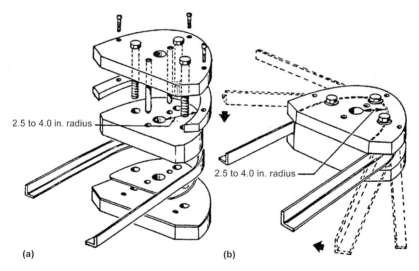

Fig. 11.4 Process for stretch forming angle shapes. Sketch illustrates (a) inboard and (b) outboard wrap techniques.

specific alloy can make a hot sizing step desirable for dimensional control. Lubricants are frequently used for room-temperature and elevated-temperature stretch forming.

Various means of evaluating formability have been developed. Figure 11.5 shows typical curves for several alloys. The formability index in this instance is based on the ductility in a 51 mm (2.0 in.) length (see Table 11.2). There is no significant change in forming ease until the working

temperatures are in the range of 430 to 650 °C (800 to 1200 °F), depending on the alloy composition.

Deep Drawing. In deep drawing, a process for forming recessed parts, a punch pulls a metal blank through a die surrounding the metal punch. A second step, generally called redraw, reduces the diameter, lengthens the part, or thins the wall. Some types of deep-drawing operations are shown in Fig. 11.6. Both single- and double-ac-

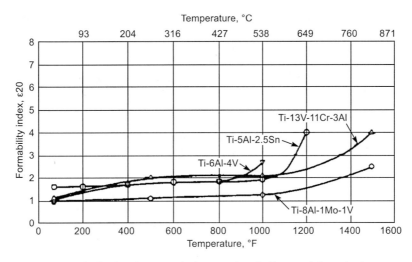

Fig. 11.5 Forming temperatures for dimpling, linear stretch, sheet stretch, and rubber stretch flange forming

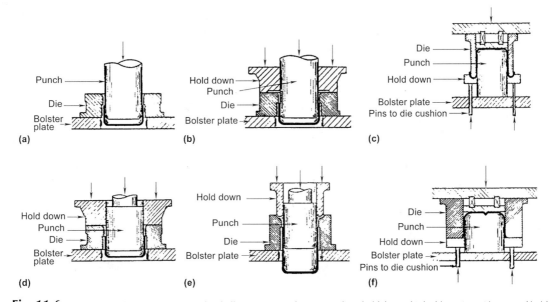

Fig. 11.6 Various deep-drawing operations. Sketch illustrates (a) single action without hold down, (b) double-action with recessed hold down, (c) single-action inverted with die-cushion hold-down reverse redraw, (d) double-action with flat hold down push-through type, (e) double-action redraw push-through type, and (f) single-action redraw with die cushion hold-down.

tion presses are adapted to the operation; double-action equipment has better control and flexibility. Hydraulic presses can be operated with slower rates of travel, and therefore, they offer an advantage over mechanical presses for deep drawing.

The deep-drawing process is very sensitive to tooling design and setup conditions; thus, it is very expensive. It is suited to production of large numbers of parts, usually 100 or more. Neither part shape nor quantity justifies its use for manufacturing many airframe components. A roll forming and welding combination is more economical in most instances. An alternative method, superplastic forming, was developed for cost-effective production of airframe components.

In typical drawing operations, the titanium blank is placed between the draw ring and pressure pad. The applied pressure must be sufficient to prevent buckling during forming but not high enough to cause excessive thinning or tearing in the wall of the part. Difficulties with buckling are minimized by sandwiching the titanium between sheets of stainless or mild steel during forming. Elevated temperatures are often used by heating the blank or the dies, or both.

Limits are related to the ratio of compressive modulus and compressive yield strength for buckling, and the tensile yield strength to compressive yield strength ratio for splitting.

Sheets for deep drawing must have uniform thickness. Dirt on the tooling affects the friction between the blank and the hold-down ring and can produce uneven drawing. Lubricants are used in the process. Tooling material includes stainless steel and other heat-resisting tool material. Speeds are rather slow. Typical drawing speeds are approximately 254 mm/min (10 in./min), with forming forces held on the part for five or more minutes after forming is complete. This holding time improves dimensional control.

Trapped-rubber forming is a pressing operation in which a rubber pad serves as a female die. Several modifications of the process use direct pressing and hydropressing on the rubber pad. Typical examples are shown in Fig. 11.7.The main advantage of this type of forming is the simplicity of the required tooling. Only the less expensive half of conventional tooling is used, with the rubber pad replacing the more expensive portion.

Several limitations are imposed by the use of the rubber pad. Typically, maximum pressure ranges from 24 to 138 MPa (3500 to 20,000 psi), depending on the equipment and setup used. Shape definition can be poor in thicker-gage and high-strength alloys. Springback is usually high, and formed parts can require a subsequent sizing operation. The process is limited to room temperature, although some high-temperature rubbers extend the range moderately for production operations. Processing temperatures are limited to a maximum of 540 °C (1000 °F).

Most tooling is made to net dimensions, although undercutting can be used to reduce the amount of sizing required after forming. Because there is relatively little rubbing action on the die during forming, the low wear enables the use of inexpensive die materials. Cold rolled steel is adequate for low-temperature service.

Tool steels could be necessary for forming at elevated temperatures. Tooling includes some means to position and hold the blank until the press is also closed. Because only male tooling is used, the workpiece must be locked on the die until the part is formed completely.

The trapped-rubber process is used extensively in the aircraft industry for flanged parts, which are formed in one step or in stages. For small parts, several are formed simultaneously. Both single-action and double-action presses are used.

Superplastic forming (SPF) of titanium alloys is commonly used in aircraft part fabrication and is an effective means of both weight and cost savings (Ref 11.18). Superplastic forming is extensively discussed in Chapter 5, "Deformation and Recrystallization of Titanium and Its Alloys," in this book.

Superplasticity is a condition in which a solid crystalline material is deformed well beyond its usual breaking point, usually more than approximately 200% during tensile deformation. This state is usually achieved at a temperature typically half of the absolute melting point ($0.5\ T_m$), which is approximately 900 °C (1650 °F) for titanium. Superplastically deformed material gets thinner in a very uniform manner, rather than forming a "neck" (a local narrowing), which leads to fracture. In general, a fine stable grain size is required for superplastic behavior to occur, with the optimum strain rate increasing with decreasing grain size, as shown in Fig. 11.8.

Requirements for superplastic behavior in titanium alloys are a fine alpha grain size (~<10 μm) and a fine dispersion of grain-boundary beta phase, which pins the alpha grain boundaries and maintains the fine alpha grains throughout SPF at elevated temperature (Ref 11.18) (Fig. 11.9). (For an explanation of how this microstructure develops, see Fig. 6.1 to 6.4 in Chapter 6, "Mechanical Properties and Testing of Titanium Alloys," in this book.) Titanium alloys that meet these criteria (Table 11.9) must also have a strain-rate sensitivity (a measurement of the way the stress on a

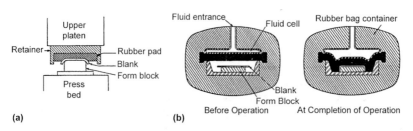

Fig. 11.7 Trapped-rubber forming techniques. Sketch illustrates (a) guerin process, and (b) wheelon process.

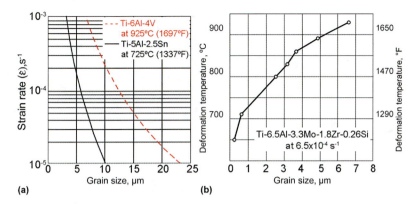

Fig. 11.8 Superplastic forming is strongly dependent on grain size. Effect of grain size on (a) strain rate of superplastic deformation for Ti-6Al-4V and Ti-5Al-2.5Sn alloys and (b) superplastic deformation temperature for Ti-6.5Al-3.3Mo-1.8Zr-0.26Si alloy

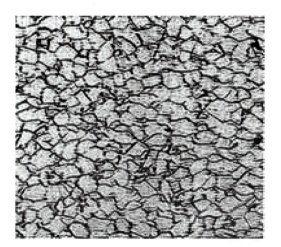

Fig. 11.9 Microstructure suitable for superplastic forming of Ti-8Al-1Mo-1V. Heat treatment: annealed at 1010 °C (1850 °F) for 1 h, furnace cooled to 595 °C (1100 °F), followed by air cooling. Original magnification: 250x

Table 11.9 Partial list of superplastic titanium alloys

Ti-6Al-4V
Ti-6Al-0.2Sn-4Zr-2Mo
Ti-3Al-2.5V
Ti-8Al-1V-1Mo
Ti-0.3Mo-0.9Ni (Ti-Code 12)
TiAl
$TiAl_3$
Ti-1100

material reacts to changes in strain rate) of >0.3 to exhibit superplastic behavior.

The precise mechanisms of superplasticity in metals are still being investigated to clearly define the phenomenon. It is generally believed that diffusion plays a part, together with the sliding of grains past each other (similar to grains of sand) (Ref 11.18).

Parts can be formed to complex configurations with complete freedom from springback. Resulting parts are also stress free and remain that way because no corrective forming is required. Elongations as high as several thousand percent are achievable. However, part thinning is usually unacceptable at such deformations. Typical elongations in use are approximately 300% maximum. These deformations far exceed those achievable by other forming methods. Further, several parts can be produced in a single die in the same cycle. The complexity of shapes obtainable enables simplifying designs that consist of several components and a number of fasteners into a single piece and no fasteners, thus providing cost and weight savings (Fig. 11.10).

Forming is performed in a sealed metal tool, or die, as shown in Fig. 11.11. The titanium sheet to be formed is located within the assembly, which is placed in a press with platens containing resistance-heating elements. The retort is flushed with argon to remove the air, providing an inert medium. Argon pressure is then applied between the top plate and the SPF sheet while heating progresses, and the sheet is formed to the contour of the tool (Fig. 11.11). Forming temperatures typically are in the range of 870 to 925 °C (1600 to 1700 °F). A superplastically formed Ti-6Al-4V part is shown being removed from a 900 °C (1650 °F) furnace in Fig. 11.12.

Superplasticity and superplastic diffusion bonding in a TiAl alloy with a fine-grained duplex microstructure were investigated to fabricate TiAl alloy products using a combination process of SPF with diffusion bonding. Superplastic tensile tests were carried out at temperatures ranging

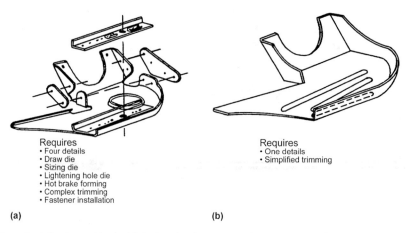

Requires
• Four details
• Draw die
• Sizing die
• Lightening hole die
• Hot brake forming
• Complex trimming
• Fastener installation

(a)

Requires
• One details
• Simplified trimming

(b)

Fig. 11.10 Superplastic forming enables the fabrication of multipiece, complex parts into a single piece. (a) Conventionally fabricated part. (b) Superplastically formed part

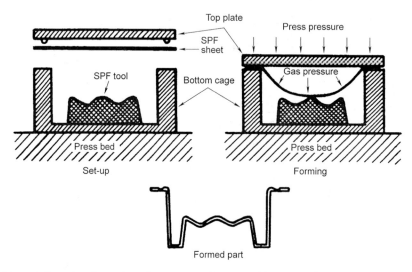

Fig. 11.11 Schematic illustration of superplastic forming (SPF) technique used for titanium alloy sheets

from 1000 to 1100 °C (1830 to 2010 °F) and at strain rates ranging from 10^{-5} to 10^{-3} s^{-1}. A low superplastic flow stress of less than 25 MPa (3.6 ksi) was observed at 1100 °C and at a strain rate of 8.3×10^{-5} s^{-1}. Under this condition, tensile elongation of 300% and a strain-rate sensitivity coefficient of over 0.5 were obtained.

Superplastic Forming Combined with Diffusion Bonding. The versatility of the SPF process can be enhanced by combining it with either diffusion bonding (DB) (see Chapter 12, "Joining Titanium and Its Alloys," in this book) or solid-state joining, because both processes require similar conditions, that is, heat, pressure, clean surfaces, and an inert environment. The combined

techniques are referred to as the SPF/DB process. Diffusion bonding can be carried out simultaneously with SPF, thereby eliminating the need for welding or brazing when producing complex parts, as shown in Fig. 11.13.

Tube Bulging. In the bulging forming method, internal pressure expands a tube or a cup to a desired shape. Methods used to apply the internal pressure include an expanding segmented mandrel, a liquid, and a rubber-type elastomer. Similar to trapped-rubber forming, tooling is simple. Free-forming can also be used. Figure 11.14 illustrates typical examples of both types. Where dies are used, the die is the female portion of the tooling.

Fig. 11.12 A Ti-6Al-4V component is removed from a 900 °C (1650 °F) furnace after superplastic forming. Courtesy of Aeromet International PLC

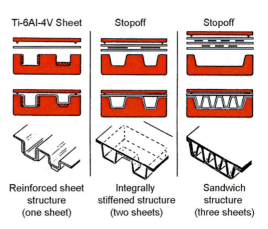

Fig. 11.13 Fabrication of Ti-6Al-4V sheet structure using a combination of superplastic forming and diffusion bonding

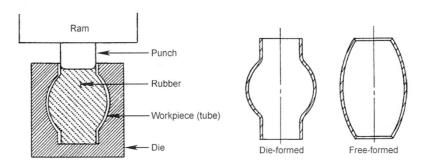

Fig. 11.14 Typical rubber bulging setup for the tube-bulging process with die-formed and free-formed configurations

Impact-energy sources (explosives, electrical discharge) are also used for bulging. Electrical sources for impact forming include electric arcs supplied by transducers, exploding bridge wires, and magnetic coils. Water, or a similar medium, transmits the energy to the workpiece, except in magnetic forming. Consequently, the systems must be sealed and trapped air eliminated. In free-forming, reflectors are used to concentrate shock waves to produce irregular shapes in tubing.

Magnetic forming stands alone in bulging. It does not require a transmission medium, and the friction characteristic of the other processes is reduced. Also, less massive tooling can contain the forces generated. The need for sealed systems is also eliminated.

Blanks to be used in bulging operations are generally in the annealed condition. Design limits must be determined for a particular material and process. Two factors that affect the forming limits in this process are ductility of the workpiece material and the tooling used.

Tube Bending. Several methods are used to shape tubes after they are fabricated. The major ones are ram or press bending, roll bending, compression bending, and draw bending. Press, roll, and compression methods are used for tubing with relatively heavy walls. Therefore, they require more powerful equipment and more expensive dies than draw bending.

In ram or press bending, rigid supports hold the tube as it is bent around a die on the press ram. In compression bending, the tube and bend die are both fixed, and a wiper die wraps the tube around the stationary bend die.

Draw bending is used to bend tubing with thinner walls. Figure 11.15 illustrates typical tooling for a draw-bending operation. The tubing is secured in the clamp die by the cleat, with the clamp plug inserted for internal support. The multiball mandrel gives internal support in the area to be formed, although in forming titanium tubing, it is not always necessary to use the ball mandrel. The pressure and clamp dies rotate around the bend

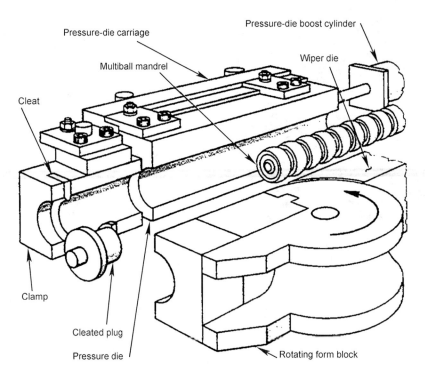

Fig. 11.15 Typical tooling for tube draw bending. The process lends itself to forming tubing for use in aerospace and other applications.

die during draw forming. Lubrication reduces galling and sliding friction.

Elevated temperatures are necessary for most bending operations with titanium alloys. Heated tooling is usually adequate, and preheating the workpiece is not commonly necessary.

Typical bending limits for unalloyed titanium appear in Table 11.10. The Ti-3Al-2.5V alloy is also widely used in tube forming in the annealed and cold-worked, stress-relieved condition. In the annealed condition, it can be cold formed to the same limits shown (Table 11.10) for commercially pure titanium tubing. In the cold-worked, stress-relieved condition, it can be cold formed to a 5T radius. Bending speeds are not shown in the table, but they are generally low (several degrees per minute or less) to prevent local deformation in the forming area.

Drop-Hammer Forming. Gravity-hammer and pneumatic drop-hammer presses form titanium by progressive deformation with repeated blows in matched dies. These processes are suited to forming recessed sheet parts from materials that are not sensitive to high strain rates. Typical parts and the associated dies are shown in Fig. 11.16, and a pneumatic hammer is shown in Fig. 11.17.

The forming energy in each stroke depends on the mass of the ram, the tooling attached to it, and the velocity at which the tooling strikes the workpiece. The hammer operator controls the forming rates.

The tooling material for the hammer is lead or other soft material that will deform to the contour of the female die. Dies are frequently capped with steel, stainless steel, and Inconel to prevent contamination of the titanium by low-melting die metals. In some instances, two punches are used: a rough punch followed by a finishing punch. Another method uses a rubber pad to equalize forming forces. In this instance, maximum rubber thickness is used where the forces are greatest during the initial forming stages. As forming progresses, the thickness is reduced by removing some of the rubber pads after each blow.

Complex parts require the use of elevated temperatures. Hot operations require ductile iron and other similar tooling, because rubber pads do not withstand the heat. Resistance heating and furnace methods are used to heat the titanium workpiece.

The severity of deformation possible is limited by the shape of the formed part and the properties of the material being formed. Table 11.2 relates

Table 11.10   Bending limitations for unalloyed and Ti-3Al-2.5V titanium tubing

| Tube diameter | | Wall thickness | | Bend radius | | | | Bend angle, degrees | |
| | | | | Minimum | | Preferred | | | |
mm	in.	mm	in.	mm	in.	mm	in.	Minimum	Preferred
Room-temperature bending									
38	1.5	0.41	0.016	57	2.25	76	3	90	120
		0.51	0.020	...	...	...	...	100	160
51	2	0.41	0.016	76	3	102	4	80	110
		0.51	0.020	...	...	...	...	100	150
64	2.5	0.41	0.016	95	3.75	127	5	70	100
		0.51	0.020	...	...	...	...	90	140
		0.89	0.035	...	...	...	...	110	180
Elevated-temperature bending									
76	3	0.41	0.016	114	4.5	152	6	90	120
		0.51	0.020	...	...	...	...	110	160
		0.89	0.035	...	...	...	...	130	180
89	3.5	0.41	0.016	133	5.25	178	7	90	120
		0.51	0.020	...	...	...	...	110	160
		0.89	0.035	...	...	...	...	130	180
102	4	0.51	0.020	152	6	203	8	110	160
		0.89	0.035	...	...	...	...	120	180
108	4.25	0.51	0.020	171	6.75	229	9	130	140
		0.89	0.035	...	...	...	...	140	140
127	5	0.51	0.020	254	10	254	10	...	110
152	6	0.51	0.020	305	12	305	12	...	100

the material properties to the forming limits. Uniform deformation in tension and minimum springback are desirable. A material that deforms elastically, not plastically, has a high amount of springback when the load is removed. It is difficult to predict the deformation in a particular part design. Consequently, drop-hammer forming tolerances are approximately 1.6 mm (0.062 in.), and the operation could be followed by a hot sizing operation.

Roll forming is a bending process in which a sequence of power-driven contoured rolls bends sheet and strip into long, shaped products. A typical roll-bending machine is shown in Fig. 11.18. Close tolerances and high-speed operation are characteristic of the process (Ref 11.2, 11.15, 11.17).

Roll forming is also used to bend strip into cylinders that are butt welded to form thin-walled tubing. Other structural shapes can also be produced. Similar products can be made on a draw bench where the workpiece is pulled through a series of heads or stands containing idling rolls instead of power-driven rolls.

Available equipment covers a very wide size range. The capacity of a particular machine is limited by its physical dimensions and the required power. A machine can contain up to 15 or 20 rolls. It is possible to roll form the more ductile titanium alloys into simple shapes at room temperature. Elevated temperature is necessary for more drastic roll-forming operations.

Elevated temperatures require controlled handling and more complex equipment. Heat-resistant bearings and rolls and extensive water cooling of the equipment may become necessary.

Lubricants are used extensively in all types of roll forming. The work blank and rolls are heated for elevated-temperature roll forming. Elevated-temperature limits of bend radii for 90° bends are in the range of $1T$ to $3T$ for the common titanium alloys.

Roll Bending. Rolls can also be used to bend a channel and other shapes to desired contours without major changes in the cross section. Equipment for roll bending can be designed for a specific job, or it can be more versatile. The distinction between roll forming and roll bending is not clearly defined, because either can be used to produce similar parts. Roll forming includes operations where contoured rolls determine the shape of the part. Bending applies to operations where adjustable cylindrical rolls are used to curve sheet, bar, and shaped sections.

Roll bending is the most economical process for producing single-contoured shapes from titanium alloys. In addition to bending flat sheet into cylindrical contours, the linear roll-bending technique is also commonly used to curve shaped cross sections such as channels. The channels ini-

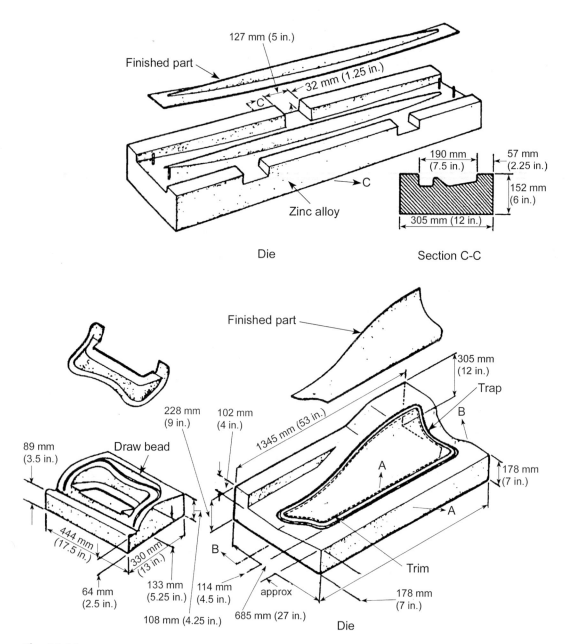

Fig. 11.16 Typical drop-hammer dies and formed parts

tially can be formed by roll forming, as previously described, or by extrusion. Linear roll-bending equipment generally is quite simple in design. One common type uses three rolls in a pyramidal configuration, the upper roll being vertically adjustable to control the bend radius.

Roll bending is a process that depends greatly on operator technique, and several trial parts generally are required to establish suitable conditions. Roll bending is done at room temperature whenever possible, but the stronger and stiffer alloys can require heating to achieve a desired radius.

Spinning and shear forming are processes for shaping sheet metal into seamless hollow parts using pressure on a rotating workpiece. Spinning involves only minor thickness changes in the sheet metal, whereas shear forming is an exten-

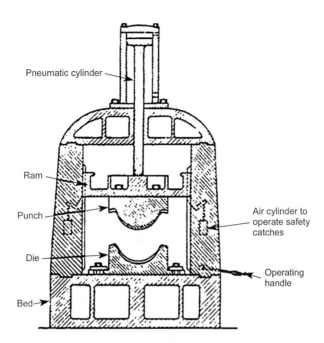

Fig. 11.17 Pneumatic hammer for forming titanium sheet

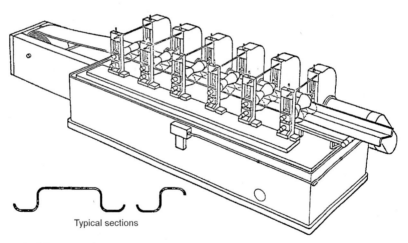

Fig. 11.18 Typical roll-forming machine

sion of spinning and causes thinning. There is no sharp delineation between the two processes.

Shapes that can be spun and shear formed include cylinders, cones, hemispheres, and modifications of these shapes, which are symmetrical around the workpiece axis of rotation. Figure 11.19 shows the progressive formation of a typical conical shear-forming operation. The forming tools in this instance are rolls that are not driven but rotate against the work on the driven mandrel, thus shaping the part.

Two typical setups for spinning are shown in Fig. 11.20. During spinning, the metal is deformed in a small, rather than a large, area. A large portion of the blank is unsupported, especially during the early stages of the process. For thin sections, the tool forces can be applied manually to the driven blank. Power spinning uses mechanical or hydraulic devices for greater tool forces when spinning thicker and stronger metals. Numerical control and automatic programming are incorporated in modem equipment.

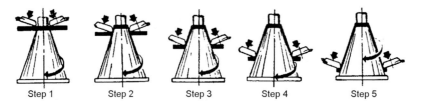

Fig. 11.19 Progressive steps in shear forming a cone

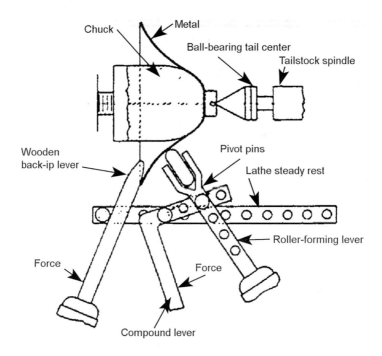

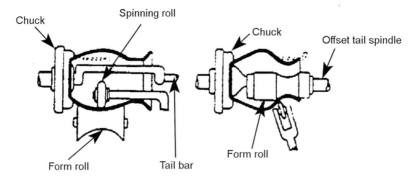

Fig. 11.20 Typical metal-spinning operations

Tooling for manual spinning is made of relatively soft materials, such as brass, to prevent surface damage to the workpiece. Rollers are commonly used in power spinning and are made of harder materials. Tool steels are necessary for high-temperature spinning. Shear spinning requires stronger tooling because of the greater forces necessary to work the metal. Roll diameters are kept small to minimize friction and power losses. Rollers are often cooled to prevent distortion and creep.

Variables to be considered include material properties of the workpiece, tooling shape, tool pressure and travel speed, lubricant, and forming temperature. Buckling and splitting, the common difficulties encountered when limits are exceeded, are shown in Fig. 11.21. Some buckling can be tolerated by using blanks large enough to allow trimming after the workpiece is formed.

Dimpling produces a conical flange around a hole in sheet metal where flush fasteners will be used. A dimpled impression has approximately the same thickness as the surrounding metal. It is commonly applied to sheets that are too thin for countersinking. A ramming-coining-dimpling process is usually used, although swaging can be used. Titanium alloys are also dimpled at elevated temperatures. Permissible deformation depends on the ductility of the alloy used. Stretching varies as fastener head size (H), rivet diameter ($2R$), and bend angle (α) vary. These parameters are shown in Fig. 11.22, and the limits are usually expressed as a graphical plot of H/R versus α.

Joggling. A joggle is an offset in a flat plane made by two parallel bends, in opposite directions, at the same angle. The bend angle for joggles is usually less than 45°, as indicated in Fig. 11.23.

This type of forming permits flush joints in sheets and structural shapes. Presses, hammers, and brakes can activate joggle dies. Presses are preferred because they provide better control. Joggling is often done at high temperatures using tool steel, heat-resistant metal, and ceramic tooling.

Because the two bends are in opposite directions and close together, there are areas of tension and compression near each other. The two types of deformation tend to compensate for each other.

Hot Sizing. The hot sizing process is used only to correct springback in parts that have been formed using other processes. In hot sizing, controlled pressure is applied at temperatures and for times sufficient to produce creep (Ref 11.2, 11.15, 11.17). Parts are formed against heated dies by pressure that approaches the yield strength of the material being formed without exceeding the yield strength. Forming temperatures are 510 °C (950 °F) or higher, and times are 10 to 30 min. Deformation occurs because the creep strength is relatively low at these temperatures. Table 11.11 shows suggested sizing conditions for several titanium alloys.

Tooling must be heat-resistant material if it is to be used repeatedly. Cast iron, tool steels, and

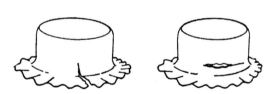

Fig. 11.21 Typical buckling and splitting in metal spinning

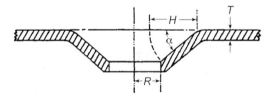

Fig. 11.22 Dimpling parameters used to determine stretching deformation limits. *H*, head size; *T*, thickness; α, bend angle; *R*, rivet diameter

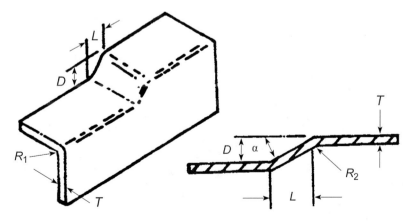

Fig. 11.23 Joggle in a flat plane showing parameters. α, joggle bend angle; *D*, joggle depth; *L*, joggle length or runout; *T*, thickness of workpiece; R_1, radius on joggling block; R_2, radius of bend on leading edge of joggle block

some ceramic materials have been used successfully. Table 11.7 lists the various tooling materials used in hot sizing. Presses supply vertical pressures, and the workpiece is restrained at the edges in rigid tooling to apply horizontal pressures. In the absence of pressing equipment, tooling can be made to use driven wedges in a rigid stand for pressure application. The locked assembly is then placed in a furnace for forming. Besides the simplicity of equipment, this system can use inert atmospheres to reduce scaling and contamination. Figure 11.24 shows several typical hot sizing tooling arrangements.

Summary

Although titanium is more difficult to form than conventional metals, such as steel and aluminum, it can be readily formed using conventional equipment. The close dimensional tolerances required for aerospace components can be achieved with titanium. However, proper preparation of the starting material, combined with careful control of the forming operation, is essential. Successful forming also relies on experience.

Titanium and titanium alloys can be hot and cold formed, but with the alloys, hot forming is preferred. Springback characteristics of the metal generally require that cold forming be followed by hot sizing if high dimensional accuracy is to be achieved.

Successful forming of titanium begins with careful preparation of the blanks. Surface and edge defects and oxygen-enriched surfaces must be removed to avoid cracking during forming.

Removal of oxide scale from hot forming can be done by mechanical and chemical methods, the latter requiring specific acid and alkaline pickling baths.

Hot forming of titanium and its alloys requires adherence to specific forming temperature ranges, depending on alloy type. Heating methods are the same as those used for more conventional hot working operations, but certain precautions regarding contamination must be observed. A variety

Table 11.11 Suggested sizing conditions for some titanium alloys

Alloy	Sizing temperature °C	°F	Time at temperature, min	Remarks
Unalloyed	482–538	900–1000	3	Blank heated by contact with heated die
Ti-5Al-2.5Sn	649–760	1200–1400	15	0.81 mm (0.032 in.) thick sheet
	...	...	20	1.60 mm (0.063 in.) thick sheet
Ti-8Al-1Mo-1V	788	1450	15	Ceramic dies used
Ti-6Al-4V	649	1200	3–15	0.81 mm (0.032 in.) thick sheet
	...	...	3–20	1.60 mm (0.063 in.) thick sheet
Ti-13V-11Cr-3Al	593	1100	10–20	1.65 mm (0.065 in.) thick sheet, solution treated; aged after sizing

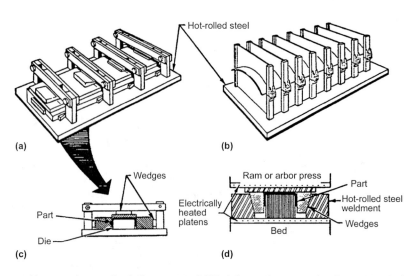

(a)

(b)

(c)

(d)

Hot-rolled steel

Wedges

Part

Die

Electrically heated platens

Ram or arbor press

Part

Hot-rolled steel weldment

Wedges

Bed

Fig. 11.24 Typical hot sizing fixtures. Sketch illustrates (a) solid-block fixture; (b) contour-plate fixture; (c) typical cross section; and (d) wedge hot-sizing tool on conventional arbor press.

of lubricants and forming-tool materials are available, and the choice of the tool material is dictated by the economics of the particular application.

Superplastic forming and DB, often in combination, are effective techniques for the cost-effective fabrication of titanium components.

Forming processes used for titanium and its alloys are the same as those commonly used in fabricating the more conventional structural metals, with the exception of the DB operation. Forming is carried out by applying mechanical and hydraulic forces to the titanium and forcing it against a rigid tool or die to produce the desired shape. The characteristics of the particular titanium alloy selected dictate the forming conditions to be observed with a given hot or cold forming process.

REFERENCES

11.1   H.E. Boyer and T.L. Gall, Ed., Forming, Chap. 26, *Metals Handbook Desk Edition,* 2nd ed., ASM International, Materials Park, OH, 1998

11.2   J.T. Benedict, C. Wick, and R.F. Veilleux, Ed., *Forming,* Vol 2, *Tool and Manufacturing Engineers Handbook,* 4th ed., Society of Manufacturing Engineers, Dearborn, MI, 1984

11.3   M.J. Donachi, *Titanium: A Technical Guide,* 2nd ed., ASM International, 2000

11.4   G. Lutjerling, U. Zwicker, and W. Bunk, *Proceedings of the Fifth International Conference on Titanium,* Metallurgical Society of AIME, Warrendale, PA, 1984

11.5   P. Lacombe, R. Tricot, and G. Beranger, Ed., *Proceedings of the Sixth International Conference on Titanium* (Nice, France), Metallurgical Society of AIME, Warrendale, PA, 1988

11.6   *Proceedings of the Seventh International Conference on Titanium* (San Diego, CA), Metallurgical Society of AIME, Warrendale, PA, 1992

11.7   *Proceedings of the Eighth International Conference on Titanium* (Birmingham, U.K.), Metallurgical Society of AIME, Warrendale, PA, 1995

11.8   *Proceedings of the Ninth International Conference on Titanium,* Metallurgical Society of AIME, Warrendale, PA, 2000

11.9   *Proceedings of the Tenth International Conference on Titanium,* Metallurgical Society of AIME, Warrendale, PA, 2004

11.10   *Proceedings of the Eleventh International Conference on Titanium,* Metallurgical Society of AIME, Warrendale, PA, 2008

11.11   *Proceedings of the Twelfth International Conference on Titanium,* Metallurgical Society of AIME, Warrendale, PA, 2012

11.12   G. Lutjering and J.C. Williams, *Titanium,* Springer, 2003

11.13   R. Boyer, E.W. Collings, and G. Welsch, Ed., *Materials Properties Handbook: Titanium Alloys,* ASM International, 1994

11.14   F.H. Froes, D. Eylon, and H.B. Bomberger, Ed., *Titanium Technology: Present Status and Future Trends,* The Titanium Development Association (now The International Titanium Association), 1985

11.15   K. Lange, Ed., *Handbook of Metal Forming,* McGraw-Hill Book Company, New York, NY, 1985

11.16   R.J. Favor and R.A. Wood, *Titanium Alloys Handbook,* MCIC-HB-02, Battelle Memorial Institute, Columbus, OH, Dec 1972

11.17   T. Lyman et al., Ed., Forming of Titanium Alloys, *Forming,* Vol 4, *Metals Handbook,* 8th ed., American Society for Metals, Metals Park, OH, 1969, p 437–445

11.18   C.H. Hamilton, Superplasticity in Titanium Alloys, *Superplastic Forming,* American Society for Metals, Metals Park, OH, 1985, p 13–22

CHAPTER 12

Joining Titanium and Its Alloys*

TITANIUM CAN BE JOINED by most methods common to the metals fabricating industry, including welding (Ref 12.1–12.13), brazing, soldering, adhesive bonding, and mechanical fastening. Welding and mechanical fastening are used in many applications, whereas brazing and adhesive bonding are used on a limited scale. Soldering operations are almost nonexistent because most applications for titanium assemblies are not compatible with this joining method. It should be noted that welding and fabricating problems are often significantly different for commercially pure titanium grades compared with many of the titanium alloys. Welding of titanium (as well as forming, machining, and drilling) requires special techniques unique to the specific material. Titanium structures fabricated by joining methods are used extensively in critical applications such as jet engines, airframes, chemical process equipment, spacecraft, deep-submergence vessels, and toxic waste disposal. Its use in such severe applications where failures are costly and often endanger lives is a tribute to titanium, its alloys, and joining methods. Titanium joining technology, with emphasis on basic fundamentals, is discussed in this chapter.

Welding

Welding operations are widely used for joining titanium and titanium alloy assemblies (Ref 12.1–12.13). Welded assemblies are fabricated from sheet, plate, bar, and forgings. With many processes, titanium welding operations are identical to those used to join other metals in similar applications. However, some welding processes require special techniques and procedures, while others are impractical. Likewise, weldments in some alloys exhibit excellent mechanical properties, while others are too brittle to be useful. The suitability of welding processes and alloys for welded assemblies is related to the chemical and metallurgical characteristics of alloys. An understanding of these characteristics is vital in designing and fabricating welded assemblies.

Chemical and Metallurgical Characteristics

Titanium and its alloys exhibit five major chemical and metallurgical characteristics that influence welding operations:

- Sensitivity to embrittlement when contaminated by small amounts of carbon, oxygen, nitrogen, and hydrogen (Ref 12.14)
- Reactivity when heated to welding temperatures
- Sensitivity to embrittlement or softening in varying degrees as a function of alloy content when subjected to the thermal cycles involved in most welding operations
- Sensitivity to embrittlement when highly alloyed with metallic elements
- Susceptibility to stress corrosion

Contamination Control

The most important precaution in welding titanium and its alloys is to avoid the embrittling effects of contamination by carbon, oxygen, nitrogen, and hydrogen from air, oil, and other foreign matter. These elements form interstitial-type solid solutions with titanium. They occur in

---

*Adapted and revised from Charles E. Hulswitt and Charles S. Young, originally from Glenn E. Faulkner, *Titanium and Its Alloys*, ASM International.

commercial titanium as residual elements, or they are added intentionally for strengthening purposes.

When carbon, oxygen, and nitrogen are present in sufficient quantities, weld-joint bend ductility and toughness are seriously impaired, as shown in Fig. 12.1 and 12.2. The loss of ductility and toughness due to carbon additions is attributed to the formation of brittle carbide networks in weld-fusion zones. Embrittlement due to oxygen and nitrogen is attributed to solid-solution hardening.

Hydrogen has different effects on welded joints, including low ductility, low toughness, and delayed weld cracking. Loss of ductility and toughness is attributed to hydride platelets in welds.

Small variations in total interstitial content of commercially pure titanium affect metal notch toughness (Ref 12.15), as shown in Fig. 12.3. These data were obtained from welds made using several different filler metals in two base metals. Filler metal B had the lowest interstitial content (total carbon, oxygen, hydrogen, and nitrogen content = 0.227%) and the highest notch toughness. Filler metal D had the highest interstitial content (total carbon, oxygen, hydrogen, and nitrogen content = 0.397%) and the lowest notch toughness. It is important to use materials with low interstitial content when good notch toughness is required.

Interstitial elements cannot be avoided entirely in welded joints. However, they can be controlled by using proper cleaning, welding procedures, and filler metals. Often, filler metals with low interstitials are used to improve weld ductility.

To control contamination, weld-fusion and heat-affected zones must be protected from contact with active elements and compounds, including all known refractories. This requirement limits the welding processes that are adaptable to titanium and its alloys. Acceptable processes include gas tungsten arc, plasma arc, gas metal arc, electron beam, friction stir, and resistance spot and seam welding methods where, depend-

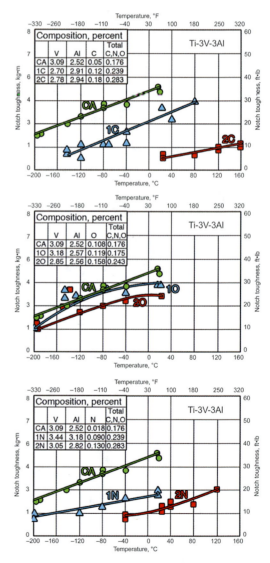

Fig. 12.2 Effects of interstitial elements carbon, oxygen, and nitrogen on Ti-3V-3Al alloy weld-metal notch toughness

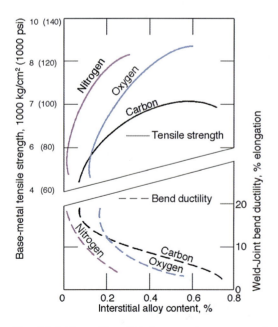

Fig. 12.1 Effects of interstitial elements carbon, oxygen, and nitrogen on strength and ductility of titanium

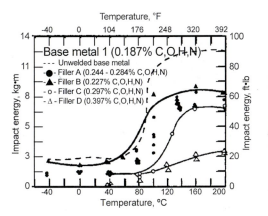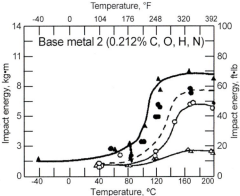

Fig. 12.3 Effects of interstitial elements carbon, oxygen, and nitrogen on commercially pure titanium weld-metal notch toughness

ing on the method, inert gas, vacuum, and/or metal contact provide the required protection.

Conversely, processes such as oxyacetylene torch and shielded metal arc with coated electrodes are unacceptable because active elements and compounds are present. A proprietary flux, which does not contaminate titanium weldments, is used as a backup to protect the root of inert gas metal arc welds from contamination. Welding processes and their adaptability to titanium alloys are listed in Table 12.1.

Alloy Selection

Alloying elements affect the mechanical properties of titanium weldments, as they do with most metals (Ref 12.1, 12.16, 12.17).

Alpha Alloys. As a class, alpha (α) alloys are readily adaptable to all applications requiring welding operations. The mechanical properties of alpha alloys are affected only slightly by heat treatment and microstructural variations. Welded joints in alpha alloys are ductile, and their strengths equal or exceed those of the base metals. Aluminum and tin are important alloying elements in alpha alloys. Their effects on the tensile strength of the base metal and weld-joint bend ductility are shown in Fig. 12.4.

Alpha-Beta Alloys. Weldability of alpha-beta (α-β) alloys varies with alloy content and welding process. Relationships between beta-stabilizing alloy content and bend ductility of welds in alpha-beta alloys are shown in Fig. 12.5.

Weld ductility decreases rapidly with increasing beta-stabilizing alloy content. Severe weld embrittlement is observed when alloy content exceeds approximately 3% (5% if vanadium).

Commercial alloys are more complex than the binary alloys represented in Fig. 12.5. Therefore, the data should be used only as a general guide in

Table 12.1 Adaptability of welding processes to titanium alloys

Welding process	Remarks
Gas tungsten arc	Requires special techniques to shield weld
Gas metal arc	Requires special techniques to shield weld
Plasma arc	Must use inert gas
Electron beam	Hard vacuum excellent
Resistance spot and seam	Excellent
Flash	Excellent
Diffusion	Excellent
Pressure	Excellent
Friction stir welding	Excellent
Oxyacetylene flame	Not applicable
Shielded metal arc	Not applicable
Submerged arc	Further development required

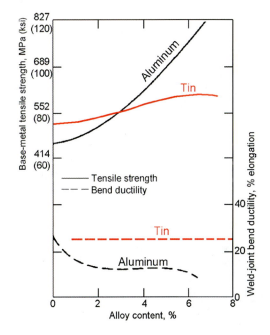

Fig. 12.4 Effects of tin and aluminum on tensile strength of alpha-alloy base metal and weld-joint bend ductility

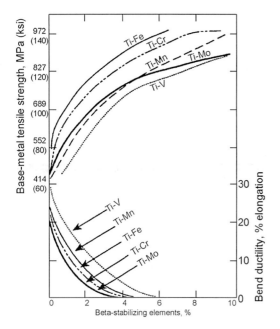

Fig. 12.5 Effects of increasing amounts of beta-stabilizing elements on the base-metal tensile strength and weld bend ductility of alpha-beta titanium alloys

assessing the weldability of a given composition. Using this guide, alloys containing only small amounts (3 to 4%) of the less potent beta-stabilizing elements combined are adaptable to most welding applications. After welding, the ductility and toughness of the weld-metal and heat-affected zones are lower than those of the base metal due to heat treatment response and microstructural variations, but they are adequate for most applications.

Aluminum and tin also are added to alpha-beta alloys. These additions strengthen the alloys but cause only slight loss of weld-joint bend ductility and notch toughness (see Fig. 12.4).

Alpha-beta alloys that contain higher beta-stabilizing content than described previously are characterized by very low ductility and toughness in weld-fusion and high-temperature heat-affected zones. Applications of these latter alloys are limited to solid-state welding processes, which produce no adverse microstructures.

Special fillers are sometimes used to alter the alloy content of weld-fusion zones, for example, commercially pure, beta-lean, alpha-beta, and alpha-alloy filler metals. The use of such filler metals lowers the beta-stabilizing alloy content of the weld-fusion zone and improves weld ductility and toughness. However, special filler metals do not alter the weld heat-affected zone. Therefore,

brittle weld heat-affected zones in some alloys limit the usefulness of this technique.

Metastable Beta Alloys. These alloys lend themselves to welding applications, especially Ti-15V-3Cr-3Sn-3Al alloy and Ti-3Al-8V-6Cr-4Mo-4Zr (Beta C) and other newer metastable beta alloys discussed in other chapters of this book. As-welded joints in commercial alloys are ductile. However, when the joints are heat treated to achieve maximum strength, ductility is severely reduced.

Thermal stability of welded joints in alpha-beta and metastable-beta alloys is important, because weld zones are not cooled under equilibrium conditions and are subject to transformation. Transformation that occurs at service temperatures generally impairs ductility. For applications requiring elevated-temperature service, thermal stability is determined by exposing weld test specimens to proposed operating temperatures for extended periods of time and determining whether or not they suffer a loss of bend or tensile ductility or fracture toughness.

Dissimilar Metals. Weldable titanium alloys are readily welded to each other. However, only limited success has been achieved in welding titanium and titanium alloys to other metals, with the exception of zirconium (now allowed by the American Society of Mechanical Engineers codes). It is almost impossible to avoid low ductility in dissimilar-metal weld-fusion zones. Low ductility is due to formation of brittle intermetallic compounds when titanium is welded to iron, nickel, aluminum, and copper (beta eutectoid elements) and to excessive solid-solution hardening when welded to vanadium, molybdenum, tantalum, and niobium (beta isomorphous elements). Embrittlement due to solid-solution hardening is less severe than that accompanying intermetallic compound formation. Therefore, metals that do not form compounds are more compatible for welding to titanium than those that do form compounds.

Susceptibility to Stress Corrosion

Titanium and titanium alloys are susceptible to stress corrosion in certain environments and under specific conditions of temperature and stress (see Chapter 14, "Corrosion," in this book). Residues containing chlorides should not be present on titanium parts subjected to elevated temperature and stress. Therefore, chlorinated cleaning solvents and cutting oils should be avoided or used only if all residues are eliminated prior to a subsequent heating operation. Also, parts should be cleaned and handled to prevent salt deposits from fingerprints before going into service. In addition, methanol causes stress-corrosion cracks when used as a pressurizing medium. Methanol should not be used as a cleaning solvent.

Surface Preparation

All metals are cleaned prior to welding critical assemblies. This removes grease, oil, dirt, and scale, which can contaminate the weld, prevent the formation of a solid-state bond, and cause weld porosity. It also produces consistent surface conditions for resistance welding. Cleaning methods for titanium alloys vary with surface conditions, welding process, and experience.

Degreasing removes dirt, oil, and foreign matter that interfere in welding and subsequent cleaning treatments. Methods of degreasing include washing and dipping in a dilute solution of sodium hydroxide; washing and dipping in a solvent such as methyl ethyl ketone, toluene, and acetone; and hand wiping with solvent immediately before welding.

Acid pickling is used to clean materials having a light scale. Typical pickling solutions contain nitric and hydrofluoric acids in concentrations ranging from 20 to 47% and 2 to 4%, respectively (it is very important to keep this ratio up during the pickling operation to avoid hydrogen pickup), at bath temperatures ranging from 25 to 80 °C (80 to 175 °F).

Titanium materials having a heavy scale are cleaned using mechanical operations (sand, grit, and vapor blasting) and molten sodium hydroxide baths. After heavy scale removal, acid pickling is required to ensure removal of all scale and subsurface contaminated metal.

Rinsing and drying treatments also are included in cleaning operations. It is important to protect cleaned surfaces from airborne contamination prior to welding.

Edges to be welded and adjacent surfaces often are abraded using draw and rotary files to control porosity. Surfaces resulting from oxyacetylene, flame, abrasive, and plasma arc cutting operations have a heavy scale and can contain microcracks due to excessive contamination. Flame-cut surfaces are best cleaned by removing the contaminated layer (and any cracks) by machining and grinding. Surfaces should be cleaned a minimum of 25 mm (1 in.) from the cut edge to remove all oxide.

Following cleaning operations, precautions must be taken in handling the parts, such as wearing of white gloves to eliminate fingerprinting, and protecting parts with lint-free and oil-free wrappings. For best results, welding should proceed without delay after cleaning operations.

Welding Procedures

Welding processes for titanium assemblies include gas tungsten arc welding (also known as tungsten inert welding); gas metal arc welding; plasma arc welding; electron beam welding; resistance (spot and seam) welding; flash, diffusion, and pressure welding; and friction stir welding (Ref 12.1–12.13, 12.18–12.21). Of these processes, only gas tungsten arc, gas metal arc, and plasma arc welding require special procedures and tooling to prevent weld contamination. However, removal of surface oxides and other foreign material is desirable for all welding processes, to ensure acceptable welds.

Gas Tungsten Arc Welding and Gas Metal Arc Welding

Gas tungsten arc welding (GTAW), gas metal arc welding (GMAW), and plasma arc welding (PAW) processes are used to join titanium assemblies. (Plasma arc welding is discussed in greater detail in the next section.) Process selection depends on shop facilities, personnel capabilities, experience, and metal thickness. For gages less than 1.5 mm (0.060 in.), GTAW and PAW welding are used exclusively. For sheet and plate more than 3 mm (0.125 in.) thick, modulated-current GMAW and conventional GMAW welding are recommended. However, GTAW welding is used by many fabricators to weld thick sections. The major drawback in GTAW welding of heavy sections is that more weld passes are required than with GMAW welding, thus increasing the possibility of weld contamination. The use of hot wire feed for GTAW and PAW can be used to increase metal-deposition rate.

Conventional welding equipment (both manual and machine) is used. Metal and ceramic gas nozzles are satisfactory, but they should be larger than for similar gages of other metals. Direct-current power sources are used with all three processes: straight polarity with GTAW and reverse polarity with GMAW. Some operators prefer straight polarity for modulated-current GMAW.

Shielding Conditions. The greatest difference between welding titanium and welding other structural metals using GTAW, GMAW, and PAW processes is the degree of shielding required (Ref 12.12). To prevent contamination, weld zones must be shielded from air while at 650 °C (1200 °F) or higher. The most critical area is the molten weld puddle, which becomes contaminated throughout its thickness. Weld zones that exceed 650 °C (1200 °F) but are not molten are subject to surface contamination. Contaminated surfaces have low ductility, which can lead to premature failures in service.

Small amounts of contaminants in the shielding gas cause reduced ductility and toughness, as shown in Table 12.2. These welds were made at a speed of 150 mm/min (6 in.) per minute with 0.25% air, oxygen, and nitrogen added to the helium shielding gas. Loss of ductility and toughness is less pronounced at higher welding speeds.

Oxygen, nitrogen, and hydrogen are introduced from the air during welding by shielding gas containing impurities, by turbulent shielding gas flow, and by inadequate gas coverage. Welding-grade argon (that is, 99.995% pure) and helium (argon with intentional oxygen additions not included) are rarely a source of contamination. Monitoring the dewpoint of the shielding gas to detect excessive moisture can avoid shielding gas problems. A dewpoint of less than −40 °C (−40 °F) measured at the torch nozzle is satisfactory.

Air is entrained when turbulence is present at the boundary of shielding gas and air. With proper gas-flow rates, standard welding torches, and proper nozzle-to-work distances, contamination from this source can be held within tolerable limits.

The most common source of weld contamination is inadequate shielding of either the face or back of the weld. Contamination from this source can be controlled by proper tooling. In some applications, special tooling is not required to shield weld faces. This condition is most common in welding thin sheets where welds cool below 650 °C (1200 °F) while they are shielded by the gas flowing from the welding torch. Methods that

improve shielding obtained from the welding torch include:

- Use of larger gas nozzles and higher gas-flow rates than are used for other metals
- Placing quench bars alongside the welds to increase weld-cooling rates
- Use of baffles to confine shielding gas to the joint area

Forms, such as V-joints in thick plate and baffles that confine the gas, allow better coverage with lower gas-flow rates. Baffles can consist only of flat bars (Fig. 12.6) and angles placed alongside the joint and small chambers that fully enclose the weld area. Baffles are especially useful in making corner-type welds, but they are not necessary in making fillet welds in T-joints.

When welding heavy sections and using fast welding speeds, trailing shields attached behind the torch provide an inert gas blanket to protect welds while cooling. Figure 12.7 illustrates some trailing shield features, including a porous diffusion plate, which minimizes turbulence and provides a uniform blanket of gas over the entire shield area, and a tight fit against the welding torch to supply a nonturbulent gas stream between the torch and shield.

Backing fixtures are used to obtain adequate shielding for the back of welds. Most fixtures of this type are designed to provide a uniform gas supply over the entire length of the fixture, but metal backing bars sometimes are used without a gas shield. These bars must have metal-to-metal contact.

Figure 12.8 shows backing fixtures used in butt welding. Similar fixtures are used for other joint

Table 12.2 Effects of impure welding atmospheres on properties of welds in commercially pure titanium

Welding atmosphere	Stock thickness		Weld-joint bend angle(a), degrees	Weld-metal notch toughness(b)		Parent-metal hardness, VHN	Maximum weld hardness, VHN	Nitrogen in weld, %
	mm	in.		kg · m	ft · lbf			
Dry helium (0.05% relative humidity)	3.2	⅛	180	...	...	160	195	0.007
	6.4	¼	180	...	...	185	230	0.038
	12.7	½	180	7.0, 7.6	50, 55	183	210	...
0.25% O, 99.75% He	3.2	⅛	180(c)	...	...	160	246	...
	6.4	¼	100(c)	...	...	187	300	...
	12.7	½	...	2.5, 3.0	18, 22	183	480	...
0.25% N, 99.75% He	3.2	⅛	180(c)	...	...	160	287	0.089
	6.4	¼	10(c)	...	...	186	304	0.380
	12.7	½	...	1.9, 1.9	14, 14	...	...	...
0.25% dry air, 99.75% He	3.2	⅛	120(c)	...	...	186	285	...
Wet helium (2.5% relative humidity)	3.2	⅛	180	...	...	160	216	...
Wet helium (5.0% relative humidity)	3.2	⅛	180	...	...	160	227	...
	6.4	¼	180(c)	...	...	185	245	...
	12.7	½	...	5.0, 5.25	36, 38	...	...	...

Note: Welds made in a chamber at speeds of 130 to 150 mm/min (5 to 6 in./min). (a) Bend radius = 2T. (b) Room-temperature notch toughness. (c) Failed during bend test

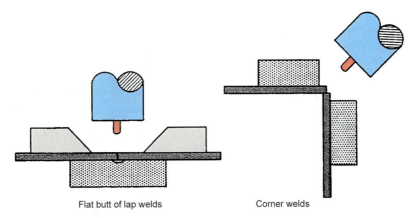

Flat butt of lap welds   Corner welds

Fig. 12.6 Baffle arrangements for improving the shielding for the face of welds. Arrangements that confine the gas enhance coverage with lower gas-flow rates.

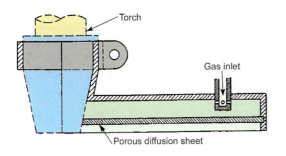

Fig. 12.7 Trailing shield for manual welding. The shield, attached behind the torch, protects the cooling welds with inert gas. Shields for machine welding are longer and often water cooled.

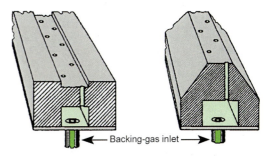

Fig. 12.8 Backing bars and butt and corner welds in titanium sheet provide inert gas to shield the back of the weld. Manifolds provide uniform flow of gas along the entire length of the bar.

designs. For fillet welds on T-joints, shielding should be supplied for two sides of the weld in addition to shielding the face of the weld. If access is not possible to the back of the joint, the structure must be enclosed so that an inert gas shield can be provided. This method is important in welding tanks, tubes, and other enclosed structures.

Inert-gas-filled welding chambers also are used to shield the face and back of welds. Welding chambers vary in size and shape. Inert atmospheres are obtained by evacuating the chamber and filling it with helium or argon, purging the chamber with inert gas, and collapsing the chamber to expel air and refilling it with an inert gas. Plastic bags are used in the collapsing-chamber technique. Inert gas is supplied through the welding torch during chamber welding operations. Advantages in using chambers are that jigs and fixtures are not required to obtain adequate shielding, and the shielding conditions are improved for manual welds.

Selection of gas-flow rates and shielding devices are based on welding tests made with applicable designs and metal thicknesses. In these tests, weld ductility is measured to determine whether or not shielding is satisfactory. Weld-joint bend and notch toughness tests are satisfactory for this purpose.

Argon is generally the most suitable shield gas because of cost and stability. More costly helium is used where shield penetration is critical. Shielding at the start and finish of welds is hard to control. Use of start and run-off tabs is recommended to minimize this problem. Also, to eliminate raising the torch to break the arc, a foot- or hand-operated switch that controls arc start and amperage is used. This tool stops the arc without moving the torch. High frequency is used to initiate the arc, minimizing start problems. Proper techniques are especially necessary in tack welding because of possible contamination in starting and stopping the arc.

Filler wire also must be shielded. To prevent contamination, the hot end of the filler wire or rod must be held within the gas shield. If not, an oxide and nitride surface scale forms on the rod and contaminates the weld.

Joint Designs. For GTAW and PAW, joint designs are similar to those used for other metals (Ref 12.12). Butt welds, fillet welds, and corner welds are commonly made. Square-butt joints with practically no root opening are preferred for thin sheets. Often, such welds are made without filler metal. When sheets exceed 1.5 mm (0.06 in.) in thickness and welding is performed from one side, a slight root opening or single V-joint design ensures complete penetration, and filler metals are added. For heavy sections, single and double U- and V-butt joints are made. For corner welds, both open- and closed-joint designs are used.

Assemblies should be designed to provide good weld shielding. Machine welds are preferred because shielding is more consistent than in manual operations. Provisions are necessary during assembly to allow access to shield the back of the joints. Also, the number of welds coming together at any one common point should be minimized.

Welding Conditions. Joint design, metal thickness, welding process, weld fixturing, and shielding gas determine welding conditions (Ref 12.1). Also, with all these variables constant, various combinations of arc current and voltage, welding speed, and filler wire feed rate produce acceptable welds. Therefore, the only reliable means for selecting welding conditions is to prepare sample weldments that simulate parts to be welded. Typical ranges for welding parameters are listed in Tables 12.3 and 12.4.

Guidelines recommended for developing welding conditions include:

- Use low welding heat input that provides consistent penetration and fusion
- Use as high as practical welding speed without sacrificing good shielding

If excessive weld porosity is encountered, increased heat input and reduced welding current can be beneficial. Also, tooling changes that reduce weld cooling rates can be beneficial in reducing the problem.

Gas Tungsten Arc and Gas Metal Arc Spot Welding

Arc spot welding (Ref 12.22) is possible by using GTAW and GMAW processes as alterna-

Table 12.3   Typical weld parameters for gas tungsten arc welds

Parameter	Ti-6Al-4V-2Sn		Ti-8Al-1Mo-1V	
Thickness, mm (in.)	1.02 (0.040)	2.03 (0.080)	1.14 (0.045)	2.29 (0.090)
Current, A	40	165	70	185
Arc voltage, V	9	13	13	12.5
Filler wire speed, mm/min (in./min)	...	150 (6)	...	150 (6)
Filler wire diameter, mm (in.)	...	0.762 (0.030)	...	0.762 (0.030)
Shielding gas	Argon	Argon	Argon	Argon
Torch, cm³ × 10⁴ (ft³/min)	42 (15)	71 (25)	42 (15)	57 (20)
Backup, cm³ × 10⁴ (ft³/min)	17 (6)	28 (10)	17 (6)	28 (10)
Training shield, cm³ × 10⁴ (ft³/min)	42 (15)	85 (30)	42 (15)	57 (20)
Welding speed mm/min (in./min)	200 (8)	200 (8)	500 (20)	127 (5)

Note: All welds made with square-butt joint design.

Table 12.4   Typical weld parameters for gas metal arc welds

Parameter	Spray arc	Pulsed arc	Short-circuiting arc(a)
Thickness, mm (in.)	25 (1)	25 (1)	25 (1)
Joint design	60° Double V 0.031	60° Double V 0.031	90° Double V 0.031
Welding current, A	315–340	...	200–240
Mean	...	75	...
Pulse	...	180	...
Arc voltage, V	315–340	22–24	17–22
Filler wire diameter, mm (in.)	1.58 (0.062)	1.58 (0.062)	0.762 (0.030)
Filler wire speed, mm/min (in./min)	14,000 (550)	4,300 (170)	16,500 (650)
Shielding gas	Argon	Argon	75Ar-25He
Welding speed, mm/min (in./min)	305–4,460 (12–175)	...	...
Weld passes	4	6	15

(a) Difficult to avoid lack of fusion defects

tives to resistance spot welding when thicknesses are not suitable for resistance welding, or one side of the joint is inaccessible. Automatic equipment is best suited, because starting and stopping conditions must be carefully controlled. Weld shielding is less critical than for conventional GTAW and GMAW, because welding is restricted to one spot. Gas tungsten arc spot welding is used for metal thickness up to approximately 6 mm (0.25 in.). Gas metal arc spot welding is used for much thicker joints. Welding conditions are similar to those used with conventional welding.

Plasma Arc Welding

Plasma arc welding provides greater welding speeds than the GTAW process. Conventional plasma arc equipment is satisfactory for welding titanium alloys. However, shielding conditions must be provided to prevent contamination.

Shielding. Argon and helium inert gases are used in PAW of titanium and its alloys to prevent contamination. Flow rates of plasma gas are low and do not shield the puddle adequately. Therefore, supplementary shielding gas is provided through an outer gas cup. In PAW, trailing shields and backing bars with functions similar to those required in GTAW and GMAW are necessary.

Welding Conditions. Greatest success is achieved in PAW if conditions for the first weld pass can be adjusted to form a hole ("keyhole") in the workpiece so the plasma jet completely penetrates the joint. Argon plasma gas is used when keyholing is desired. As the plasma jet progresses over the workpiece, molten metal flows together behind the keyhole to form the weld bead. Keyholing is possible with square-butt joints ranging in thickness from 2.4 to 13 mm (0.1 to 0.5 in.). For thicker metals, U- and V-joints are used to enhance keyholing.

Observing the keyhole during welding indicates complete weld penetration. Welding conditions for square-butt joints in titanium are shown in Table 12.5. Conditions determined for joining various thicknesses of stainless steel can often be used as a starting point for welding the same thicknesses of titanium.

In PAW, careful consideration of the orifice size and plasma gas-flow rates is necessary. The equipment manufacturer should be consulted to establish these settings.

Electron Beam Welding

Electron beam welding (EBW) is suitable for joining titanium and its alloys. The process is applicable to a range of thicknesses from approxi-

Table 12.5 Typical weld parameters for plasma arc welds

Parameter(a)	Thickness	
	3.2 mm (⅛ in.)	4.8 mm (³⁄₁₆ in.)
Welding speed, mm/min (in./min)	500 (20)	380 (15)
Arc current, A	185	190
Arc voltage, V	21	26
Gas flow rate, cm³/h × 10⁴ (ft³/h)		
Nozzle	23 (8)	34 (12)
Shielding	34 (12)	130 (45)

(a) Electrode set back 3.2 mm (⅛ in.) from face of torch nozzle, with nozzle-to-work distance 4.8 mm (³⁄₁₆ in.)

mately 0.04 to 50 mm (0.0015 to 2 in.). Only hard-vacuum welding has been adapted to titanium alloys. In such welds, there is no contamination from external sources. Soft-vacuum techniques have been used, and there is some contamination, but the extent is uncertain. Electron beam welds have high depth-to-width ratios. Such narrow welds cause less distortion than is obtained with GMAW, GTAW, and PAW. Conventional equipment is used to weld titanium and its alloys using low and high voltage. Fixtures that hold parts in position for welding are not as heavy as for other welding methods.

Joint Designs. Characteristics that affect joint design in the electron beam process include:

- Beam is focused in small area.
- Beam penetrates at the angle that it strikes the plate.
- Filler metal generally is not added.

These characteristics dictate the use of flat, tightly abutting surfaces, including square- and scarf-butt joints, closed-corner welds, and melt-through T-welds. With the exception of melt-through joints, the beam typically is focused parallel to the joint interface.

Welding Conditions. Electron beam welding conditions depend on the material thickness and type of electron gun. For a given thickness of material, various combinations of accelerating voltage, beam current, and travel speed are satisfactory. Electrical parameters do not adequately describe weld heat input of electron beams because of the importance of beam focus. Beam diameter measurements are complex, so the transfer of welding parameters between different units is very difficult. Instead, suitable welding parameters must be developed for each machine. This can be done with only a few trials.

Electron beam weld passes that completely penetrate joints sometimes are undercut along both edges of the weld metal. Undercutting is eliminated by making a second weld pass at lower energy levels and with a slightly defocused beam. Table 12.6 shows typical EBW conditions for joining titanium alloys.

Resistance Spot and Seam Welding

Resistance spot welding (Ref 12.23) is used extensively to join titanium and titanium alloy sheet metal assemblies. Seam welding is also adaptable, but it is used to a lesser extent. Except for some seam welding operations, special shielding is not required. In resistance welds, the molten and heated titanium is surrounded by solid metal, and the closeness of the surfaces being joined, the short welding cycles, and the electrodes pressed tightly against the sheets cut down contamination of the joints. Therefore, conventional spot and seam welding equipment is adequate if the controls provide consistent welding time and current. However, removal of surface oxide is necessary for a quality weld nugget.

Joint Designs. Sheet overlap and spot spacing are controlled to maintain consistent and reliable mechanical properties. Sheet should overlap sufficiently so that a spot weld nugget is pulled out in tension tests rather than end tearing of the sheet. Spacing of spots must be sufficient to prevent excessive shunting of the current. However, shunting is not a major problem due to the high electrical resistivity of titanium.

Welding Conditions. Spot and seam welding conditions that have the greatest effect on weld quality are welding current and time. Weld diameter, strength, penetration, and indentation can vary significantly with variations in these two conditions. Electrode force also affects these properties, but not by a large amount if force is maintained within acceptable limits.

Spherical-faced copper-alloy electrodes are most satisfactory for spot welding titanium and its alloys. They provide higher strengths, larger weld nuggets, better penetration control, and less sheet separation and indentation for given welding conditions than other electrodes. Electrodes are cooled internally and externally. The Resistance Welder Manufacturers Association class 2 and 3 alloys are used. Class 3 electrodes provide the greatest life for high production rates.

To establish spot and seam welding schedules, welding current and time are adjusted on production welding machines to obtain specified nugget diameters and tension-shear strengths without excessive penetration, indentation, and sheet separation. Penetration is usually high, but it can be controlled by proper heat input. Examples of typical spot welding conditions are listed in Table 12.7.

Table 12.6 Typical weld parameters for electron beam welds

Parameter	Ti-6Al-6V-2Sn		Ti-8Al-1Mo-1V	
Thickness, mm (in.)	1.02 (0.040)	2.04 (0.080)	1.14 (0.045)	2.28 (0.090)
Voltage, V	13,000	18,500	13,000	19,000
Current, mA	50	90	55	100
Welding speed, mm/s (in./min)	2,095 (82.5)	1,905 (75)	1,700 (68)	1,800 (70)
Focusing current, %	55	65	55	65
Gun-to-work distance, mm (in.)	25 (1)	25 (1)	25 (1)	25 (1)

Table 12.7 Spot welding schedules for equal-gage combinations of Ti-6Al-4V sheet

| Material thickness | | | Electrode | | | | Force | | Weld time, | Weld current, | Nugget diameter | |
| | | | Diameter | | Radius | | | | | | | |
mm	in.	Class	mm	in.	mm	in.	kg	lb	cycles	A	mm	in.
0.508	0.020	II	12.7	½	250	10	550	1,200	5	...	3.81	0.150
0.635	0.025	III	...	...	80	3	260	575	2	...	...	...
0.889	0.035	II	15.8	⅝	80	3	300	600	7	5,500	5.72	0.225
1.270	0.050	III	...	...	100	4	400	900	4	...	...	...
1.575	0.062	II	15.8	⅝	80	3	680	1,500	10	10,600	9.11	0.359
1.600	0.063	II	12.7	½	250	10	680	1,500	12	...	8.89	0.350
1.778	0.070	II	15.8	⅝	80	3	770	1,700	12	11,500	9.93	0.391
2.362	0.093	II	15.8	⅝	80	3	1,090	2,400	16	12,500	10.95	0.431
3.175	0.125	II	12.7	½	250	10	1,050	2,300	14	...	10.80	0.425

Flash Welding

Flash welding is commonly used to weld titanium rings for jet engines. It is also used to weld many prototype assemblies, including propeller blades and other complex assemblies.

Flash welding is a resistance butt-welding process in which two workpieces are clamped in suitable current-carrying fixtures holding them lightly end to end. A current is passed through the workpieces, producing a flashing or arcing that heats the workpiece ends to the fusion point. At the proper temperature, the workpieces are suddenly brought together with sufficient force to cause an upsetting action.

Flash welding is better adapted to high-strength, heat treatable alloys than arc, electron beam, spot, and seam welding for two reasons. First, molten metal is not retained in the joint, so cast structures are not present. Second, the hot metal in the joint is upset, and the upsetting operation improves the ductility of the heat-affected zone. Flash welds with mechanical properties approaching those of the base metals are produced regularly in conventional machines.

Shielding. Solid cross sections are welded without inert gas shielding, but shielding sometimes is used. When used, fiberglass enclosures are placed around the joint, and inert gas is introduced into the enclosure. For joints in tubing and assemblies with hollow cross sections, inert gas is introduced into the assemblies.

Joint designs for flash welds are also similar to those used for other metals. Flat edges are satisfactory for welding sheet and plate up to approximately 6 mm (0.25 in.) thick. For thicker sections, the edges are sometimes beveled slightly. Figure 12.9 shows metal allowances used in making titanium flash welds. The allowances include metal loss in the flashing and upsetting operations.

Welding Conditions. The transformer capacity required to weld titanium alloys does not differ much from that required for steel. The upset pressure capacity for making titanium weldments is not as high as that required for steel. Figures 12.10 and 12.11 show the transformer and upset capacity required for welds of different cross-sectional area.

The most important flash welding conditions are flashing current, speed and time, and upset pressure and distance. With proper control of these variables, molten metal, which may be contaminated, is not retained in the joint, and the metal at the joint interface is at the proper temperature for upsetting.

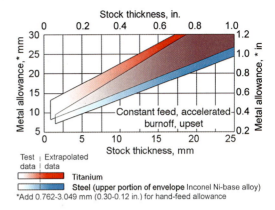

Fig. 12.9 Comparison of total metal allowance as a function of stock thickness in flash welding titanium and steel. Allowances include metal loss in the flashing and upsetting operations.

Generally, fast flashing feeds and short flashing times are used to weld titanium and titanium alloys. These conditions are desirable to minimize weld contamination, and they are possible because of the low electrical and thermal conductivities of these metals. Also, the use of a parabolic flashing curve is more desirable than the use of a linear flashing curve, because maximum joint efficiency can be obtained with a minimum loss of metal at low-to-intermediate upset pressures (500 to 140 kg/cm², or 7 to 2 ksi).

Flash welding conditions differ from machine to machine and application to application. Table 12.8 lists some conditions that work for Ti-6Al-4V alloy. Welding current is not given, but welding current and arc voltage depend on the transformer tap that is used.

Solid-State Welding

Titanium and its alloys are readily joined without melting. Welds form in these processes through deformation and diffusion across the interface of the two parts. Deformation results in intimate contact, while diffusion forms a metallurgical bond between the parts. The absence of melting makes these processes less restrictive to weldable alloys, but most applications involve weldable alloys. Processes that use little or no deformation are classified as diffusion bonding. Processes (Ref 12.1) using significant deformation are termed pressure and deformation welding.

Diffusion bonding (DB) is done with just enough pressure to bring the surfaces to be joined into intimate contact (Ref 12.1). Some deforma-

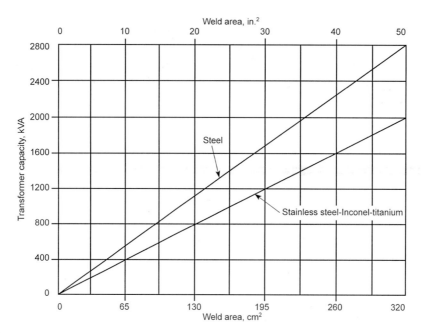

Fig. 12.10 Transformer capacity as a function of weld area in flash welding titanium and other materials

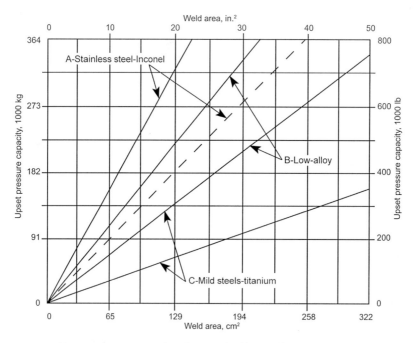

Fig. 12.11 Maximum machine upset pressure required as a function of weld area in flash welding. The upset pressure capacity required for titanium is much less than for stainless and high-strength, low-alloy steels.

tion is required to obtain adequate contact, but generally it is not measurable.

Diffusion bonding can be done in several types of equipment. Resistance spot-welding equip-

ment can be used to form solid-state spot welds. However, conditions must be controlled much closer than in conventional spot welding. Also, large surfaces can be joined in presses and auto-

Table 12.8   Flash welding conditions

	Ti-6Al-4V	
Condition	89 mm (3.5 in.) round	6.4 mm (0.25 in.) plate
Flashing voltage, V	5.0	5.0
Total metal allowance, mm (in.)	15.67–23.88 (0.617–0.940)	17.78–20.32 (0.700–0.800)
Upset, mm (in.)	5.08–12.7 (0.200–0.500)	5.72-6.99 (0.225–0.275)
Current cutoff, mm (in.)	3.81–10.16 (0.150–0.400)	5.21-6.45 (0.205–0.255)
Upset pressure, MPa (ksi)	105 (15)	125 (18)
Sheet higher, mm (in.)	28 (1.10)	16 (0.63)
Atmosphere	Argon	Argon

claves where external pressure is applied and in evacuated envelopes where pressure differential provides intimate contact between the surfaces. Diffusion bonding is often done in conjunction with superplastic forming (SPF) of sheet metal structures, known as SPF/DB (see Chapter 11, "Forming of Titanium Plate, Sheet, Strip, and Tubing," in this book). For both types of operations, a fine, equiaxed alpha microstructure is desirable (in alpha and alpha-beta alloys) for good flow characteristics.

Shielding. Theoretically, auxiliary shielding is not required because the intimate contact between the metal surfaces prevents contamination. This condition is achieved in solid-state spot welds. Where large surfaces are joined, such as in SPF/DB, additional precautions are taken, such as placing assemblies in inert gas enclosures.

Welding Conditions. Various combinations of time, temperature, and pressure apply to welding titanium and its alloys. Welding time decreases with increasing temperature. In all operations, pressure must be sufficient to ensure intimate contact. Temperatures are high in resistance solid-state spot-welding operations, enabling welding in very short times. A relatively long time is required to join larger areas. Temperatures, times, and pressures used to join titanium alloy assemblies with large areas are 820 to 1040 °C (1500 to 1900 °F), 30 min to 6 h, and 350 to 1050 kg/cm^2 (5 to 15 ksi).

Pressure Welding

Pressure welding is done to provide relatively large deformation at the joint interface. The deformation ensures intimate contact and promotes diffusion across the joint interface.

Presses and rolling mills are used for pressure welding. Presses join structural parts such as plates, bars, formed tubes, and hemispheres. Rolling mills assemble structural members, such as sandwich structures. Oxyacetylene flame and induction heating are used for structural parts

joined in presses. Parts for roll welding are heated in furnaces.

Shielding. Structural parts with solid cross sections are pressure welded without external shielding. Metal that is contaminated is expelled from the weld and removed by machining. Inert gas is supplied to the inside hollow structural parts to minimize contamination during welding. Some contamination occurs using this technique, and it should be removed for best conditions. With induction heating, the entire weld area can be shielded if desired.

In roll-welding operations, parts are assembled in a pack, which is evacuated and outgassed to prevent scale formation and subsequent poor welding.

Welding Conditions. Deformation and temperature or pressure are controlled in pressure welding. In joining structural shapes in presses, upset pressure of approximately 175 kg/cm^2 (2500 psi) is maintained throughout the heating cycle. When a preset amount of upset occurs, the operation is stopped. In roll welding, temperature is controlled, and after a certain reduction is achieved, the operation is stopped. For most titanium alloys, rolling temperatures of 760 to 980 °C (1400 to 1800 °F) are used, which do not impair the mechanical properties of the alloys being welded.

Friction Stir Welding

Friction stir welding (FSW) is a nonfusion technique for joining sheet and plate material, invented by The Welding Institute (TWI), United Kingdom, in 1991 and licensed by TWI (Ref 12.18–12.21). The basic form of the process uses a cylindrical (nonconsumable) tool consisting of a flat, circular shoulder with a smaller probe protruding from its center. The tool is rotated and plunged into the joint line (between two rigidly clamped plates) so the shoulder sits on the plate surface and the probe is buried in the workpiece, as shown in Fig. 12.12.

Friction between the rotating tool and the plate material generates heat, and the high normal pressure from the tool causes a plasticized zone to form around the probe. The tool is then traversed, frictionally heating and plasticizing new material as it moves along the joint line. As the tool traverses, the probe stirs the locally plasticized area and forms a solid-phase joint.

The development and incorporation of FSW over the past two decades was very rapid. Nearly all current uses involve joining aluminum alloys for use in applications including airframes and aircraft components, ship decking and structures, rail carriages, automotive components, bridge components, and space launch systems. Other materials being investigated for joining using FSW include magnesium alloys, copper, steels, and titanium alloys.

Friction Stir Welding of Titanium Alloys. Although the majority of common titanium alloys are generally weldable using conventional means, problems with workpiece distortion and poor weld quality occur. In addition, some of the more advanced titanium alloys (such as Ti-6246 and Ti-17) are difficult to weld using fusion processes. The development of FSW offers the possibility of

a cost-effective method to produce high-quality, low-distortion welds in titanium alloy sheet and plate (Ref 12.18–12.21).

The first trials on FSW of titanium were carried out at TWI as early as 1995. The initial welds were conducted on commercially pure (grade 2) titanium and proved the potential of applying FSW to titanium alloys. A section from one of these initial trials is shown in Fig. 12.13.

The darker areas in the section show where the material has been heated to above the beta transus (approximately 900 °C, or 1650 °F, for this material). The lighter areas of the weld are untransformed but significantly refined.

The success of the early trials led to the formation of a TWI Group-Sponsored Project (GSP) in 1996. Under this GSP, several TWI member companies jointly funded a research program on FSW of titanium alloys, which ran from 1996 through 2002.

Friction Stir Welding of Ti-6Al-4V. The majority of work during TWI's GSP was carried out on 6.35 mm (0.25 in.) Ti-6Al-4V plate. After identifying a suitable tool material, welding trials were carried out to develop effective tool designs and processing conditions to weld the plate. This ultimately led to the production of fully formed, high-quality friction stir welds in Ti-6Al-4V, as shown in Figs. 12.14 and 12.15.

Fig. 12.12 Schematic of the friction stir welding process. Source: Ref 12.18

Fig. 12.14 Surface appearance of a good-quality friction stir weld in 6.35 mm (0.25 in.) thick Ti-6Al-4V. Source: Ref 12.18

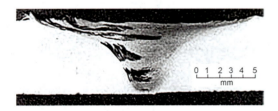

Fig. 12.13 Cross section of a friction stir weld in commercially pure (grade 2) titanium produced at The Welding Institute in 1995. Source: Ref 12.18

Fig. 12.15 Cross section of a good-quality friction stir weld in 6.35 mm (0.25 in.) thick Ti-6Al-4V. Source: Ref 12.18

Friction stir welding of other alloys and thicknesses was investigated in addition to the main body of work on the development of FSW for 6.35 mm (0.25 in.) thickness Ti-6Al-4V.

Friction Stir Welding of 3 mm (0.125 in.) Thick Ti-6Al-4V. Trials were conducted on 3 mm (0.125 in.) thick Ti-6Al-4V sheet. As in the work on 6.35 mm (0.25 in.) thick material, FSW tool designs and welding conditions were adjusted to achieve a good-quality weld, as shown in Fig. 12.16.

Very little distortion was experienced using optimized welding conditions. Minimal workpiece distortion is a significant advantage of the FSW process.

Friction Stir Welding of 6.7 mm (0.250 in.) Thick Ti-15V-3Al-3Cr-3Sn. In a study on the application of FSW to 6.7 mm (0.250 in.) thick Ti-15V-3Al-3Cr-3Sn plate, researchers found this beta-phase titanium alloy was significantly more formable than alpha-beta Ti-6Al-4V. Friction stir welding of Ti-15V-3Al-3Cr-3Sn generated lower peak temperatures (approximately 800 °C, or 1470 °F) than those observed in Ti-6Al-4V (approximately 1000 to 1200 °C, or 1830 to 2190 °F), and excellent weld surface quality was achieved, as shown in Fig. 12.17.

Microstructure of Friction-Stir-Welded Ti-6Al-4V. Weld sections were evaluated using optical and scanning electron microscopy. The sections showed the area heated and stirred by the FSW tool was surrounded by a very narrow heat-affected zone, which highlighted the low thermal conductivity of titanium. The weld root in many cases was outside the hot deformed zone, and, in early trials, voids in the area were relatively common. A typical example of this weld structure is shown in Fig. 12.18.

Microstructural characterization was carried out on the three different zones (A, B, and C) identified in Fig. 12.18.

Zone A (parent material) consisted of a rolled microstructure of elongated grains of alpha (light) in a matrix of alpha and beta (dark), as shown in Fig. 12.19.

Zone B (deformed zone) shows evidence of alpha-beta transformation, which occurs at a temperature of approximately 990 °C (1815 °F) in Ti-6Al-4V. Significant grain growth occurs at this elevated temperature, producing large, equiaxed beta grains in the weld center. The beta phase reverts on cooling, and the resultant weld microstructure consists of large alpha grains with a smaller amount of retained beta, as shown in Fig. 12.20. The extent of the grain growth in this region suggests there is potential to reduce the heat input to this area of the weld.

Fig. 12.18 Cross section of an early friction stir welding trial in 6.35 mm (0.25 in.) thick Ti-6Al-4V shows three different weld zones: parent metal (A), deformed surface region (B), and partially transformed weld-root zone containing voids (C). Source: Ref 12.18

Fig. 12.16 Cross section of a good-quality friction stir weld in 3 mm (0.125 in.) thick Ti-6Al-4V sheet. Source: Ref 12.18

Fig. 12.17 Surface appearance of a friction stir weld in 6.7 mm (0.25 in.) thick Ti-15V-3Al-3Cr-3Sn. Source: Ref 12.18

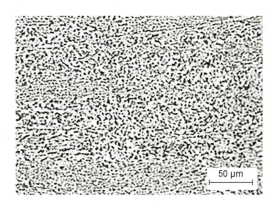

Fig. 12.19 Microstructure of parent metal (zone A in Fig. 12.18). Source: Ref 12.18

Zone C (partially transformed weld-root zone) shows that only partial transformation occurred in this region. Grain growth is also much less than that in the weld, leaving a fairly fine-grained structure with small areas of transformed beta, as shown in Fig. 12.21. This partially transformed structure confirms that lower temperatures were experienced in this area of the weld, which probably accounts for the voids observed in the region.

Hardness tests were also conducted across the centerline of the weld (in the fully transformed zone). Hardness increased from a base of approximately 305 HV to approximately 340 HV in the weld zone. The increase is consistent with an increased proportion of alpha phase in the transformed weld structure.

Tensile Properties of Friction-Stir-Welded Ti-6Al-4V. Table 12.9 lists results of transverse tensile tests on selected good-quality welds produced during the later stages of the program. Tensile strengths were higher than those of the parent material in some cases. An increase in strength in the weld zone was expected (due to the higher proportion of alpha phase in the transformed region), and the results confirm that high-quality friction stir welds were produced.

Advantages and Disadvantages of FSW. Advantages of FSW over conventional fusion welding processes include:

- Good mechanical properties in the as-welded condition
- Improved safety due to the absence of toxic fumes and spatter
- No consumables. A threaded pin made of conventional tool steel (e.g., hardened H13) can weld over 1 km (~3300 ft) of aluminum without the use of filler material and shielding gas.
- Easily automated on simple milling machines, translating to lower setup costs and less training
- Operation possible in all positions (horizontal, vertical, etc.), because there is no weld pool
- Generally good weld appearance and minimal thickness under/overmatching, thus reducing the need for expensive machining after welding
- Low environmental impact

Disadvantages of the process include:

- Exit hole remains when tool is withdrawn

Fig. 12.20 Microstructure of transformed material (zone B in Fig. 12.18). Source: Ref 12.18

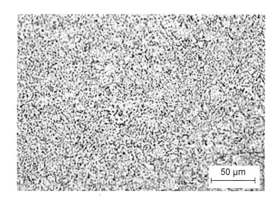

Fig. 12.21 Microstructure of partially transformed weld root (zone C in Fig. 12.18). Source: Ref 12.18

Table 12.9 Tensile properties of selected weld in 6.35 mm (0.25 in.) thickness Ti-6Al-4V

Sample number	Section area		Maximum load		Maximum stress		Elongation, %
	mm²	in.²	kN	tonf	N/mm²	ksi	
Parent	...	...	...	...	1035	150	14
W5.7A	66.0	0.102	67.3	6.86	1020	148	8.5
W5.7B	65.7	0.102	64.8	6.61	986	143	8.5
W5.18A	60.2	0.093	64.3	6.56	1068	155	8
W5.18B	61.2	0.095	65.1	6.64	1064	154	7.5

- Large downforces required, with heavy-duty clamping necessary to hold plates together
- Less flexible than manual and arc processes (difficulties with thickness variations and nonlinear welds)
- Often slower traverse rate than some fusion welding techniques (this can be offset if fewer welding passes are required)

Weld Defects

In general, the types of weld defects observed in titanium and titanium alloys are similar to those encountered with other metals. Major weld defects in all welding processes used to weld titanium assemblies are porosity, tungsten inclusions, lack of penetration and fusion, and cold cracks. Tungsten inclusions, penetration, and fusion are controlled by adjusting welding conditions.

Porosity in titanium welding is directly related to the cleanliness (removal of oxides) of the base metal and filler metal, as well as to the welding speed and heat input. Base metals cleaned using a hot salt bath descale and then pickled and welded within a short time period appear water-white in an x-ray. Recommended procedures to minimize porosity include:

- Improve cleaning and handling to prevent foreign materials, including fingerprints and oxides, from contaminating the weld
- Examine filler wire to eliminate embedding foreign materials in the surface
- Abrade sheared and machined edges, or prepare surfaces with draw filing
- Use smooth, square joints and good fit-up: use tightly butted joints if using filler metal, and use root opening if using filler metal
- Reduce weld cooling by adjusting welding heat input and using mild steel rather than copper quench bars, high current, helium as the shielding gas, and multipass welds
- Use slower welding speeds to allow porosity to surface
- Do not allow weld oxide to remain for extended time between passes for multiple-pass welding; all welds should be cleaned during the same work shift as welded.

Weld cracks in titanium and its alloys are attributed to excessive contamination by hydrogen, oxygen, and nitrogen and to the heat treatment response of certain alloys. To control crack-ing due to oxygen and nitrogen contents, the base metal surface must not contain excessive amounts of these dissolved elements. Good shielding procedures are required. Cracking due to hydrogen content can be overcome by vacuum annealing the base metals to remove hydrogen prior to welding. Cracking due to the heat treatment response of the alloys can be avoided by selecting a more suitable alloy. If such a change is not practical, annealing treatments immediately after welding are helpful.

Major defects in plasma arc and electron beam welds are the same as for GMAW and GTAW welds. Porosity in GMAW is controlled by increasing the voltage.

Major defects in resistance spot and seam welds include "dud" welds, excessive penetration, sheet separation, metal expulsion, indentation, and small nugget diameters. These problems are controlled by welding current, time, and electrode force and shape. If fine-tuning the weld settings does not minimize or eliminate these problems, joint fit-up and machine operation should be carefully evaluated.

Major defects in flash and solid-state welding processes are unwelded areas. Such defects are extremely hard to detect and must be controlled by cleaning procedures, welding equipment, shielding devices, and weld settings.

Dissimilar-Metal Joints

As mentioned previously, titanium is difficult to weld to other metals because the weld becomes brittle when titanium is alloyed with other metals. Techniques used to produce a dissimilar-metal joint with titanium include:

- Join using GTAW with pure vanadium or niobium on the surface or in an insert for attaching titanium alloys to steel
- Coat the surface of steel with vanadium for spot welding
- Melt only low-temperature material, such as aluminum, in resistance, spot, and GTAW operations
- Use intermediate material such as vanadium (a beta-isomorphous element) with solid-state processes
- Use 99.99% fine silver as a braze-type joint with the GTAW process

Note that these methods should not be used on pressure-containing parts.

Quality Control

Quality-control procedures for titanium weldments are similar to those used for other metals. Dye penetrant, ultrasonic, and radiographic inspection are used to locate cracks, porosity, and incomplete weld penetration. Likewise, visual examinations for undercut, penetration, and weld reinforcement are used. In addition, metallographic examinations, mechanical tests, and hardness surveys are made on representative samples to check weld quality.

The greatest problem in quality control is to determine whether or not the shielding obtained in inert-gas-shielded metal arc welds is adequate. Bend tests on representative samples provide a good measure of shielding, but nondestructive tests are not as reliable as desired. Ultrasonic testing can be used for hardness testing. Also, weld hardness is compared to base metal hardness using portable hardness testers.

The only widely used nondestructive method for evaluating weld contamination is surface appearance (which indicates oxygen level). With this method, bright surfaces with the appearance of newly polished silver are desired. Also, shiny surfaces with straw or peacock blue colors are acceptable. Dull blue and gray surfaces are unacceptable. This method is fairly reliable if shielding conditions are well controlled. An indication of well-controlled conditions is the variation in surface appearance of a given weld or several similar welds. If large variations are observed, better controls are needed.

A major drawback for this type of nondestructive test is that highly contaminated welds can be rewelded and provide the desired appearance while being totally contaminated. Highly skilled, qualified operators are a must for a successful welding program.

In chamber welding, the shield atmosphere should be monitored during welding using an oxygen analyzer. No welding should be initiated with oxygen levels above 3900 ppm. The level should rapidly decrease to below 100 ppm during the welding sequence. Superior weld ductility can be achieved at or below these oxygen levels. Again, shield gas must be maintained on the torch at all times during welding.

Another check on weld shielding is obtained by welding bend-test specimens before starting and after welding an assembly. If the bend specimens are satisfactory, then it is assumed that the welds in the assembly are satisfactory. Also, if the welds become discolored while welding the assembly, bend-test specimens are welded to determine if the atmosphere in the chamber has become excessively contaminated.

Repair Welding

In making repair welds, defects must first be completely removed by machining and/or grinding, and the repair area dye penetrant tested to ensure removal of cracks. The repair area is cleaned and welded using the same procedures that are typically used. For electron beam welds, a portion of the structure including the weld is often removed and another piece welded into the part. Also, electron beam welds are sometimes repaired merely by remelting the defective area.

Repair welding is not desirable. Therefore, the cause of defects should be determined before making repair welds, so they will not be defective. Cracks can be caused by contamination, residual stresses, or stress corrosion. If due to contamination, all contaminated metal should be removed prior to repair welding. If due to residual stresses, the parts should be stress relieved prior to repair welding. Also, it may be necessary to upset the material in the repair weld area to provide excess metal to allow for weld shrinkage. If due to contamination or stress corrosion, parts may not be reparable unless the defective area is localized.

Stress Relieving

Stress-relieving operations are commonly used to reduce residual welding stresses and promote dimensional stability. They are performed after completion or at various stages of production. Data on stress-relaxation annealing times and temperatures are presented in Table 12.10. Cleaning procedures outlined previously are used after stress-relieving treatments.

Mechanical Properties

Tensile and compressive strengths of titanium and titanium alloy weldments are excellent. With the exception of some alloys, they are equal to or exceed those of the base metal regardless of testing temperature. Most specifications allow design stresses as high as 90% of base metal strength.

When using titanium alloys for strength as opposed to corrosion resistance, as-welded strength must be evaluated. Commercially pure titanium-grade weldments are usually within the minimum tensile ranges specified by ASTM International and the American Society of

Table 12.10   Stress-relaxation annealing temperature and time

Nominal composition, wt%	Stress-relieving temperature °C	°F	Time at temperature, h
Commercially pure	427	800	8
	482	900	0.75
	538	1000	0.5
Ti-5Al-2.5Sn	482	900	20
	538	1000	6
	593	1100	2
	649	1200	1
Ti-6Al-4V, annealed	482	900	20
	538	1000	2
	593	1100	1
Ti-6Al-4V, solution treated	482	900	15
	538	1000	4
Ti-6Al-4V, solution treated and aged	482	900	15
	538	1000	5

Mechanical Engineers. Proper selection of filler and base metal is also necessary.

Tensile ductility and toughness of weldments are often lower than those of the base metal. For most applications, ductility and toughness of weldable alloys are adequate to meet all service requirements. However, in solid rocket motor cases that are subjected to high strain rates and biaxial state of stress, techniques used to increase weldment ductility and toughness at the expense of tensile strength include:

- Fully heat treat parts prior to welding, and only partially heat treat weldments
- Specify minimum possible interstitial content in filler wires
- Use filler wire with lower alloy content than that of base metal

When these techniques are used, parts are machined with thickened weld bands to match the load-carrying capacity of the weld with that of the thinner base metal. Spot-welded joints are evaluated on the basis of tensile shear and cross tensile strength. Tension shear strengths are very high, ranging from 133 to 445 N per 0.28 mm (30 to 100 lbf per 0.011 in.) of single sheet thickness. The lower values are for thinner-gage materials. Cross tension strength is approximately average. Therefore, tension-to-shear ratio is lower than that normally encountered with other metals. To compensate for the low tension-to-shear ratio, rivets sometimes are placed at the ends of long spot-welded joints to minimize tensile loads.

Under prolonged exposure to elevated temperatures, the properties of weldments in Ti-6Al-4V, Ti-5Al-2.5Sn, and Ti-8Al-1Mo-1V are comparable to those of the base metal.

Brazing

There are fewer applications for brazed titanium structures than for welded structures. This is due to the inherent characteristics of brazing and welding. Compared with weldments, brazed joints generally are:

- Heavier for given strength requirements
- Less resistant to corrosion
- Less desirable for elevated-temperature application
- More difficult to inspect
- More difficult to process

Most of these characteristics stem from the need to use a brazing filler metal of completely different composition than the base metal. However, there are some applications where brazing is more adaptable than welding operations. These applications include fabricating composites, such as honeycomb sandwich structures, and dissimilar-metal joints.

Chemical and Metallurgical Characteristics

Brazing of titanium alloy structures is limited by the chemical and metallurgical characteristics described in the welding section, including:

- Sensitivity to embrittlement when contaminated by small amounts of carbon, oxygen, nitrogen, and hydrogen
- Reactivity when heated to brazing temperatures
- Sensitivity to embrittlement or softening in varying degrees as a function of alloy content when subjected to the thermal cycles involved in brazing operations
- Sensitivity to embrittlement when highly alloyed with metallic elements

The effects of these characteristics on welding operations should be reviewed because they also have similar effects on brazing. The effects include:

- Parts must be protected from contamination during brazing operations. Vacuum is most often used for this purpose, but inert gases and special fluxes also are applicable.

- Titanium alloys that can undergo brazing thermal cycles without severe mechanical property deterioration must be used. This requirement limits most brazing operations to weldable alloys. Attempts to develop brazing filler metals that melt at temperatures compatible with some of the alloys with poor weldability have not been successful.
- Very short brazing times or special alloys that do not readily react with titanium alloys must be used to prevent embrittlement of brazed joints and titanium base metals.

Surface Preparation

The preparation of clean, scale-free surfaces is as important in brazing operations as it is in welding. Cleaning operations are necessary to remove surface scale, oil, and other foreign matter that contaminate titanium and prevent braze metal wetting. Cleaning procedures discussed in the welding sections are satisfactory for brazing operations.

Sometimes, special surface preparation is required when titanium is brazed to other metals. This consists of applying an overlay of brazing filler metal on the titanium or other metal prior to the brazing operations. Overlays are especially beneficial in brazing operations requiring the use fluxes or involving metals that are wetted by the brazing filler metal, as is titanium.

Shielding

Special fluxes, inert gas atmospheres, and vacuum promote the formation of sound brazed titanium joints. Fluxes attack and float away surface scale, and they protect the underlying surfaces from additional oxidation. Inert gas atmospheres and vacuum do not remove surface scale, but they prevent its formation.

Fluxes are available for brazing titanium and titanium alloys, but they are not recommended for most applications because reliability and quality are lower compared with the use of vacuum or inert gas protection. Also, fluxes are extremely corrosive and must be removed entirely after welding. Their corrosive nature is due to their components, including NaCl, KCl, AgCl, KF, and LiF.

Inert gases and vacuum are most satisfactory for protecting titanium alloys during brazing operations. Compared with the use of flux, these methods provide advantages, including:

- Entire assemblies are protected from contamination.
- Flux inclusions in the joint are not a problem.
- Special flux-removal operations are not required.

The last advantage is especially important in brazing assemblies in which flux-removal operations are not possible.

The main disadvantage of using vacuum and inert atmospheres lies in the equipment required. Special furnaces, retorts (gastight envelopes), and vacuum pumps are required, and tooling costs are high. However, the advantages obtained by using these methods overshadow the difficulties, and inert gas or vacuum protection is recommended for most applications.

Induction, furnace, torch, and resistance brazing methods are adaptable to titanium alloys.

Induction Brazing

Induction heating is satisfactory. Its use allows shorter brazing cycles than are generally possible with other brazing methods. Consequently, reactions between the base metal and filler metals are minimized.

Furnace Brazing

Furnace heating is satisfactory for brazing titanium and titanium alloys, and it can be adapted to most joint designs. However, brazing alloys generally are molten longer in furnace brazing processes than with other processes. Therefore, furnace-brazed joints exhibit greater reaction zones than joints brazed using other methods.

Torch Brazing

Both oxyacetylene and helium arc torches are used in brazing titanium. Short brazing cycles are possible, but temperature control is difficult. Skilled operators are required, especially for oxyacetylene brazing, to obtain sound, strong joints.

In oxyacetylene brazing operations, use of a flux promotes braze-metal wetting. Skilled operators are required with this process, because it is difficult to feed filler metal into the joint. When the joint is overheated, a crest forms on the flux and makes it almost impossible to add more braze metal. This problem can be overcome by placing the filler metal in the joint prior to brazing.

Helium arc torch brazing operations do not require the use of fluxes. A standard tungsten arc torch is used, and inert gas flowing through the torch prevents formation of surface scale. However, care must be taken to prevent heating a larger area than can be shielded by the torch. In this operation, low current is used to prevent melting the base metal.

Resistance Brazing

Conventional spot-welding equipment is used for resistance brazing titanium. High-conductivity electrodes are preferred over graphite and other electrode materials that have high electrical resistance. In this process, the brazing filler metal (in the form of foil) is placed between the faying surfaces of the joint.

Electrode pressure, current density, and heating time are particularly important in resistance brazing operations. Too high of a pressure can squeeze the brazing alloy out of the joint. If brazing current or time is too low, temperature distribution is poor, and the brazing alloy may not melt completely. If current or time is too high, the brazing alloy is superheated, reacting with the titanium and squeezing out of the joint.

Brazing Filler Metals

Special silver-base brazing alloys are used in brazing titanium alloys. They wet and flow on titanium and melt within compatible, but not optimum, temperature ranges. Short brazing cycles using these alloys can preclude excessive penetration and embrittlement.

Silver-Lithium Alloys. Joints brazed using silver-base alloys containing 0.5 and 3.0% Li are degraded by oxidation when heated to 430 °C (800 °F) in air. Also, they have poor corrosion resistance in salt spray environments.

Silver-Aluminum-Manganese Alloys. The most promising brazing filler metals are the Ag-Al-Mn alloys, such as Ag-5Al-1Mn. These alloys have better oxidation and corrosion resistance than silver-lithium alloys.

Silver-cadmium-zinc alloys were developed for use in oxyacetylene brazing applications.

Aluminum. Pure aluminum is widely used for brazing aircraft structural components.

Experimental alloys are being developed, such as a palladium-base alloy containing 15.4% Ag and 3.5% Si. The alloy flows well and melts at 695 °C (1280 °F). Joint penetration is low, and joint strengths are high.

Dissimilar-Metal Joints

Brazing is an excellent method to join titanium alloys to other metals. Most applications require joining titanium to stainless steel. Several silver-base alloys and the experimental palladium-base alloy are used for such joints.

Major defects in brazed joints are unbrazed areas, poor fillet geometry, excessive penetration, and embrittlement of the base metals. Effective quality control is possible through process development and control. To achieve this, optimum brazing conditions should be developed and all known variables studied. This enables controlling variables within close limits in production.

Brazed joints are examined using nondestructive testing methods, including visual examination and radiographic, ultrasonic, and dye penetrant techniques.

Mechanical Properties of Brazed Joints

Brazing filler wire selection, joint design, brazing process, and brazing cycles have the greatest effect on joint properties. Shear strengths ranging from 103 to 620 MPa (15 to 90 ksi) are developed with most alloys. Joints that fail in the base metal are produced by adjusting the size of the braze overlap and using short brazing cycles to prevent joint and base metal embrittlement.

The effect of the brazing thermal cycle on base metal properties is also important. For best results, brazing temperature should not exceed the beta transus. Otherwise, ductility and toughness of the base metal can be impaired. Also, if the brazing alloy is not compatible with the heat treatment of the heat treatable alloys, the full potential of the alloy will not be used in brazed joints.

Soldering

Soldering operations on titanium are limited, and data are not available on properties of soldered joints. The likely reason for this is because the properties and temperature capabilities of soldered joints are not compatible with applications for titanium structures.

Titanium has been successfully soldered by coating the titanium with silver, coating the titanium with solder, or using a tungsten arc torch to solder directly to the titanium.

Coating titanium with silver prior to soldering enables the coated surfaces to be readily soldered using conventional methods. Coating methods include using a tungsten arc torch, sprinkling a silver-depositing flux on the titanium and heating it, and plating. However, a joint with a thin coating must not be overheated, because the silver coating is removed, resulting in an unsound joint.

Titanium surfaces coated with solder also are joined using conventional methods. A method used to apply a coating is to "load" a grinding wheel with solder and lightly grind the surfaces to be joined. Solder "rubs on" in this fashion and coats hard-to-wet surfaces. Overheating solder-coated surfaces also removes the solder coating. Using a tungsten arc torch for heating makes it possible to solder directly to titanium.

Adhesive Bonding

Adhesive bonding is used extensively to join aluminum and steels, but its use in joining titanium alloy structures is limited. However, all titanium alloys and dissimilar metals can be economically joined using adhesive bonding, providing good strength, mechanical damping, and good appearance. Therefore, the use of adhesive bonding is expected to increase.

The metallurgical and chemical characteristics of titanium, which affect welding and brazing operations, have little effect on adhesive bonding, except for surface treatment requirements.

Surface Preparation

As is true for all metals, proper surface preparation is critical in adhesive bonding operations. Surface treatments for titanium include cleaning and surface conditioning. Typical surface-preparation steps include:

1. Degreasing
2. Acid etching and alkaline cleaning
3. Rinsing
4. Drying
5. Surface conditioning
6. Rinsing
7. Drying
8. Priming and bonding

Degreasing, alkaline cleaning, and acid-pickling techniques were discussed previously in the welding section but are reviewed here.

Surface-conditioning treatments form a corrosion-resistant film of controlled thickness and composition on the surface to be joined (adhered). Films used are complex mixtures of phosphates, fluorides, chromates, sulfates, and nitrates. The composition and thickness of the film is important.

Water rinsing using hot and cold tap, demineralized, and distilled water removes residue from cleaning and surface-conditioning treatments. The best procedure is to try existing tap water, hot and cold, over random time periods to determine bonding results. If results are not satisfactory, demineralized or distilled water should be used.

No more than 8 h should elapse before cleaned parts are bonded. If longer time between cleaning and bonding are expected, surfaces should be primed in accordance with the manufacturer's instructions, to prevent deterioration of the cleaned surfaces.

Many surface-preparation procedures are used successfully, but comparisons between them are not possible. The most commonly used test for cleanliness is the water-break test, where a droplet of distilled water will wet and spread over clean surfaces, and the resulting thin film over the entire surface will not break up into droplets.

Bonding Procedures

All commonly available adhesives are satisfactory for joining titanium. Techniques for applying adhesives depend on their form. Liquids are applied by roller, brushing, dipping, troweling, flow coating, and spraying, depending on viscosity. Hand application is satisfactory for adhesive film and tapes.

After adhesive application, parts are laid up. For some adhesive systems, external pressure is required for curing. In this case, laid-up assemblies are loaded with dead weights or placed in vacuum bags, presses, and autoclaves for curing, depending on pressure requirements. For solid adhesives such as epoxy, pressure is not necessary (except in critical applications) to control bond thickness. Simple clamping and self-aligning joints are used with these adhesives.

Manufacturer's instructions are followed for curing. Some adhesives cure at room temperature. Shelf life of the adhesives is limited. Highest strengths are obtained using adhesives that cure at elevated temperatures. Laid-up assemblies are placed in furnaces, heated autoclaves, and heated molds to cure the adhesives.

Joint design in adhesive bonding is very important. Bonds have the maximum strength in shear and the lowest in peel. Therefore, lap joints loaded parallel to the adhesive bond line are preferred. If tension loads normal to the bond line are necessary, the joints should be designed to minimize peel.

Quality Control. Unbonded and poorly bonded areas are major defects encountered in adhesive-bonded joints. They may be due to cleaning, bonding, lay-up tolerances, and adhesive quality. Guarding against such defects is accomplished through process development and control, the main steps of which include:

1. Document important processing variables
2. Test all incoming materials
3. Monitor all bonding operations

Nondestructive tests are ineffective for adhesive-bonded joints. Unbonded areas can be determined using sonic bond-testing equipment and by an experienced inspector using a coin or light hammer to test the part. However, low-strength bonds cannot be detected.

Mechanical Properties

Mechanical properties vary with adhesive system and processing variables. Tensile shear strengths of adhesive-bonded joints range from approximately 14 to 48 MPa (2 to 7 ksi).

Mechanical Fastening

The first operational titanium alloy structures were joined by mechanical fastening. In the early stage of titanium development, when only commercially pure titanium was used, this was the most easily adapted joining process, and little information on heat treatment or shielding was required. However, as the stress-carrying capabilities of such joints increased through the use of high-strength fasteners and alloys, mechanical fastening technology became more complex (Ref 12.2–12.10).

Metallurgical and Chemical Properties

Metallurgical and chemical characteristics that affect welding and brazing have little effect on mechanical fastening operations. However, titanium characteristics that affect mechanical joining include:

* Brittle nature of oxide scale
* Hot forming requirements
* Machining characteristics
* Susceptibility to halogen stress corrosion

Methods used to compensate for these characteristics in mechanical fastening procedures are discussed in subsequent sections.

Surface Preparation

Oxide scale on titanium alloys is brittle and hard and dulls machine tools. Therefore, it must be removed before machining. Acid pickling and molten salt bath techniques described in the welding section remove this scale.

Forming

When titanium and its alloys are formed at room temperature, springback is high and unpredictable. To achieve satisfactory forming results, elevated-temperature forming generally is used. Elevated temperatures require equipment and dies that maintain pressure and dimensions at temperature.

Commercially pure titanium, for the most part, can be severely formed at temperatures in the range of 300 to 500 °C (600 to 1000 °F). Springback is a function of grade purity. For example, grade 1 has considerably less springback than grade 4.

Rivet holes are countersunk by dimpling, an operation that requires high deformation in the sheet. Dimpling is carried out at temperatures to 530 °C (1000 °F). Operations, such as the use of a hydraulic press, are used to form sheets so overlap joints are smooth on one surface. Heating parts to 590 °C (1100 °F) for joggling maintains good dimensional tolerances.

Machining

Drilling is preferred for preparing rivet and bolt holes. Punching is also used, depending on the application. Recommended guidelines to overcome troublesome machining characteristics (see Chapter 13, "Machining and Chemical Shaping of Titanium," in this book) include:

* Liberally use cutting fluids with good cooling capacity to minimize temperature rise, which causes titanium chips to gall and weld to the machine tool, thus breaking off small pieces from the cutting
* Maintain sharp drills to obtain holes of required dimensional tolerances
* Use short drills with large flutes, proper relief angles, and solid fixturing to achieve maximum drill life
* Use low speeds and positive feeds to maintain high tool-bit pressures

After drilling, holes are reamed and deburred on the back side. Reaming is easily done using spiral-fluted reamers having a 5 to 10° relief angle and a 0.254 mm (0.010 in.) margin to reduce chatter.

Tapping drilled holes in titanium is achieved using automatic equipment that taps in reverse when a critical torque is exceeded. Also, the equipment should feature threads reduced to 55 to 60% of full depth. The number of threads cut should be an absolute minimum to overcome tapping problems. Special lubricants are available to aid tapping operations. Consulting a tool specialist is recommended.

Holes for rivets and bolts are often drilled with the joint fitted up and temporarily clamped for precise alignment. Such alignment ensures maximum joint strength.

Stress-Corrosion Susceptibility. This characteristic also imposes limits on machining operations. Although chlorinated or sulfochlorinated cutting oils result in the highest machining rate, nonchlorinated fluids are recommended to avoid stress corrosion. Oil-water emulsions should be used instead.

Riveting

Riveting is used extensively to fabricate titanium assemblies, using conventional joint designs to join sheets and plates and to attach stiffeners such as T-shapes and angles. Typical minimum edge distances for rivet holes are listed in Table 12.11. Rivet hole sizes are listed in Table 12.12.

Rivet holes are carefully prepared to eliminate sharp corners, tool marks, and other defects that could cause premature failures. Rivet holes are prepared by drilling and reaming. For machined countersunk holes, special tooling is used to develop a radius 0.050 to 0.180 mm (0.002 to 0.007 in.) at the bottom of the countersink. Holes are deburred after drilling and reaming. Special tooling is used to blend the corners where the edge of the hole intersects the surface of the sheet. Typical radii are listed in Table 12.13.

Upsetting Rivets. Rivet shank length is selected to provide specified upset dimensions without the projection exceeding the nominal rivet diameter. Typical upset dimensions for commercially pure titanium (and A-286 and Fe-Ni-Cr-Mo alloy) rivets are listed in Table 12.14.

Rivets are upset both hot and cold and in several different ways. Cold riveting is preferred on the basis of speed, efficiency, and potential thermal damage to rivets and parts. Air-powered,

Table 12.11 Minimum edge distance in riveting

Rivet diameter		Minimum edge distance			
		Protruding head		Flush head	
mm	in.	mm	in.	mm	in.
2.4	3/32	5.6	0.22	6.4	0.25
3.2	1/8	7.1	0.28	7.9	0.31
4.0	5/32	8.6	0.34	8.6	0.34
4.8	3/16	10	0.41	10	0.41

Table 12.12 Rivet hole size

Rivet diameter		Hole size			
		Minimum		Maximum	
mm	in.	mm	in.	mm	in.
2.4	3/32	2.393	0.0942	2.431	0.0957
3.2	1/8	3.188	0.1255	3.239	0.1275
4.0	5/32	3.980	0.1567	4.044	0.1592
4.8	3/16	4.775	0.1880	4.857	0.1910

Table 12.13 Typical radii for deburring rivet holes

Sheet thickness		Radius adjacent to manufactured head		Radius between sheets and adjacent to upset head	
mm	in.	mm ± 0.13	in. ± 0.005	mm	in.
0.508–0.762	0.020–0.030	0.254	0.010	0.076	0.003
0.81–1.52	0.032–0.060	0.381	0.015	0.102	0.004
1.63–3.18	0.064–0.125	0.508	0.020	0.127	0.005

Table 12.14 Upset dimension of rivets

Rivet diameter		Flat upset			
		Diameter		Height	
mm	in.	mm	in.	mm	in.
2.4	3/32	2.921–3.429	0.115–0.135	0.762–1.524	0.030–0.060
3.2	1/8	3.810–4.318	0.150–0.170	1.02–1.78	0.040–0.070
4.0	5/32	4.826–5.558	0.190–0.220	1.27–2.29	0.050–0.090
4.8	3/16	5.842–6.858	0.230–0.270	1.52–2.79	0.060–0.110

squeeze-type tools are preferred when access permits. With this technique, the steady squeeze bulges the rivet into the hole, and tolerance for oversized holes and misalignment is increased. When the contour of the part or edge distance of the rivet does not permit use of squeeze- type tools, rivet sets and an impact hammer are used. Care is required to use backing bars of proper contours to avoid damaging the part. Sometimes, spin riveting is used. With this technique, a bead is formed by rotating a tool against the rivet. The technique does not expand the rivet into the hole.

Rivet sets must have proper die shape and length to produce specified upsets. Cup-shaped universal and conical die sets are used when non-flush upsets are required on exterior surfaces. Flat sets are used for nonflush upsets on interior surfaces, but conical sets are preferred to reduce cracking, especially for larger rivet sizes. The use of knurled flat sets is not recommended. Also, rivets are never upset into a countersink.

Before installing rivets, parts must be tightly clamped with the holes properly aligned. Clamping devices should be placed as close as possible to the rivet being upset. When titanium rivets are upset, they tend to separate the sheets rather than pull them together. Also, the rivet set must be located directly in line with the rivet and parallel to the rivet axis. Otherwise, offset or eccentric upsets can be formed. As is true for most manufacturing operations, it is desirable to install rivets in test pieces before starting on production parts. Test rivets are inspected to ensure proper rivet sets, air pressure, rivet length, rivet quality, and clamping force.

Quality Control. Rivet installations are inspected visually for upset dimensions, contour and quality, sheet separation, and flushness to determine acceptability. Major defects include spiral and burst cracks in the upset, oversized and undersized upset, offset and eccentric heads, concentric ring heads, cut and stepped heads, swelled rivets, open and incompletely driven rivets, and flush heads below the adjacent flat surface. All defects are due to mechanical causes, including improper hole preparation, rivet length, die set size, clamping pressure, and installation.

Threaded Fastening

Threaded fasteners also are used extensively to join titanium assemblies, including joining sheets and plates, attaching stiffeners, and joining structural members with flange-type joints. Conventional joint designs are used.

Installation of Threaded Fasteners. Selection of threaded fastener length is specified in engineering drawings. Evidence of adequate length is the extension of threaded fasteners through the nut. Flat-end fasteners should extend through the nut at least 0.8 mm (0.03 in.). With round- and chamfered-end fasteners, the full round and chamfer should extend through the nut. Restrictions are sometimes placed on loading threads in bearing. With A-286 alloy fasteners, for example, no threads are permitted in bearing unless the sheet next to the nut is over 2.4 mm (0.10 in.) thick, and the threads in bearing must not exceed 25% of the sheet thickness.

The operation sequence is similar to that used for riveted joints. Parts are positioned to align the holes, clamped to maintain alignment if necessary, and the fastener is inserted in the hole. If the fastener does not enter the hole, alignment and hole and fastener size should be checked and corrected, if necessary. Enlarging holes or driving a drift pin to correct alignment is not permitted. After all fasteners are in place, they are tightened by hand to pull the mating parts together before torquing.

Pretension in a threaded fastener is important in determining the strength of the joint. Torque wrenches are used to achieve uniform pretension. Typical torque values are shown in Table 12.15.

In bolt assemblies where the bolt is turned in tightening, the torque required to overcome shank friction is added to the required torque value. Fasteners and nuts must be removed and discarded if they are torqued in excess of the specified amounts.

Assembly of seal-type joints involving flanged fittings, housings, and similar mating surfaces requires rotation or sequence tightening. For circular flanges, for example, two diametrically opposite fasteners constitute a pair. The fasteners

Table 12.15 Typical torque values for fasteners

Bolt or screw size(a), in.	Self-locking nuts	
	N · m	lbf · in.
10–32	3–5	30–40
¼–28	10–11	85–100
⁵⁄₁₆–24	23–25	200–225
³⁄₈–24	40–44	350–390
⁷⁄₁₆–20	51–62	450–550
½–20	70–78	620–690
¾–16	260–283	2300–2500
1–12	565–621	5000–5500

(a) These are standard sizes. Metric bolts and screws are available in similar but not equivalent sizes. Therefore, conversion factors are not provided.

are first seated by tightening pairs to approximately 20% of the required torque values. After they are seated, successive pairs are tightened in ¼-turn increments to approximately 75 to 100% of required torque. Final torque is applied in clockwise order.

Quality Control. Threaded fastener installations are inspected for flushness of flush-head fasteners, extension of fastener through the nut, and gaps between the head and adjacent surface of the part. Heads of flush fasteners must never be below the adjacent flat surface but instead should extend 0.050 to 0.30 mm (0.002 to 0.012 in.) above the surface, depending on the application. Small gaps, typically 0.1 mm (0.004 in.) between fastener head and sheet surface, are permitted partially around fastener heads if the number of such fasteners is only a small percentage (typically 10%) and no two adjacent fasteners have this condition. Extension of fastener through the nut must be at least 0.8 mm (0.031 in.) for flat fasteners and full round and chamfer for round- and chamfer-end fasteners.

Repair

An advantage of mechanically fastened joints is that defective rivets and threaded fasteners are readily removed and replaced. Rivets are removed by filing a flat surface on the head, center-punching, and drilling through the head with a slightly smaller drill than the rivet shank. The head can then be snapped off by placing a punch in the drilled hole and applying torque. After the head is removed, a hole is drilled through the remainder of the shank, and the rivet is driven out.

For threaded fastener installations, the individual threaded fasteners are unscrewed ⅙ turn to remove maximum torquing stress. Then, following a rotation sequence, each fastener is unscrewed ¼ turn until all retention and bearing forces are removed. After this, fasteners may be removed in any sequence desired.

Properties

Mechanically fastened joints are loaded either predominantly in tension or in shear. Rivets are generally used only in shear, because tensile loads tend to loosen the rivets. In long joints, where uneven loads may cause tension, bolts often are placed at each end to minimize failure of rivets one by one. Threaded fasteners are used for both shear and tensile loads.

A number of fastener materials are used to join commercially pure titanium, titanium alloys, Monel, and A-286 alloy. The mechanical properties of these fasteners affect joint design and strength. These data and allowable design loads are included in most design manuals, so extensive load calculations are not strictly necessary in designing joints.

When rivet materials other than titanium are used, the effects of galvanic corrosion and thermal expansion must be considered in design. Monel fasteners present very little problem from galvanic corrosion, but steel and aluminum fasteners are rapidly attacked when used with titanium in a corrosive environment.

Summary

Titanium and its alloys are significant materials of construction in a wide range of applications. Basically, commercially pure titanium is the prime material for industrial, chemical, and related commercial uses, although some titanium alloys are being used where higher strengths are required. Titanium alloys such as Ti-6Al-4V are widely used in aircraft and aerospace applications.

Titanium and titanium alloys can be joined by welding, brazing, soldering, solid state diffusion bonding, and mechanical fastening. The different welding procedures available include GTAW, GMAW, PAW, EBW, flash welding, and FSW. The suitability of a joining process is related to the chemical and metallurgical characteristics of the materials used and the specified requirements of the finished product.

Familiarity with titanium in a wide spectrum of applications has grown rapidly since the late 1950s. It is important to emphasize that one basic tenet is common to all fabricated products: cleanliness and good housekeeping are a requisite for high-quality products.

ACKNOWLEDGMENT

The section "Friction Stir Welding" in this chapter was adapted from "Friction Stir Welding of Titanium Alloys—A Progress Update" by M.J. Russell of TWI Ltd., Cambridge, U.K., which was presented at the Tenth World Conference on Titanium on July 13 to 18, 2003, in Hamburg, Germany (Ref 12.18).

REFERENCES

12.1 F.H. Froes, D. Eylon, and H.B. Bomberger, Ed., *Titanium Technology: Present Status and Future Trends,* Titanium

Development Association, Dayton, OH, 1985

12.2 H. Kimura and O. Izumi, *Proceedings of the Fourth International Conference on Titanium: Titanium '80 Science and Technology,* Metallurgical Society of AIME, 1980, p 2337–2431

12.3 G. Lutjerling, U. Zwicker, and W. Bunk, *Proceedings of the Fifth International Conference on Titanium,* Metallurgical Society of AIME, Warrendale, PA, 1984

12.4 P. Lacombe, R. Tricot, and G. Beranger, Ed., *Proceedings of the Sixth International Conference on Titanium* (Nice, France), Metallurgical Society of AIME, Warrendale, PA, 1988

12.5 F.H. Froes and I.L. Caplan, Ed., *Proceedings of the Seventh International Conference on Titanium* (San Diego, CA), Metallurgical Society of AIME, Warrendale, PA, 1992

12.6 *Proceedings of the Eighth International Conference on Titanium* (San Birmingham, U.K.), Metallurgical Society of AIME, Warrendale, PA, 1995

12.7 *Proceedings of the Ninth International Conference on Titanium,* Metallurgical Society of AIME, Warrendale, PA, 2000

12.8 *Proceedings of the Tenth International Conference on Titanium,* Metallurgical Society of AIME, Warrendale, PA, 2004

12.9 *Proceedings of the Eleventh International Conference on Titanium,* Metallurgical Society of AIME, Warrendale, PA, 2008

12.10 *Proceedings of the Twelfth International Conference on Titanium,* Metallurgical Society of AIME, Warrendale, PA, 2012

12.11 G. Lutjering and J.C. Williams, *Titanium,* Springer, 2003

12.12 *Welding, Brazing, and Soldering,* Vol 6, *ASM Handbook,* ASM International, 1993

12.13 M.J. Donachi, *Titanium: A Technical Guide,* 2nd ed., ASM International, 2000

12.14 D.C. Martin, Effects of Carbon, Oxygen, and Nitrogen on Welds in Titanium, *Weld. J.,* Vol 32 (No. 3), 1953, p 139s–154s

12.15 D.M. Daley, Jr. and C.E. Hartbower, Notch Toughness of Weld Deposits in Commercial Titanium Alloys, *Weld. J.,* Vol 35 (No. 9), 1956, p 447s–456s

12.16 G.E. Faulkner, The Effects of Alloying Elements on Welds in Titanium: Part II, *Weld. J.,* Vol 34 (No. 6), 1955, p 295s–321s

12.17 G.E. Faulkner, G.B. Grable, and C.B. Voldrich, The Effects of Alloying Elements on Welds in Titanium, *Weld. J.,* Vol 32 (No. 10), 1953, p 481s–496s

12.18 M.J. Russell, "Friction Stir Welding of Titanium Alloys—A Progress Update," TWI Ltd., Cambridge, U.K., Paper presented at Tenth World Conference on Titanium, July 13–18, 2003 (Hamburg, Germany)

12.19 R.S. Misra and M.W. Mahoney, *Friction Stir Welding and Processing,* ASM International, 2007

12.20 N.R. Deb and T. Badeshi, Recent Advances in Friction Stir Welding—Process, Weldment Structure, and Properties, *Prog. Mater. Sci.,* Vol 53 (No. 6), 2008, p 980–1023

12.21 D. Lohwasser and Z. Chen, *Friction Stir Welding—From Basics to Applications,* Woodhead Publishing, 2010

12.22 J.C. Barrett and I.R. Lane, Jr., Effect of Arc Welds in Titanium, *Weld. J.,* Vol 33 (No. 3), 1954, p 121s–128s

12.23 R.K. Nolen, J.F. Rudy, H. Schwarztbart, and H.D. Kessler, Spot Welding of Ti-6Al-4V Alloy, *Weld. J.,* Vol 37 (No. 4), Research Supplement, 1958, p 129s–137s

CHAPTER 13

Machining and Chemical Shaping of Titanium*

THIS CHAPTER INTRODUCES the terminology of conventional and nonconventional machining processes and details general guidelines for successful handling of titanium and its alloys using these machining processes. The physical properties of titanium and their effects on machining are also discussed. The machining recommendations provided in this chapter are guidelines for instructional purposes. In practice, the processes used and their results may vary from those described here.

Machinability

Factors affecting the machinability of metals include tool material, tool geometry, cutting fluid, machine settings, and the properties of the workpiece material (Ref 13.1–13.16). The machinability of a metal affects machining costs from the standpoints of tool life, surface finish, power consumption, and chip form.

The relative ease of metal removal for equal tool lives can be expressed in terms of relative cutting speeds and machinability ratings. The machinability of unalloyed titanium is similar to that of annealed austenitic stainless steels, while that of titanium alloys is more comparable to that of ¼-hard and ½-hard stainless steels. This comparison is also valid from another viewpoint; both materials produce tough, stringy chips. However, austenitic stainless steel usually requires heavier feeds to penetrate below a heavily strain-hardened skin, whereas titanium, a material with low strain hardenability, does not necessarily require heavy feeds. Table 13.1

shows the approximate machinability ratings of titanium alloys, austenitic stainless steel, and other alloys. Other newer alloys of similar types behave in the same manner as those listed.

Machining Behavior. Generally, machining problems for titanium originate from three sources: high cutting temperatures, chemical reactivity, and a relatively low modulus of elasticity. However, a built-up edge does not form on tools, which accounts for the characteristically good finish on machined surfaces, but it also leaves the cutting edge naked to the abrading action of the chip peeling off the work (Ref 13.1, 13.2, 13.13–13.16).

In addition, titanium produces a thin chip, which flows at high velocity over the tool face on a small tool-chip contact area. This situation, plus the high strength of titanium, produces high contact pressures at the tool-chip interface. This combination of events and the poor heat conductivity of titanium result in unusually high tool-tip temperatures.

Thermal Problems. Cutting temperature achieved at the tool point depends partly on the rate at which heat is generated from forces at the tool point and partly on the rate at which it is removed by the chip, the workpiece, the cutting fluid, and by conduction through the tool.

The heat-transfer characteristics of the chip and work material are important and depend on thermal diffusivity, which is a function of density, specific heat, and thermal conductivity. Because titanium exhibits poor thermal diffusivity, as indicated in Table 13.2, tool-chip interface temperatures are higher than they would be when machining other metals at equal tool

---

*Adapted and revised from Walter E. Herman, originally from Carl T. Olofson, *Titanium and Its Alloys*, ASM International.

Table 13.1 Machinability ratings of titanium and its alloys compared with those of other materials

Alloy	Type	Condition	Hardness, HB	Rating(a)
6061	Aluminum alloy	Solution heat treated and artificially aged	...	200
B1112	Resulfurized steel	Hot rolled	...	100
1020	Carbon steel	Cold drawn	...	70
4340	Alloy steel	Annealed	...	45
Titanium	Commercially pure	Annealed	200	30
302	Stainless steel	Annealed	178	40
Ti-5Al-2.5Sn	Titanium alloy	Annealed	310	25
Ti-6Al-4V	Titanium alloy	Annealed	320	22
Ti-8Al-1Mo-1V	Titanium alloy	Annealed	320	22
Ti-6V-2Sn	Titanium alloy	Annealed	...	20
Ti-6Al-2Sn-4Zr-2Mo	Titanium alloy	Solution treated and aged	320	18
Ti-6Al-2Sn-4Zr-6Mo	Titanium alloy	Solution treated and aged	...	18
Ti-6Al-4V	Titanium alloy	Solution treated and aged	365	18
Ti-6Al-6V-2Sn	Titanium alloy	Solution treated and aged	365	16
Ti-10V-2Fe-3Al	Titanium alloy	Solution treated and aged	...	12
Ti-13V-11Cr-3Al	Titanium alloy	Annealed	...	16
Ti-13V-11Cr-3Al	Titanium alloy	Solution treated and aged	365	12
HS25	Cobalt base	Annealed	...	10
Inconel 718	Nickel base	Solution treated plus aged	385	6

(a) Based on AISI B1112 steel as 100

Table 13.2 Properties of commercially pure titanium, 7075 aluminum alloy, and AISI 1020 steel

Property	Commercially pure titanium	AISI 1020 steel	7075 age-hardened aluminum
Density, g/cm^3 ($lb/in.^3$)	4.5 (0.163)	8.03 (0.290)	2.8 (0.101)
Thermal conductivity, $W/m \cdot °C$ ($Btu/ft^2 \cdot h \cdot °F \cdot in.$)	15.1 (105)	56.2 (390)	122 (845)
Specific heat, $cal/g \cdot °C$ ($Btu/lb \cdot °F$)	0.13 (0.13)	0.117 (0.117)	0.021 (0.021)
Volume specific heat	0.021	0.031	0.021
Thermal diffusivity	4.950	11.500	39.800

stresses. The higher temperatures in the cutting zone lead to rapid tool failure unless efficient cooling is provided by suitable cutting fluids.

In addition, titanium may shrink on steel drills, reamers, and taps because of differences in the thermal expansion of these two materials.

Chemical Reactivity. The strong chemical reactivity of titanium with tool materials at high cutting temperatures and pressures induces galling, welding, and smearing, because an alloy is continuously formed between the titanium chip and the tool material. This alloy passes off with the chip, producing tool wear. Titanium reactivity also shows up when the tool dwells in the cut even momentarily, as in drilling. In this instance, the cutting temperature drops, causing the chip to fuse to the tool. When cutting is resumed, this chip breaks off, leaving a layer of titanium on the cutting edge. This layer then picks up additional titanium to form an "artificial" built-up edge, which spalls off, taking part of the tool edge with it. This undesirable situation can be prevented by not dwelling in the cut or by dressing the tool to

remove the titanium layer before cutting is resumed.

Alpha Case. The reactivity of titanium shows up in another way. Previous thermal processing in air, such as heat treating, forging, extruding, casting, and hot forming, can cause scaling and surface contamination up to 0.31 mm (0.012 in.) deep, but normally on the order of 0.13 mm (0.005 in.) deep (Fig. 13.1). This contamination is caused by the diffusion of oxygen into the surface of the highly reactive titanium, and it is hard and abrasive (Ref 13.1). Continuous cuts through this layer (such as turning) present few, if any, difficulties. However, interrupted cuts (such as milling) are generally troubled by tool spalling at the entrance and exit portions of the cut.

This condition is eliminated by a chemical etch or chemical mill in a bath consisting of HNO_3-HF and water (10%HNO_3-3%HF is commonly used) applied to a scale-free surface before machining.

A simple test for alpha case is to draw a sharp scriber over the surface to be machined. A high-pitched squeak may indicate the presence of alpha case. A low-pitched sound should indicate the uncontaminated surface.

Modulus of Elasticity. The stiffness of a part, determined by the shape and the modulus of elasticity of the workpiece material, is important when designing fixtures and selecting machining conditions. This is so because the thrust force that deflects the part being machined is considerably greater for titanium than for steel. Because the modulus of elasticity for titanium is only approximately half that of steel, a titanium part may deflect several times as much as a similar steel part during machining, creating tolerance and tool-robbing problems.

General Machining Requirements

Successful machining of titanium and its alloys requires high-quality machine tools and cutting tools, minimum vibration, rigid setups, and faithful observance of recommended machining practices (Ref 13.1, 13.13–13.16).

Machine Tools. Selection of machine tools is a primary factor; that is, just any machine will not do. Tools used for machining titanium must be in excellent condition and have attributes that ensure vibration-free operations. These attributes include dynamic balance of rotating elements; true-running spindles; snug bearings, slides, and screws;

sturdy frames; wide ranges of speed and feed; and ample power to maintain speed throughout cutting. Undersized and underpowered machines should be avoided. Certain locations of machines near or adjacent to heavy aisle traffic also can induce vibration and chatter during machining.

Rigidity of operation is also important. Generally, it is obtained through the use of adequate clamping and by minimizing deflection of work and tool during machining. In milling, this means strong, short tools, machining close to the table, rigid fixturing, frequent clamping of long parts, and the use of backup support for thin walls and delicate workpieces. Rigidity in turning is achieved by machining close to the spindle, gripping the work firmly in the collet, and providing steady or follow rests for slender parts. Drilling requires short drills, positive clamping of sheet, and backup plates on through-holes.

Cutting speed is important in all machining operations and is a very critical variable for titanium. As shown in Fig. 13.2, cutting speed has a pronounced effect on tool-chip temperature. Excessive speeds could easily cause overheating and short tool life. Consequently, speeds are limited to relatively low values, unless adequate cooling can be supplied at the cutting site. However, all machining variables should be carefully selected for optimum machining rates.

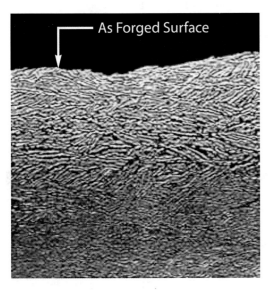

Fig. 13.1 Typical alpha case on titanium alloy surface. Original magnification: 100×. Courtesy of Sikorsky Aircraft

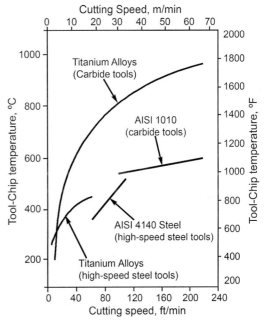

Fig. 13.2 Effect of cutting speed on cutting temperature for carbide and high-speed steel

All machining operations require a positive, uniform, mechanical feed. The cutting tool should never dwell or ride in the cut without removing metal. As an added precaution, all cutters should be retracted when they are returned across the work. The cutter must be up to speed and must maintain this speed as it takes the load.

Cutting Tools. High-quality cutting tools, properly ground, are needed for all machining operations (Ref 13.1, 13.13–13.16). The face of the tool should be smooth and the cutting edges sharp and free of feather burrs. Milling cutters, drills, and taps should be mounted to run true. Lathe tools should usually cut on dead center. In a multiple-tooth cutter such as a mill or a drill, all teeth should cut the same amount of material.

Carbide, cast alloy, and high-speed steel tools are used to machine titanium, the choice depending on the process parameters, including:

- Conditions of the machine tool
- Rigidity of the system
- Type of cut to be made
- Surface condition of the titanium
- Amount of metal to be removed and removal rate
- Skill of the operator
- Chip-recycling (remelt) considerations

Carbide cutting tools are usually selected for high-production items, extensive metal-removal operations, and scale removal. The so-called nonferrous or cast iron grades of carbides are used for titanium. These have been identified as grades C-1 to C-4, inclusive, by the Carbide Industry Standardization Committee. A partial list of companies producing these grades of carbide cutting tools is given in Table 13.3. For reference, Table 13.4 shows American National Standards Institute standard insert identification.

Although competitive brands of cutting tools classified under the same grade are similar, they are not necessarily identical. Variations in tool life should be expected from carbides produced by different manufacturers and even between lots made by the same producer. However, in all instances, carbide tools require heavy-duty, amply powered, vibration-free machine tools and rigid tool-work setups to prevent chipping. If these two basic conditions cannot be met, then high-speed steel tools give better results.

High-speed steel tools are used at low production rates, but tool life is low by usual plant standards. Both tungsten and molybdenum types of conventional high-speed steel are used. Cobalt was added to these steels to increase their red hardness above 540 °C (1000 °F), because conventional high-speed steels normally become too soft to cut effectively above this cutting temperature. Newer superhigh-speed steels (M41 to M44) produce good results.

Certain precautions must be observed when cobalt high-speed steels and superhigh-speed steels are used. They are sensitive to checking and cracking from abrupt temperature changes, such as those that may occur during grinding. Consequently, any kind of sharp, localized overheating or sudden heating or cooling of these steels must be avoided. They are more brittle than conventional high-speed steels; thus, they are not usually suitable for razor-edged quality tools. In addition, precautions must be taken to protect these high-speed steels from excessive shock and vibration in service.

Cast Co-Cr-W alloys are used for metal cutting at speeds intermediate between carbide and high-speed steel. The main constituents of these alloys are combined in various proportions to produce different grades.

Cutting Fluid Considerations. Cutting fluids are used in machining titanium to increase tool life, improve surface finish, minimize welding, and reduce residual stresses in the workpiece. A list of cutting fluids and the corresponding code numbers by which they are referenced appears in Table 13.5. Soluble oil-water emulsions, water-soluble waxes, and chemical coolants are usually used at the higher cutting speeds (23 to 30.5 m/min, or 75 to 100 ft/min, and higher) where cooling is important.

Low-viscosity sulfurized oils and chlorinated and sulfochlorinated oils are used at lower cutting speeds (below 23 m/min, or 75 ft/min) to reduce tool-chip friction and to minimize welding to the tool. Cutting oils have either mineral oil or mineral oil/lard oil bases. All cutting fluids have been identified for use in subsequent machining tables.

For many machining operations on conventional materials, it is possible to specify cutting fluids by using class designations, such as soluble oil, sulfurized oil, and sulfochlorinated oil. However, for some of the difficult-to-machine alloys, class designations sometimes are inadequate. Many fluids that improve machinability are complex, often proprietary, and sometimes contain unidentifiable active compounds.

Flood application through multiple nozzles to cover the cutting tool and immediate cutting area can be used for oil-based fluids. This form of

Table 13.3 Guide for carbide tool materials

Classification system								Partial list of carbides made by various manufacturers(a)							
CISC(b)		ISO	Adamas	Carboloy	Carmet	Excello	Greenleaf	Iscar	Kennametal	Newcomer	Sandvik	Seco	Teledyne	Valenits	VR/Wesson
C-1	Roughing	K-30 K-40	B	820 44A	CA-3 CA-12	E8	G10	IC-28	K1	N10	...	G27	H	VC-1 VC-111	2A68
C-2	General purpose	K-20 K-25	AM A	883	CA-443 CA-4	E6	G02	IC-2	K6 K8735	N22 N25	H20	SU41 H13 HX	HTA HA	VC-24 VC-2	2A5 VR-54
C-3	Finishing	K-10 K-15	PWX	905	CA-7	E5 XL028	G30	IC-20	...	...	H10	...	...	VD-3	2A7 VR-82
C-4	Precision finishing	K-01 K-05	AAA	999	CA-8	E3	G40	IC-4	K11	N40	...	...	HF	VC-4	...
C-5	Roughing	P-40 P-50	499 434	390	CA-740	CL85 10A	G50	IC-54	...	N52	535 56 R4 R1P	S25M S-2 5-6	TO4	VD-76 VD-5	...
C-6	General purpose	P-25 P-35	495	370	CA-720 CA-721	XL86 XL061 606 8A	G52 G53	IC-50M IC-50	K420 K21	N60 N65	S4	S4 SIG SIF	TXH NTA	VC-6 VC-55 VC-125	...
C-7	Finishing	P-10 P-35	548	350	CA-711	6AX 509 6A	G74, G70	IC-70	K45 K4H K2884	N72 N70	SIP S2	...	T25	VD-7	VR-71
C-8	Precision finishing	P-01 P-05	490	320	R03	XL88	G80	IC-80T	K7H	N95 N93 N80	F02 HIP	...	SA8	VC-83 VC-8	VR-65

(a) For the same Carbide Industry Standardization Committee (CISC) grade, there are no truly equivalent carbides of different brands. Where two carbide grades from the same manufacturer are shown for the same CISC grade, the first is sometimes recommended. (b) The following chip-removal applications apply for the Carbide Industry Standardization Committee (CISC) grade indicated. It will be noted that some grades specify the type of metal removal for which they are best suited: C-1, roughing cuts; C-2, general purpose; C-3, light finishing; and C-4, precision boring for cast iron and nonferrous materials; C-5, roughing cuts; C-6, general purpose; C-7, finishing cuts; and C-8, precision boring for steel. Note: This chart can function only as a guide. The so-called "best grade" may differ for each specific job even if the material being machined is the same. The final selection can be made only by trial and error. Instructions regarding the specific use and application of any competitive grade should be obtained from the manufacturer.

Table 13.4   American National Standards Institute (ANSI) standard insert identification

1. Insert shape

A: parallelogram 85°
B: parallelogram 82°
C: diamond 80°
D: diamond 55°
E: diamond 75°
H: hexagon
K: parallelogram 55°
L: rectangle
M: diamond 86°
O: octagon
P: pentagon
R: round
S: square
T: triangle
V: diamond 35°
W: trigon

2. Clearance (secondary facet angle may vary by ±1°)

N: 0°; G: 30°
A: 3°; G: 0 to 11°
B: 5°; J: 0 to 14°
C: 7°; K: 0 to 17°
P: 11°; L: 0 to 20°
O: 15°; M: 11 to 14°
E: 20°; R: 11 to 17°
F: 25°; S: 11 to 20°

5. Inscribed circle size

For rhombic or regular polygon inserts, number of 3.175 mm (0.125 in., or eighths of in.) of inscribed circle.(c) On intermediate fractions, number is carried to one decimal place, e.g., 7/16 in. is 3.5; 7/32 in. is 1.8.

6. Thickness

Number of 1.59 mm (0.0625 in., or sixteenths of in.) of thickness, carried to one decimal place on intermediate fractions, e.g., 1.5 for 3/32 in.

7. Cutting point configuration

0: sharp corner 0.0762 mm (0.003 in.) or less
1: 0.397 (1/64 in.) radius
2: 0.794 mm (1/32 in.) radius
3: 1.19 mm (3/64 in.) radius
4: 1.59 mm (1/16 in.) radius
6: 2.38 mm (3/32 in.) radius
8: 3.175 mm (1/8 in.) radius
A: square insert with 45° chamfer
D: square insert with 30° chamfer
E: square insert with 15° chamfer
F: square insert with 3° chamfer
K: square insert with 30° doubled chamfer
L: square insert with 15° doubled chamfer
M: square insert with 3° doubled chamfer
N: truncated triangle insert
P: flatted corner triangle

3. Tolerance class (Tolerances shown are plus and minus from nominal.)

Cutting point (B), mm (in.)	Insert inscribed circle, mm (in.)	Thickness (T), mm (in.)
A(a): 0.0051 (0.0002)	0.0254 (0.001)	0.0254 (0.001)
B: 0.0051 (0.0002)	0.0254 (0.001)	0.127 (0.005)
C: 0.0127 (0.0005)	0.0254 (0.001)	0.0254 (0.001)
D: 0.0127 (0.0005)	0.0254 (0.001)	0.127 (0.005)
E: 0.0254 (0.001)	0.0254 (0.001)	0.0254 (0.001)
G: 0.0254 (0.001)	0.0254 (0.001)	0.127 (0.005)
M(b): 0.0508–0.254 (0.002–0.010) min	0.0508–0.1016	0.127 (0.005)
U(b): 0.127–0.305 (0.005–0.012)	(0.002–0.004)	0.127 (0.005)
	0.127–0.254 (0.005–0.010)	

R: blank with grind stock on all surfaces
S: blank with grind stock on top and bottom surfaces only

4. Insert style

A: with hole
B: with hole and one countersink (70 to 90°)
C: with hole and two countersinks (70 to 90°)
D(c): smaller than 6.35 mm (0.25 in.) inscribed circle with hole
E(c): smaller than 6.35 mm (0.25 in.) inscribed circle without hole
F: with hole-chip grooves on two rake faces
G: with hole and chip grooves on two rake faces
H: with hole, one countersink (70 to 90°), and chip grooves on two rake faces
M: with hole and chip groove on one rake face
P(c): 10° positive land with hole and chip grooves on two rake faces
R(c): with hole and extrawide chip grooves on two rake faces
Z(c): high positive land with hole-chip grooves on two rake faces
X(c): negative land with hole-chip grooves on two rake faces

8. Edge Preparation

A: 0.0127–0.0762 mm (0.0005–0.003 in.) rounded edge (honed)
B: 0.0762–0.127 mm (0.003–0.005 in.) rounded edge (honed)
C: 0.127–0.1718 mm (0.005–0.007 in.) rounded edge (honed)
E: rounded edge (honed)
F: sharp edge
J: rake surface finish
T: chamfered edge
S: chamfered and rounded (honed)

(a) Tolerances normally apply to inserts with facets (secondary cutting edges). (b) Tolerance depends on size and shape of insert and should be as shown in the standards for corresponding shapes and sizes (see ANSI 894.25). (c) TRW identification only

application is not recommended for those high cutting speeds that would throw off the fluids.

The mist system provides cooling or lubrication to inaccessible areas, or both, visibility of the cutting zone, and better tool life (or lower costs) in some instances. Water-based fluids are preferred over oil-based fluids because of possible health hazards related to oil mists.

Fluids, whether applied as flood or mist, must be directed to give maximum cooling or lubrication, or both, to the tool-work interface. Care must be taken not to direct the fluid directly onto the chip, thereby blocking the flow of fluid to the zone of maximum heat.

Chlorinated oils are used in some instances on titanium and its alloys. However, they are generally avoided if nonchlorinated fluids satisfy the machining requirements. Residual chloride residues from these fluids can lead to stress-corrosion cracking of titanium alloy parts during subsequent annealing or in service. (Chloride residues can cause stress corrosion at temperatures above 205 °C, or 400 °F, in titanium alloys but not necessarily in commercially pure titanium.) If chlorinated fluids are used on titanium, the residues must be removed promptly with a nonchlorinated degreaser such as methyl ethyl ketone or by acid cleaning. Fundamentally, it is always good prac-

tice to completely remove all cutting fluid and lubricant residues from workpieces, especially before any heating operations. Furthermore, due consideration should be given to the difficulties of washing complex assemblies.

Many shops have well-defined machining and cleaning procedures relating to the use of chlorine-type cutting fluids for titanium alloys.

Scrap Prevention

Waste should be avoided because titanium is a relatively expensive metal. Table 13.6 lists common sources of revert scrap, their importance in different machining operations, and ways of preventing excessive scrap.

Any scrap-prevention program requires adherence to the basic recommendations previously stated for machining titanium. In addition to those practices, parts should be handled and transported with care. Nicks and scratches must be avoided, both on parts in process and on finished parts. Therefore, suitable containers and paper separators should be used for parts in process to prevent damage in handling and storage.

The machining and grinding of titanium normally requires closer supervision than do operations on other metals, not only to prevent scrapping of parts but also to detect defective parts early in the processing schedule.

Table 13.5   Types of cutting fluids used in machining

Code No.(a)	Type of cutting fluid
1	Soluble oil-water emulsion (1:10)
2	Water-soluble waxes
3	Chemical coolants (synthetics, barium hydroxide)
4	Highly chlorinated oil
5	Sulfurized oil
6	Chlorinated oil
7	Sulfochlorinated oil
8	Rust-inhibitor types (such as nitrite amine)
9	Heavy-duty soluble oil (such as chlorinated, barium sulfonated, extreme-pressure additive types)

(a) Code numbers are used in tables of machining data.

Table 13.6   Sources of scrap in various machining operations and corrective actions needed

	Burned surfaces	Rough finish	Chatter marks	Dimensional discrepancies	Residual stresses	Distortion	Broken tools	Handling scratches
Incidence of scrap for machining operation shown								
Operation								
Turning	...	X	X	X	X	X	...	...
Milling	...	X	X	X	...	...	...	X
Drilling	...	...	...	X	...	...	X	X
Tapping	...	X	...	X	...	...	X	X
Grinding	X	X	...	...	X	X	...	X
Belt grinding	X	...	...	...	X	...	...	X
Cut-off	X	...	...	...	...	...	...	...
Sawing	...	...	...	X	...	...	...	X
Corrective action needed to avoid defects indicated above								
Action								
Strong, sharp tools	...	X	...	X	X	X	X	...
Dressed wheels	X	...	...	...	...	X	...	...
Positive chip removal	...	X	X	...	...	X	...	...
More rigid setups	...	X	X	X	...	...	...	...
Modern machine tools	...	--	X	X	...	...	...	...
Speed/feed/cutting fluid	...	X	...	X	X	X	X	...
Careful handling	...	...	...	...	X	...	...	...
Stress relief	...	...	...	...	X	X	...	...

Hazards and Safety Considerations

Titanium by itself is not particularly hazardous. However, a potential explosion hazard exists if very finely divided titanium is present in air in the proper proportions. A fire hazard also exists in that fine chips and turnings can ignite under certain conditions. Titanium turnings also can ignite when the metal is cut at high speeds without adequate use of coolants. In the same manner, dry grinding can cause trouble due to the intense spark stream. Finally, chip accumulations from poor housekeeping habits and improper storage produce likely sites for titanium fires.

From the health viewpoint, no adverse physiological reaction from titanium has been reported. However, barium compounds such as barium hydroxide used in cutting fluids can be hazardous to personnel unless suitable precautions are taken to protect machine operations. Barium compounds can cause acute and chronic toxicity if inhaled at high concentrations. Consequently, positive measures must be taken to exhaust all fumes and mist from the machining area. Recommended maximum atmospheric concentration per 8 h day is 0.15 kg/m^3 (0.15 oz/ft^3) of air.

Safety considerations are concerned with both preventive and emergency measures. Preventive measures generally mean that good housekeeping practices must be maintained at all times. Specifically, they involve:

- Regular chip collection and immediate storage in covered containers (once a day)
- Removal of containers to an outside location
- Keeping machine ducts and working areas clean of titanium dust, chips, and oil-soaked sludge
- Cleaning area and equipment of all oil and grease, and removal of rags and waste subject to spontaneous combustion

If a fire starts, it should be smothered with dry powders suitable for combustible-metal fires. These include graphite powder, talc, powdered limestone, absolutely dry sand, dry salt (NaCl), and dry compound extinguisher powder for magnesium fires. Carbon tetrachloride and carbon dioxide extinguishers should not be used. Water and foam should never be applied directly to a titanium fire. Water accelerates the burning rate and can cause hydrogen explosions. However, water can be applied to the surrounding area up to the edge of the fire to cool the unignited material below the ignition point.

Milling Titanium

Milling is a process for removing metal using a rotary cutter. The process lends itself to both unit and mass production. Milling operations can produce titanium parts of various shapes and sizes to aircraft standards of surface finish and dimensional accuracy. A surface finish of 1.6 μm (63 μin.) or better is readily attainable, and a 0.43 μm (17 μin.) surface is possible in finishing cuts.

This section deals with milling machines, milling cutters, machine settings, and cutting fluids required for titanium. The following sections provide general milling information and describe face-milling operations and peripheral-milling techniques.

Milling Behavior of Titanium. Milling is an intermittent cutting operation that can be difficult to control because of the large number of variables involved. Welding and edge chipping are the basic problems when milling titanium. The amount of titanium welded on the cutter edges is proportional to the chip thickness as each tooth leaves the cut. The weld metal and part of the underlying cutting edge then chips off as each tooth reenters the cut, thus starting a wear land. Edge welding increases progressively as cutting continues, and the wear land grows until the tool fails suddenly. This progressive chipping and wear phenomenon also produces gradual deterioration of the surface finish and loss of tolerance. Both factors can become serious unless the worn or damaged tool is replaced promptly.

Other problems include heat, deflection, and abrasion. High cutting temperatures soften chips, which tend to clog and load milling cutters. Deflection of thin parts and slender milling cutters promote rubbing and heat. Abrasive surface oxide of titanium, if present, can notch the cutter at the depth-of-cut line.

Another problem in milling titanium, particularly in extrusions, is distortion originating from the release of stresses originally imposed by the basic mill-processing operation. Distortion can occur when unequal amounts of metal containing residual stresses are removed from opposite surfaces, or by the machining operation itself.

The welding-chipping behavior can be minimized by providing thin exit chips characteristic of down (climb) milling. Slower speeds and light feeds also reduce chipping and permit lower cutting temperatures. Water-based coolants also reduce cutting temperatures and therefore minimize galling. Chemical removal of any oxide skin before machining alleviates the abrasion problem. Stress relieving in fixtures before final machining overcomes the distortion problem.

Milling Machines. Heavy-duty horizontal or vertical knee-and-column milling machines are required for face milling, end milling, and pocket milling titanium. Heavy-duty, fixed-bed milling machines also are used for face milling and end milling large titanium workpieces. Numerically controlled, vertical-profile milling machines or tracer-controlled milling machines are recommended for profile and pocket milling operations. Milling machines used on titanium must be free of backlash (climb milling) and must have snug table gibs.

Generally, 10 to 15 hp machines are usually sufficient for milling titanium. Examples are a No. 2 heavy-duty and No. 3 standard knee-and-column milling machine. However, large machines that often need to accommodate large parts have as much as 25 to 50 hp available.

Milling Cutters. The choice of milling cutter depends on the type of machining to be done. Face mills, rotary face mills, plain milling cutters, and slab mills are used for milling plane surfaces. End mills are used for light operations, such as profiling and slotting. Form cutters and gang-milling cutters produce shaped cuts. All cutters need adequate body sections and tooth sections to withstand the cutting loads. Helical cutters are preferred for smoother cutting action. The use of the smallest-diameter cutter with the largest number of teeth without sacrificing necessary chip space minimizes chatter and deflection.

All cutters should be ground and mounted to run absolutely true, with all teeth cutting the same amount of material. The total runout should be no more than 0.025 mm (0.001 in.) total indicator heading.

Cutter Design. Tool angles of a milling cutter should be chosen to promote unhampered chip flow and immediate ejection of the chip. The controlling angles in this regard include the axial rake, radial rake, and corner angles. These angles should be chosen to provide a positive angle of inclination to lift the chip from the machined surface.

Rake angles are not especially critical. Tool life progressively improves as the radial rake is reduced from +6 to 0° and down to −10°. Positive rake angles are generally used on high-speed steel cutters, but occasionally it is necessary to reduce the rake to zero to overcome a tendency for a cutter to "dig in," or to chip prematurely. Carbide cutters often use negative rake angles.

The use of a corner angle, plus a small nose radius, provides a longer cutting edge. This dis-tributes cutting forces over a greater area, thus causing less tool pressure. It also dissipates the heat of cutting. A 30 to 45° chamfer also can produce a longer cutting edge and a wider, thinner chip. However, a corner angle is usually more effective than a chamfer.

Relief angles are probably the most critical of all tool angles when milling titanium. Relief angles of approximately 12° give longer tool life than do the standard relief angles of 6 and 7°. If chipping occurs, the 12° relief angles should be reduced toward the standard values. In general, relief angles less than 10° lead to excessive smearing along the flank, while angles greater than 15° weaken the tool and encourage digging in, as well as chipping of the cutter edge.

Tool Materials. The choice of the proper tool material for a milling cutter is not a simple matter. Conventional high-speed steel cutters, for example, T1, M1, and M2, are popular mainly because they are readily available. Applications of these tools on titanium include:

- Low-volume production of small parts
- Slots and form cuts
- Milling under conditions of insufficient rigidity
- End mills, form mills, narrow side-cutting slitting saws, and large-radius cutters

The T4 and T5 cobalt grades are used for high-production milling of small parts. Tool life of high-speed steel cutters is low by the usual standards and quite sensitive to speed. Furthermore, some differences in the performance of high-speed steel cutters exist between cutters of the same type and geometry but obtained from different suppliers. This difference can be attributed to composition or heat treatment of the tool, or both. Therefore, high-speed cutters should be purchased to the specifications covering the grade and appropriate heat treatment of the steel.

Carbide milling cutters are especially useful for high-production and extensive metal-removal operations, particularly in face-milling and slab-milling applications. Carbide milling is done extensively in the aircraft industry, and it is recommended whenever possible because of higher production rates.

Feeds. Feed rates are measured in millimeters per tooth (mm/t), or inches per tooth (in./t). Feed rates for milling titanium generally lie in the range of 0.51 to 0.203 mm/t (0.02 to 0.008 in./t) to avoid overloading the cutters, fixtures, and milling machine. Light feeds at slow speeds also reduce premature chipping. Delicate types of cutters and flimsy and nonrigid workpieces require smaller feeds. A positive, uniform feed must be maintained. Positive gear feeds without backlash are sometimes preferred over hydraulic feed mechanisms. Cutters should not dwell or stop in the cut.

Down-milling (climb-milling) techniques are usually used for carbide and cast alloy cutters to encourage formation of a thin chip. Conventional milling is usually more suitable for high-speed steel tools and for removing scale.

Depth of Cut. Selection of cut depth depends on the rigidity of the part, the tolerances required, and the type of milling operation undertaken. For skin milling, light cuts (0.25 to 0.50 mm, or 0.010 to 0.020 in.) cause less warpage than deeper cuts (1.02 to 1.52 mm, or 0.040 to 0.060 in.). When cleaning up extrusions, a 1.27 mm (0.050 in.) depth is usually allowed. However, depths of cut up to 3.81 mm (0.15 in.) can be used in other situations if sufficient power is available. When forging scale is present, the nose of each tooth must be kept below the scale to avoid rapid tool wear.

Cutting speed is a critical factor in milling titanium. Excessive speeds overheat the cutter edges and cause rapid tool failure. Consequently, speeds listed in tables in this chapter should not be exceeded. When starting on a new job, it is advisable to try cutting speeds in the lower portion of the recommended range for each alloy.

Sufficient flywheel-assisted spindle power should be present to maintain constant cutting speed as the cutter takes the cutting load.

Cutting Fluids. A wide variety of cutting fluids is used to reduce cutting temperatures and to inhibit galling. Sulfurized mineral oils are used extensively and are usually flood-applied. Water-based cutting fluids are also widely used and are either flood- or mist-applied. Tool life improves significantly (up to 300%) when a 5% barium hydroxide-water solution is used as a spray mist. However, the toxic fumes must be exhausted from the cutting area to protect the operator.

Some shops prefer the spray-mist technique for all water-based coolants because the air blows the chips free of the cutter. The mist should be applied ahead of a peripheral milling cutter (climb cutting) and at both the entrance and exit of a face-milling-type cutter. Pressurizing the fluid in an aspirator system enables improved penetration to the tool-chip area, as well as better cooling, chip removal, and tool life by a factor of 2. With a flood coolant, chips tend to accumulate behind the cutter and occasionally are carried through the cutter. Flood application is preferred when cutting oils are used.

General Milling Techniques. Machining titanium requires reasonably close supervision. The supervisor should check all new milling setups before cutting operations begin. Thereafter, parts should be spot checked for nicks and scratches to prevent defective workpieces from being processed too far.

Fixtures should hold and support the workpiece as close to the machine table as possible. The solid part of the fixture (rather than the clamps) should absorb the cutting forces. Fixtures should be rugged enough to minimize distortion and vibration.

The selection of speeds, feeds, and depths of cut in any setup should take into account the rigidity of the setup, the optimum metal-removal rate/tool-life values, and the surface finish and tolerances needed on the finished part.

Milling cutters should be sharp, and they should be examined for early indications of dulling. If a dull-red chip starts to form, the tool should be replaced. Some shops have at least two cutters for a given operation. Minimal downtime usually occurs when the entire cutter is replaced by a new one, rather than waiting for a dull cutter to be resharpened. The normal criterion of wear for replacing a cutter is a wear land of 0.25 mm (0.010 in.) for a carbide cutter and 0.38 mm (0.015 in.) for a high-speed steel cutter.

Surface contamination breaks down cutters prematurely. If this is a problem, the surface can be removed by acid pickling.

Face-milling operations use the combined action of cutting edges located on the periphery and face of the cutter. The milled surface is generally at right angles to the cutter axis and is flat, except when milling to a shoulder. Face mills and end mills are used in this operation. Face mills are suitable for facing workpieces more than 127 mm (5 in.) wide. End mills are used for facing narrow surfaces and for profiling and slotting. Table 13.7 shows the type of mills used in various operations.

Face or Skin Milling. Conventional face mills of normal design are suitable for machining relatively wide, flat surfaces usually wider than 127 mm (5 in.). Special face mills are also used, including the rotating insert and conical types.

Face-mill diameter is important. Diameters can exceed 152 mm (6 in.) but should not be appreciably greater than the width of the cut. If a smaller-diameter cutter can perform the operation and still overhang the cut by 10%, then a larger cutter should not be used. It is not good practice to bury the cutter in the work.

A good surface finish and freedom from distortion are desirable. Surface finish in milling becomes considerably better with decreasing feed and slightly better with increasing speed. Light cuts (0.25 to 0.50 mm, or 0.010 to 0.020 in.) on sheet metal cause less warpage than deeper cuts (1 to 1.5 mm, or 0.040 to 0.060 in.).

Table 13.8 contains data on feeds, speeds, depths of cut, tool design, and other important variables when face milling titanium alloys. Figure 13.3 explains the tool nomenclature codes.

End milling, a form of face milling, uses the cutting action of teeth on the circumferential surface and one end of a solid-type cutter. End-milling cutters are used for facing, profiling, slotting, and end-milling operations and include the standard end mills and two-lip end or slotting mills.

End mills should have wide flutes to permit unrestricted chip flow. Helical-type end mills perform better than the straight-tooth designs.

In profile and pocket milling, cutters are fed gradually into the work to keep them from grabbing and breaking. Chip crowding, chip disposal, and tool deflections can be problems during this machining operation. Pocket milling is done best with the cutter axis in a horizontal plane to avoid recutting of chips and to give better chip-removal conditions.

The proper combinations of direction of helix and direction of cut should be considered to avoid deflection of the cutter in the direction of an increasing depth of cut. The choice depends on the type of milling being done. For example, when machining slots where the end of the cutter is in contact with the work, the direction of the helix and the direction of the cut should be the same. This means a right-hand helix for a right-hand cut, and a left-hand helix for a left-hand cut. When profile milling where the periphery of the cutter is doing the cutting, the opposite is true: left-hand helix for a right-hand cut and vice versa.

Cutter diameter in pocket milling depends on the radius required on the pockets. Due to an inherent lack of rigidity, end mills should be as short as practicable, and their shank diameters should equal their cutting diameters.

High-speed steel cutters are typically used for end-milling, slotting, and profile-milling operations. The shank of end mills should be softer than the cutter flutes to avoid breakage between shank and flutes.

Tables 13.9 to 13.12 provide cutting data on peripheral milling, slotting, profile milling, and pocketing using high-speed steel and solid carbide cutters. Other newer alloys of similar types behave in the same manner as those listed. Figures 13.3 and 13.4 illustrate the tool nomenclature and codes used.

Peripheral-milling operations use the cutting action of teeth located on the periphery of the cutter body. Arbor-mounted cutters used for such operations include plain mills, helical mills, slab mills, form-relieved cutters, formed profile cutters, side mills, and slotting cutters. However, face mills are usually more efficient in removing metal from flat surfaces, and they produce flat surfaces more accurately than plain-milling cutters.

Faster feed rates are also possible with face mills because they are more rugged. In addition, the complicated supports usually required for arbor-mounted cutters are unnecessary when face mills are used.

Spar and slab milling can be used to bring extrusions into aircraft tolerance and to provide a good surface finish. The operation is performed on a heavy-duty fixed-bed mill. However, large bed mills may not have adequate feed ranges.

Spars and similar sections, being relatively long and thin, require special handling. As-received extrusions could require straightening before machining because extrusion straightness tolerances exceed mill fixture and part tolerances.

Rigid setups are necessary, but spar extrusions should not be forced into a fixture. Spars can be

Table 13.7 Types of mills

Mill type	Diameter		Application
	mm	in.	
Face mills(a)	152 and greater	6 and greater	Roughing and finishing
Shell end mills	25–152	1–6	Facing wide surfaces
End mills	13–51	0.5–2	Facing narrow surfaces, end milling, profiling, slotting
Slotting mills	13–51	0.5–2	Slots

(a) Indexable face-milling cutters using throwaway carbide inserts are available in positive or negative rakes with lead angles up to 45°.

Table 13.8 Milling titanium alloys with helical face mills

Depth of cut: 0.64–6.4 mm (0.025–0.25 in.)

Titanium alloy (condition)(a)	Hardness, HB	C-2 carbide tools(b)					High-speed steel tools(b)				
		Tool geometry(c)	Cutting fluid(d)	Cutting speed, m/min (ft/min)(e) Brazed	Throwaway	Feed, mm/t (in./t)	Tool material	Tool geometry(c)	Cutting fluid(d)	Cutting speed, m/min (ft/min)	Feed, mm/t (in./t)
Commercially pure, grade 1 (annealed)	110–170	C	1, 3, 5	122–162 (400–530)	134–178 (440–585)	0.10–0.20 (0.004–0.008)	M1, M2, M7, T1	A	1, 3, 5	38–53 (125–174)	0.08–0.20 (0.003–0.008)
Commercially pure, grades 2, 3 (annealed)	140–220	C	1, 3, 5	91–122 (300–400)	101–134 (330–440)	0.10–0.15 (0.004–0.006)	M1, M2, M7, T1	A	1, 3, 5	31–43 (100–140)	0.08–0.15 (0.003–0.006)
Commercially pure, grade 4 (annealed)	200–275	C	1, 3, 5	61–91 (200–300)	67–95 (220–310)	0.10–0.15 (0.004–0.006)	M1, M2, M7, T1	A	1, 3, 5	23–34 (75–110)	0.08–0.15 (0.003–0.006)
Ti-8Al-1Mo-1V (annealed)	320–370	B, D	1, 3	34–46 (110–150)	37–49 (120–160)	0.10–0.15 (0.004–0.006)	M3, M33, T5, T15	A, G	1, 3	8–12 (25–40)	0.08–0.15 (0.003–0.006)
Ti-5Al-2.5Sn (annealed) Ti-6Al-2Nb-1Ta-1Mo	300–340	A, B, E, F	1, 2, 3	52–64 (170–210)	58–75 (190–245)	0.05–0.15 (0.002–0.006)	M2, M3, M10, M33, T5, T15	A, C, E, F	1, 2, 5	9–18 (30–60)	0.05–0.20 (0.002–0.008)
Ti-6Al-2Sn-4Zr-2Mo (annealed)	310–350	A, B, C, E, F	1, 3, 4	31–52 (100–170)	58–75 (190–245)	0.10–0.15 (0.004–0.006)	M3, M33	A, C, E	1, 4, 5	12–15 (40–50)	0.05–0.15 (0.002–0.006)
Ti-6Al-4V (annealed)	310–350	A, B, C, E, F	1, 3, 4	31–52 (100–170)	58–75 (190–245)	0.10–0.15 (0.004–0.006)	M3, M33	A, C, E	1, 4, 5	12–15 (40–50)	0.05–0.15 (0.002–0.006)
Ti-6Al-4V (STA) Ti-6Al-2Sn-4Zr-2Mo Ti-6Al-2Sn-4Zr-6Mo	350–400	B, C, E	1, 4	24–44 (80–145)	37–49 (120–160)	0.10–0.15 (0.004–0.006)	T15	A, C, E	6, 7	6–14 (20–45)	0.05–0.15 (0.002–0.006)
Ti-6Al-6V-2Sn (annealed) Ti-5Al-2Sn-2Zr-4Mo-4Cr	320–370	B, C, E	1, 3, 4	34–46 (110–150)	37–53 (120–175)	0.10–0.15 (0.004–0.006)	M2, M3, M10, M33	A, C, E	1, 4	8–18 (25–60)	0.05–0.15 (0.002–0.006)
Ti-5Al-2Sn-2Zr-4Mo-4Cr (STA)	375–420	C, E	1, 4	24–32 (80–105)	27–35 (90–115)	0.10–0.15 (0.004–0.006)	M2, M3, M10, M33	A, C, E	1, 4	8–18 (25–60)	0.05–0.15 (0.002–0.006)
Ti-13V-11Cr-3Al (annealed) Ti-10V-2Fe-3Al Ti-3Al-8V-6Cr-4Zr-4Mo	310–350	B, C, E	1, 3, 4	31–38 (100–125)	34–44 (110–145)	0.08–0.15 (0.003–0.006)	T15, M33	C, E	4	5–11 (15–35)	0.10–0.18 (0.004–0.007)
Ti-13V-11Cr-3Al (STA) Ti-10V-2Fe-3Al Ti-3Al-8V-6Cr-4Zr-4Mo	375–440	B, C, F, E	1, 4	18–24 (60–80)	20–27 (65–90)	0.08–0.15 (0.003–0.006)	T15, M33	C, E	4	5–8 (15–25)	0.10–0.18 (0.004–0.007)

(a) STA, solution treated and aged (b) Carbide Industry Standardization Committee designations for carbides; American National Standards Institute designations for high-speed steels. (c) See Fig. 13.3 and 13.4 for tool angles involved. (d) See Table 13.5 for specific types. Remove all chlorinated oil residues with methyl ethyl ketone. (e) The lower speed in each range is used for the heavier feeds and depths of cut. Also, some shops may prefer these lower speeds for general operations, and somewhat higher speeds for numerical-control machining

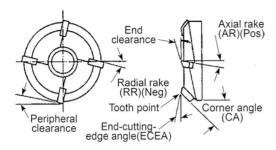

Fig. 13.3 Tool geometry and nomenclature for facing mills

straightened mechanically if the distortion is not too severe. Otherwise, they should be hot straightened in fixtures. Straightening of spars must be rigidly controlled to prevent significant loss in properties due to stress redistribution.

When using arbor-mounted cutters, the arbor should be of the largest possible diameter and should be supported on each side of the cutter with over-arm supports. Furthermore, the arbor should have just the proper length required for the number of cutters mounted and the arbor support used. Arbor overhang beyond the outer support should be avoided, because it is conducive to chatter and vibration. Finally, cutters should be mounted as close to the column face of the milling machine as the work will permit.

Slab-milling cutters should be mounted so cutting forces are absorbed by the spindle of the machine. This is done by using cutters with a left-hand helix for a right-hand cut, and vice versa. When two milling cutters are used end-to-end of the arbor, cutters having helixes of opposite direction to the cut involved should be used. This setup neutralizes cutting forces that tend to push the cutters away from the arbor.

Carbide cutters are preferred for spar milling because of the higher production rates attainable, except under conditions where the inherent brittleness of the carbide precludes its use. Helical cutters are recommended because they provide wider, thinner chips than do the corresponding straight-tooth types. In slab milling, six cutting edges per 25 mm (1 in.) diameter enables heavier feeds and longer tool lives than the conventional three cutting edges per 25 mm diameter.

Unit feeds range from 0.10 to 0.30 mm/t (0.004 to 0.012 in./t), depending on the finish desired. However, too light a feed can produce red-hot chips, which can cause a fire.

When cleaning up extrusions, only approximately 1.27 mm (0.05 in.) depth of cut can be taken to reduce material costs. Therefore, for long

spans, a feed-speed combination should be used that gives the most economical metal removal based on machines available and cutting-tool inventory.

Table 13.13 gives machining data for slab milling commercially pure titanium and several titanium alloys. Other newer alloys of similar types behave in the same manner as those listed.

Turning, Facing, and Boring

Turning, facing, and boring operations are essentially similar, and no unusual difficulties are experienced with any of them. They give less trouble than milling, especially when cutting is continuous rather than intermittent. The same speeds used for turning can be used for boring and facing cuts. However, in most instances, the depths of cuts and feeds must be reduced when boring because of an inherent lack of rigidity of the operation.

The problems to be minimized in turning-type operations include high tool-tip temperatures and the galling and abrasive properties of titanium toward tool materials. They can be avoided by following the precautions listed earlier in the section "General Machining Requirements" and the suggestions that follow. The conditions given here for turning should be suitable for boring using single-point tools.

Turning Machines. Machine tools for turning should be heavy duty and powerful enough to cut without the spindle slowing down. Thus, a modern lathe in good condition provides production rates 5 to 10 times those possible using older machines. Vibration and lack of rigidity are common problems with older equipment. Furthermore, the overall range of spindle speeds available on many existing lathes is not broad enough to cover some of the lower speeds required for titanium.

In this regard, lathes should have either a variable-speed drive for the spindle or the spindle-gear train should have a geometric progression of 1.2 or less to provide speed steps of 20% or less for more precise speed selections. The trend in new lathes is toward variable-speed drives. Rigidity, backlash elimination, dimensional accuracy, rapid indexing of tools, and flexibility are additional features being emphasized.

The application of numerical control in turning is rapidly spreading. On lathes equipped with tracer control and numerical control, variable speed and feed features can be added so speed and feed can be optimized during contouring operations.

Table 13.9 Peripheral milling titanium alloys with helical end mills

Depth of cut: 1.3–6.4 mm (0.05–0.25 in.)

Titanium alloy (condition)(a)	Hardness, HB	C-2 carbide tools (tool geometry: E)(b)				M2 high-speed steel tools (tool geometry: D)(b)			
		Cutting speed(c), m/min (ft/min)	Feed(d), mm/t (in./t); mill diameter, mm (in.)			Cutting speed(c), m/min (ft/min)	Feed(d), mm/t (in./t); mill diameter, mm (in.)		
			9.5 (%)	19 (%)	25–51 (1–2)		9.5 (%)	19 (%)	25–51 (1–2)
Commercially pure, grade 1 (annealed)	110–170	78–114 (250–375)	0.051 (0.002)	0.127–0.152 (0.005–0.006)	0.176–0.254 (0.007–0.010)	31–53 (100–175)	0.051–0.127 (0.002–0.005)	0.127–0.178 (0.005–0.007)	0.178–0.229 (0.007–0.009)
Commercially pure, grades 2, 3 (annealed)	140–200	69–114 (225–375)	0.051 (0.002)	0.127–0.152 (0.005–0.006)	0.178–0.254 (0.007–0.010)	27–48 (90–155)	0.051–0.127 (0.002–0.005)	0.127–0.178 (0.005–0.007)	0.178–0.229 (0.007–0.009)
Commercially pure, grade 4 (annealed)	200–275	46–66 (150–215)	0.025–0.076 (0.001–0.003)	0.102–0.152 (0.004–0.006)	0.152–0.178 (0.006–0.007)	18–26 (60–85)	0.0254–0.076 (0.001–0.003)	0.102–0.127 (0.004–0.005)	0.152–0.178 (0.006–0.007)
Ti-8Al-1Mo-1V (annealed)	320–370	38–53 (125–175)	0.025–0.076 (0.001–0.003)	0.102–0.152 (0.004–0.006)	0.152–0.178 (0.006–0.007)	15–21 (50–70)	0.0254–0.076 (0.001–0.003)	0.102–0.127 (0.004–0.005)	0.152–0.178 (0.006–0.007)
Ti-5Al-2.5Sn (annealed)	300–340	41–61 (135–200)	0.025–0.076 (0.001–0.003)	0.102–0.152 (0.004–0.006)	0.152–0.178 (0.006–0.007)	17–24 (55–80)	0.0254–0.076 (0.001–0.003)	0.102–0.127 (0.004–0.005)	0.152–0.178 (0.006–0.007)
Ti-6Al-2Nb-1Ta-1Mo (annealed)	300–340	41–61 (135–200)	0.025–0.076 (0.001–0.003)	0.102–0.152 (0.004–0.006)	0.152–0.178 (0.006–0.007)	17–24 (55–80)	0.0254–0.076 (0.001–0.003)	0.127–0.152 (0.005–0.006)	0.152–0.178 (0.006–0.007)
Ti-6Al-4V (annealed)	310–350	38–58 (125–190)	0.025–0.076 (0.001–0.003)	0.102–0.152 (0.004–0.006)	0.152–0.178 (0.006–0.007)	15–23 (50–75)	0.0254–0.076 (0.001–0.003)	0.127–0.152 (0.005–0.006)	0.152–0.178 (0.006–0.007)
Ti-6Al-4V (STA) Ti-6Al-2Sn-4Zr-2Mo Ti-6Al-2Sn-4Zr-6Mo	350–400	31–50 (100–165)	0.0203–0.076 (0.0008–0.003)	0.076–0.127 (0.003–0.005)	0.127–0.152 (0.005–0.006)	12–20 (40–65)	0.0203–0.076 (0.0008–0.003)	0.076–0.127 (0.003–0.005)	0.127–0.152 (0.005–0.006)
Ti-6Al-6V-2Sn (annealed)	320–370	38–53 (125–175)	0.0254–0.076 (0.001–0.003)	0.102–0.152 (0.004–0.006)	0.152–0.178 (0.006–0.007)	15–21 (50–70)	0.0254–0.076 (0.001–0.003)	0.102–0.127 (0.004–0.005)	0.127–0.152 (0.005–0.006)
Ti-6Al-6V-2Sn (STA) Ti-5Al-2Sn-2Zr-4Cr-4Mo	375–420	31–50 (100–165)	0.0203–0.0508 (0.0008–0.002)	0.076–0.127 (0.003–0.005)	0.127–0.152 (0.005–0.006)	12–20 (40–65)	0.0203–0.051 (0.0008–0.002)	0.038–0.102 (0.0015–0.004)	0.076–0.102 (0.003–0.004)
Ti-13V-11Cr-3Al (annealed) Ti-10V-2Fe-3Al Ti-3Al-8V-6Cr-4Zr-4Mo	310–350	15–21 (50–70)	0.025–0.076 (0.001–0.003)	0.102–0.127 (0.004–0.005)	0.127–0.152 (0.005–0.006)	15–21 (50–70)	0.0254–0.076 (0.001–0.003)	0.102–0.127 (0.004–0.005)	0.127–0.152 (0.005–0.006)
Ti-13V-11Cr-3Al (STA) Ti-10V-2Fe-3Al Ti-3Al-8V-6Cr-4Zr-4Mo	375–440	9–17 (30–55)	0.0152–0.051 (0.0006–0.002)	0.0381–0.076 (0.0015–0.003)	0.076–0.127 (0.003–0.005)	9–17 (30–55)	0.0152–0.051 (0.0006–0.002)	0.0381–0.076 (0.0015–0.003)	0.076–0.127 (0.003–0.005)

(a) STA, solution treated and aged. (b) Carbide Industry Standardization Committee designations for carbides, American National Standards Institute designations for high-speed steels. See Fig. 13.3 and 13.4 for tool angles. (c) The higher speeds correlate with higher feeds and smaller depths of cut. (d) Lower feeds are needed for 3.2 mm (0.125 in.) end mills.

Table 13.10 Slotting titanium alloys with helical end mills

Depth of cut: 0.38–6.4 mm (0.015–0.25 in.)

Titanium alloy (condition)(a)	Hardness, HB	C-2 carbide tools (tool geometry: E)(b)				M2 high-speed steel tools (tool geometry: D)(b)			
		Cutting speed(c), m/min (ft/min)	Feed(d), mm/t (in./t); mill diameter, mm (in.)			Cutting speed(c), m/min (ft/min)	Feed(d), mm/t (in./t); mill diameter, mm (in.)		
			9.5 (⅜)	19 (¾)	25–51 (1–2)		9.5 (⅜)	19 (¾)	25–51 (1–2)
Commercially pure, grade 1 (annealed)	110–170	61–114 (200–375)	0.038–0.051 (0.0015–0.002)	0.076–0.102 (0.003–0.004)	0.127–0.152 (0.005–0.006)	24–46 (80–150)	0.038–0.076 (0.0015–0.003)	0.0102–0.152 (0.0004–0.006)	0.015–0.178 (0.0006–0.007)
Commercially pure, grades 2, 3 (annealed)	140–200	53–114 (175–375)	0.038–0.076 (0.0015–0.003)	0.076–0.102 (0.003–0.004)	0.127–0.152 (0.005–0.006)	21–46 (70–150)	0.038–0.076 (0.0015–0.003)	0.0102–0.152 (0.0004–0.006)	0.015–0.178 (0.0006–0.007)
Commercially pure, grade 4 (annealed)	200–275	31–58 (100–190)	0.0178–0.076 (0.0007–0.003)	0.076–0.152 (0.003–0.006)	0.127–0.203 (0.005–0.008)	12–23 (40–75)	0.0178–0.076 (0.0007–0.003)	0.076–0.127 (0.003–0.005)	0.102–0.152 (0.004–0.006)
Ti-8Al-1Mo-1V (annealed)	320–370	23–62 (75–200)	0.0178–0.076 (0.0007–0.003)	0.076–0.152 (0.003–0.006)	0.127–0.203 (0.005–0.008)	9–15 (30–50)	0.0178–0.076 (0.0007–0.003)	0.076–0.127 (0.003–0.005)	0.102–0.152 (0.004–0.006)
Ti-5Al-2.5Sn (annealed)	300–340	27–53 (90–175)	0.0178–0.076 (0.0007–0.003)	0.076–0.152 (0.003–0.006)	0.127–0.203 (0.005–0.008)	10.5–21 (35–70)	0.0178–0.076 (0.0007–0.003)	0.076–0.127 (0.003–0.005)	0.102–0.152 (0.004–0.006)
Ti-6Al-4V (annealed)	310–350	23–50 (75–165)	0.0178–0.076 (0.0007–0.003)	0.076–0.152 (0.003–0.006)	0.127–0.203 (0.005–0.008)	9–20 (30–65)	0.0178–0.076 (0.0007–0.003)	0.076–0.127 (0.003–0.005)	0.102–0.152 (0.004–0.006)
Ti-6Al-4V (STA) Ti-6Al-2Sn-4Zr-2Mo	350–400	18–35 (60–115)	0.0152–0.076 (0.0006–0.003)	0.076–0.127 (0.003–0.005)	0.102–0.178 (0.004–0.007)	7.5–13.5 (25–45)	0.0152–0.076 (0.0006–0.003)	0.051–0.102 (0.002–0.004)	0.102–0.152 (0.004–0.006)
Ti-7Al-4Mo (annealed) Ti-6Al-6V-2Sn	320–370	23–43 (75–140)	0.0178–0.076 (0.0007–0.003)	0.076–0.152 (0.003–0.006)	0.127–0.203 (0.005–0.008)	9–17 (30–55)	0.0178–0.076 (0.0007–0.003)	0.076–0.127 (0.003–0.005)	0.102–0.152 (0.004–0.006)
Ti-7Al-4Mo (STA) Ti-6Al-6V-2Sn	375–420	18–35 (60–115)	0.0152–0.076 (0.0006–0.003)	0.076–0.127 (0.003–0.005)	0.102–0.178 (0.004–0.007)	7.5–13.5 (25–45)	0.0127–0.102 (0.0005–0.004)	0.0025–0.102 (0.0001–0.004)	0.051–0.152 (0.002–0.006)
Ti-13V-11Cr-3Al (annealed)	310–350	23–43 (75–140)	0.0178–0.076 (0.0007–0.003)	0.076–0.152 (0.003–0.006)	0.127–0.203 (0.005–0.008)	9–17 (30–55)	0.0178–0.076 (0.0007–0.003)	0.076–0.127 (0.003–0.005)	0.102–0.152 (0.004–0.006)
Ti-13V-11Cr-3Al (STA)	375–440	18–31 (60–100)	0.0102–0.051 (0.0004–0.002)	0.051–0.102 (0.002–0.004)	0.051–0.152 (0.002–0.006)	6–13.5 (20–45)	0.0102–0.051 (0.0004–0.002)	0.0025–0.076 (0.0001–0.003)	0.051–0.152 (0.002–0.006)

(a) STA, solution treated and aged. (b) Carbide Industry Standardization Committee designations for carbides; American National Standards Institute designations for high-speed steels. See Fig. 13.3 and 13.4 for tool angles. (c) The higher speeds correlate with higher feeds and smaller depths of cut. (d) Lower feeds are needed for 3.2 mm (0.125 in.) end mills. Feeds for the high-speed steel end mills are conservative and may be increased in some instances.

Table 13.11 Profiling titanium alloys with helical end mills

Depth of cut: 0.3810–1.3 mm (0.015–0.05 in.)

Titanium alloy (condition)(a)	Hardness, HB	C-2 carbide tools (tool geometry: A)(b)				M2 and M7 high-speed steel tools (tool geometry: A)(b)			
		Cutting speed(c), m/min (ft/min)	Feed(d), mm/t (in./t); mill diameter, mm (in.)			Cutting speed(c), m/min (ft/min)	Feed(d), mm/t (in./t); mill diameter, mm (in.)		
			6.3 (¼)	12.7 (½)	19 (¾)		6.3 (¼)	12.7 (½)	19 (¾)
Commercially pure, grade 1 (annealed)	110–170	99–114 (325–375)	0.025–0.051 (0.001–0.002)	0.076 (0.003)	0.102 (0.004)	38–46 (125–150)	0.038–0.051 (0.0015–0.002)	0.102 (0.004)	0.152 (0.006)
Commercially pure, grades 2, 3 (annealed)	140–200	91–114 (300–375)	0.025–0.051 (0.001–0.002)	0.076 (0.003)	0.102 (0.004)	37–46 (120–150)	0.038–0.051 (0.0015–0.002)	0.102 (0.004)	0.152 (0.006)
Commercially pure, grade 4 (annealed)	200–275	49–58 (160–190)	0.025–0.051 (0.001–0.002)	0.051–0.076 (0.002–0.003)	0.127–0.152 (0.005–0.006)	18–23 (60–75)	0.025–0.038 (0.001–0.0015)	0.051–0.076 (0.002–0.003)	0.102–0.127 (0.004–0.005)
Ti-8Al-1Mo-1V (annealed)	320–370	14–17 (45–55)	0.025–0.038 (0.001–0.0015)	0.051–0.076 (0.002–0.003)	0.127–0.152 (0.005–0.006)	14–17 (45–55)	0.025–0.038 (0.001–0.0015)	0.051–0.076 (0.002–0.003)	0.102–0.127 (0.004–0.005)
Ti-5Al-2.5Sn (annealed)	300–340	43–53 (140–175)	0.025–0.038 (0.001–0.0015)	0.051–0.076 (0.002–0.003)	0.127–0.152 (0.005–0.006)	17–21 (55–70)	0.025–0.038 (0.001–0.0015)	0.051–0.076 (0.002–0.003)	0.102–0.127 (0.004–0.005)
Ti-6Al-4V (annealed)	310–350	15–20 (50–65)	0.025–0.038 (0.001–0.0015)	0.051–0.076 (0.002–0.003)	0.127–0.152 (0.005–0.006)	15–20 (50–65)	0.025–0.038 (0.001–0.0015)	0.051–0.076 (0.002–0.003)	0.102–0.127 (0.004–0.005)
Ti-6Al-2Nb-1Ta-1Mo									
Ti-6Al-4V (STA)	350–400	11–14 (35–45)	0.025–0.038 (0.001–0.0015)	0.051–0.076 (0.002–0.003)	0.102–0.152 (0.004–0.006)	14–17 (45–55)	0.025–0.038 (0.001–0.0015)	0.051–0.076 (0.002–0.003)	0.076–0.102 (0.003–0.004)
Ti-6Al-2Nb-1Ta-1Mo									
Ti-6Al-2Sn-4Zr-2Mo	320–370	15–20 (50–65)	0.025–0.038 (0.001–0.0015)	0.051–0.076 (0.002–0.003)	0.127–0.152 (0.005–0.006)	14–17 (45–55)	0.025–0.038 (0.001–0.0015)	0.051–0.076 (0.002–0.003)	0.102–0.127 (0.004–0.005)
Ti-6Al-4Zr-6Mo									
Ti-7Al-4Mo (annealed)									
Ti-6Al-6V-2Sn (STA)	375–420	11–14 (35–45)	0.025–0.038 (0.001–0.0015)	0.051–0.076 (0.002–0.003)	0.102–0.127 (0.004–0.005)	11–14 (35–45)	0.025–0.038 (0.001–0.0015)	0.051–0.076 (0.002–0.003)	0.076–0.102 (0.003–0.004)
Ti-5Al-2Sn-2Zr-4Cr-4Mo									
Ti-6Al-6V-2Sn (annealed)	310–350	11–14 (35–45)	0.025–0.038 (0.001–0.0015)	0.051–0.076 (0.002–0.003)	0.127–0.152 (0.005–0.006)	14–17 (45–55)	0.025–0.038 (0.001–0.0015)	0.051–0.076 (0.002–0.003)	0.102–0.127 (0.004–0.005)
Ti-5Al-2Sn-2Zr-4Cr-4Mo									
Ti-13V-11Cr-3Al (annealed)	375–440	11–14 (35–45)	0.0178–0.025 (0.0007–0.001)	0.038–0.051 (0.0015–0.002)	0.076–0.102 (0.003–0.004)	11–14 (35–45)	0.0178–0.025 (0.0007–0.001)	0.038–0.051 (0.0015–0.002)	0.051–0.076 (0.002–0.003)
Ti-3Al-8V-6Cr-4Zr-4Mo									
Ti-10V-2Fe-3Al									

(a) STA, solution treated and aged (b) Carbide Industry Standardization Committee designations for carbides; American National Standards Institute designations for high-speed steels. See Fig. 13.3 and 13.4 for tool angles. (c) The higher speeds correlate with higher feeds and smaller depths of cut. (d) Lower feeds are needed for 3.2 mm (0.125 in.) end mills. Feeds for the high-speed steel end mills are conservative and may be increased in some instances.

Table 13.12 Pocket milling titanium alloys with helical end mills

Axial depth of cut: 25 mm (1 in.); radial depth of cut: 0.76–3.185 mm (0.03–0.125 in.)

Titanium alloy (condition)(a)	Hardness, HB	Cutting speed(c), m/min (ft/min)	M2 or M10 high-speed steel tools (tool geometry: H)(b)		
			Feed(d), mm/t (in./t); cutter diameter, mm (in.)		
			13 (½)	16 (⅝)	19 (¾)
Ti-5Al-2.5Sn (annealed) Ti-6Al-2Nb-1Ta-1Mo	310	12–15 (40–50)	0.051–0.076 (0.002–0.003)	...	...
		12–18 (40–60)	...	0.076–0.102 (0.003–0.004)	...
		14–20 (45–65)	...	...	0.102–0.127 (0.004–0.005)
Ti-6Al-4V (STA)(e) Ti-6Al-2Sn-4Zr-2Mo Ti-6Al-2Sn-4Zr-6Mo	365	7.6–9.1 (25–30)	0.051–0.102 (0.002–0.004)	...	...
		7.6–12 (25–40)	...	0.076–0.102 (0.003–0.004)	...
		11–14 (35–45)	...	...	0.102–0.127 (0.004–0.005)
Ti-7Al-4Mo (STA)(e) Ti-6V-2Sn Ti-5Al-2Sn-2Zr-4Cr-4Mo	365	9.1–12 (30–40)	0.051–0.102 (0.002–0.004)	...	...
		11–15 (35–50)	...	0.102 (0.004)	...
		12–17 (40–55)	...	...	0.102–0.127 (0.004–0.005)
Ti-13V-11Cr-3Al (STA)(e) Ti-3Al-3V-6Cr-4Zr-4Mo Ti-10V-2Fe-3Al	365	3.0–6.0 (10–20)	0.051–0.102 (0.002–0.004)	...	...
		6.0–9.1 (20–30)	...	0.051–0.102 (0.002–0.004)	...
		9.1–12 (30–40)	...	...	0.102–0.127 (0.004–0.005)

(a) STA, solution treated and aged. (b) Carbide Industry Standardization Committee designations for carbides; American National Standards Institute designations for high-speed steels. See Fig. 13.3 and 13.4 for tool angles. (c) The higher speeds correlate with higher feeds and smaller depths of cut. (d) Lower feeds are needed for 3.2 mm (0.125 in.) end mills. Feeds between 0.127 and 0.203 mm (0.005 and 0.008 in.) are used for 25.4 mm (1 in.) mills. For the high-speed steel, end mills are conservative and may be increased in some instances. (e) Speed and feed recommendations for these heat treated alloys can be increased 30% when machining these alloys in the annealed condition.

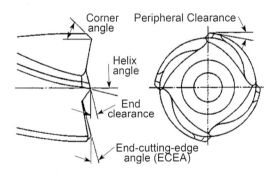

Fig. 13.4 Tool geometry and nomenclature for end mills

Lathes with 10 hp ratings should be ample for most turning operations. Workpieces ranging between 25 and 250 mm (1 and 10 in.) in diameter can be turned on a standard or heavy-duty 1610 engine lathe. The number 1610 is the lathe-industry designation for a 406 mm (16 in.) swing-over bed and a 254 mm (10 in.) swing-over cross slide. These lathes have a range of spindle speeds that closely approximate the requirements previously described.

Cutting Tools. Standard lathe tools are used for turning titanium. They are available in a variety of shapes, sizes, tool angles, and tool materials. High-speed steel, carbide, and cast alloy tools can be used.

Tool angles are important for controlling chip flow to minimize smearing and chipping and for maximum heat dissipation. Rake angles and side-cutting-edge angle determine the angle of inclination and chip flow. Relief angles, together with rake angles, control chipping and smearing.

Positive, zero, and negative rake angles are used depending on the heat treated condition of the alloy, the tool material, and the machining operation. The side rake is the important angle, with positive rakes being best for finish turning and high-speed steel tools, and negative rakes for carbide tools at heavier feeds.

The side-cutting-edge angle influences the temperature near the cutting zone. Larger angles reduce cutting pressure and present longer tool edges. The reduced pressure minimizes heat formation, and the longer cutting edges dissipate more heat. Therefore, higher values of the side-

Table 13.13 Slab milling titanium alloys with high-speed steel peripheral mills

Titanium alloy (condition)(a)	Hardness, HB	Rough machining depth of cut: 3.8 mm (0.150 in.)		Finish machining depth of cut: 0.64 mm (0.025 in.)		High-speed steel type
		Cutting speed, m/min (ft/min)	Feed, mm/t (in./t)	Cutting speed, m/min (ft/min)	Feed, mm/t (in./t)	
Commercially pure, grade 1 (annealed)	110–170	34 (110)	0.203 (0.008)	47 (155)	0.102 (0.004)	M2, M7
Commercially pure, grades 2, 3 (annealed)	140–200	27 (90)	0.152 (0.006)	38 (125)	0.102 (0.004)	M2, M7
Commercially pure, grade 4 (annealed)	200–275	20 (65)	0.152 (0.006)	31 (100)	0.102 (0.004)	M2, M7
Ti-8Al-1Mo-1V (annealed)	320–370	7.6 (25)	0.152 (0.006)	11 (35)	0.102 (0.004)	T15, M33
Ti-5Al-2.5Sn (annealed)	300–340	14 (45)	0.152 (0.006)	17 (55)	0.102 (0.004)	T15, M33
Ti-7Al-2Nb-1Ta-1Mo (annealed)	300–340	14 (45)	0.152 (0.006)	17 (55)	0.102 (0.004)	T15, M33
Ti-6Al-4V (annealed)	310–350	11 (35)	0.152 (0.006)	14 (45)	0.102 (0.004)	T15, M33
Ti-6Al-4V (STA)	350–400	9.1 (30)	0.152 (0.006)	12 (40)	0.102 (0.004)	T15, M33
Ti-6Al-2Sn-4Zr-2Mo	350–400	9.1 (30)	0.152 (0.006)	12 (40)	0.102 (0.004)	T15, M33
Ti-6Al-2Sn-4Zr-6Mo	350–400	9.1 (30)	0.152 (0.006)	12 (40)	0.102 (0.004)	T15, M33
Ti-7Al-4Mo (annealed)	320–370	7.6 (25)	0.152 (0.006)	11 (35)	0.102 (0.004)	T15, M33
Ti-7Al-4Mo (STA)	375–420	6.1 (20)	0.152 (0.006)	9.1 (30)	0.102 (0.004)	T15, M33
Ti-6Al-6V-2Sn (annealed)	320–370	7.6 (25)	0.152 (0.006)	11 (35)	0.102 (0.004)	T15, M33
Ti-6Al-6V-2Sn (STA)	375–420	6.1 (20)	0.152 (0.006)	9.1 (30)	0.102 (0.004)	T15, M33
Ti-5Al-2Sn-2Zr-4Cr-4Mo (STA)	375–420	6.1 (20)	0.152 (0.006)	9.1 (30)	0.102 (0.004)	T15, M33
Ti-13V-11Cr-3Al (annealed)	310–350	6.1 (20)	0.152 (0.006)	9.1 (30)	0.102 (0.004)	T15, M33
Ti-13V-11Cr-3Al (STA)	375–440	6.1 (20)	0.152 (0.006)	7.6 (25)	0.102 (0.004)	T15, M33
Ti-3Al-8V-6Cr-4Zr-4Mo (annealed)	310–350	6.1 (20)	0.152 (0.006)	9.1 (30)	0.102 (0.004)	T15, M33
Ti-3Al-8V-6Cr-4Zr-4Mo (STA)	375–440	6.1 (20)	0.152 (0.006)	7.6 (25)	0.102 (0.004)	T15, M33
Ti-10V-2Fe-3Al (annealed)	310–350	6.1 (20)	0.152 (0.006)	9.1 (30)	0.102 (0.004)	T15, M33
Ti-10V-2Fe-3Al (STA)	375–440	6.1 (20)	0.152 (0.006)	7.6 (25)	0.102 (0.004)	T15, M33

(a) STA, solution treated and aged

cutting-edge angle generally permit greater feeds and speeds, unless chipping occurs as the cutting load is applied or removed from the work.

Relief angles between 5 and 12° are used in cutting titanium. Angles less than 5° increase smearing of titanium on the flank of the tool. Relief angles of approximately 10° are better, although some edge chipping can occur. Chipbreaking devices should be used for good chip control.

Figure 13.5 explains the nomenclature for single-point cutting tools. Table 13.14 shows tool geometries for turning titanium. Cutting tools should be carefully ground and finished before use. Typically, this means that tool surfaces over which chips pass should have a good finish, with the direction of finishing corresponding to the chip-flow direction. A rough surface can cause a properly designed tool to deteriorate rapidly. The life of a carbide tool can be extended if the sharp cutting edge is slightly relieved by honing. Carbide inserts are available with this feature.

Tool Materials. High-speed steel, cast alloy, and cemented carbide cutting tools are suitable for lathe turning titanium. Ceramic tools have not proved successful for machining titanium. High-speed steel tools are best suited for form cutting heavy plunge cuts, interrupted cutting, and minimally rigid setup conditions. Carbide tools are

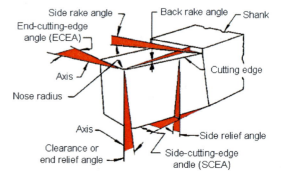

Fig. 13.5 Tool geometry and nomenclature for single-point cutting tools

typically used for continuous cutting situations, high-production items, extensive metal-removal operations, and scale removal. Nonferrous cast alloy tools are suitable for severe plunge cuts, machining to dead center, and producing narrow grooves.

High-speed steel and cast alloy tools should be ground on a tool grinder to the required tool geometry. The same is true for carbide tools. However, off-the-shelf brazed and throwaway carbide tools that fit the rake, lead, and relief-angle requirements are convenient to use.

Brazed tools come in standard sizes and styles, as shown in Table 13.15. However, mechanically clamped inserts perform as well as brazed tools, so they cost less per cutting edge.

Throwaway carbide inserts are designed for mechanical clamping in either positive- or negative-rake tool holders. The general coding system for mechanical tool holders is explained in Table 13.16. The tool geometries for solid-base tool holders suitable for titanium are shown in Table 13.17.

Use of throwaway tooling is reported to substantially reduce costs. Factors contributing to these savings include:

- Reduced tool-grinding costs
- Reduced tool-changing costs
- Reduced scrap
- Increased use of harder carbides for longer tool life and increased metal-removal rates
- Savings through tool standardization
- Maximum carbide use per tool dollar

Disposable-type carbide (C-2) inserts in heavy-duty, negative-rake tool holders provide maximum metal removal at minimum cost in turning titanium.

Carbide cutting tools are most sensitive to chipping. Therefore, they require "overpowered," vibration-free lathes and more rigid tool-work setups. If these conditions cannot be met, high-speed steels must be used.

Feeds. Three important rules for feeding when turning titanium are:

- Always use constant, positive feeds
- Avoid dwelling in the cut
- Never stop or slow down in the cut

Metal-removal rate and surface-finish requirements determine the amount of feed to be taken:

heavy feeds for greater metal-removal rates, light feeds for better surface finishes.

Depth of Cut. The selection of cut depth depends on the amount of metal to be removed and the metal-removal rate desired. In removing scale, the tool must get under the scale and cut at least 0.050 mm (0.0020 in.) deeper than the tool radius. For rough cuts, the nose of the tool must get below any hard skin and oxide remaining from previous processing operations. In finish turning, light cuts must be used for the best finish and the closest tolerances.

Cutting Speed. Tool life is more sensitive to cutting speed than to any other machining variable when turning titanium. However, high speeds are not necessary to produce good finishes on titanium. Therefore, relatively low cutting speeds compared with those associated with conventional metals are used to obtain reasonable tool life for titanium.

Cutting fluids are almost always used during turning and boring operations to cool the tool and to aid in chip disposal. Dry cutting is done in only a few instances, usually where chip contamination is objectionable. Dry cutting is not recommended for semifinishing and finishing operations on titanium.

Water-based coolants are the most satisfactory cutting fluids for turning titanium. Specifically, a 5% solution of sodium nitrite in water gives the best results, while a 1:20 soluble oil in water

Table 13.15 Tool geometries of brazed carbide tools

	Tool geometry				
Tool style	A	B	C	D	E
Back rake	0	0	0	0	0
Side rake	+7	+7	0	0	0
End relief	7	7	7	7	7
Side relief	7	7	7	7	7
End-cutting-edge angle	8	15	...	50	60
Side-cutting-edge angle	0	15	...	40	30

Table 13.14 Tool geometry data

	Tool angles (in degrees) and nose radius for indicated tool geometry code								
	A	B	C	D	E	F	G	H	I
Back rake	−5	+5 to −5	+5 to −5	0	0	0 to +5	0 to +10	+6 to +10	+5 to +15
Side rake	−5	+6 to 0 0 to −6	+5	5 or 6	15	+5 to +15	0 to +10	+0 to +15	+10 to +20
End relief	5	5–10	8–10	5	5	5–7	6–8	6–10	5–8
Side relief	5	5–10	8–10	5	5	5–7	6–8	6–10	5–8
End-cutting edges	15–45	5–20	0–45	15	15–45	15–20	0–30	0–45	0–30
Side-cutting edge (lead)	15–45	5–20	0–45	15	15–45	15–20	0–30	0–45	
Nose radius, mm (in.)	0.79–1.2 (1/32–3/64)	0.76–1.0 (0.03–0.04)	0.76–1.0 (0.03–0.04)	1.2 (3/64)	1.2 (3/64)	0.51–0.76 (0.02–0.03)	...	0.76–1.0 (0.03–0.04)	0.25–1.5 (0.01–0.06)

Table 13.16   General coding system for mechanical tool holders

Company identification	Shape of insert	Lead angle	Rake angle	Cut type(c)
(a)	T = triangle	B	(b)	R or L
(a)	R = round	A	(b)	R or L
(a)	P = parallelogram	A	(b)	R or L
(a)	S = square	B	(b)	R or L
(a)	L = rectangle	B	(b)	R or L

Lead angle or tool style

A = 0° turning
B = 15° lead
D = 30° lead
E = 45° lead
F = facing
G = 0° offset turning

(a) Some producers place a letter here for company identification. (b) Some companies use the letter "T" for negative rake, "P" for positive rake, and sometimes add "S" to indicate solid-base holders. For example, a TATR designation denotes a tool holder for a triangular insert mounted in such a way to give a 0° lead angle and an S° negative rake. The "R" denotes a right-hand cut. (c) R = right hand; L = left hand; N = neutral

Table 13.17   Tool geometries of solid-base tool holders for throwaway inserts

	Negative-rake tools				Positive-rake tools			
	Back rake: 5° Side rake: 5° End relief: 5° Side relief: 5°				Back rake: 0° Side rake: 0° End relief: 5° Side relief: 5°			
Tool-holder geometry(a)	Insert type(a)	ECEA(b), degrees	SCEA(c), degrees		Tool holder geometry(a)	Insert type(a)	ECEA(b), degrees	SCEA(c), degrees
A	T	5	0		A	T	3	0
A	T	3	0		A	T	5	0
A	R	8	0		...	...	...	...
B	T	23	15		B	T	23	15
B	T	18	15		B	S	15	15
B	S	15	15		B	T	20	15
B	T	20	15		...	...	...	...
D	T	35	30		D	T	35	30
E	S	45	45		...	...	...	...
F	T	0	0		F	T	0	0
F	S	15	0		F	S	15	0
G	T	3	0		G	T	3	0

(a) See Tables 13.15 and 13.16 for explanation. (b) ECEA, end-cutting-edge angle. (c) SCEA, side-cutting-edge angle

emulsion is second best. Sulfurized oils are used at low cutting speeds, but precautions must be taken to avoid possible fires. A full, steady flow of cutting fluid must be maintained at the cutting site, particularly when carbide tools are used.

Control and Inspection. When setting up a turning operation, the workpiece must be firmly chucked in the collet of the spindle and supported by the tailstock, using a live center. The tool must be held firmly in a flat-base holder and set to cut on dead center. Machining should be done as closely as possible to the spindle for minimum work overhang. A steady rest or follow should be used to add rigidity to slender parts. While machining, chips should be removed from the

work area promptly, particularly during boring. Chips lying on the surface tend to produce chatter and poor surface finishes.

The tool should be examined frequently for nicks or worn flanks. These defects promote galling, increase cutting temperature, accelerate tool wear, and increase residual stresses in the machined surface.

Arbitrary tool-changing schedules are desirable. Usually this means replacing carbide tools after 0.38 mm (0.015 in.) wear land in rough turning and 0.25 mm (0.010 in.) wear land in finish turning. High-speed steel tools are replaced after a wear land of 0.76 mm (0.030 in.) develops. When periodic interruptions are made in a machining operation before the maximum wear land occurs, remove any welded-on metal, nicks, and crevices by honing before resuming operations.

Sharp edges of turned titanium surfaces are potential sources of failure. Therefore, they should be "broken" with a wet emery. This operation should not be performed dry or using oil because of a potential fire hazard.

After certain turning operations, parts may require stress relieving. Recommended treatments include:

* Annealing after rough machining, machining, and chemical shaping
* Stress relieving thin-walled parts after semi-finishing operations
* Stress relieving all finished parts

Data on speeds, feeds, and depths of cut for carbides and high-speed steel tools are shown in Tables 13.18 and 13.19. Other newer alloys of similar types behave in the same manner as those listed.

Drilling Titanium

Titanium is difficult to drill using techniques considered as conventional for other materials. Thin chips flowing at high velocities fold and clog in the flutes of the drill. Also, the usual galling action of titanium, accentuated by high cutting temperatures and pressures, causes rapid tool wear. Out-of-round holes, tapered holes, and smeared holes result, with subsequent tap breakage, if the holes are to be threaded.

Methods to minimize these problems include:

* Using short, sharp drills
* Supplying cutting fluids to the cutting zone
* Using low speeds and positive feeds
* Supplying solid support to the workpiece, especially on the exit side of the drilled hole where burrs otherwise would form

Machine Tools for Drilling. Drilling machines must be sturdy and rigid enough to withstand the thrust and torque forces built up during cutting.

Table 13.18 Finish turning of titanium alloys
Depth of cut: 0.64–2.5 mm (0.025–0.10 in.); feed: 0.13–2.5 mm/rev (0.005–0.10 in./rev)

| | C-3, C-2 carbide tools(b) | | | High-speed steel tools(b) | | |
| | | Cutting speed(d), m/min (ft/min) | | | | |
Titanium alloy (condition)(a)	Tool geometry(c)	Brazed tools	Throwaway	AISI steels	Tool geometry(c)	Cutting speed, m/min (ft/min)
Commercially pure, grade 1 (110–170 HB) (annealed)	A, D	11.43–12.70 (37.5–41.7)	12.70–13.97 (41.7–45.8)	T15, M3	D	5.08–6.35 (16.7–20.8)
Commercially pure, grade 2, 3 (140–200 HB) (annealed)	D	9.53–10.80 (31.3–35.4)	10.80–12.07 (35.4–39.6)	T15, M3	D	4.06–4.57 (13.3–15.0)
Commercially pure, grade 4 (200–275 HB) (annealed)	D	6.35–6.99 (20.8–22.9)	7.87–8.89 (25.8–29.2)	T15, M3	D	2.54–2.79 (8.3–9.2)
Ti-8Al-1Mo-1V (annealed)	A, D	2.29–3.94 (7.5–12.9)	4.19–4.70 (13.8–15.4)	M3, T5, T15	D, F, G	0.51–1.27 (1.6–4.2)
Ti-5Al-2.5Sn Ti-7Al-2Nb-1Ta Ti-6Al-2Nb-1Ta-1Mo (annealed)	A, B, D	4.19–5.46 (13.8–17.9)	5.59–6.35 (18.3–20.8)	M3, T5, T15	D, F	1.02–2.03 (3.3–6.7)
Ti-6Al-4V (annealed)	A, D, G	3.81–4.32 (12.5–14.2)	4.57–5.33 (15.0–17.5)	M3, T5, T15	D, F	1.14–1.78 (3.8–6.3)
Ti-6Al-4V Ti-6Al-2Sn-4Zr-2Mo Ti-6Al-2Sn-4Zr-6Mo (STA)	A, D	3.05–3.68 (10.0–12.1)	4.06–4.70 (13.3–15.4)	M3, T15	D, F	0.64–1.65 (2.1–5.4)
Ti-7Al-4Mo Ti-6Al-6V-2Sn (annealed)	A	3.30–3.94 (10.8–12.9)	4.19–4.70 (13.8–15.4)	M3, T15	D, F	1.27–1.52 (4.2–5.0)
Ti-6Al-6V-2Sn Ti-5Al-2Sn-2Zr-4Cr-4Mo (STA)	A	2.54–3.05 (8.3–10.0)	3.05–3.81 (10.0–12.5)	T15	E, F	0.76–1.27 (2.5–4.2)
Ti-13V-8V-6Cr-4Zr-4Mo Ti-13V-11Cr-3Al Ti-10V-2Fe-3Al (annealed)	A	2.54–3.18 (8.3–10.4)	3.04–3.81 (10.0–12.5)	M3, T15	D, F	0.64–0.89 (2.1–2.9)
Ti-13V-8Al-6Cr-4Zr-4Mo Ti-13V-11Cr-3Al Ti-10V-2Fe-3Al (STA)	A	2.03–2.54 (6.7–8.3)	2.54–3.05 (8.3–10.4)	T15	E, F	0.51–0.89 (1.7–2.9)

(a) STA, solution treated and aged. (b) Carbide Industry Standardization Committee designations for carbides; American National Standards Institute designations for high-speed steels. (c) See Table 13.14 for tool angles. (d) Higher speeds are associated with lower feeds and lower depths of cut.

Table 13.19 Rough turning of titanium alloys

Depth of cut: 2.54–6.35 mm (0.10–0.25 in.); feed: 0.254–0.38 mm/rev (0.010–0.015 in./rev)

Titanium alloy (condition)(a)	C-2 carbide tools(b)			High-speed steel tools(b)		
	Tool geometry(c)	Cutting speed(d), m/min (ft/min)				
		Brazed tools	Throwaway	AISI steels	Tool geometry(c)	Cutting speed, m/min (ft/min)
Commercially pure, grade 1 (110–170 HB) (annealed)	A, F, G	102–114 (335–374)	114–127 (374–417)	T15, M3	D	4.45–5.08 (14.6–16.7)
Commercially pure, grade 2, 3 (140–200 HB) (annealed)	A	82.6–95.2 (271–312)	95.3–108 (313–354)	T15, M3	D	3.56–4.06 (11.7–13.3)
Commercially pure, grade 4 (200–275 HB) (annealed)	A	57.2–83.5 (188–274)	69.9–78.7 (229–258)	T15, M3	D	2.29–2.54 (7.5–8.3)
Ti-8Al-1Mo-1V (annealed)	A, F, G	17.8–35.6 (58–117)	38.1–50.8 (125–167)	M3, T5, T15	B, E, F, G	0.51–1.52 (1.7–5.0)
Ti-5Al-2.5Sn Ti-6Al-2Nb-1Ta-1Mo (annealed)	A, F, G, H	35.6–45.7 (117–150)	35.6–55.9 (117–183)	M3, T5, T15	B, E, K, F	0.76–2.03 (2.5–6.7)
Ti-6Al-4V (annealed)	A, F, G, I	35.6–38.1 (117–125)	45.7–50.8 (150–167)	M3, T5, T15	B, E, L	0.89–1.78 (2.9–5.8)
Ti-6Al-4V Ti-6Al-2Sn-4Zr-2Mo Ti-6Al-2Sn-4Zr-6Mo (STA)	A, F, G	25.4–30.5 (83–100)	38.1–40.6 (125–133)	M3, T5	B, E	0.76–1.40 (2.5–4.6)
Ti-7Al-4Mo Ti-6Al-6V-2Sn (annealed)	A	27.9–33.0 (92–108)	3.81–4.19 (13–14)	M3, T15	B, F	1.02–1.52 (3.3–5.0)
Ti-6Al-6V-2Sn Ti-5Al-2Sn-2Zr-4Cr-4Mo (STA)	A	20.3–30.5 (67–100)	25.4–30.5 (83–100)	T15	C, E, F	0.64–1.02 (2.1–3.3)
Ti-13V-11Cr-3Al Ti-3Al-8V-6Cr-4Zr-4Mo Ti-10V-2Fe-3Al (annealed)	A	20.3–25.4 (67–83)	25.4–30.5 (83–100)	M3, T15	B, F	0.51–0.64 (1.7–2.1)
Ti-13V-11Cr-3Al Ti-3Al-8V-6Cr-4Zr-4Mo Ti-10V-2Fe-3Al (STA)	A	15.2–25.4 (50–83)	20.3–25.4 (67–83)	T15	D, F	0.51–0.64 (1.7–2.1)

(a) STA, solution treated and aged. (b) Carbide Industry Standardization Committee designations for carbides; American National Standards Institute designations for high-speed steels. (c) See Table 13.14 for tool angles. (d) Higher speeds are associated with lower feeds and lower depths of cut.

Therefore, spindle overhang should be no greater than necessary for a given operation. In addition, excessive clearances in spindle bearings cannot be tolerated. Radial and thrust bearings should be good enough to minimize runout and end play. Finally, the feed mechanism should be free of backlash to reduce the strain on the drill when it breaks through the workpiece on through-holes.

Machines for drilling operations are made in many different types and sizes. Size and capacity can be expressed in horsepower. The horsepower rating is that usually needed to drill cast iron using the maximum drill diameter. Suitable sizes of machines for drilling titanium include:

- Upright drill No. 3 and 4
- Upright drill, production: 533 mm (21 in.), heavy duty, 5 hp
- Upright drill, production: 610 mm (24 in.), heavy duty, 7.5 hp
- Upright drill, production: 711 mm (28 in.) heavy duty, 10 hp

Industry also has requirements for drilling parts at assembly locations. These needs are ful-filled using portable power-feed, air-drilling machines. Modern units incorporate positive mechanical-feed mechanisms, depth control, and automatic return. Some are self-supporting and self-indexing. Slow-speed, high-torque drill motors are required. Spindle speeds between 230 and 550 rpm at 620 kPa (90 psi) air pressure are appropriate for high-speed drills, while speeds up to 1600 rpm have been used for carbide drills. Thrusts between 145 and 454 kg (320 and 1000 lb) are available on some portable drilling machines.

Drills and Drill Design. The choice of drills depends on the drilling operation. Heavy-duty, stub-type screw machine drills instead of jobber-length drills are recommended for drilling opera-tions on workpieces other than sheet. For deep-hole drilling, oil-feeding drills or a series of short drills of various lengths are used in sequence. Drill jigs and bushings are used when added rigidity is needed.

Aircraft drills such as National Aerospace Stan-dards (NAS) 907 types C, D, and E are usually used on sheet metal. The NAS 907 type B drill can be used where the type C drill may be too short because of bushing length and hole depth.

Large flutes reduce the tendency for chips to clog. Drill length should be kept as short as feasible (not much longer than the intended hole depth) to increase columnar rigidity and decrease torsional vibration that causes chatter and chipping.

Drills having conventional drill geometry and special-point grinds are used. They have a normal helix of 29° (just enough relief to prevent rubbing and pickup), a thinned web to reduce drilling pressure, a correct point angle with its apex held accurately to the centerline of the drill, and cutting lips of the same slope and of equal length. Special-point grinds include crankshaft notch-type drills and split points with positive-rake notching.

Relief angles are extremely important to drill life. Small angles cause excessive pickup of titanium, while very large angles weaken the cutting edge. Relief angles between 7 and 12° are used.

The web is often thinned to reduce drilling pressure. However, when doing so, the effective rake angle should not be altered.

Point angles have a marked effect on drill life. The selection of 90, 118, or 135° depends on feed, drill size, and workpiece. Therefore, it is advisable to try all three angles to find which is best suited for the job. Generally, blunt points (135 and 140°) are superior on small-sized drills (No. 40 to 31) and on sheet metal, while 118 and 90° and the double angle (140 and 118° + 90° chamfer) are best on larger sizes and bar stock.

Figures 13.6 and 13.7 illustrate typical nomenclature and data for standard and NAS 907-type drills. Table 13.20 lists some drill specifications used in the aircraft industry. Drill geometry should conform to recommendations. If necessary, drills should be reground accurately on a drill grinder, and the point angle, relief angle, and web thickness rechecked. Drills should never be sharpened by hand.

The apex of the point angle should be held accurately to the centerline of the drill, and the cutting lips should have the same slope. This combination avoids uneven chip formation, drill deflection, and oversized holes.

When dull drills are reconditioned, resharpening the point alone is not always adequate. The entire drill should be reconditioned to ensure conformance with recommended drill geometry.

Machine-ground points with fine finishes give the best tool life. A surface treatment, such as chromium plating or black-oxide coating of the flutes, minimizes welding of chips to the flutes.

Drill Materials. Conventional molybdenum-tungsten high-speed steel drills are usually used in production. Cobalt high-speed steels provide 50% more tool life, but their costs are 1½ to 2 times higher than standard high-speed steels. Table 13.21 indicates drilling applications for various grades of high-speed steel.

Drilling Feeds. When drilling titanium, it is important to keep the drill cutting. The drill should never be allowed to ride in the hole without cutting metal. The best technique is to use a positive mechanical feed. Even assembly drilling of sheet should be done using portable power drills having positive feed arrangements.

Hand drilling can be done only if sufficient thrust can be applied to ensure a heavy chip throughout drilling. However, the high axial thrust required to keep the drill cutting, especially in heat treated titanium alloys, can cause rapid operator fatigue. Furthermore, allowing the drill to advance rapidly on breakthrough, as is generally the case with hand feeding, will seriously shorten drill life by chipping the corners of the drill.

The selection of feeds depends largely on the size of the drill being used. Generally, a feed range of 0.025 to 0.127 mm/rev (0.001 to 0.005 in./rev) is used for drills up to 6.35 mm (0.25 in.) in diameter. Drills 6.35 to 19.0 mm (0.25 to 0.75 in.) in diameter require a heavier feed range of 0.051 to

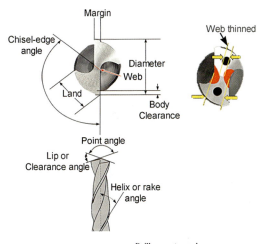

	Drill geometry code		
Drill elements	**X**	**Y**	**Z**
Drill diameter, mm (in.)	<3.15 (<0.125)	3.175–6.35 (0.125–0.25)	6.35 (0.25) and greater
Helix angle, degrees	29	29	29
Clearance angle, degrees	7–12	7–12	7–12
Point angle, degrees	135	118, 135	118, 90, or double angle
Type point		Crankshaft (split)	

Fig. 13.6 Drill nomenclature and tool angles

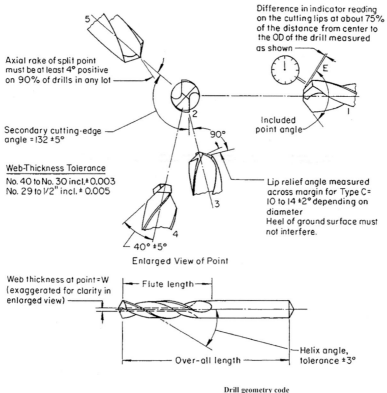

Fig. 13.7 Drill nomenclature and geometry for NAS 907 aircraft drills (type C illustrated)

Drill elements	Drill geometry code			
	C	D	B	E
Notch rake angle, degrees	4–7	20	4–7	10
Helix angle, degrees	23–30	28–32	23–30	12
Clearance angle, degrees	10–14	6–9	10–14	6–9
Point angle, degrees	118±5	135±5	135±5	135±5
Type point	P-5	P-1	P-3	P-2
NAS 907 drill types	C	D	B	E
Drilling application	Sheet	Hand drilling sheet	Fixed feed	Fixed feed (dry)

Table 13.20   Specifications for drills used on titanium alloys

Drill diameter, mm (in.)	Overall length, mm (in.)	Flute length, mm (in.)	Helix angle, degrees	Lip relief angle, degrees	Point angle, degrees	Web thickness (below split), mm (in.)	Chisel-edge angle, degrees	Chisel thickness (after splitting), mm (in.)
2.49 (0.098)	41.3 (1⅝)	15.9 (⅝)	28 ± 2	7–10	135 ± 3	0.74–0.81 (0.029–0.032)	115–125	0.10–0.20 (0.004–0.008)
3.26 (0.1285)	49.2 (1⁹⁄₁₆)	33.3 (1⁵⁄₁₆)	28 ± 2	7–10	135 ± 3	0.97–1.1 (0.038–0.042)	115–125	0.10–0.20 (0.004–0.008)
4.04 (0.1590)	54.0 (2⅛)	28.6 (1⅛)	30 ± 2	12 ± 2	135 ± 3	1.19–1.32 (0.047–0.052)	115–125	0.10–0.20 (0.004–0.008)
4.70 (0.1850)	55.6 (2³⁄₁₆)	30.2 (1³⁄₁₆)	30 ± 2	12 ± 2	135 ± 3	1.4–1.5 (0.055–0.060)	115–125	0.10–0.20 (0.004–0.008)
4.91 (0.1935)	57.2 (2¼)	31.8 (1¼)	30 ± 2	12 ± 2	135 ± 3	1.4–1.6 (0.055–0.063)	115–125	0.10–0.20 (0.004–0.008)
6.25 (0.246)	63.5 (2½)	38.1 (1½)	30 ± 2	12 ± 2	135 ± 3	1.5–1.6 (0.060–0.063)	115–120	0.10–0.20 (0.004–0.008)
6.35 (0.250)	63.5 (2½)	34.9 (1⅜)	30 ± 2	12 ± 2	135 ± 3	1.5–1.6 (0.060–0.063)	115–120	0.10–0.20 (0.004–0.008)

Note: All drills have point angles of 135 ± 3°, crankshaft (split) points, and are made from M33 high-speed steel.

Table 13.21 High-speed steels used for drills in drilling titanium alloys

AISI grade of high-speed steel	Commercially pure titanium	Titanium alloy			
		Ti-5Al-2.5Sn	Ti-8Al-1Mo-1V	Ti-6Al-4V	Ti-13V-11Cr-3Al
M1	S	S	S	…	G
M2	…	…	…	…	G
M2	S	S	S	…	…
M7	G, D	G, D	…	G, D	…
M10	G, D, S	G, D, S	G, D, S	G, S	…
M33	G, D	G, D	G, D	…	G
M34	G, D	G, D	G, D	…	
M36	…	…	…	S	…
T4	G, D, S	G, D, S	G, D, S	…	…
T5	G, D, S	G, D, S	G, D, S	G, S	…

Note: S = sheet drilling; G = general drilling; D = deep-hole drilling

0.178 mm/rev (0.002 to 0.007 in./rev). Suggested feed rates for 1.6 to 12.7 mm (0.0625 to 0.5 in.) diameter drills are given in Fig. 13.8.

Drilling Speeds. Because its cutting zone is confined, drilling requires low cutting speeds for minimum cutting temperatures. Further, the selection of speed depends largely on the strength level of the titanium material and the nature of the workpiece. Thus, speeds up to 30 m/min (~100 ft/min) are used for commercially pure titanium, while only 5.5 to 7.3 m/min (18 to 24 ft/min) should be used on aged Ti-13V-11Cr-3Al and Ti-3Al-8V-6Cr-4Zr-4Mo and other aged metastable beta alloys.

Speeds should remain constant throughout drilling. This requires an "overpowered" drilling machine. Slow-speed, high-torque drill motors should be used for portable power drills.

Cutting Fluids. Drilling titanium usually requires cutting fluids. Although holes in single sheets with thicknesses up to twice the drill diameter can be drilled dry, sulfurized oils and sulfurized lanolin paste are recommended for low speeds and for drills less than 6.35 mm (0.25 in.) in diameter. Good coolants, such as soluble oil-water emulsions, are used for higher drilling speeds.

A steady, full flow of fluid, externally applied at the cutting site, is used, but the use of a spray mist seems to give better tool life. However, a two-diameter depth limit exists for external applications. Therefore, oil-feeding drills work best for deep holes.

General Drilling Techniques and Inspection. Setup conditions for drilling titanium should provide overall setup rigidity and sufficient spindle power to maintain drilling speeds. Successful drilling of titanium also depends on the ability to

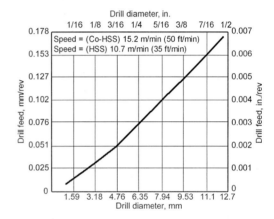

Fig. 13.8 Feed rate versus drill diameter for high-speed drills. Feed rates are based on the use of adequate supply of coolant, heavy-duty drills having point angles of 118 or 135°, and on speeds of 15.2 m/min (50 ft/min) for cobalt high-speed steel (Co-HSS) or 10.7 m/min (35 ft/min) for conventional HSS.

reduce the temperature at the cutting lips. Methods used to accomplish this include:

- Using low cutting speeds
- Reducing the feed rate
- Supplying adequate cooling at the cutting site

Thin sheet-metal parts must be properly supported at the point of thrust. This is done with a backup block of AISI 1010 or 1020 steel. Where this is not possible because of part configuration, a low-melting alloy can be cast around the part.

When drilling stacked sheet, they should be clamped securely with clamping plates to eliminate gaps between sheets.

When starting a drilling operation, the drill should be up to speed and under positive feed as it advances toward the work. Drilling should never be started with a dull drill, and a triangular center punch should be used to mark the hole location on the part. Holes should be drilled to size in one operation when possible. Center drills and undersized starting drills usually are not recommended. The use of drill bushings is desirable for close-tolerance holes.

The margin of the drill should be examined periodically for smearing to prevent oversized holes. Also, the outer corner of the lips should be inspected for the presence of possible breakdowns. An arbitrary drill replacement schedule should be established to prevent spoilage of work and drills.

The nature of the chips indicates the condition of the drill during drilling. A sharp drill usually produces tight-curling chips without difficulty. As the drill dulls progressively, cutting temperature rises, and the titanium begins to smear on the lips and margins. The subsequent appearance of feather-type chips in the flutes is a warning signal that the drill is dulling and should be replaced. Afterward, any appearance of irregular and discolored chips indicates that the drill has failed. Out-of-round holes, tapered holes, and smeared holes result from poor drilling action, with subsequent reaming problems, or even tap breakage, when the holes are threaded. Chips should be removed at periodic intervals unless the cutting fluid successfully flushes them away.

When drilling holes more than one diameter deep, the drill should be retracted once for each half diameter of drill advance, to clear the flutes. The drill should be retracted simultaneously with the stop of the feed to minimize dwell. The drill is quickly (but carefully) reengaged with the drill up to speed and under positive feed.

When drilling through-holes, it is sometimes advisable not to drill all the way through on a continuous feed. Instead, the drill is retracted before breakthrough, and the drill and the hole are flushed to remove the chips. The drill is then returned under positive feed, and the hole is drilled through carefully to avoid any "feed surge" at breakthrough.

All assembly drilling should be done using portable, fixed-feed, jig-mounted drilling machines. Hand drilling is used, but the practical limit appears to be the No. 40 drill. Above this diameter, insufficient feed results with consequent heat buildup and short drill life. Hand drilling should not be used if the hole is to be tapped. Drilled holes could require reaming to meet specific tolerances, unless a bushing is used immediately adjacent to the part. Drilled holes in sheet probably require deburring on the exit side. Table 13.22 presents general operating data for drilling. Other newer alloys of similar types behave in the same manner as those listed.

Surface Grinding

Recommended abrasives are silicon carbide wheels for cut-off and portable grinding, and aluminum oxide wheels for cylindrical and surface grinding. Other recommendations include:

- Use a sharply dressed wheel
- Use the largest wheel diameter and thickness that is feasible
- Use harder wheels
- Use ample power at the spindle for grinding

Reduced wheel speeds, compared with those used with steel, aid in grinding performance. In surface grinding, for example, 914 surface m/min (3000 surface ft/min) causes minimum surface stresses and distortion. The basic reasons for using slower wheel speeds in grinding titanium are to:

- Avoid high temperatures developed at the chip-grit interface
- Compensate for the abrasive character of titanium
- Avoid wheel loading
- Avoid smearing
- Compensate for the heat sensitivity of titanium
- Prevent fire hazard

The following downfeeds should be used to minimize residual stresses in ground surfaces of titanium parts: 0.025 mm (0.001 in.) per pass to last 0.051 mm (0.002 in.), then 0.0127 mm (0.0005 in.), 0.010 mm (0.0004 in.), 0.0076 mm (0.0003 in.), 0.0051 mm (0.0002 in.), and 0.0025 mm (0.0001 in.) per pass, no sparkout.

Nitride amine-base fluids are used in a majority of operations. For form grinding, straight grinding oil is recommended.

Because grinding titanium in the presence of a grinding oil, especially at increased speeds, pres-

Table 13.22 Drilling titanium alloys with high-speed drills

Titanium alloy (condition)(a)	Tool material(b)	Cutting speed, m/min (ft/min)	Drill diameter, mm (in.)	Power feed(c), mm/rev (in./rev)
Commercially pure (annealed)	M1, M2, M10	10.2–20.3 (34–67)	3.18 (⅛)	0.025–0.051 (0.001–0.002) 0.0127–0.025 (0.0005–0.001)
Ti-8Al-1Mo-1V (annealed)	M1, M2, M10	5.1 (17) 10.2 (34) for sheet	6.35 (¼)	0.025–0.127 (0.001–0.005) 0.025–0.127 (0.001–0.005)
Ti-5Al-2.5Sn (annealed)	M1, M2, M10	7.6–10.2 (25–34)	12.7 (½)	0.051–0.152 (0.002–0.006) 0.025–0.102 (0.001–0.004)
Ti-6Al-4V (annealed)	M1, M2, M10	7.6–10.2 (25–34)	50.8 (2)	0.127–0.330 (0.005–0.013) 0.076–0.203 (0.003–0.008)
Ti-6Al-4V (STA)	T15, M33	5.1–7.6 (17–25)	76.2 (3)	0.127–0.381 (0.005–0.015) 0.102–0.229 (0.004–0.009)
Ti-6Al-2Sn-4Zr-6Mo (STA)	T15, M33	5.1–7.6 (17–25)	50.8 (2)	0.127–0.381 (0.005–0.015) 0.076–0.229 (0.003–0.009)
Ti-7Al-4Mo (annealed)	M1, M2, M10	5.1 (17)	…	…
Ti-6Al-6V-2Sn (annealed)	M1, M2, M10	5.1 (17)	…	…
Ti-6Al-6V-2Sn (STA)	T15, M33	5.1–6.4 (17–21)	…	…
Ti17 (5Al-2Sn-2Zr-4Cr-4Mo) (STA)	T15, M33	5.1–6.4 (17–21)	…	…
Ti-13V-11Cr-3Al (annealed)	M1, M2, M10	5.1–7.6 (17–25)	…	…
Ti-3Al-8V-6Cr-4Zr-4Mo (STA)	T15, M33	3.8–5.1 (13–17)	…	…

Tool geometry: For general drilling operations, choose drill geometry X, Y, or Z, depending on drill size (see Fig. 13.6). For drilling sheet, use drill geometry C, D, or B according to application (see Fig. 13.7). Cutting fluids used: Use a soluble oil-water emulsion or sulfurized oil, the latter at lower speeds and for small drills 6.35 mm (0.25 in.). Chlorinated oils are also used, provided oil residues are promptly removed by methyl ethyl ketone. Holes in single sheets up to twice the drill diameter can be drilled dry. (a) STA, solution treated and aged. (b) AISI designation. (c) Use the lower feeds for the stronger or aged alloys. Data are for holes deeper than 12.7 mm (0.5 in.). Somewhat higher feeds may be used for holes less than 12.7 mm (0.5 in.) deep. Medium and steady hand pressure is needed for hand drilling.

ents a fire hazard, precautions should be taken, including:

- Installing extra cutting fluid lines to quench sparking as much as possible
- Installing filters where possible to remove fine titanium particles from the cutting fluid
- Frequent cleaning of titanium dust from external surfaces of machines, and changing oil more often than is customary with steels
- Having available material such as soapstone in the vicinity of the machine to quench fires

Table 13.23 lists general grinding recommendations for a variety of titanium alloys.

Broaching

As in other titanium machining operations, it is essential that the entire machine-tool setup and the titanium component be rigid, to ensure a high-quality broaching job. Recommended tool geometry for broaching is illustrated in Fig. 13.9. It is also recommended that broaches be wet-ground, to improve tool finish, thereby giving better tool performance. During the broaching operation, vapor blasting with coolant helps lengthen broach life and reduce the tendency for smearing.

Titanium chips tend to weld to the tool on an interrupted cut such as broaching, which increases as the wear land develops. Both broach and broach slots should be examined regularly for signs of smearing, because this and chip welding are indications of wear. Watching for signs of smearing can minimize poor finish, rapid tool wear, and loss of tolerances. Table 13.24 lists general recommendations for broaching titanium.

Tapping

It is essential that straight, clean holes are drilled to ensure good tapping results, because variations in diameter and tapered holes are detrimental to this work. Sound threads can be ensured by reducing the tendency of the titanium to smear on the lands of the tap and by providing for a free flow of chips in the flutes. Failure to do this will result in poor threads, undersized holes, seizures, and, consequently, broken taps.

The use of nitrided taps also helps to reduce adherence of titanium to the lands of the tap. Relieving the land, or an interrupted tap, also helps minimize smearing.

Chip clogging is reduced by using spiral-pointed taps, which push the chips ahead of the tool. More chip clearance is provided by sharply

Table 13.23 Surface grinding

Material (condition)(a)	Type of cut(b)	Wheel designation		Table speed, m/min (ft/min)	Downfeed, mm/pass (in./pass)	Crossfeed, mm/pass (in./pass)
		Aluminum oxide(c)	Silicon carbide(d)			
Grade 1 (110–170 HB) (annealed)	R	A 46 JV	C 46 JV	12.2 (40)	0.025 (0.001)	1.57 (0.062)
	F	A 60 LV	C 70 LV	12.2 (40)	0.013 (0.0005) max	1.27 (0.050)
Grade 2, 3, 7 (140–200 HB) (annealed)	R	A 46 JV	C 46 JV	12.2 (40)	0.025 (0.001)	1.57 (0.062)
	F	A 60 LV	C 70 LV	12.2 (40)	0.013 (0.0005) max	1.27 (0.050)
Grade 4 (200–275 HB) (annealed)	R	A 46 JV	C 46 JV	12.2 (40)	0.025 (0.001)	1.57 (0.062)
	F	A 60 LV	C 70 LV	…	0.013 (0.0005) max	1.27 (0.050)
Ti-3Al-2.5V (200–260 HB) (annealed)	R	A 45 JV	C 46 JV	12.2 (40)	0.025 (0.001)	1.57 (0.062)
	F	A 60 LV	C 70 LV	…	0.013 (0.0005) max	1.27 (0.050)
Ti-5Al-.2.5Sn	R	A 46 JV	C 46 JV	12.2 (40)	0.025 (0.001)	1.57 (0.062)
Ti-5Al-.2.5Sn ELI Ti-8Al-2Nb-1Ta-1Mo (300–340 HB) (annealed)	F	A 60 LV	C 70 LV	12.2 (40)	0.013 (0.0005) max	1.27 (0.050)
TI-6Al-4V	R	A 46 JV	C 46 JV	12.2 (40)	0.025 (0.001)	1.57 (0.062)
Ti-6Al-4V ELI (310–350 HB) (annealed)	F	A 60 LV	C 70 LV	…	0.013 (0.0005) max	1.27 (0.050)
Ti-7Al-4Mo	R	A 46 JV	C 46 JV	12.2 (40)	0.025 (0.001)	1.57 (0.062)
Ti-8Al-1Mo-1V Ti-6Al-6V-2Sn (320–370 HB) (annealed)	F	A 60 LV	C 70 LV	…	0.013 (0.0005) max	1.27 (0.050)
Ti-6Al-4V	R	…	C 46 JV	12.2 (40)	0.025 (0.001)	1.57 (0.062)
Ti-6Al-2Sn-4Zr-2Mo Ti-6Al-2Sn-4Zr-6Mo (350–400 HB) (STA)	F	…	C 60 KV	…	0.013 (0.0005) max	1.27 (0.050)
Ti-6Al-6V-2Sn	R	…	C 46 JV	12.2 (40)	0.025 (0.001)	1.57 (0.062)
Ti-7Al-4Mo Ti-5Al-2Sn-2Zr-4Cr-4Mo (375–420 HB) (STA)	F	…	C 60 KV	…	0.013 (0.0005) max	1.27 (0.050)
Ti-13V-11Cr-3Al	R	A 46 JV	C 46 JV	12.2 (40)	0.025 (0.001)	1.57 (0.062)
Ti-3Al-8V-6Cr-4Zr-4Mo (310–350 HB) (STA)	F	A 60 LV	C 70 LV	12.2 (40)	0.013 (0.0005) max	1.27 (0.050)
Ti-13V-11Cr-3Al	R	…	C 46 JV	12.2 (40)	0.025 (0.001)	1.57 (0.062)
Ti-3Al-8V-6Cr-4Zr-4Mo Ti-10V-2Fe-3Al (375–440 HB) (STA)	F	…	C 60 KV	12.2 (40)	0.013 (0.005) max	1.27 (0.050)

(a) ELI, extralow interstitial; STA, solution treated and aged. (b) R = roughing; F = finishing. (c) 457–762 m/min (1500–2500 ft/min). (d) 914–1676 m/min (3000–5500 ft/min). Nominal grinding ratio (15) to be expected for the recommended surface-grinding conditions for aluminum oxide and silicon carbide wheels. Grinding ratio = mm³ (in.³) of metal removed by grinding/mm³ (in.³) of wheel loss. Courtesy of RMI Co.

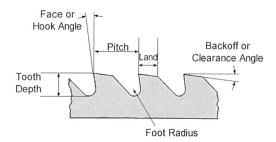

Fig. 13.9 Tool geometry and nomenclature for broaching. Reprinted with permission from Ref 13.17

grinding away the trailing edges of the flutes. To give proper clearance, two-fluted spiral point taps are recommended for diameters up to 7.9 mm (0.312 in.). Three-fluted taps are recommended for larger sizes. General tap geometry is shown in Fig. 13.10. Table 13.25 gives general recommendations for tapping titanium. Other newer alloys of similar types behave in the same manner as those listed.

Recent Advances in Machining

The high titanium use on newer commercial aircraft with more composites structure is a strong driver for increased efforts to reduce machining costs. The quantity of titanium chips generated during the manufacture of the Boeing 787 is estimated at ~68,500 kg (151,000 lb) per airframe. Titanium machining represents, on the average, over 38% of total part fabrication cost. Titanium metal-removal rates (MRRs) average 10 to 20 times slower than traditional aluminum MRRs used on heritage airframe structures. Original equipment manufacturers, fabricators, cutting-tool manufacturers, and others have invested heavily in improving the MRR of titanium components since the 1980s, studying all aspects, such as cutter materials and design, cutter coatings, machine design, machining software, and so on. Very substantial MRR gains have been made since that time; for roughing operations, MRRs have increased by more than a factor of 4. The

Table 13.24 Broaching

| Material (condition)(a) | Type of cut(b) | High-speed steel tool | | | |
		Speed, m/min (ft/min)	Chip load, mm/t (in./t)	Tool material	Tool geometry hook/clearance angle
Grade 1 (110–170 HB) (annealed)	R	10.7 (35)	0.127–0.203 (0.005–0.008)	T5	8–10°/3–4°
	F	16.8 (55)	0.051–0.127 (0.002–0.005)	T5	8–10°/2–3°
Grade 2, 3, 7 (140–200 HB) (annealed)	R	9.14 (30)	0.127–0.203 (0.005–0.008)	T5	8–10°/3–4°
	F	13.7 (45)	0.051–0.127 (0.002–0.005)	T5	8–10°/2–3°
Grade 4 (200–275 HB) (annealed)	R	6.1 (20)	0.102–0.178 (0.004–0.007)	T5	8–10°/3–4°
	F	9.14 (30)	0.051–0.102 (0.002–0.004)	T5	8–10°/2–3°
Ti-5Al-2.5Sn Ti-5Al-2.5Sn ELI Ti-6Al-2Nb-1Ta-1Mo (300–340 HB) (annealed)	R	4.6 (15)	0.076–0.152 (0.003–0.006)	T5	8–10°/3–4°
	F	6.7 (22)	0.038–0.076 (0.0015–0.003)	T5	8–10°/2–3°
Ti-6Al-4V Ti-6Al-4V ELI Ti-8Mn (310–350) (annealed)	R	3.7 (12)	0.076–0.152 (0.003–0.006)	T5	8–10°/3–4°
	F	5.5 (18)	0.038–0.076 (0.0015–0.003)	T5	8–10°/2–3°
Ti-7Al-4Mo Ti-8Al-1Mo-1V Ti-6Al-6V-2Sn (320–370 HB) (annealed)	R	3.0 (10)	0.076–0.152 (0.003–0.006)	T5	8–10°/3–4°
	F	4.8 (16)	0.038–0.076 (0.0015-0.003)	T5	8–10°/2–3°
Ti-6Al-4V Ti-6Al-2Sn-4Zr-2Mo Ti-6Al-2Sn-4Zr-6Mo (350–400 HB) (STA)	R	2.4 (8)	0.051–0.127 (0.002–0.005)	T15	8–10°/3–4°
	F	3.7 (12)	0.025–0.051 (0.001–0.002)	T15	8–10°/2–3°
Ti-6Al-6V-2Sn Ti-7Al-4Mo Ti-5Al-2Sn-2Zr-4Cr-4Mo (375–420 HB) (STA)	R	2.1 (7)	0.051–0.102 (0.002–0.004)	T15	8–10°/3–4°
	F	3.0 (10)	0.025–0.051 (0.001–0.002)	T15	8–10°/2–3°
Ti-13V-11Cr-3Al Ti-3Al-8V-6Cr-4Zr-4Mo 10V-2Fe-3Al (310–350 HB) (solution annealed)	R	3.4 (11)	0.076–0.152 (0.003–0.006)	T5	8–10°/3–4°
	F	5.2 (17)	0.038–0.076 (0.0015–0.003)	T5	8–10°/2–3°
Ti-13V-11Cr-3Al Ti-3Al-8V-6Cr-4Zr-4Mo (375–440 HB) (STA)	R	1.8 (6)	0.051–0.102 (0.002–0.004)	T15	8–10°/3–4°
	F	2.7 (9)	0.025–0.052 (0.001–0.002)	T15	8–10°/2–3°

Due to the complexity of most broaching tools and the configurations machined, general predictions of broach life are not practical. (a) ELI, extralow interstitial; STA, solution treated and aged. (b) R = roughing; F = finishing. Courtesy of RMI Co.

rate of improvement has decreased over the past few years, but recent technology developments are beginning to show promise toward further MRR improvements. More recent gains are attributed primarily to cutting-tool technology advancements, design of higher-performance machine tools, and advances in machining optimization software (Fig. 13.11). Boeing anticipates that further improvements in machinability may be gained by the use of alternative alloys, such as Timetal Ti-54M (Ti-5Al-4V-0.75Mo-0.5Fe).

BAE Systems opened a titanium-machining facility at its Samlesbury site in Lancashire, United Kingdom, that will manufacture, detail, and assemble titanium components of the aft fuselage and the vertical and horizontal tails for the F-35 Lightning II Joint Strike Fighter. Improvements in machining are anticipated from a highly automated computer-integrated machining system working on a "just-in-time" basis.

The machining of the equiatomic titanium aluminide (TiAl) is generally considered to be difficult. However, Moeller Manufacturing Co. Inc. of Wixom, Wisconsin, has gained expertise in handling TiAl such that it achieves defect-free yields and machining costs comparable with equivalent parts made of Ti-6Al-4V titanium and stainless steel. This is the result of implementing unique fixturing concepts of the company's own design along with proven combinations of machining parameters and tooling.

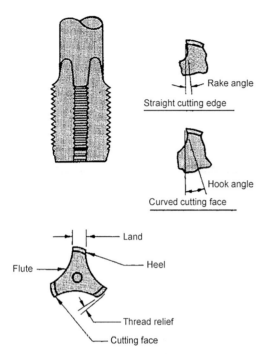

Rake angle

Straight cutting edge

Hook angle

Curved cutting face

Land

Flute

Heel

Thread relief

Cutting face

Fig. 13.10 Tool geometry and nomenclature for taps. Reprinted with permission from Ref 13.17

Thus, it appears that improvements in the machining of titanium are coming from better, more efficient overall setups with extensive use of computer control.

Flame Cutting

Metal removal and forming by use of oxyacetylene torch has long been used by the construction trades and machine shops in tool fabrication of nonstainless steels. It was soon learned that titanium could also be easily torch cut, which enabled fast removal of excess stock, trimming of thick plate, and cutting of simple contours.

An example of the savings possible is the use of flame cutting on a large titanium ring, where circumferential cuts on a 127 mm (5 in.) thick section only require 12 min, including contouring of a large boss. Removal of the same stock by conventional methods would have required approximately 50 h of machining time.

However, there are a number of precautions that must be taken when flame cutting. Caution must be exercised because large volumes of nox-

Table 13.25 Tapping

Tool geometry—Two-fluted spiral point for 7.93 mm (⁵⁄₁₆ in.) tap and smaller; three-fluted spiral point for >7.93 mm (⁵⁄₁₆ in.) tap

Material(a)	Condition	Hardness, HB	Speed, cm/min (in./min)	High-speed steel tool material
Grade 1	Annealed	110–170	12.7 (5.0)	Nitrided
	…	…	…	M1, M10
Grade 2	Annealed	140–200	10.2 (4.0)	Nitrided
Grade 3	…	…	…	M1, M10
Grade 7	…	…	…	…
Grade 4	Annealed	200–275	76.2 (30)	Nitrided
	…	…	…	M1, M10
Ti-5Al-2.5Sn	Annealed	300–340	63.5 (25)	…
Ti-5Al-2.5Sn ELI	…	…	…	Nitrided
Ti-6Al-2Nb-1Ta-1Mo	…	…	…	M1, M10
Ti-6Al-4V	Annealed	310–350	50.8 (20)	Nitrided
Ti-6Al-4V ELI	…	…	…	M1, M10
Ti-7Al-4Mo	Annealed	320–370	38.1 (15)	Nitrided
Ti-8Al-1Mo-1V	…	…	…	M1, M10
Ti-6Al-6V-2Sn	…	…	…	…
Ti-6Al-4V	Solution treated	350–400	25.4 (10)	Nitrided
Ti-6Al-2Sn-2Zr-2Cr-4Mo	Solution treated and aged	…	…	M1, M10
Ti-6Al-2Sn-2Zr-4Cr-4Mo	…	…	…	…
Ti-6Al-6V-2Sn	Solution treated	375–420	25.4 (10)	Nitrided
Ti-7Al-4Mo	Solution treated and aged	…	…	M1, M10
Ti-5Al-2Zr-4Cr-4Mo	…	…	…	…
Ti-13V-11Cr-3Al	Solution treated	310–350	38.1 (15)	Nitrided
Ti-3Al-8V-6Cr-4Zr-4Mo	Annealed	…	…	M1, M10
Ti-10V-2Fe-3Al	…	…	…	…
Ti-5V-11Cr-3Al	Solution treated	375–440	17.8 (7.0)	Nitrided
Ti-3Al-8V-6Cr-4Zr-4Mo	Solution treated and aged	…	…	M1, M10
Ti-10V-2Fe-3Al	…	…	…	…

Nominal tool life to be expected for the recommended tapping conditions for high-speed steel taps: 75 holes for 2:1 depth-to-diameter ratio. (a) ELI, extralow interstitial

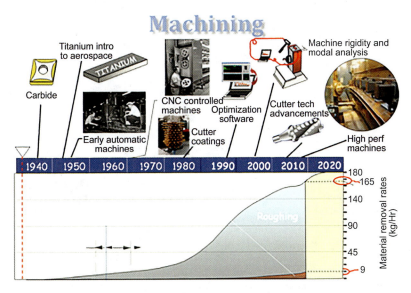

Fig. 13.11 Advances in machining. The gray area represents rough machining, the brown area finish machining, and the arrows indicate the 10 year lag between technology introduction and widespread use. Courtesy of Boeing

ious (but nontoxic) white smoke (TiO_2) are evolved from the molten slag resulting from flame cutting. To reduce the smoke nuisance, the area must be very well ventilated. Another successful, less costly way of dealing with the smoke is to do the cutting over a water table so that the time of transit through air is minimized. The water level should be within 50 to 100 mm (2 to 4 in.) of the underside of the material being cut.

Because the reaction between the hot titanium slag and water results in the liberation of hydrogen, which, under improper conditions, will detonate, additional precautions that must be observed include:

- Blowing air between the water and the underside of the workpiece when trimming large sheets and plates to dilute, vent, and otherwise prevent the accumulation of hydrogen gas. It is not unusual to see a blue flame on the surface of the water from the burning of hydrogen.
- Using a 100 to 300 mm (4 to 12 in.) deep water table at a temperature below 38 °C (100 °F) and circulating it so hydrogen gas is not allowed to saturate the area where the hot titanium slag is being deposited under the cut

Hot slag buildup and hydrogen generation in water become more of a problem with heavy sections because the quantity of slag being deposited in a given location increases. At ≥50 mm (2 in.)

thick, there is little difficulty with smoke from the slag. As the section increases up to 250 mm (10 in.) thick, the water in the table should be turned over every few minutes to prevent hydrogen saturation of the water to the critical point of detonation. Typical setups for flame cutting are given in Table 13.26 for the use of liquid-petroleum gases.

Cutting speed should be appropriate to the quality of cut required. A rule of thumb is never to dwell during the flame-cutting operation. Tip sizes and pressures are also varied.

It should be noted that the cut surface will be coated with an alpha case and slag deposit that must be removed by grinding or other technique prior to machining or use.

The depth of surface contamination is quite shallow, under 0.51 mm (0.020 in.), while the depth of affected microstructure is under 3.1 mm (0.120 in.) deep. As a safeguard against torch blowout or dwell, which may cause undercutting, a 6.4 mm (0.250 in.) envelope should be maintained over finish-part dimensions.

Flame cutting is not limited to oxyacetylene and gas torches. Plasma torches are also widely used for flame cutting, and even lasers are used for cutting sheet and thin plate.

Chemical Machining

Chemical machining is the oldest of the chipless machining processes. The first recorded

Table 13.26   Flame-cutting setups

Thickness				Preheat O₂ pressure		
mm	in.	Cutting tip	Gas, psig	MPa	ksi	Cutting O₂, psig
25	1	AFS-52	10	170	25	20
127	5	AFS-49	4–6	170–241	25–35	70–90
203	8	AFS-38	8–10	276–345	40–50	70–90
254	10	AFS-20	8–12	276–345	40–50	80–100

work was observed in approximatively 1000 A.D., mostly involving various forms of art. It has also been used for many years in the production of engraved plates for printing of printed circuit boards. In its use as a machining process, it is applied to parts ranging from the very small, such as microelectronic circuits, to very large parts, up to 18 m (59 ft) long.

In chemical machining, material is removed from selected areas of a workpiece by exposing it to a chemical reagent. The material is removed by microscopic electrochemical cell action, as occurs in corrosion and chemical dissolution of a metal; no external circuit is involved. Surfaces not to be attacked are protected by an adhering mask that is chemically inert to the reactive chemicals.

The process is also known as chemical milling, photofabrication, and chemical blanking. Photofabrication is similar to photographic processes used to produce the necessary mask that shields portions of the workpiece from which no material is to be removed. Chemical blanking is applied when the resulting parts otherwise may have been produced by the use of blanking dies. (In most instances, the size and detail of the parts are such that they could not be made by ordinary blanking processes.)

Five process steps of industrial chemical machining or chemical milling are:

1. Precleaning to ensure uniform adhesion of the maskant and a uniform etch rate
2. Masking to protect the areas not to be etched
3. Scribing to define the required material-removal pattern
4. Etching to remove the material
5. Demasking to remove the residual mask

Chemical milling is generally used to produce shallow step-cuts over a relatively large surface area. However, it is used to produce much deeper cuts. Step-cuts deeper than 13 mm (0.5 in.) can require special workpiece and tank control to maintain tolerances. It is also used for tapering and for uniform gage reduction of extrusions, forgings, and sheet metal.

Most titanium alloys are homogeneous and fine grained and can be chemically milled in the annealed and solution-treated and aged condition. Generally, the annealed condition is preferred due to the selective relaxation of surface stresses in the area being relieved by the etching process. A surface finish of 1.6×10^{-6} mm (63 μin.) is usually specified, with actual surface finishes of 5.8×10^{-7} mm (20 μin.) or better. Castings are a general exception to this rule, with a surface finish of approximately 3.125×10^{-6} mm (125 μin.).

Generally, chemical machining of titanium is more uniform than for aluminum and steel. The standard linear tolerance of ±0.76 mm (±0.030 in.) is easily produced. Thickness tolerance of ±0.025 mm (±0.001 in.) for light removals to ±0.127 mm (±0.005 in.) for heavy removals are readily attainable, and a tolerance of ±0.013 mm (±0.0005 in.) or better is possible by using local masking and etching techniques on hand-selected, close-tolerance titanium. When close-tolerance chemical milling is used for weight reduction, a considerable cost-savings can be obtained by milling to a maximum average/absolute minimum tolerance rather than to the conventional absolute range tolerance (Table 13.27).

The fillet produced at the edge of a chemically milled titanium step is either circular with a radius approximately equal to the depth of cut (Fig. 13.12) or beveled at approximately 45°. The type of fillet produced is controlled by the composition of the chemical milling etchant, the maskant, and the part orientation during etching. Hand routing to increase the fillet radius is not practical on titanium.

The transition zone is the distance from the edge of a milled step to the point where the thickness of the chemically milled area becomes constant (Fig. 13.13). Because titanium is a poor conductor of the heat generated during chemical milling, a transition zone of 10 times the depth of cut, or at least 12.7 mm (0.5 in.), must be allowed.

Chemical milling does not introduce surface stresses but causes a redistribution of any residual stresses from forming. In some instances, chemi-

Table 13.27 Thickness tolerances in chemical machining

	Thickness tolerances			
	±0.127 mm (0.005 in.)	±0.051 mm (0.002 in.)	Alternate method, ±0.051 mm (0.002 in.)	
Typical drawing callout, mm (in.)	1.14 (0.045)	0.99 (0.039)	0.94 (0.037)	Maximum
	1.02 (0.040)	0.94 (0.037)	0.91 (0.036)	Average
	0.89 (0.035)	0.89 (0.035)	0.89 (0.035)	Absolute minimum
Chemical milling costs (relative), U.S. dollars/kg (U.S. dollars/lb)	$3.60 ($8) removed	$7.20 ($16) removed	$5.40 ($12) removed	
Final weight, kg/m² (lb/ft²)	4.66 (0.955)	4.25 (0.870)	4.25 (0.870)	

Note: A weight savings of approximately 0.342 kg/m² (0.07 lb/ft²) can be realized by decreasing the thickness tolerance of chemically milled parts from ±0.127 mm (0.005 in.) to the weight equivalent of ±0.051 mm (0.002 in.). A comparison of these tolerances for thin-web, integrally stiffened titanium aircraft skins chemically milled from 914 × 2438 × 3.18 mm (36 × 96 × 0.125 in.) sheet is shown above.

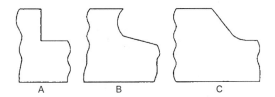

Fig. 13.12 Fillet-edge geometry. (a) Not possible by chemical machining. (b) More usual geometry achieved. (c) Chamfered edge possible but difficult with special etchants

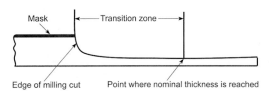

Fig. 13.13 Transition zone on chemically milled workpiece, that is, the distance from the edge of a milled step to the point where the thickness of the milled area becomes constant

cal milling is used as a surface stress-relieving operation. Warpage occurs in formed, severely machined, quenched, and welded titanium parts unless they are stress relieved prior to chemical milling. As an example, a titanium plate rolled flat and chemically machined from only one side bows inward from the opposite side. This is caused by an unbalance of the residual surface stresses remaining after chemical milling.

Particular care when handling titanium before heat treatment is advised; fingerprints that have been baked on during heat treatment are difficult to remove prior to chemical milling.

Precleaning. All surfaces must be absolutely clean, even if only one surface is going to be etched. Surfaces must be free from oil, grease, primer coatings, markings and identification inks, scale, and other foreign matter. This is a prime requirement in ensuring consistent adhesion of the maskant and a uniform etch rate.

Precleaning begins with a solvent wipe or emulsion cleaner dip. An alkaline cleaner follows, with a thorough rinse in cold water. While still wet, the surface should be checked for water breaks; if any are observed, the cleaning should be repeated. A descaling operation may be required using a formulation such as 25% solution of sodium hydroxide at approximately 95 °C (205 °F). This softens the scale to a degree that

the sides are lifted away from the surface by a nitric-hydrofluoric acid pickle solution at room temperature.

Masking. The maskant must have sufficient adhesion to cling tightly to the part when it is immersed in the etchant solution. The maskant must also have sufficient adhesion and inherent strength to protect the edges of the etched areas, as well as the areas not to be etched. This can require special taping and additional maskant layers on surfaces and sharp edges that could be rubbed during the process. The adhesion level must not be too high, or the maskant will be difficult to hand-strip prior to etching. It is for these reasons that chemical cleanliness prior to masking is most important.

Rubberized polymer maskants are the preferred peelable coatings for titanium because of their chemical resistance and uniform film properties. Water-based maskant materials are being developed and are the preferred maskants because of the ease of application. Two or more mask coatings are generally required, depending on the amount of suspended solids in the maskant solution. The coats should be applied at an angle of 90 to 180° to each other to achieve a total dry-film coating thickness of 0.152 to 0.381 mm (0.006 to 0.015 in.). When required, entrapped air bubbles and subsequent pinholes are minimized by spraying each wet mask coat with toluene. Minimum drying times between coating applications should be 20 min between first and second coats, 30 min between second and third coats, 45 min between subsequent coats, and 1 h before oven cure. These times are highly dependent on relative humidity, ambient temperature, and the type of maskant being applied.

The cured mask should be free of all holes and damaged areas, should provide complete coverage, and should have an adhesion of 0.268 to 0.714 g/cm (0.0015 to 0.004 lb/in.). Small defects in the mask are repaired by reapplying liquid air-curing maskant and by using a masking tape specially designed for this purpose. Parts having low mask adhesion should be completely stripped, recleaned, and remasked.

Scribing. Recessed patterns are produced by scribing an outline through the mask, peeling the mask from the outlined areas, and exposing the bared areas to the etching solution. Scribing should be done using a thin, sharp-pointed knife held as nearly vertical as possible. Care should be taken to cut completely through the mask without damaging the remaining mask. Excess heavy scribing pressures result in deep scores, which cause channeling in the final product.

After the mask is completely scribed, the waste mask should be peeled slightly away from the scribe line in a direction as nearly parallel to the part surface as possible. Multiple step-cuts are produced by scribing and milling the deepest step to the difference between its required linear dimensions and depth and those of the next deepest step, scribing this step, and following through similarly (Fig. 13.14).

Etching. With etching, the clean, bare metal surfaces are exposed to the etching solution, and the titanium is dissolved at a controlled rate until the required depth of cut has been reached. Three variables involved in this operation are the depth of cut required, the rate of etching (titanium usually etches between 0.013 and 0.076 mm/min, or 0.0005 and 0.003 in./min), and time. To determine the correct time of immersion in the etchant for the first items presented to the etching line, the rate of etch must be determined at the beginning of the working period and especially after any additions to the solution are made. Etch rate is determined by the amount of cross section removed divided by the time it took.

A 0.007 to 0.703 mm/m (0.001 to 0.09 in./ft) continuous taper can be produced by slowly immersing and withdrawing the part in the solution. Multiple steps are used to produce irregular and unusual tapers and when the etching tank is not sufficiently deep to permit continuous tapering. Multiple steps can be reasonably produced as close together as 1.27 mm (0.050 in.) and as shallow as 0.076 mm (0.003 in.).

The part should be racked for etching so as to limit contact to trim and masked areas and to avoid placing surfaces to be milled in close parallel position. Parts should be positioned to eliminate air pockets and racked vertically or face up to permit good circulation of the solution over all significant surfaces. The part is removed from the solution after final etching is complete and rinsed in cold water. The remaining mask is removed by peeling, and the sharp milled edge is either mechanically or chemically removed.

Etchants used for chemical milling titanium are water-based solutions generally containing nitric and hydrofluoric acids. Hydrofluoric acid is the etching component, while the nitric acid limits hydrogen absorption by the titanium and moderates the etch rate. A wetting agent minimizes the size of the reactant gas bubble to prevent gas channeling and rough fillets. Table 13.28 gives an example of a typical etching solution.

Hydrogen is absorbed on the surface of titanium parts during etching and diffuses inward at a rate determined by the temperature and the metallurgical phases present in the titanium alloy. Hydrogen diffusion in the alpha phase is quite slow at lower temperatures (below 705 °C, or

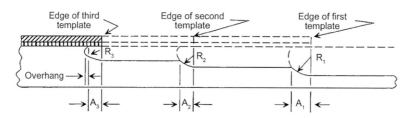

Fig. 13.14 Effect of multistep chemical milling of titanium. A_3 is approximately equal to R_3; A_2 and A_1 are less than R_2 and R_1, respectively, depending on their depths of cut. Overhang is diminished as depth of cut becomes shallower.

Table 13.28   Makeup of nitric hydrofluoric acid chemical milling solution

Ingredient or property	Makeup per 380 L (100 gal)	Control range
Nitric acid	38 L (10 gal) of 60–70% acid	283–425 g/3.8 L (10–15 oz/gal) as HNO_3
Hydrofluoric acid	11 L (3 gal) of 70% acid	As required to maintain etch rate
Wetting agent	42.5 g (1.5 oz)	29–35 dynes/cm at 21 °C (70 °F) (31 dynes/cm optimum)
Dissolved titanium	1.4 kg (3 lb)	2.2–67 g/L (0.3–9 oz/gal)
Etch rate	…	0.015–0.028 mm/side/min
		(0.0006–0.0011 in./side/min)

Temperature 32–51 °C (90–125 °F) (46 °C, or 115 °F) optimum; maintain desired temperature within ±3 °C (5 °F)

1300 °F) and tends to precipitate out as hydrides. The majority of the hydrogen diffusion at lower temperature occurs in the beta phase at a diffusion rate that increases rapidly as the continuity of the beta phase is increased.

Excessive hydrogen pickup is minimized by selecting a suitable treatment after qualifying each titanium alloy and heat treating condition with the etching solution, and by maintaining the proper oxidizer-to-fluoride ratio in the solution. Etching should not be permitted to increase the hydrogen content of parts by more than 20 ppm for alpha alloys, 30 ppm for alpha-beta alloys, and 50 ppm for beta alloys. In some instances, customer material specifications allow a variance from recommended limits.

The solution in Table 13.28 etches most alpha and alpha-beta titanium alloy conditions without excessive hydrogen pickup. Beta annealing (and, to a lesser extent, solution treating) results in increased hydrogen pickup during etching of many alpha-beta alloys. Hydrogen pickup during etching of the more sensitive alloys is minimized by increasing the nitric-to-fluoride ratio to 10:1, and to 14:1 or higher for beta alloys.

Chemical Milling Equipment. The chemical milling tank can be made of mild steel, concrete, and marine plywood, but it must be protected with a liner that resists oxidizing acids at temperatures up to 65 °C (150 °F). Commonly used linings are welded unplasticized polyvinyl chloride (PVC) and a coating of epoxy fiberglass. Tanks constructed entirely of molded polyethylene are the most desirable. Temperature must be controlled within ±3 °C (5 °F) of the set value to prevent nonuniform etching of pan surfaces. Slow solution agitation is used to minimize temperature and concentration gradients. Lipboard tank ventilation with appropriate air scrubbing should be used.

Corrosion-resistant racks such as polyethylene, polypropylene, synthetic fluorine-containing resin, PVC, 8S25, and Hastelloy X should be used. Corrosion-resistant steel racks and wires cause notching of the part surface if they are in point contact with titanium during chemical milling. Under certain conditions, metal racking can create an undesirable galvanic action on the pan surface, resulting in pits or discoloration. Plastic and plastic-coated racking is preferred.

Conventional scribing knives and templates are used with titanium. Scribing templates for titanium alloys should allow for a mask undercut during chemical milling of 40 to 80% of the depth of cut.

Electrochemical Machining

An electrochemical process for machining difficult-to-machine alloys was introduced in 1959. Known as the ECM process, it is not affected by the hardness and toughness of the material, does not destroy the working surfaces of the tooling used, and does not induce stresses (work hardening and distortion by working pressure) on the surface of the workpiece.

Figure 13.15 shows a simplified electrochemical circuit in which a high-amperage, low-voltage (5 to 20 V) current flows through an electrolyte-filled gap between the work and the tool. The workpiece is the anode and the tool is the cathode in this electrolytic cell. Electrolytes used in machining titanium alloys are sodium chloride (NaCl) and potassium chloride (KCl) at a concentration of 120 g/L (1 lb/gal) of water. The electrolyte is pumped through the gap between the tool and workpiece.

The process is sometimes described simply as the reverse of electroplating. Metal removal occurs by the ionization effect of the current in the cutting zone working on the electrolytic chemical and the metal workpiece. The configuration desired in the workpiece is achieved by designing the cathode to a required shape and confining the working action by insulators and fluid dams. The proper activity of the electrolyte is maintained by supplying a sufficient flow of high-pressure fluid. It replaces the expended ions, carries off the products of decomposition (both

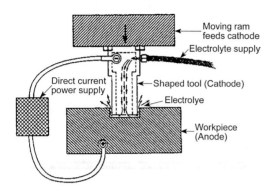

Fig. 13.15 Schematic of electrochemical machining setup. The workpiece is the anode (positive charge), while the tool is the cathode (negative charge). An electrolyte (electrically conductive solution) is pumped under high pressure between the tool and workpiece while a direct current of several thousand amperes at over 11 V is applied to the circuit.

solids and gases), and removes heat produced in the operation.

Movement of the cathode maintains the proper spacing in the gap. Equilibrium is reached when the inward motion of the cathode just matches the electrochemical removal rate. The tool never touches the workpiece, and there is no friction, no damage from heat or sparking, and no tool wear.

The removal rate for titanium alloys is 0.017 mm³/min/1000 A (0.10 in.³/min/1000 A) at 11 V minimum. The area between the cathode tool and workpiece and the rate at which the tool is fed determine the metal-removal rate.

Electrochemical machining is an economical way of fabricating aerospace hardware where the metal is hard to machine and in instances where conventional machining methods cannot be applied advantageously.

Summary

This chapter discusses the basic principles of conventional and nonconventional machining methods as they apply to titanium and titanium alloys. Special requirements for the handling of titanium alloys in machining and surface-finishing processes are covered, with descriptions of precautions to avoid health and safety hazards.

Correct machining setups for titanium require strong, sharp cutting tools, positive feeds, relatively low cutting speeds, and certain types of cutting fluids. Improper cutter rigidity or geometry, or both, contributes to vibration. Spindle speeds and feeds should be verified on each

machine to ensure correct cutting conditions. Small changes in cutting conditions can produce large changes in tool life. All machining variables should be carefully selected for optimum machining rates. Recommended speeds and feeds suggested here only serve as guides to assist in the selection of proper cutting conditions.

Successful machining of titanium depends on the use of machining equipment in excellent condition, operating in a vibration-free environment. Titanium and its alloys are processed by milling, turning, facing, boring, drilling, surface grinding, broaching, tapping, and flame cutting. Other successful techniques are chemical and electrochemical machining.

REFERENCES

13.1 F.H. Froes, D. Eylon, and H. B. Bomberger, Ed., *Titanium Technology: Present Status and Future Trends,* Titanium Development Association, Dayton, OH, 1985

13.2 G. Lutjering and J.C. Williams, *Titanium,* Springer, 2003

13.3 G. Lutjering, U. Zwicker, and W. Bunk, *Proceedings of the Fifth International Conference on Titanium,* Metallurgical Society of AIME, Warrendale, PA, 1984

13.4 P. Lacombe, R. Tricot, and G. Beranger, Ed., *Proceedings of the Sixth International Conference on Titanium* (Nice, France), Metallurgical Society of AIME, Warrendale, PA, 1988

13.5 *Proceedings of the Seventh International Conference on Titanium* (San Diego, CA), Metallurgical Society of AIME, Warrendale, PA, 1992

13.6 *Proceedings of the Eighth International Conference on Titanium* (San Birmingham, U.K.), Metallurgical Society of AIME, Warrendale, PA, 1995

13.7 *Proceedings of the Ninth International Conference on Titanium,* Metallurgical Society of AIME, Warrendale, PA, 2000

13.8 *Proceedings of the Tenth International Conference on Titanium,* Metallurgical Society of AIME, Warrendale, PA, 2004

13.9 *Proceedings of the Eleventh International Conference on Titanium,* Metallurgical Society of AIME, Warrendale, PA, 2008

13.10 *Proceedings of the Twelfth International Conference on Titanium,* Metallurgical Society of AIME, Warrendale, PA, 2012

13.11 R. Boyer, E.W. Collings, and G. Welsch, Ed., *Materials Properties Handbook: Titanium Alloys,* ASM International, 1994

13.12 M.J. Donachi, *Titanium: A Technical Guide,* 2nd ed., ASM International, 2000

13.13 G. Schneider, Machinability of Metals, *Am. Mach.,* Dec 2009

13.14 S. Kalpakjian and S.R. Schmid, *Manufacturing Processes for Engineering Materials,* Pearson Education, 2003, p 437–440

13.15 Supra Alloys, Inc., Camarillo, CA, www.supraalloys.com

13.16 American Machinist, www.americanmachinist.com

13.17 "Facts about Machining Titanium," RMI Company

CHAPTER 14

Corrosion*

CORROSION (Ref 14.1, 14.2) is of major concern in our present-day world, and no study of titanium is complete without an understanding of its corrosion resistance. Scientists and engineers recognized the exceptional corrosion resistance of titanium in natural water, oxidizing environments, acidic brines, and many other environments since it was commercialized in late 1948 (Ref 14.3–14.6). This chapter discusses corrosion resistance of titanium.

Titanium is, by nature, an active-passive metal. A titanium dioxide film that forms on the surface of titanium metal has exceptional thermodynamic stability. The film protects the reactive metal underneath in environments that routinely destroy other passive metals. However, as with any metal, titanium is not immune to all environments. In fact, it exhibits some rather spectacular failures when misapplied (Ref 14.3, 14.5, 14.7–14.14).

The sections that follow explore the corrosion performance of titanium and the effect of alloying on corrosion resistance.

Corrosion Behavior of Titanium

The types of corrosion that can occur are shown in Table 14.1. Due to its tenacious oxide film, titanium has excellent corrosion resistance (better than stainless steels, copper, and aluminum) in oxidizing environments, such as solutions containing chlorine ions, seawater, bleach, and hypochlorates.

Titanium corrosion resistance decreases in reducing environments, such as sulfuric, hydrochloric, and phosphoric acids. The general corrosion behavior of titanium can be extended into the "reducing-acid" region by adding a small amount (0.2%) of palladium (Pd), as shown in Fig. 14.1 and Table 14.2 (Ref 14.3). The addition of 0.2% Pd induces anodic passivation, increasing it by 3 orders of magnitude. Corrosion resistance in this region is also improved by means of anodic protection at the power source or with a more noble metal and by ion plating (physical vapor deposition) a thin (1 μm) coating of platinum (Pt).

Titanium is susceptible to crevice corrosion in areas of restricted circulation. However, crevice corrosion resistance is enhanced by alloying with palladium and by ion plating a thin coating of platinum.

Titanium Grades. The various titanium grades, as defined by ASTM International and the American Society of Mechanical Engineers, are numbered from 1 and upwards, where all

Table 14.1 Types of corrosion

Corrosion type	Description	Conditions
General	Uniform attack, expressed as mils/year	Titanium can passivate if certain oxidizing agents and metal ions are present (e.g., HCl with ferric, cupric ions).
Crevice	Localized in tight crevices (under deposits or structural, i.e., gaskets)	Oxidizing species depleted, metal in crevice becomes anode, HCl formed in crevice
Stress-corrosion cracking	Cracking under stress, abbreviated as stress-corrosion cracking (SCC)	Only in extreme cases (absolute methanol, red fuming nitric) for commercially pure titanium. Ti-6Al-4V can exhibit SCC in chloride.
Pitting (anodic breakdown)	Localized corrosion that can pit or lead to holes in the material	When potential exceeds breakdown for oxide layer; increase in temperature and acidity lower breakdown potential

* Adapted and revised from David E. Thomas, *Titanium and Its Alloys*, ASM International.

numbers except 6 and 8 are represented (Table 14.3).

Most of the grades are alloys with various additions of aluminum (Al), vanadium (V), nickel (Ni), ruthenium (Ru), molybdenum (Mo), chromium (Cr), and zirconium (Zr) to improve and/or combine various mechanical characteristics, heat resistance, conductivity, microstructure, creep, ductility, and corrosion resistance.

Titanium is alloyed with palladium, ruthenium, nickel, and molybdenum to obtain a significant improvement in corrosion resistance, particularly in slightly reducing environments where conditions are not conducive to the formation of the necessary protective oxide film on the metal surface.

Alpha Alloys. Expensive Grade 7 and 11 alloys are traditionally used in uninhibited reducing acids and hot acidic (and oxidizing) halide environments. Other platinum-group metals and

lower-cost titanium-palladium alloy grade 16 can be used in less aggressive environments. Interest in the addition of ruthenium, a metal similar to palladium, is increasing due to its lower cost (approximately 7 times less expensive than palladium). A Ti-0.1%Ru alloy is 20% less expensive than a Ti-0.05%Pd alloy and 40% less expensive than a Ti-0.1%Pd alloy. This opens the use of titanium-ruthenium alloy grades to replace Fe-Ni-Cr-Mo and Ni-Cr-Mo alloys in dilute reducing acids, hot aqueous halides, oxidizing-acid halides, and wet halogen environments.

Potential-pH Diagram. An introduction to metal corrosion should include a discussion about the thermodynamic nature of the metal. The easiest way to do this is to look at a potential-pH (also called Pourbaix) diagram of a metal (Ref 14.1, 14.2).

These diagrams show the thermodynamically stable phase regions as a function of electrochemical potential and solution pH. Used as a starting point, these diagrams can be very helpful in predicting corrosion problems. Care must be exercised in their use, because they do not predict the magnitude of corrosion attack. In addition, potential-pH diagrams do not always account for highly localized attack, such as crevice corrosion, pitting, and stress-corrosion cracking.

Electrochemical potential is related to the change in free energy (G) or work capacity of a system. In the case of corrosion, the system is in the metal/environment combination (Ref 14.1, 14.2). A positive increase in potential indicates that electrons are lost and that the metal-ion valence state is increased (i.e., oxidation), given by the expression:

$$M \rightarrow M^+ + e^-$$

where M represents the metal, and e^- represents electrons.

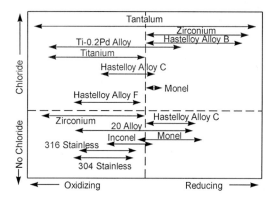

Fig. 14.1 General corrosion behavior of commercially pure titanium and titanium-palladium alloys compared with other metals and alloys in oxidizing and reducing environments, with and without chloride ions. Each metal or alloy can generally be used in those environments below its respective solid lines.

Table 14.2   Cost-optimized titanium-ruthenium and titanium-palladium alpha alloys

| Alloy | ASTM grade | Minimum ASTM strength | | | | Approximate cost ratio(a) |
| | | Yield strength | | Ultimate tensile strength | | |
		MPa	ksi	MPa	ksi	
Unalloyed Ti	2	275	40	345	50	1.00
Ti-0.15Pd	7	275	40	345	50	1.90
Ti-0.05Pd	16	275	40	345	50	1.38
Ti-0.1Ru	26	275	40	345	50	1.12
Ti-0.15Pd(b)	11	170	25	240	35	...
Ti-0.05Pd(b)	17	170	25	240	35	...
Ti-0.1Ru(b)	27	170	25	240	35	...
Alloy C-276 (Ni-Cr-Mo)	...	...	...	...	...	1.90

(a) For 6.3 mm (0.25 in.) plate, corrected for density. (b) Lower interstitial, soft grade

Table 14.3 Common ASTM International/ American Society of Mechanical Engineers titanium grades

Grade	Description
Grade 1	Unalloyed titanium, low oxygen, low strength
Grade 2	Unalloyed titanium, standard oxygen, medium strength
Grade 3	Unalloyed titanium, medium oxygen, high strength
Grade 4	Unalloyed titanium, high oxygen, extra high strength
Grade 5	Titanium alloy (6% Al, 4% V)
Grade 7	Unalloyed titanium plus 0.12–0.25% Pd, standard oxygen, medium strength
Grade 9	Titanium alloy (3% Al, 2.5% V), high strength; mainly aerospace applications
Grade 11	Unalloyed titanium plus 0.12–0.25% Pd, low oxygen, low strength
Grade 12	Titanium alloy (0.3% Mo, 0.8% Ni), high strength
Grade 13	Titanium alloy (0.5% Ni, 0.05% Ru), low oxygen
Grade 14	Titanium alloy (0.5% Ni, 0.05% Ru), standard oxygen
Grade 15	Titanium alloy (0.5% Ni, 0.05% Ru), medium oxygen
Grade 16	Unalloyed titanium plus 0.04–0.08% Pd, standard oxygen, medium strength
Grade 17	Unalloyed titanium plus 0.04–0.08% Pd, low oxygen, low strength
Grade 18	Titanium alloy (3% Al, 2.5% V, plus 0.04–0.08% Pd)
Grade 19	Titanium alloy (3% Al, 8% V, 6% Cr, 4% Zr, 4% Mo)
Grade 20	Titanium alloy (3% Al, 8% V, 6% Cr, 4% Zr, 4% Mo) plus 0.04–0.08% Pd
Grade 21	Titanium alloy (15% Mo, 3% Al, 2.7% Nb, 0.25% Si)
Grade 23	Titanium alloy (6% Al, 4% V, extralow interstitial, or ELI)
Grade 24	Titanium alloy (6% Al, 4% V) plus 0.04–0.08% Pd
Grade 25	Titanium alloy (6% Al, 4% V) plus 0.3–0.8% Ni and 0.04–0.08% Pd
Grade 26	Unalloyed titanium plus 0.08–0.14% Ru, standard oxygen, medium strength
Grade 27	Unalloyed titanium plus 0.08–0.14% Ru, low oxygen, low strength
Grade 28	Titanium alloy (3% Al, 2.5% V) plus 0.08–0.14% Ru
Grade 29	Titanium alloy (6% Al, 4% V, with ELI elements) plus 0.08–0.14% Ru

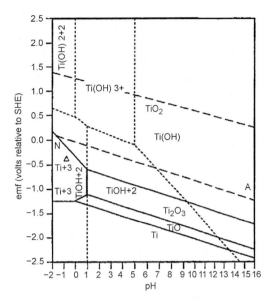

Fig. 14.2 Potential-pH diagram for titanium-water system at 85 °C (185 °F). Chloride ion activity is 10. All other ionic activities are 10^5. emf, electromotive force; SHE, standard hydrogen electrode

If, on the other hand, the potential decreases, electrons are gained and the ion-valence state decreases (i.e., reduction), given by the expression:

$$M^+ + e^- \rightarrow M$$

In any metal/environment system, it is possible to measure electrochemical potential with respect to a known reference and to predict whether corrosion can or cannot occur. The term *pH* indicates the hydrogen ion activity.

The potential-pH diagram for titanium is shown in Fig. 14.2. Titanium oxide is stable over a wide range of pH and potential, with the exception of a small region on the highly acidic (i.e., low pH) side of the diagram. It is also important to note that this region falls below the hydrogen evolution line (indicated as "A" in the diagram). This means that environments with limited oxi-

dizing power (those with limited ability to remove electrons from the metal) can lead to poor oxide stability. Because titanium relies on its oxide for corrosion resistance, these environments can lead to undesirable attack.

Another region of the diagram to be aware of is the high-pH, moderate-potential region where Ti_2O_3 is stable. This condition occurs just occasionally. At elevated temperatures, attack in this region usually manifests itself as excessive hydrogen pickup, which leads to embrittlement, as opposed to the normal thinning caused by uniform corrosion.

In general, the key to avoiding corrosion attack in titanium is to keep it out of environments typical of these two thermodynamic regions.

Corrosion Reactions. As with any electrochemical process, corrosion occurs by the exchange of electrons (Ref 14.1, 14.2). As mentioned previously, electrons are lost by the corroding metal in a process known as oxidation, and electrons are gained by an ion in solution in a process known as reduction. Both reactions result in the corrosion of a metal.

In the corrosion of titanium, two oxidation reactions occur, depending on the potential of the metal environment. In weak oxidizing environments, the oxidation reaction is:

$$Ti \rightarrow Ti^{3+} + 3e^{-1}$$

while in oxidizing environments, the reaction is:

$$Ti \rightarrow Ti^{4+} + 4e^-$$

The surface that loses electrons, thereby making its valence state greater, is said to be oxidized and is called the anode.

Likewise, the reduction reaction, which receives the lost electrons, varies depending on the environment. In weak oxidizing (or reducing) environments, the primary reactions are:

$$H^+ + e^- \rightarrow H_{ads}$$

$$2H_{ads} \rightarrow H_2$$

$$H_{ads} \rightarrow H_{abs}$$

where H_{ads} is an adsorbed (i.e., surface) hydrogen atom, and H_{abs} is an absorbed hydrogen atom.

In oxidizing environments, an oxidant gains electrons and is reduced. Examples include:

$$O_2 + 2H_2O + 4e^- \rightarrow 4OH^-$$

$$Fe^{3+} + e^- \rightarrow Fe^{2+}$$

$$Cu^{2+} + e^- \rightarrow Cu^+$$

In each case, a species is reduced by gaining electrons at a surface called the cathode. Note that in receiving electrons, the cathode valence state is reduced, thus the term *reduction*.

Therefore, it is necessary to have an anode to lose electrons and a cathode to receive electrons for the overall corrosion process to take place. If either electrode is absent or altered, the corrosion process is affected.

Because oxidizing environments enhance the stability of titanium dioxide, promoting the reduction process by adding oxidants also reduces the chance that titanium will corrode. As discussed later in this chapter, this is precisely what happens.

Mixed-Potential Theory. Because it is possible to separate the corrosion process into two distinct reactions, it is convenient to examine the effects of variations in anodic (oxidizing) and cathodic (reducing) reactions. Figure 14.3 is a diagram of the mixed-potential theory (Ref 14.1, 14.2).

Mixed-potential theory uses the relationship between potential and current density (i.e., corrosion rate) to study the possibility and the magnitude of corrosion. The two parameters are graphically presented in two-dimensional plots such as those in Fig. 14.3.

Behavior for an active metal is depicted by Fig. 14.3(a) and for a passive metal in Fig. 14.3(b). The anode reaction is represented by the line with upward slope; the cathode reactions are represented by lines with downward slope. The point at which these lines cross gives the potential

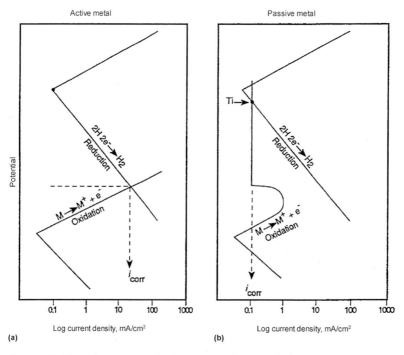

Fig. 14.3 Mixed-potential behavior of an active metal and a passive metal in an acid solution

and current density at which corrosion occurs. Because current density is directly related to corrosion rate, the active metal corrodes at a higher rate than the passive metal (comparing i_{corr}). Moving either of these lines changes the corrosion response. What causes these lines to move and how corrosion performance is affected is discussed later in this chapter. In general, titanium behavior falls into the passive-metal category.

Polarization. The term *polarization* is used to indicate a change in electrochemical potential as a result of a change in the surface characteristics of a conductor (usually metal) in an electrolyte (Ref 14.1, 14.2). Polarization results from electrolyte concentration, species variations, and artificial (i.e., external) application of potential. Polarization effects are especially pronounced in titanium because of its highly reactive surface and passive oxide.

A good example of polarization is when an oxidizing ion, such as the ferric ion (Fe^{3+}) is added to a hydrochloric acid (HCl) solution in which titanium is actively corroding. The ferric ion (an oxidizing agent) polarizes titanium in the anodic direction, a process called cathodic depolarization. If enough ferric ion is added, polarization is sufficient to promote the formation of a passive film on the titanium, stopping active corrosion. The potential-pH diagram in Fig. 14.2 shows why this occurs. The titanium electrochemical potential (at low pH) moves vertically out of the corrosion region into the passive region.

Similarly, it is possible to polarize a surface in the cathodic direction. One method to do this is to remove oxygen from the environment. Unfortunately, for titanium, cathodic polarization is not typically effective as a method to reduce corrosion attack. However, it is often used with other metals, such as carbon steel, in a process known as cathodic protection.

Forms of Corrosion

The forms of corrosion attack associated with titanium and the effect various environments have on each form of corrosion (Ref 14.1, 14.2) are discussed in this section.

Uniform Corrosion

Uniform corrosion, the most common form of corrosion attack, is characterized by general thinning of a metal. It is uniform because attack generally proceeds at the same rate on all exposed surfaces. Because titanium relies on the oxide

film on its surface for corrosion resistance, it can suffer dramatic uniform attack if exposed to inappropriate environments. Several classes of environments and their effect on titanium are discussed subsequently.

Acids. In general, acidic environments are separated into two categories: oxidizing and reducing. Titanium exhibits excellent resistance in most oxidizing acids, notably nitric acid and chromic acid. However, in reducing acids, such as hydrochloric acid and sulfuric acid, titanium exhibits moderate resistance, and only at low concentration.

Nitric Acid. Titanium has been used successfully for many years in nitric acid production and storage. It is capable of handling high temperatures over a wide concentration range with little or no corrosion attack. In heat-exchanger equipment, such as cooler condensers and reboilers, finite corrosion attack has been observed, with evidence of preferential attack of the weld zone.

Laboratory studies indicate that this type of attack is specific to heat-exchanger tubing, because oxidizers (in this case Ti^{4+}) do not build to a level sufficient to promote passivity. As a result, titanium performance is, at best, erratic. Therefore, care should be exercised when using titanium in heat-exchanger equipment.

The overall corrosion performance of titanium in nitric acid is shown in Fig. 14.4. This is a good first approximation in determining the suitability

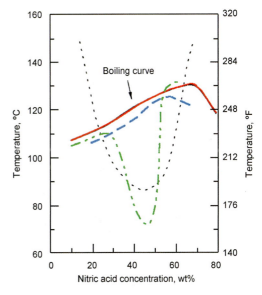

Fig. 14.4 Isocorrosion diagram for titanium in nitric acid. Curves represent corrosion rates of 0.12 mm (0.005 in.) per year. Courtesy of RMI Company

of titanium in a particular nitric acid environment. However, it should be used carefully because it is based on laboratory data rather than generated under the precise circumstances typical of a given application.

The diagram in Fig. 14.4, termed an isocorrosion diagram, enables depicting a great amount of corrosion data on one two-dimensional plot. Contour lines are placed on the diagram to indicate the range of temperature and concentration where unacceptable corrosion attack occurs.

Titanium also corrodes in red fuming nitric acid (RFNA). It can react so violently that it ignites, resulting in serious accidents. Attack can be avoided either by keeping RFNA away from titanium, by adding at least 1.5 wt% water, or by keeping the NO_2 content below 6 wt%. This situation is illustrated in Fig. 14.5.

Chromic Acid Titanium also exhibits excellent corrosion resistance in chromic acid at all concentrations up to boiling. Typical performance in chromic acid service is shown in Table 14.4.

Hydrochloric acid (HCl) is one of the most common reducing acids. Titanium exhibits limited corrosion resistance in anything but dilute HCl. This range of performance is illustrated in Fig. 14.6. Concentrations higher than 1% HCl are

corrosive to titanium. Also illustrated in this diagram are the effects of alloying on corrosion attack. These effects are discussed later in the section "Alloying for Corrosion Prevention" in this chapter.

Sulfuric acid, another reducing acid, attacks titanium at nearly all concentrations. The isocorrosion diagram for titanium in sulfuric acid is shown in Fig. 14.7.

Phosphoric Acid. Like sulfuric and hydrochloric acid, phosphoric acid is a reducing acid and generally is not suitable for titanium. An isocorrosion diagram for phosphoric acid is shown in Fig. 14.8.

Hydrofluoric acid (HF) is the most corrosive acid for titanium because of its ability to easily dissolve titanium dioxide films. Because titanium has little inherent corrosion resistance without its oxide film, corrosion rates could be expected to be dramatic. Corrosion rates of titanium in HF are measured in mm/min (mils/min) instead of the normal mm/yr (mils/yr).

Two types of additions that reduce the corrosive effects in very dilute HF acid solutions are an

Table 14.4 Corrosion of unalloyed titanium in chromic acid

Acid concentration, %	Temperature, °C (°F)	Corrosion rate, mm/yr (mils/yr)
20	21 (70)	4.0 (158) max
10	Boiling	Nil
10	Boiling	<5.0 (197)
20	Room	Nil

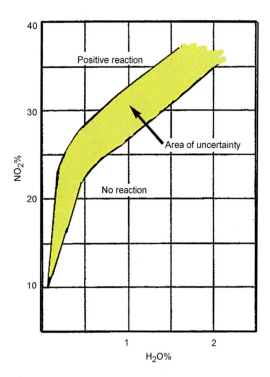

Fig. 14.5 Acid-composition limits to avoid rapid pyrophoric reactions of titanium with red fuming nitric acid

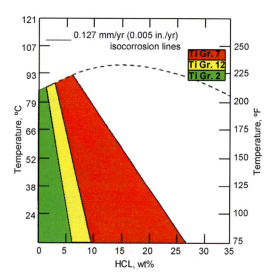

Fig. 14.6 Isocorrosion diagram for titanium alloys in pure, naturally aerated hydrochloric acid (HCl) solutions

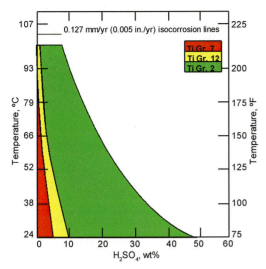

Fig. 14.7 Isocorrosion diagram for titanium alloys in pure, naturally aerated sulfuric acid (H_2SO_4) solutions

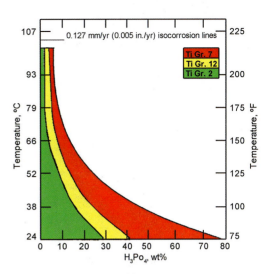

Fig. 14.8 Isocorrosion diagram for titanium alloys in pure, naturally aerated phosphoric acid (H_3PO_4) solutions

oxidizing agent (e.g., HNO_3) and complexing metal-ion corrosion inhibitors. The complexing metal ion takes the detrimental F^- ion out of the environment as an insoluble complex. Inhibitors in this category include Al^{3+}, Fe^{3+}, and Cr^{6+}. Probably the most effective agent is Al^{3+} because it produces a strong AlF^{6-} complex.

Anodic protection, discussed later in the section "Corrosion Inhibitors" in this chapter, is somewhat effective because it promotes reformation of the oxide. However, because HF dissolves the oxide directly, anodic protection merely slows the corrosion process; it does not eliminate it.

Other reducing acids produce uniform corrosion attack similar to that found with HCl and H_2SO_4, including hydrobromic, hydriodic, sulfamic, oxalic, trichloroacetic, and formic acids.

Weakly reducing acids, such as boric acid, carbonic acid, aqueous hydrogen sulfide, and sulfurous acid do not attack titanium. Most organic acids do not attack titanium, with the exceptions of oxalic, trichloroacetic, and formic acids. Corrosion rates typical of these acids are shown in Table 14.5.

Corrosion Inhibitors. The corrosion resistance of titanium in sulfuric acid and all other reducing acids changes dramatically when inhibitors (Ref 14.1, 14.2), often as inadvertent impurities, are present. Tables 14.6 and 14.7 list the important oxidizing agents and other corrosion-inhibiting compounds, as well as the effects these agents have in reducing attack in normally corrosive mineral acids.

In general, the characteristic that distinguishes these compounds is that they are oxidizing agents. What this means to titanium is illustrated by the Pourbaix diagram in Fig. 14.2. The area where corrosion is thermodynamically possible (low-pH reducing environments) is represented by reducing acids such as hydrochloric, sulfuric, and phosphoric acids. Adding an oxidizing agent shifts the environment upward at constant pH to a point where corrosion no longer occurs. This effect is also illustrated in the mixed-potential diagram in Fig. 14.9.

The more potent the oxidizing agent, the more effective it is at enhancing titanium corrosion resistance. Artificial oxidizing agents are produced by applying an anodic potential to a material with an external power source. This technique, termed anodic protection, is also very effective in reducing corrosion attack on titanium in acids such as HCl, H_2SO_4, and, as mentioned previously, HF. This effect is illustrated in Table 14.8.

Halide Solutions/Metal Salts. Titanium is unique in its corrosion resistance in most neutral-pH, halide-containing solutions (except fluoride). This resistance is illustrated in Table 14.9.

With the exception of concentrated, hydrolyzable chloride salts, such as aluminum chloride and zinc chloride, titanium exhibits little or no corrosion attack. This attribute extends to the highly oxidizing metal salts, such as ferric chloride, that typically cause severe pitting and localized attack in other passive-alloy systems, such as the stainless steels.

Table 14.5 Resistance of titanium alloys to organic acids

Acid	Concentration, wt%	Alloy grade	Temperature °C	Temperature °F	Corrosion rate mm/yr	Corrosion rate mils/yr
Acetic	0–99.5	2, 7, 12	Boiling		0.00	0.00
Adipic	67	2	240	465	0.00	0.00
Citric, aerated	10–50	2	100	212	0.01	0.4
Citric	50	2	Boiling		0.35	14
	50	7, 12	Boiling		0.01	0.4
Di-and mono-chloroacetic	100	2	Boiling		<0.013	<0.5
Formic, aerated	25–90	2	100	212	0.001	0.04
	25	2	Boiling		2.4	95
Formic	45	2	Boiling		11.0	433
	45	7, 12	Boiling		0.00	0.00
	10	2	Boiling		0.00	0.00
Lactic, aerated	10	2	Boiling		0.014	0.56
Lactic	10	2	100	212	0.048	1.9
	25	2	Boiling		0.028	1.11
	85–100	2	Boiling		0.01	0.4
Oxalic	0.5	2	60	140	2.4	95
	1	2	35	95	0.15	6
	10	7	Boiling		32.3	1275
Stearic	100	2	180	355	0.003	0.12
Tartaric	10–50	2	100	212	0.013	0.5
Terephthalic	77	2	225	435	0.00	0.00
Trichloroacetic	100	2	Boiling		14.6	575

Table 14.6 Species that inhibit corrosion of titanium alloys in reducing acids

Inhibitor category	Species	Relative inhibitor potency
Oxidizing metal cations	Ti^{4+}, Fe^{3+}, Cu^{2+}, Hg^{4+}, Sn^{4+}	High
	VO_2^+, Te^{4+}, Te^{6+}, Se^{4+}, Se^{6+}	High
	Ni^{2+}	Low
Oxidizing anions	ClO_4^{2-}, $Cr_2O_7^{2-}$, MoO_4^{2-}, MnO_4^{2-}	Very high
	WO_4^-, IO_3^-, VO_4^{3-}, VO_3^-	Very high
	NO_3^-, NO_2^-, $S_2O_3^{2-}$	Moderate
Precious metal ions	Pt^{2+}, Pt^{4+}, Pd^{2+}, Ru^{3+}, Ir^{3+}	High
	Rh^{3+}, Au^{3+}	High
Oxidizing organic compounds	Picric acid, o-dinitrobenzene, 8-nitroquinoline, m-nitroacetanilide, trinitrobenzoic acid, and certain other nitro-, nitroso-, and quinone-organics	Moderately high
Others	O_2, H_2O_2, ClO_3^-, OCl^-	Moderate

Table 14.7 Effect of multivalent metal ions on corrosion resistance of titanium in boiling reducing acids

Inhibiting ion	Boiling 5 wt% HCl ion concentration, mm/yr (mils/yr) 0 ppm	100 ppm	500 ppm	Boiling 10 wt% H_2SO_4 ion concentration, mm/yr (mils/yr) 0 ppm	100 ppm	500 ppm
Fe^{3+}	29.0 (1142)	0.025 (1.0)	0.020 (0.8)	>76.2 (3000)	0.208 (8.3)	0.069 (2.7)
Cu^{2+}	29.0 (1142)	0.033 (1.3)	0.030 (1.2)	>76.2 (3000)	0.419 (16.7)	0.361 (14.4)
Mo^{6+}	29.0 (1142)	0.000	0.000	>76.2 (3000)	0.001 (0.04)	0.000
Cr^{6+}	29.0 (1142)	0.000	0.000	>76.2 (3000)	0.001 (0.04)	0.001 (0.04)
V^{5+}	29.0 (1142)	0.020 (0.8)	0.008 (0.3)	>76.2 (3000)	0.005 (0.20)	0.005 (0.20)

Titanium is also highly resistant to chlorine compounds, as shown in Table 14.10. While titanium resistance is excellent in most halide-containing environments, localized corrosion in the form of crevice corrosion can occur under certain conditions. This phenomenon is discussed in the section "Localized Attack" in this chapter.

Dry halide gases (chlorine, bromine, and iodine) react rapidly with titanium. The reaction can be pyrophoric under severe circumstances. However, water is a powerful inhibitor to this reaction, even in concentrations as low as 1%.

Alkaline Environments. Titanium exhibits excellent resistance to alkaline environments at moderate temperatures. However, corrosion occurs in hot, concentrated sodium and potassium hydroxide. In addition, hydrogen embrittlement can accompany this attack, leading to unexpected brittle fracture. Corrosion rates for several alkaline environments are shown in Table 14.11.

Natural Waters. Titanium is immune to corrosion attack in all natural waters. Titanium is also resistant to steam at temperatures as high as 600 °C (1110 °F). Titanium resistance is typified by passive film thickening during exposure, leading to slight surface discoloration.

In seawater as well as brackish and polluted water, titanium exhibits no uniform corrosion at temperatures as high as 500 °C (930 °F). Localized attack is not observed in titanium in ambient seawater, even under the most severe biological fouling. No evidence of microbiologically

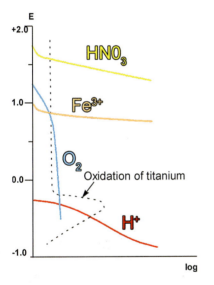

Fig. 14.9 Effect of oxidizing agents (HNO_3, Fe^{3+}, O_2, H^+) on mixed-potential diagram for titanium

induced corrosion has ever been identified on titanium. Titanium is so resistant to attack in seawater that it has become the material of choice in critical seawater applications in the chemical, petrochemical, desalination, and power-generation industries.

Organic Compounds. Titanium alloys are resistant to most organic compounds, including alcohols, aldehydes, ketones, ethers, and hydrocarbons, because of the presence of trace amounts of water. Typical corrosion performance in several organic environments is shown in Table 14.12. In totally anhydrous organics, care should be exercised because passive film formation is impeded and localized attack and stress-corrosion cracking can occur. This phenomenon is primarily restricted to anhydrous methanol and is discussed in the section "Environmental Cracking" in this chapter.

Hydrocarbons that contain halides, especially chlorides, do not typically pose a threat to titanium. They are used effectively as degreasers and cleaning agents. However, careless use of these hydrocarbons can lead to the formation of HCl if water is allowed to accumulate at elevated temperature.

Organic acids, as discussed previously, promote corrosion attack if they are sufficiently reducing in nature. However, most organic acids do not attack titanium. Alloys, such as Ti-0.15Pd and Ti-0.3Mo-0.8Ni are significantly more resistant in reducing organic acids, such as oxalic acid. These effects are discussed later in the section "Alloying for Corrosion Prevention" in this chapter.

Gaseous Environments. Most gases do not attack titanium at temperatures as high as 175 °C (350 °F). As a result, titanium is not attacked in marine, industrial, urban, and farm atmospheres. Oxygen and sulfur-bearing gases do not affect titanium up to 315 °C (600 °F). At temperatures

Table 14.8 Effect of impressed anodic potential on corrosion resistance of unalloyed titanium in reducing acids

Acid rate	Temperature		Applied potential (volts vs. H_2)	Corrosion rate		Reduction of corrosion, times
	°C	°F		mm/yr	mils/yr	
40% sulfuric	60	140	+2.1	0.005	0.2	11,000
	90	195	+1.4	0.07	2.8	896
	114	237	+2.6	1.8	71.5	189
60% sulfuric	60	140	+1.7	0.035	1.4	662
	90	195	+3.0	0.10	4	163
37% hydrochloric	60	140	+1.7	0.068	2.7	2,080
60% phosphoric	60	140	+2.7	0.018	0.72	307
	90	195	+2.0	0.50	20	100
50% formic	Boiling		+1.4	0.083	3.3	70
25% oxalic	Boiling		+1.6	0.25	9.9	350
20% sulfamic	90	195	+0.7	0.005	0.2	2,710

Table 14.9 Corrosion rates for unalloyed titanium in inorganic environments

Environment(a)	Weight, %	Temperature °C	°F	Corrosion rate mm/yr	mils/yr
Aluminum chloride	5, 10	100	212	<0.005	<0.2
	25	21	70	0.001	0.04
	25	100	212	6.6–50 pitting	260–2000 pitting
Barium chloride	5, 20	100	212	0.0025	0.1
Calcium chloride	5, 20, 40	100	212	<0.025	<1
	70	175	350	>25 pitting	>1000 pitting
Calcium hypochlorite	2, 6	100	212	0.001	0.05
Cupric chloride	1–20	100	212	0.001	0.05
	55	115	240	0.0025	0.1
Cuprous chloride	50	32	90	Nil	
Ferric chloride	1–30	100	212	Nil	
	50	150	302	0.0025	0.1
Hydrogen peroxide	5	21	70	Nil	
	30	24	75	1.14	45.0
Hydrogen sulfide + water	...	21	70	Nil	
Mercuric chloride	1, 10, 55	100	212	Nil to 0.1	Nil to 4.0
Magnesium chloride	5, 42	100	212	Nil	
	50	200	393	Pitting	
Nickel chloride	20	93	200	Nil	
Potassium chloride	36	110	232	0.01	0.5
Potassium bromide	Saturated	21	70	Nil	
Potassium iodide	Saturated	21	70	Nil	
Sodium chloride	1 to saturated	Boiling		Nil	
Sodium carbonate	20	Boiling		Nil	
Sodium cyanide	Saturated	21	70	Nil	
Sodium hydroxide	10	Boiling		0.02	0.84
	28	21	70	0.0025	0.1
	40	80	176	0.127	5.0
Sodium hypochlorite	2, 10	100	212	Nil	
Sodium nitrate	Saturated	21	70	Nil	
Sodium phosphate	Saturated	21	70	Nil	
Sodium sulfite	Saturated	21	70	Nil	
Sodium sulfide	10	Boiling		Nil	
Sodium sulfate	Saturated	21	70	Nil	
Stannic chloride	25, 100	32	90	Nil	
Zinc chloride	5, 20	Boiling		Nil	
	50, 75	150	302	Nil	
	75, 89, 90	150	302	0.6–76.5 pitting	24–3000 pitting

(a) Not aerated

Table 14.10 Corrosion of unalloyed titanium in solutions containing oxidizing chlorine compounds

Reagent	Concentration, wt%	Temperature °C	°F	Corrosion rates mm/yr	mils/yr
Cl$_2$-saturated water	...	75	167	0.003	0.12
	...	88	190	0.002(a)	0.08(a)
	...	97	206	0.07	2.8
NaOCl	6	25	77	0.000	0.000
ClO$_2$ + HOCl	15	43	109	0.000	0.000
ClO$_2$ + steam	5	100	212	0.005	0.2
Ca(OCl)$_2$	2	100	212	0.001	0.04
	6	100	212	0.001	0.04
	18	25	77	0.000	0.000
HOCl + ClO$_2$ + Cl$_2$	17	38	100	0.000	0.000

(a) Welded sample

above 370 °C (700 °F), surface degradation can eventually occur, leading to embrittlement. The higher the temperature, the more rapid the attack.

Highly oxidizing gases such as dry chlorine, dry bromine, and nitrogen tetroxide can produce violent, exothermic reactions that lead to ignition

Table 14.11 Corrosion of unalloyed titanium in alkaline solutions

Media	Concentration, wt%	Temperature		Corrosion rate	
		°C	°F	mm/yr	mils/yr
Ammonium hydroxide	28	26	79	0.002	0.08
	70		Boiling	0.000	0.000
Sodium carbonate	20		Boiling	0.000	0.000
Sodium hydroxide	28	25	77	0.003	0.12
	10		Boiling	0.02	0.8
	40	66	150	0.038	1.5
	40	93	200	0.064	2.5
	40	121	250	0.13	5.2
	50	66	150	0.018	0.72
	50–73	188	370	<1.1	<44
	73	110	230	0.05	2
	73		Boiling	0.13	5.2
Potassium hydroxide	10		Boiling	0.13	5.2
	25		Boiling	0.3	12
	50	25	77	0.010	0.4
	50		Boiling	2.7	107

Table 14.12 Corrosion of unalloyed titanium in organic compounds

Medium	Concentration, wt%	Temperature		Corrosion rate	
		°C	°F	mm/yr	mils/yr
Acetic anhydride	99–99.5	20 to boiling	68 to boiling	<0.13	<5.2
Adipic acid	0–57	204	400	<0.05	<2
Adipic acid + 200/0 glutaric acid + 5% acetic acid	25	200	390	0.000	0.000
Aniline hydrochloride	5–20	35/100	95/212	<0.001	<0.04
Benzene + HCl + NaCl	Vapor + liquid	80	176	0.005	0.2
Carbon tetrachloride	99		Boiling	0.003	0.12
	100		Boiling	0.003	0.12
Chloroform	100		Boiling	0.000	0.000
Chloroform + water	50		Boiling	0.12	4.8
Cyclohexane + traces of formic acid	...	150	302	0.0003	0.12
Ethyl alcohol	95		Boiling	0.013	0.52
Ethylene dichloride	100		Boiling	<0.13	<5.2
Formaldehyde	37		Boiling	<0.13	<5.2
Tetrachloroethylene	100		Boiling	0.000	0.000
Tetrachloroethylene + water	...		Boiling	0.13	5.2
Tetrachloroethane	100		Boiling	0.000	0.000
Trichloroethylene	99		Boiling	<0.13	<5.2

of titanium. A minimum water content is necessary to preclude this occurrence. The critical amount of water depends on service conditions but is typically less than 0.5%.

Localized Attack

Localized attack is characterized by corrosion at discrete sites on a metal surface. These sites are randomly distributed throughout the surface and are associated with a deposit or crevice.

Two types of localized attack are pitting and crevice corrosion. Localized attack is of particular importance because it leads to premature failure and often unpredictable service life.

Pitting. Titanium is resistant to pitting in nearly all environments. The exception to this is in hot,

hydrolyzable metal salts, such as magnesium, aluminum, and zinc chlorides.

Typical pitting potentials, that is, oxidizing levels at which the passive film of titanium breaks down, are given in Table 14.13. The pitting potential of titanium is well above that associated with naturally occurring environments. Pitting potential decreases with increasing temperature but remains well above 1.0 V (Ag/AgCl). Titanium alloys typically have poorer resistance to pitting than commercially pure titanium.

Crevice Corrosion. Titanium is susceptible to crevice corrosion in chloride environments above 80 °C (175 °F). Crevice corrosion is the most often overlooked form of attack on titanium. In general, pH and temperature are the most important variables that influence crevice-corrosion

Table 14.13 Anodic breakdown potentials for titanium alloys in chloride solutions

Alloy	Solution	pH	Temperature °C	Temperature °F	E_b(a), V
Titanium grade 2	1N NaCl	7	25	77	+11.0
Titanium grade 5	1N NaCl	7	25	77	52
Titanium grade 2	Saturated NaCl(b)	1 and 7	25	77	9.6
Titanium grade 12	Saturated NaCl(b)	1 and 7	25	77	9.6
Titanium grade 7	Saturated NaCl(b)	1 and 7	25	77	9.6
Titanium grade 5	Saturated NaCl	1 and 7	25	77	8.9
Titanium grade 2	Saturated NaCl	1 and 7	95	203	5.0–6.5
Titanium grade 12	Saturated NaCl	1 and 7	95	203	5.0–5.7
Titanium grade 7	Saturated NaCl	1 and 7	95	203	5.2–7.0
Titanium grade 5	Saturated NaCl	1 and 7	95	203	2.5–3.4
Titanium grade 2	1N NaCl		125	257	−4.4
		7	150	302	−2.2
			175	347	−1.2
			200	392	−1.2
Titanium grade 12	Seawater	8	245	473	2.3
	C_2 saturated seawater	8	245	473	3.3
Titanium grade 2	1N KCl + 0.2M H_2SO_4	...	25	77	80.0

(a) E_b vs. Ag/AgCl reference electrode. Ag/AgCl is a standard reference electrode used to measure electrochemical potential. The potential indicated is thus referenced to the Ag/AgCl electrode. (b) Similar values measured for synthetic seawater, pH 8

attack. Figure 14.10 illustrates this effect for three grades of titanium. Chloride concentration has little effect on crevice-corrosion susceptibility.

Alloys such as Ti-0.15Pd (Grade 7) and Ti-0.3Mo-0.8Ni (Grade 12) extend titanium resistance to crevice corrosion at higher temperature and lower pH. At pH levels above 3.0, Ti-0.3Mo-0.8Ni exhibits crevice-corrosion resistance at temperatures to 290 °C (550 °F). The alloy Ti-0.15Pd is useful at pH levels below 1.0 and is generally recognized as the most crevice-corrosion-resistant titanium alloy.

Crevice geometry is also critical to titanium crevice-corrosion resistance. Crevices produced by flexible polymers such as polytetrafluoroethylene aggravate this type of attack by producing very tight crevices. Metal-to-metal crevices are generally less severe, because they allow the area to be refreshed from the bulk environment. Possibly the most effective crevice protection is that produced by carbonaceous scales. These deposits form very tight interfaces, often incorporating chlorides.

Oxidizing agents often exacerbate the problem by promoting the crevice-corrosion mechanism. The oxidizing agent can decrease initiation or incubation time, which is the most critical aspect of titanium crevice attack.

Galvanic Corrosion

Corrosion attack enhanced by contact with dissimilar metals is called galvanic corrosion (Ref 14.1, 14.2). This type of corrosion is rarely observed for titanium, except in environments where the metal corrodes rapidly anyway (i.e.,

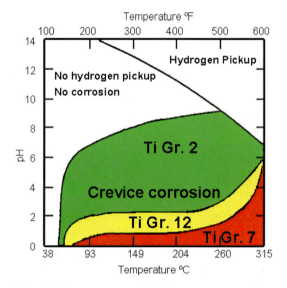

Fig. 14.10 Temperature-pit limits for crevice corrosion of titanium grades 2, 7, and 12 in concentrated sodium chloride (NaCl) brine

strongly reducing acids). This is illustrated in the galvanic series of Table 14.14.

Metals at the top of the list are generally cathodes while metals at the bottom are generally anodes. As discussed in the section "Mixed-Potential Theory" in this chapter, the anode corrodes and the cathode does not. Thus, titanium (normally a cathode) is not affected by dissimilar-metal galvanic attack. However, the other metal in the galvanic couple can suffer significant accelerated attack.

Galvanic attack in seawater for several alloys in contact with titanium is shown in Fig. 14.11.

This figure illustrates an important point about galvanic corrosion. If the surface area of titanium (the cathode) is appreciably lower than the surface area of the other metal (the anode), galvanic effects are minimal. However, if the surface areas are reversed, galvanic corrosion can become significant.

Because titanium, as the cathode, supports the cathodic portion of the corrosion reaction, it is possible for titanium to absorb hydrogen and become embrittled. This potential occurrence is almost always associated with temperatures above 80 °C (180 °F) when titanium is coupled to actively corroding metal. In such cases, care should be exercised to avoid galvanic contact, if at all possible.

Methods used to avoid galvanic attack include:

- Removing less-noble metals
- Insulating all joints
- Using controlled cathodic protection of the anodic member of the couple
- Coating the cathode

Erosion-Corrosion

When corrosive environments are in motion, metal attack can increase. This is a phenomenon known as erosion-corrosion. Titanium, by virtue of its hard, dense oxide, is highly resistant to erosion-corrosion in all but the most abrasive environments. For example, seawater velocities in excess of 30 m/s (100 ft/s) do not attack titanium.

In sand-laden seawater, titanium is resistant to velocities up to 6 m/s (20 ft/s). Environments that cause erosion-corrosion of titanium are so abrasive that it is questionable whether corrosion is involved at all.

Environmental Cracking

Environmental cracking typically shows itself as either stress-corrosion cracking or hydrogen embrittlement. Stress-corrosion cracking occurs as a result of the presence of some species, usually chlorides, that produce or promote cracking in a highly localized area. Hydrogen embrittlement occurs as a result of hydrogen absorption

Table 14.14 Galvanic series in flowing water (4 m/s, or 13 ft/s) at approximately 25 °C (75 °F)

Material	Steady-state electrode potential (saturated calomel half-cell), V
Graphite	+0.25
Platinum	+0.15
Zirconium	−0.04
Type 316 stainless steel (passive)	−0.05
Type 304 stainless steel (passive)	−0.08
Monel 400	−0.08
Hastelloy C	−0.08
Titanium	−0.10
Silver	−0.13
Type 410 stainless steel (passive)	−0.15
Type 316 stainless steel (active)	−0.18
Nickel	−0.20
Type 430 stainless steel (passive)	−0.22
Copper alloy 715 (70-30 cupro-nickel)	−0.25
Copper alloy 706 (90-10 cupro-nickel)	−0.28
Copper alloy 442 (admiralty brass)	−0.29
G bronze	−0.31
Copper alloy 687 (aluminum brass)	−0.32
Copper	−0.36
Alloy 4S4 (naval rolled brass)	−0.40
Type 410 stainless steel (active)	−0.52
Type 304 stainless steel (active)	−0.53
Type 430 stainless steel (active)	−0.57
Carbon steel	−0.61
Cast iron	−0.61
Aluminum 3003-H	−0.79
Zinc	−1.03

Source: Timet

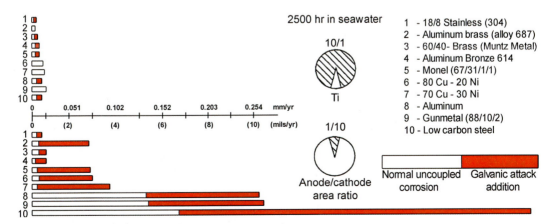

Fig. 14.11 Effect of galvanic coupling of titanium to less corrosion-resistant metals in seawater

and subsequent transformation of titanium to brittle titanium hydride.

Titanium has excellent resistance to stress-corrosion cracking in most aqueous environments, while titanium alloys typically exhibit somewhat lower resistance. Fortunately, stress-corrosion cracking is almost exclusively confined to the laboratory. Environments known to promote stress-corrosion cracking are presented in Table 14.15 and are discussed subsequently.

Oxidizers. Highly oxidizing environments, such as nitrogen tetroxide and red fuming nitric acid, produce severe cracking in all titanium alloys. However, water is a powerful stress-corrosion-cracking inhibitor in these environments.

Methanol. Anhydrous methanol, which contains chlorides, promotes stress-corrosion cracking in most titanium alloys. Alloys containing aluminum contents above 6% are particularly susceptible. As with oxidizing environments, small water additions (as little as 1.5%) greatly reduce stress-corrosion-cracking susceptibility.

Other Alcohols. Most higher-order alcohols, such as butanol, ethanol, or propanol, do not pro-

Table 14.15   Environments that promote stress-corrosion cracking of unalloyed titanium and titanium alloys

Oxidizers	Temperature		Susceptible titanium materials
	°C	°F	
Medium			
Red fuming nitric acid	Room temperature		Ti, Ti-8Mn, Ti-6Al-4V, Ti-5Al-2.5Sn, Ti-2Fe-2Cr-2Mo
Nitrogen tetroxide (no excess NO)	30–75	85–165	Ti-6Al-4V
Organic compounds			
Methanol	Room temperature		Ti-6Al-4V
Chloride, bromide			Ti-6Al-4V, Ti-8Al-1Mo-1V
			Ti-75Al, Ti-6Al-4V, Ti-8Al-1Mo-1V,
			Ti-5Al-2.5Sn, Ti-4Al-3Mo-1V
Methyl chloroform	370	700	Ti-8Al-1Mo-1V, Ti-6Al-4V, Ti-5Al-2.5Sn,
			Ti-13V-11Cr-3Al
Ethyl alcohol	Room temperature		Ti-8Al-1Mo-1V, Ti-5Al-2.5Sn
Ethylene glycol	Room temperature		Ti-8Al-1Mo-1V
Trichloroethylene	370	700	Ti-8Al-1Mo-1V, Ti-5Al-2.5Sn
	620	1150	
	815	1500	
Trichlorofluoroethane (Freon PCA, DuPont Fluoroproducts)	788	1450	Ti-8Al-1Mo-1V, Ti-5Al-2.5Sn, Ti-6Al-4V, Ti-13V-11Cr-3Al
Chlorinated diphenyl (Aerochlor 1262)	315–370	600–700	Ti-5Al-2.5Sn
Hot salt			
Various chloride salts, residues	288–426	550–800	All commercial alloys
Metal embrittlement			
Cadmium	>321	>610	Ti-4Al-4Mn
	329–400	625–750	Ti-8Mn
Mercury	Room temperature		Ti-75Al, Ti-6Al-4V
	370	700	Ti-13V-11Cr-3Al
Silver			
Ag plate	470	875	Ti-7Al-4Mo, Ti-5Al-2.5Sn
AgCl	370–480	700–900	Ti-7Al-4Mo, Ti-5Al-2.5Sn
Ag-5Al-2.5Mn	340	650	Ti-6Al-4V, Ti-8Al-1Mo-1V
Seawater			
	Ambient		Unalloyed Ti (with high oxygen content, i.e., 0.3178)
			Ti-8Mn
			Ti-2.5Al-1Mo-11Sn-5Zr-0.2Si (IMI 679)
			Ti-3Al-11Cr-13V
			Ti-4Al-4Mn
			Ti-5Al-2.5Sn
			Ti-6Al-4V
Miscellaneous			
Chlorine	288	550	Ti-8Al-1Mo1V
Hydrochloric acid, 10%	Room temperature		Ti-5Al
	35	95	Ti-5Al-2.5Sn
	340	650	Ti-8Al-1Mo-1V
Sulfuric acid, 7–60%	Room temperature		Ti-5Al

mote stress corrosion. However, ethanol induces stress-corrosion cracking in Ti-8Al-1Mo-1V, a particularly sensitive alloy.

Hot Salt. Titanium is susceptible to stress-corrosion cracking in halides at temperatures higher than 260 °C (500 °F); sodium chloride is particularly aggressive. This phenomenon, known as hot salt corrosion, is typically associated with laboratory creep testing and stress-relief heat treatment and rarely occurs in service. (This is interesting, considering all the titanium flying over the oceans at temperatures well within the hot salt corrosion range.) Alloys containing appreciable amounts of aluminum are particularly sensitive. Commercially pure titanium is considered immune to this type of attack.

Seawater. As mentioned previously, titanium and its alloys are highly resistant to stress-corrosion cracking in aqueous environments. However, under laboratory conditions, the more highly alloyed titanium alloys can crack. This is most often presented as a reduction in material fracture toughness in seawater compared with air.

Fortunately, laboratory conditions have not been observed in service. However, laboratory observations indicate the most important alloying additions that promote stress-corrosion cracking are aluminum and oxygen. Molybdenum, niobium, tantalum, and vanadium generally improve cracking resistance.

In addition, heat treatment of aluminum-bearing alpha and alpha-beta alloys in the temperature range of 480 to 815 °C (900 to 1500 °F) reduces stress-corrosion-cracking resistance.

Beta titanium alloys, especially those containing molybdenum, are nearly immune to stress-corrosion cracking in seawater.

Hydrogen embrittlement (Ref 14.3) is caused by excessive absorption of hydrogen and transformation of the titanium alpha phase to brittle titanium hydride. The phenomenon is associated with either hot caustic environments or with cathodic polarization in hot, aqueous environments. The latter phenomenon, cathodic polarization, leading to cathodic hydrogen pickup, is more frequently observed. Reduction of hydrogen on the cathode leads to hydrogen adsorption and diffusion into the bulk metal, leading to hydriding.

Mechanisms by which titanium is polarized cathodically include inadvertent cathodic protection, galvanic coupling, and continuous mechanical surface damage. Temperature, cathodic potential, solution chemistry, surface area, surface roughness, and stress state all influence the degree to which embrittlement occurs.

When hydrogen discharge is the dominant reduction reaction (>0.75 V [Ag/AgCl] in seawater), commercially pure titanium absorbs hydrogen regardless of temperature. At low current densities, hydrogen remains primarily at the surface as a hydride film having little effect on structural integrity. At high current densities, bulk hydriding occurs, leading to severe embrittlement.

Higher temperatures (especially above 80 °C, or 175 °F) promote diffusion of hydrogen inward, increasing the rate and severity of embrittlement. Acidic pH levels and sulfides, arsenates, and antimony compounds (hydrogen recombination poisons) also promote hydrogen absorption and subsequent embrittlement.

Alloys having greater beta-phase proportions absorb hydrogen more rapidly than commercially pure titanium. However, alloys with a high proportion of beta phase, such as beta alloys, resist hydriding and subsequent embrittlement more effectively.

Surface roughness also influences hydrogen absorption. In general, smoother surfaces absorb hydrogen more slowly. In addition, surfaces with artificially thick oxides produced by thermal oxidation are more resistant to hydrogen absorption.

Liquid/Solid Metal Cracking. Titanium alloys suffer embrittlement and/or cracking when in contact with cadmium, zinc, mercury, and certain silver brazing alloys.

Alloying for Corrosion Prevention

Despite the excellent corrosion performance of titanium, several alloys have been developed to improve crevice corrosion and corrosion resistance in a reducing environment. Two approaches are used to alloy titanium for improved corrosion resistance:

- Use elements that facilitate cathodic depolarization by forming low-hydrogen overvoltage sites, such as platinum-group metals and nickel
- Add large amounts of alloying elements with inherent corrosion resistance, such as tantalum or molybdenum

Table 14.16 depicts the increased resistance of titanium in oxidizing environments compared with stainless steel. Note that neither material performs particularly well in reducing environments (last two listed in the table).

Cathodic alloying is successfully used in alloys Ti-0.15Pd (ASTM grade 7) and Ti-0.3Mo-0.8Ni (ASTM grade 12). Both alloys have substantially improved resistance to reducing acids and crevice corrosion, as shown in Tables 14.16 and 14.17 and Fig. 14.12. Replacement of palladium with platinum, ruthenium, or iridium produces similar results.

This type of alloying is easily shown using a mixed-potential diagram. Figure 14.12 shows a typical anodic curve for titanium and two different reduction curves in a reducing acid. Adding palladium (or nickel) to titanium changes the potential at which hydrogen reduces (i.e., hydrogen overvoltage). In doing this, the corrosion current (i_{corr}) shifts, effectively reducing corrosion attack (titanium-palladium).

Other alloys that take advantage of this alloying concept are the high-strength molybdenum-containing alloys, such as Ti-6Al-2Sn-4Zr-6Mo and Ti-3Al-8V-6Cr-4Zr-4Mo. Corrosion resistance of these alloys is shown in Fig. 14.13.

The second alloying technique is not commercialized, although several alloys have been examined. Molybdenum was the first element to be examined in detail. Investigators found that molybdenum additions greater than 25% appreciably enhanced corrosion resistance in reducing acids. At approximately 50% Mo, the titanium-molybdenum alloy has essentially the same corrosion resistance as pure molybdenum. However,

oxidizing-acid corrosion increases at the same time that reducing-acid corrosion decreases. Representative corrosion performance of titanium-molybdenum alloys in a number of environments is shown in Table 14.18.

Tantalum is another alloying element that fits into the second alloying concept. However, tantalum is unique in that it enhances corrosion in both oxidizing and reducing acids. Reductions in corrosion attack in HNO_3, an oxidizing acid, by as much as an order of magnitude are observed in titanium alloys with 8% Ta (Table 14.19).

In reducing acids, corrosion improvements are observed for tantalum additions greater than 20%. In general, higher concentrations of reducing acid require greater percentages of tantalum. At approximately 50% Ta, the alloy is resistant to most concentrations of HCl, H_3PO_4, H_2SO_4, and oxalic acids. This effect is shown in Fig. 14.14.

Chromium was studied to some extent in binary 11-Cr alloys and is detrimental. However, in alloys containing both chromium and molybdenum, improved reducing-acid corrosion without sacrificing the inherent oxidizing-acid corrosion resistance of titanium is observed.

If both alloying concepts are used in, for example, Ti-15Mo-0.2Pd, corrosion resistance is improved in both oxidizing and reducing acids. Typical corrosion rates for this alloy are shown in Table 14.18.

Table 14.16 Comparison of titanium alloys in boiling acid solutions

Solution	Corrosion rate, mm/yr (mils/yr)			
	316 stainless steel	Titanium grade 2	Titanium grade 12	Titanium grade 7
65% nitric acid	0.41 (16)	0.254 (10)	0.15 (6)	0.3 (12)
30% FeCl₃	Severe pitting	Nil	Nil	Nil
15% chromic acid	>10 (>400)	12.7 (500)	Nil	Nil
20% acetic acid	0.1 (4)	Nil	Nil	Nil
5% sulfuric acid	>2.5 (>100)	25 (1000)	>2.5 (>100)	0.5 (20)
5% hydrochloric acid	>2.5 (>100)	25 (1000)	>2.5 (>100)	>0.025 (>1)

Table 14.17 Resistance of titanium alloys to crevice corrosion in boiling salt solutions

Boiling solution	pH	Titanium grade 2	Titanium grade 12	Titanium grade 7
Saturated ZnCl₂	3.0	Failed	Resistant	Resistant
10% MgCl₂	4.2	Failed	Resistant	Resistant
10% CaCl₂	3.0	Failed	Resistant	Resistant
10% KCl	3.0	Failed	Resistant	Resistant
Saturated NaCl	3.0	Failed	Resistant	Resistant
Saturated NaCl + Cl₂	1–2	Failed	Failed	Resistant
10% NH₄Cl	4.1	Failed	Resistant	Resistant
10% FeCl₃	0.6	Failed	Failed	Resistant
10% Na₂SO₄	2.0	Failed	Resistant	Resistant

Note: Tight metal-to-gasket crevices were tested.

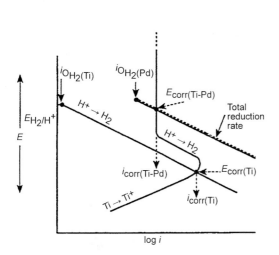

Fig. 14.12 Mixed-potential diagram showing the reduction in corrosion of titanium by coupling to palladium

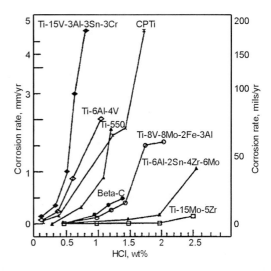

Fig. 14.13 General corrosion profiles of aged titanium alloys in naturally aerated hydrochloric acid (HCl) solutions. CP, commercially pure

Table 14.18   Corrosion of unalloyed titanium and certain titanium alloys

	Temperature		Corrosion rates, mm/yr (mils/yr)				
Environment	°C	°F	Unalloyed titanium	Ti-0.2Pd	Ti-15Mo	Ti-0.2Pd-15Mo	Ti-30Mo
25% AlCl$_3$	Boiling		51 (2,000)	0.025 (1)	...	...	Nil
30% FeCl$_3$	Boiling		<0.025 (1)	Nil	...	...	0.13 (5)
5% HCl	Boiling		28 (1,100)	0.18 (7)	...	...	Nil
20% HCl	21	70	0.64 (25)	0.1 (4)	...	...	Nil
	Boiling		>127 (5,000)	20 (770)	25 (1,000)	4.1 (160)	0.13–0.25 (5–10)
20% HCl + 1% FeCl$_3$	Boiling		2.8 (110)	2.9 (115)	...	...	38 (1,500)
5% H$_2$SO$_4$	Boiling		48 (1,900)	0.5 (20)	...	...	...
40% H$_2$SO$_4$	21	70	1.7 (65)	0.23 (9)	...	...	Nil
	Boiling		>330 (13,000)	...	15 (600)	2.4 (95)	0.05–0.25 (2–10)
40% H$_2$SO$_4$ + 1% Fe$_2$(SO$_4$)$_3$	Boiling		2.8 (110)	2.3 (90)	...	...	19 (750)
65% HNO$_3$	21	70	Nil	Nil	Nil	Nil	Nil
	Boiling		0.05–0.5 (2–20)	0.64 (25)	1.0 (40)	1.0 (40)	1.3 (50)
1% oxalic	100	212	46 (1,800)	1.1 (45)	...	...	Nil
10% oxalic	100	212	89 (3,500)	0.02 (0.8)	...	...	0.04 (1.5)
100% trichloroacetic	Boiling		15 (591)	...	...	...	...

Chemical and Related Applications

Titanium is used in many applications where its corrosion resistance is important (see Chapter 15, "Applications of Titanium," in this book). One major chemical company found that the metal can be used in approximately 95% of their environments, and this wide acceptance reduced the number of spare parts required for routine maintenance. In most chemical applications, solid titanium metal is used. However, in certain applications, liners (loose or otherwise) or titanium-clad steel parts are applied where economics justify this type of construction.

The first important applications in the chemical industries were in environments highly corrosive to conventional materials and in which major cost-savings could be realized. One company reported that titanium thermowells lasted 10 times longer than previously used materials in hot nitric acid, and that each unit saved the company more than $10,000. While initial costs of titanium can exceed those for more common materials, experience from wide use shows that trouble-free service, with no costly maintenance and downtime, can quickly recover initial costs.

Some important applications where the corrosion resistance of titanium is advantageous are presented as follows.

Table 14.19 Effect of alloying elements on the corrosion resistance of titanium alloys in boiling nitric acid solutions

	Corrosion rate					
	40% HNO₃, 4 h at 6 °C (−14 °F)		65% HNO₃, 4 h at 6 °C (−14 °F)		65% HNO₃, 96 h(a)	
Alloying element, %	mm/yr	mils/yr	mm/yr	mils/yr	mm/yr	mils/yr
Unalloyed Ti	0.74	29	1.43	56.9	0.07	2.78
Al 1.05	0.94	37	1.76	70	...	...
Si 0.43	1.22	48	1.90	75.6	0.005	0.199
V 1.27	0.73	29	1.09	43.3	...	...
Mn 3.89	...	...	...	...	0.13	5.17
Mn 8.73	...	...	...	...	0.19	7.56
Zr 1.06	0.92	36.5	1.79	71.2	0.06	2.39
Nb 0.90	0.75	30	1.25	49.7	0.09	3.58
Mo 1.00	1.04	41.4	0.72	28.6	...	...
Ta 0.98	0.61	24	0.58	23	0.09	3.58
Ta 4.67	0.14	5.6	0.20	7.95	...	...
Ta 7.97	0.05	2.0	0.07	2.78	0.01	0.40

(a) Test solution renewed every 4 h

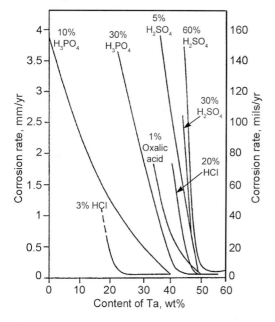

Fig. 14.14 Effect of tantalum on the corrosion resistance of titanium-tantalum alloys in boiling acid solutions

grids for packing in absorption towers. The metal is also popular in organic chlorinators. Titanium has good resistance to most chlorinated hydrocarbons at moderate temperatures, and chlorine inhibits attack by the more aggressive acid solutions. Restricted crevices should be avoided in all chlorine applications, and the metal must not be used in the dry gas.

Various titanium components have been used in sodium and calcium hypochlorite service for many years without noticeable effects. Components include heat exchangers, coolers, tanks, filter presses, atomizers, and dryers.

Chlorine dioxide is a popular bleach in the paper industry because it improves whiteness with less fiber degradation. Titanium equipment made this process practical; it has been in constant trouble-free service for more than 12 years, whereas previously used materials failed in a few weeks or months. The metal is also used in chlorine-dioxide-generating units as baffle plates, thermowells, tubing, and valves.

The metal has excellent resistance to metallic chlorides and is used extensively for handling brine, and iron, nickel, and copper chlorides. The metal is also used to process sodium, potassium, ammonium, calcium, aluminum, magnesium, and zinc chlorides under certain conditions. Crevices should be avoided in equipment for the acidic, light-metal chlorides. At high temperatures and concentrations, it is best to use an alloy such as Ti-0.2%Pd, which offers improved corrosion resistance. The Ti-1%Ni alloy shows promise for scrubbers in boiling 42% magnesium chloride.

Water Purification. Because of the good resistance of titanium to chloride solutions, the metal

Chlorine and Chlorides. Titanium is a key material of construction in the chlorine and chloride industries. Nearly all new chlorine coolers are made of titanium. They cost less than glass units and require only approximately one-eighth the space. Because of better heat transfer, more efficient design, and no maintenance, titanium coolers pay for themselves quickly.

In addition to coolers, the chlorine industry uses titanium in headers, duets, cell covers, anodes, sparger pipes, compressors, and support

is functional for condensers, tubing, pumps, filters, and other components in water-purification systems. However, at temperatures above approximately 95 °C (200 °F), alloys such as Ti-1%Ni and Ti-12%Pd with resistance to crevice corrosion are recommended. A desalination plant on the Virgin Islands uses 156 km (97 miles) of commercially pure titanium tubing. The unit has been in continuous trouble-free service. It operates at significantly higher efficiency than anticipated because of the unexpected low fouling factor. In some instances, the fouling factor was 75% lower than expected.

Marine Applications. The metal is functional in many marine applications because of the good corrosion resistance and high fracture toughness. Unalloyed metal has been used as heat exchangers, tubing, and pumps. The Ti-6Al-4V and Ti-6Al-2Nb-1Ta-1Mo alloys are used as heavy plate in experimental deep-submergence vehicles and are being studied for use as hydrofoils, propellers, shafts, and hulls in high-performance marine craft.

Acid Solutions. Titanium is useful for handling hot nitric acid and inhibited reducing acids. Reactors, tube bundles, heaters, pumps, and thermowells are used in nitric acid at temperatures up to 315 °C (600 °F) and concentrations up to 70%. As previously noted, hazards can be involved in using the metal in the fuming acids.

Titanium is especially functional for certain mixed acids and inhibited reducing acids because most common metals corrode rapidly under these conditions. The original nickel-cobalt leaching plant at Moa Bay, Cuba, contained approximately 16,820 kg (37,000 lb) of titanium heat exchangers, valves, pipes, pumps, and reactor parts. The process involves treating a thick ore slurry with 98% sulfuric acid at pressures above 3400 kPa (500 psi) and between 750 and 930 °C (1380 and 1700 °F). The acid is consumed rapidly to a residual under 3%, yielding a highly erosive-corrosive liquor containing 36% fine iron ore in suspension. Titanium is superior to other materials in this application because of its naturally good erosion and corrosion resistance and the inhibiting effect of the cobalt, nickel, and iron.

Another important application involves the use of 45 heating coils in a 25 to 30% sulfuric acid solution containing metallic sulfates at 215 °C (420 °F). Each coil is made from 51 mm (2 in.) diameter by 36.5 m (120 ft) long welded tubing.

Organic Petroleum and Petrochemicals. Titanium is being used for fluid ends in large pumps handling brackish waters. It is also functional as pipes, pumps, and valves for gas and oil wells containing corrosion sand-laden chlorides, sulfides, sulfur dioxide, and organic halides.

The metal is used in urea reactors and other ammoniated environments as tube bundles and strippers. However, not all urea applications performed well. In some instances, hydrogen embrittlement was encountered, which appears to be associated with high pressures, high temperatures, high-velocity flow, and other operating conditions.

A large amount of titanium is used for the production of acetaldehyde by air oxidation of ethylene in aqueous chlorides. Much of the metal is unalloyed and is used as liners for steel vessels up to 3 m (10 ft) in diameter, process piping, heat-transfer equipment, wire-mesh eliminators, and cast pumps and valves. Virtually all mill shapes with extensive welding were used.

Foods, Drugs, and Medical Implants. Titanium is not affected by the products of the food processing industry and the strongest cleansing agents it uses. Titanium equipment (unlike other metal components) can be used for seasonal foods without maintenance. Neither foods nor drugs are contaminated by titanium.

Another application of growing importance is the use of titanium and titanium alloys for orthopedic implants such as nails, hip joints, plates, heart valves, screws, and external braces on human beings. Complete inertness and body compatibility are essential, and high strengths and light weight are desirable in such applications.

Electrochemical Applications. The first large industrial application for titanium was in the aluminum anodizing industry, where the metal is used extensively as support racks. The metal resists the chemicals and the anodizing treatment, and it can be used indefinitely. The thin anodic oxide film reduces current leakage but permits good electrical contact with the aluminum parts.

The application of a thin film of platinum to titanium results in an excellent, low-cost, insoluble anode, even if the platinum is discontinuous. Anodes of this type have been in continuous service over two years for electroplating gold, platinum, rhodium, copper, nickel, silver, and other metals. The purity of deposits is of high order. Anode baskets are widely used in nickel plating, enabling 100% use of the nickel, resulting in significant cost-savings.

Platinum-titanium anodes are also used in cathodic-protection systems on ships, harbor installations, chemical equipment, water heaters, and pumping systems. Such anodes are also used in a low-cost power system in a 200 kV direct-current power link between Sardinia and the Italian mainland. The durable anode system enables using the sea as a conductor for one direction, while cables are used in the other direction. Platinized titanium electrodes are used in small water-purification systems.

Platinized-titanium anodes are particularly functional for the production of chlorine and caustic. Several full-scale pilot cells (both the mercury and diaphragm types) have been in operation for a few years and indicate production feasibility. Advantages cited for these anodes are lower power requirements, less maintenance, constant electrode spacing, purer products, and greater design flexibility than is possible with conventional graphite anodes.

Miscellaneous Applications. Several soda-ash plants are using titanium equipment, which is expected to last at least five years without maintenance. Heat exchangers in some ammonia stills contain as many as 800 tubes per bundle. One company reported a production increase of more than 25% after retubing a still with titanium. The improvement resulted from better heat transfer. Unfortunately, not all soda-ash applications have performed well. In at least one instance, embrittlement of the tubes was reported after three years of operation. Other soda-ash plants report satisfactory service with titanium for long periods. Presumably, performance is related to operating variables.

Titanium valve plates and springs provide good service in gas compressors for ammonia synthesis. Lighter-weight Ti-5Al-2.5Sn alloy springs and plates operate at higher speeds and are said to be less prone to corrosion fatigue than components made of other metals.

Titanium is also used successfully as large fans and stack liners to resist hot gases containing hydrochloric acid, sulfur dioxide, hydrogen sulfide, and various metallic and organic chlorides.

Titanium alloys perform well as dies in the zinc die-casting industry. Regular die materials do not experience early cracking failures from repeated thermal cycling.

Summary

Titanium has excellent corrosion resistance in oxidizing environments, including virtually all natural waters and in many commercial aqueous chemicals. It is attacked by reducing mineral acids, such as sulfuric, hydrochloric, and oxalic acid, and by ionizable fluoride compounds. The corrosion resistance of titanium is improved in reducing environments by the addition of platinum-group metals, with ruthenium being particularly attractive because it is 7 times less expensive than palladium. Improved corrosion resistance is also achieved by adding a more noble metal (e.g., silver) and by adding compound formers that promote cathode reactions, such as nickel.

Titanium corrodes slowly in hot caustic solutions, often accompanied by hydrogen absorption and embrittlement. Most organic compounds are compatible with titanium, with the exception of anhydrous methanol and reducing organic acids, such as trichloroacetic acid.

Strong anhydrous oxidizers, such as red fuming nitric acid, chlorine, and nitrogen tetroxide, can react violently with titanium and can lead to fires and/or environmental cracking.

Oxidizing agents commonly encountered in commercial environments, such as Fe^{3+}, Cu^{2+}, and Cr^{6+}, are powerful corrosion inhibitors in reducing acids. Even air in a solution can be an effective corrosion inhibitor.

Environmental cracking occurs in halide salts (Cl^-, Br^-, I^-) at approximately 290 °C (550 °F) under controlled laboratory conditions but rarely occurs in operating practice. Certain titanium alloys can crack in anhydrous methanol, nitrogen tetroxide, and in contact with cadmium.

In seawater, titanium alloys suffer a loss in environmental cracking resistance, as measured by a reduction in fracture roughness.

REFERENCES

14.1 M.G. Fontana and N.D. Greene, *Corrosion Engineering,* McGraw-Hill, New York, 1978

14.2 W.D. Callister, Jr. and D.G. Rethwisch, *Materials Science and Engineering: An Introduction,* John Wiley & Sons, Inc., Hoboken, NJ, 2012

14.3 F.H Froes, D. Eylon, and H.B. Bomberger, Ed., *Titanium Technology: Present Status and Future Trends,* Titanium Development Association, Dayton, OH, 1985

14.4 R. Boyer, E.W. Collings, and G. Welsch, Ed., *Materials Properties Handbook: Titanium Alloys,* ASM International, 1994

14.5 G. Lutjerling and J.C. Williams, *Titanium,* Springer, 2003

14.6 M.J. Donachi, *Titanium: A Technical Guide,* 2nd ed., ASM International, 2000

14.7 G. Lutjerling, U. Zwicker, and W. Bunk, *Proceedings of the Fifth International Conference on Titanium,* Metallurgical Society of AIME, Warrendale, PA, 1984

14.8 P. Lacombe, R. Tricot, and G. Beranger, Ed., *Proceedings of the Sixth International Conference on Titanium* (Nice, France), Metallurgical Society of AIME, Warrendale, PA, 1988

14.9 F.H. Froes and I.L. Caplan, Ed., *Proceedings of the Seventh International Conference on Titanium* (San Diego, CA), Metallurgical Society of AIME, Warrendale, PA, 1992

14.10 *Proceedings of the Eighth International Conference on Titanium* (San Birmingham, U.K.), Metallurgical Society of AIME, Warrendale, PA, 1995

14.11 *Proceedings of the Ninth International Conference on Titanium,* Metallurgical Society of AIME, Warrendale, PA, 2000

14.12 *Proceedings of the Tenth International Conference on Titanium,* Metallurgical Society of AIME, Warrendale, PA, 2004

14.13 *Proceedings of the Eleventh International Conference on Titanium,* Metallurgical Society of AIME, Warrendale, PA, 2008

14.14 *Proceedings of the Twelfth International Conference on Titanium,* Metallurgical Society of AIME, Warrendale, PA, 2012

CHAPTER 15

Applications of Titanium*

THE BIRTH of the commercial titanium industry occurred when U.S. companies TIMET and Rem-Cru Titanium were established in 1950. With the founding of these two companies, and of those that followed, titanium mill products began to be produced in sufficient quantities to enable aerospace and industrial designers to exploit the unique engineering properties of titanium: high strength, low density, and corrosion resistance. By 1951, the titanium industry, with government support, produced 440 metric tons (990,000 lb) of sponge and produced and shipped 67 metric tons (150,000 lb) of mill products. The production history of the U.S. titanium industry in both sponge and mill products has seen numerous peaks and valleys, but the overall trend has been steady growth.

Early Applications

The first commercial applications of titanium were jet engine parts for the J-57 engine produced by Pratt & Whitney Aircraft (P&WA) for the B-52 bomber. Titanium replaced steel parts whose weight prevented the engine (and therefore the aircraft) from achieving its design potential. Titanium was quickly applied in the manufacture of airframe structural parts, following the successful early use of titanium in engine parts. By 1952, Douglas Aircraft Company was using titanium engine nacelles and firewalls on their new DC-7. Alcoa's use of titanium in 1951 for anodizing racks was the first industrial application. In the early 1950s, titanium, "the wonder metal," was successfully tried in applications as diverse as shovels, frying pans, railroad spikes,

and mirrors, but these applications were quickly discarded due to high cost.

Why were the designers and engineers so eager to use titanium? Previously, designers had to choose between metals that had strength but were heavy, such as steel, copper-nickel, and nickel-base alloys, and metals that were lightweight but had very low strengths, such as aluminum and magnesium. Titanium offered the engineer lower density than steel, higher strength than aluminum and magnesium at all temperatures, and the fatigue, creep, and fracture toughness properties required for jet engine and airframe service (Ref 15.1–15.12). In addition to these attractive mechanical and physical properties, titanium and its alloys exhibited resistance to corrosion and erosion from a wide range of industrial chemicals, natural waters, and gases. This corrosion and erosion resistance resulted the use of titanium in myriad nonaerospace industrial, commercial, and medical applications.

Material Availability

Is titanium readily available? This is a question that engineers apply to every material. Engineering materials must be available in mill product forms at a stable, competitive price to be considered for use in most applications. Titanium is the fourth most abundant structural metal and ninth most abundant element in the Earth's crust. The known commercially mineable deposits of titanium ore are available throughout the world and are not isolated to geopolitically unstable regions. Also, the industry's capacity to produce sponge and melt ingots

---

*Adapted and revised from Walter E. Herman, originally by Edward E. Mild and Ronald W. Schutz, *Titanium and Its Alloys*, ASM International.

is nearly double the demand. Therefore, the availability of titanium is not foreseen to be limited by either ingot or sponge capacity.

Sufficient manufacturing capacity exists to produce mill products to meet demand for the foreseeable future. See Table 15.1 (world demand forecast), Fig. 15.1 (engines), Fig. 15.2 (commercial airframes), Fig. 15.3 (industrial market outlook), Fig. 15.4 (Chinese titanium market), and Fig. 15.5 (Japanese titanium market). Titanium

Table 15.1 World demand forecast for titanium mill products

	European Union and North America	China	Rest of world	Total
2012, kt				
Industrial applications	16	44	27	87
Aerospace	45	5	10	59
Consumer and other	12	4	4	19
Total	72	53	40	165
Average annual growth rate, %				
Industrial applications	2.5	8.0	4.5	6.0
Aerospace	2.5	5.0	4.0	3.0
Consumer and other	2.0	5.0	2.0	2.7
Total	2.4	7.5	4.2	4.6
2018, kt				
Industrial applications	19	70	35	123
Aerospace	52	7	12	71
Consumer and other	13	5	4	22
Total	83	82	51	216

Courtesy of Roskill Information Services

mill product availability is only limited by mill lead time and available inventories.

Material Cost Table. The cost of titanium is frequently higher than other common metals on a per kilogram (per pound) basis due to the complexity of the extraction and melting processes (Table 15.2) (see Chapter 1, "History and Extractive Metallurgy," and Chapter 8, "Melting, Casting, and Powder Metallurgy," in this book). However, when density and engineering properties are considered and optimized in a design, titanium is cost-effective (Ref 15.1, 15.2, 15.4–15.12). When properly used, titanium is cost-effective, and where life-cycle cost, density-corrected cost, and reliability are considered, titanium is often the most economic material available.

Aerospace Applications

The histories of the jet engine and use of titanium have been intertwined since the beginning of the titanium industry (Ref 15.1–15.3). The use of Ti-6Al-4V by Pratt & Whitney in their J-57 jet engine for blades and disks saved approximately 95 kg (210 lb) per engine, or 945 kg (2080 lb) per B-52. This represented the first significant use of titanium in the aerospace industry. The first airframe applications of titanium were the nacelle skins and firewall webs of the B-52 made of commercially pure grade 4 titanium. Both these applications were weight-reduction designs that look

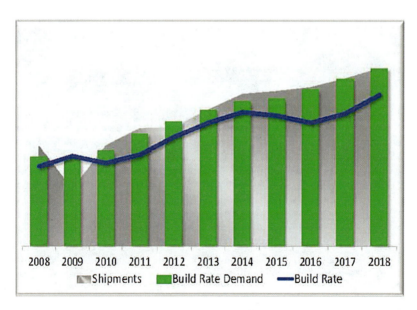

Fig. 15.1 Engine demand for titanium during the mid- to late-2010s. Courtesy of the International Titanium Association

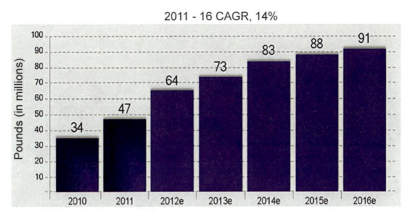

Fig. 15.2 Commercial airframe demand for titanium through the mid-2010s. CAGR, compound annual growth rate. Courtesy of the International Titanium Association

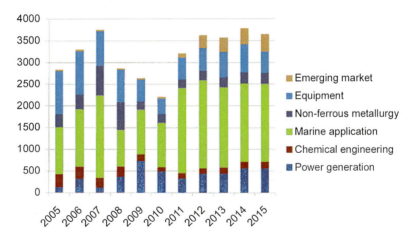

Fig. 15.3 The U.S. use of titanium is approximately 70% for aerospace applications, while other countries use much less in nonaerospace applications. Courtesy of the International Titanium Association

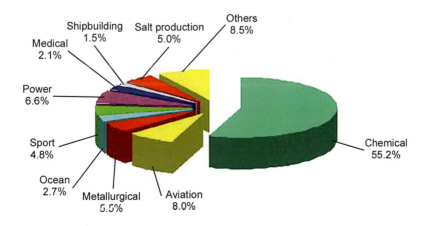

Fig. 15.4 Chinese titanium market in 2011. Domestic demand was 56,000 tons for sponge and 44,500 tons for mill products.

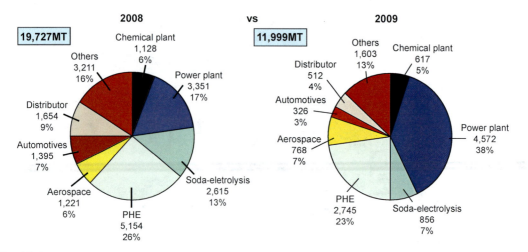

Fig. 15.5 Titanium use in Japan. PHE, plate heat exchangers. Courtesy of the International Titanium Association

Table 15.2 Cost comparison of titanium

| | Material(a), $/lb | | |
Item	Steel	Aluminum	Titanium
Ore	0.02	0.10	0.22 (rutile)
Metal	0.10	1.10	5.44
Ingot	0.15	1.15	9.07
Sheet	0.30–0.60	1.00–5.00	15.00–50.00

Note: Contract prices. The high cost of titanium compared to aluminum and steel is a result of high extraction and processing costs. The high processing costs relate to the relatively low processing temperatures used for titanium and the conditioning (surface regions contaminated at the processing temperatures, and surface cracks, both of which must be removed) required prior to further fabrication. (a) U.S. dollars in 2014

Table 15.3 Titanium buy weights for early military and commercial airframes

| | Titanium buy weight | |
Airframe	kg	lb
Military		
F-14	18,814	41,442
F-15	23,154	51,000
F-16	804	1,770
F/A-18A-D	3,884	8,554
C-5B	6,870	15,130
B-1B	82,707	182,175
KC-10A	10,773	23,730
C-17	19,068	42,000
ATB (Stealth)	61,835	136,200
AH-64A	452	995
CH-53E	7,860	17,315
S/UH-60	1,910	4,200
Commercial		
737-200	1,730	3,810
737-300/400	1,820	4,005
757	8,900	19,600
MD-80	1,378	3,036
A-320	7,462	16,437
747	18,247	40,191
767	5,397	11,887
A-300/310	2,765	6,090
MD-11	10,773	23,730

advantage of the high strength-to-weight ratio of titanium. (The strength-to-weight ratio is calculated as tensile strength divided by density.)

As thermomechanical processing improved and new alloys were developed, the use of titanium in aircraft grew. The SR-71 Blackbird (reconnaissance plane), which was 93% Ti, was the proving ground for sophisticated alloys and design configurations. Aerospace applications comprise approximately 70% of all titanium shipments, and jet engines use well over half of the total aerospace consumption. Even though it is expected that the industrial use of titanium will grow significantly, the aerospace market is forecast to remain the single largest application for many years. Table 15.3 lists titanium buy weights for military and commercial airframes. Figure 15.6 shows the increasing use of titanium over the years on commercial airframes (Ref 15.1–15.3).

Gas turbine engines historically have been the largest single consumer of titanium. In addition, the widest variety of alloys is used in the gas turbine engine. Table 15.4 shows typical titanium uses in a jet engine, listed by alloy and application.

Fan blades, compressor blades, and disks are produced from forged alloys such as Ti-8Al-1V-1Mo, Ti-6Al-4V, Ti-6Al-2Sn-4Zr-2Mo, Ti-6Al-2Sn-4Zr-2Mo-0.1Si, Ti-6Al-2Sn-4Zr-6Mo, and Ti-17 IMI829 and 834, Ti-1100 (see Table 6.9 in this book), and most recently TiAl compositions (see Chapter 6, "Mechanical Properties and Testing of Titanium Alloys" in this book). These forged and heat treated alloys are used for their high-temperature properties and high fatigue strength, as well as their low modulus, high

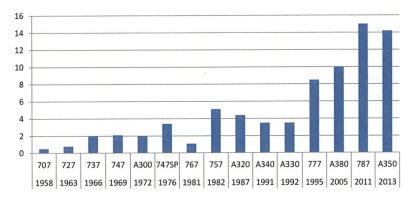

Fig. 15.6 Growth in titanium use as a percentage of total gross empty weight on Boeing and Airbus aircraft. The decreased use on the 767 was due to a perceived shortage of titanium when the plane was designed, and other materials, such as steel and aluminum, were substituted for titanium.

Table 15.4   Early titanium use in a fan-jet engine

Part	Material	Alternate titanium alloy(a)	Alternate material(a)
Inlet case	5Al-2.5Sn	Ti(b), 8Al-1Mo-1V	12Cr stainless steel
Fan blades	8Al-1Mo-1V	6Al-4V, 6Al-6V-2Sn	Composites(c)
Fan disks	8Al-1Mo-1V	6Al-4V, 6Al-6V-2Sn, Ti 679	Low-alloy steel, 12Cr steel
Fan exit struts	6Al-4V	Ti(b), 5Al-2.5Sn	12Cr steel
Fan duct	Ti-5Al-2.5Sn	8Al-1Mo-1V	Al alloy
Fan duet fairings	Ti(b)	5Al-2.5Sn	Al alloy
Front compressor blades	8Al-1Mo-1V	6Al-4V, 6Al-6V-2Sn	12Cr steel
Front compressor, disk	8Al-1Mo-1V	6Al-4V, 6Al-6V-2Sn	Low-alloy steel, 12Cr
Multistage disk	6Al-2Mo-4Zr-2Sn	6Al-4V	None
Front compressor case	6Al-4V	5Al-2.5Sn, 6Al-1Mo-1V	12Cr steel
Intermediate case	5Al-2.5Sn	8Al-1Mo-1V	12Cr steel
Rear compressor blades	8Al-1Mo-1V	6Al-2Mo-4Zr-2Sn, Ti 679	12Cr steel, Ni alloy

(a) Depending on temperature, stress, corrosion resistance, cost, and weight requirements. (b) Commercially pure titanium A-55 or A-70. (c) Such as graphite-fiber-reinforced epoxy and boron-reinforced aluminum. Source: Simmons and Wagner

strength-to-weight ratio, and high corrosion resistance. Also, in static (nonrotating) parts such as vanes, cases, and ducts where fatigue is less critical, titanium castings are successfully used. In addition to being weight critical, engines use high-speed rotating components, such as blades and disks, which encounter high centrifugal loading and vibrating (cyclic) stresses. These conditions place a premium on weight reduction to reduce the stress that must be carried. Titanium, with its unique combination of engineering properties, provides the jet engine designer with an efficient way to meet these challenges. Table 15.5 lists early titanium buy weights for military and commercial aircraft engines (Ref 15.3).

Continuing development of larger thrust engines (which use larger amounts of nickel-base and titanium alloys) resulted in the development of General Electric's GEnx engine for use on the Boeing 787 and 747-8. The GEnx is the first "all-titanium engine," with the use of new titanium

alloys and the introduction of cast gamma TiAl (Ti-48Al-2Nb-2Cr at.%) (see Chapter 8, "Melting, Casting, and Powder Metallurgy," in this book) for use in the fifth- and sixth-stage compressor. Heavier nickel-base superalloys were previously used in these high-temperature environments. The GEnx is a quieter engine with decreased fuel consumption (compared with the CF6). Gamma alloys are also being evaluated for similar use by P&WA and Rolls-Royce. Figure 15.7 shows a fully assembled GEnx engine on display at the Paris airshow, and Fig. 15.8 shows the historical demand of titanium for jet engines.

Titanium jet engine components are manufactured in numerous ways. Blades and disks are closed-die forged or precision forged and machined to final configuration or dimension. Precision forging produces a near-net shape and reduces the machining loss. Ring rolling is used to produce seals, spools (multistage disks), and engine cases. Casting, followed by hot isostatic pressing (Fig. 15.9), is used to produce compressor cases and structural housing members. Ducts, fan cases, and engine cases typically are fabricated plate and sheet metal structures. As a result

of the titanium engine disk failure, which led to an uncontained engine explosion and subsequent crash of a Boeing 737 in Sioux City, Iowa, melting of titanium for engine applications has turned to cold-hearth melting (see Chapter 8, "Melting, Casting, and Powder Metallurgy," in this book). Cold-hearth melting minimizes or eliminates the type of low-density inclusion (enriched in oxygen and/or nitrogen) that was the root cause of the catastrophic engine failure and crash.

Competitive Comparison. Titanium alloys compete primarily with stainless steel, precipitation-hardening steel, high-strength steel, and nickel-base alloys in various sections of the jet engine. Titanium also competes with aluminum alloys in the jet engine lower-temperature section.

Titanium has a significant design advantage over steels on the basis of its high strength and low density. This is particularly true for rotating components. As temperature increases, creep strength becomes the critical design parameter.

Compressor blades also encounter high cyclic stresses (vibration and deflection). The fatigue strength of many titanium alloys at room and elevated temperatures is equivalent or superior to most steels and nickel-base alloys (Ref 15.1, 15.2). In many cases, titanium alloy compressor blades operate at lower stresses than steel blades. This is because of the lower modulus and density of titanium, which allows dissipation of centrifu-

Table 15.5 Early titanium buy weights for military and commercial aircraft engines

Engine	Titanium buy weight	
	kg	lb
Military		
TF-30	3,800	8,380
F-110	2,363	5,204
F-100	2,744	6,044
F-404	1,816	4,000
PW1120	2,744	6,044
TF-39	4,441	9,781
F-101	1,949	4,293
F-108 (CFM 56)	1,500	3,300
F-117	3,860	8,500
T-700	73	160
T-64	321	706
Commercial		
JT8D-217/219	1,040	2,290
CFM-56	1,500	3,300
PW2037	3,860	8,500
RB211-535	2,036	4,485
V-2500	1,500	3,300
JT9D-7R4	5,345	11,774
CF6-50/80	3,147	6,931
PW4000	4,858	10,700
RB211-S24	2,724	6,000

Fig. 15.7 General Electric's GEnx engine, which makes use of the gamma intermetallic TiAl alloys in its compressor. The GEnx is an all-titanium engine. Source: Ref 15.13

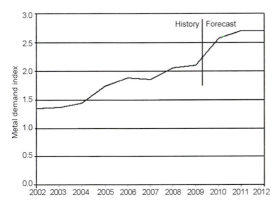

Fig. 15.8 Titanium consumption in commercial jet engines. Courtesy of the International Titanium Association

Fig. 15.9 Titanium support structure for a jet engine thrust reverser from a Rolls-Royce RB199 engine. Source: Ref 15.14

gal and vibrating stresses. This combination of low density with high creep and fatigue strength makes titanium an excellent engineering material for temperatures up to 540 °C (1000 °F). Much of the aluminum once used in the forward section of jet engine compressors has been displaced by titanium. In many instances, titanium is used not because of weight considerations but because of its superior corrosion and erosion resistance.

Improved jet engine designs result in increased operating temperature and stress levels. Thus, it has been necessary to continuously improve the creep strength of titanium alloys in the range of 370 to 595 °C (700 to 1100 °F). For example, early titanium applications in jet engines involved alloys in the annealed condition. Later designs involved large rotating fan components operating below 315 °C (600 °F), where the onset of creep occurs in titanium alloys. For such components, a higher ultimate tensile strength is advantageous, and annealed titanium alloys are often superseded by higher strength, heat treated (solution-treated and aged) titanium alloys.

While the use of titanium in jet engines is based on its advantageous combination of strength, density, fracture toughness, fatigue, and corrosion resistance, most alloys currently in use are limited to a maximum temperature of 575 °C (1070 °F). This is due to a sharp drop in creep strength and an increase in embrittling surface oxidation effects above this temperature. However, alloys such as Ti-1100 and IMI 834 and intermetallic compounds, particularly gamma titanium aluminide, TiAl (see Chapter 6, "Mechanical Properties and Testing of Titanium Alloys," in this book), have pushed the useful temperature range up to 595 °C (1100 °F) for the terminal alloys and 750 °C (1380 °F) for the

intermetallic. Also, fabrication techniques, such as superplastic forming coupled with diffusion bonding (see Chapter 11, "Forming of Titanium Plate, Sheet, Strip, and Tubing"), are capable of producing hollow blades, thereby drastically reducing the weight of fan and compressor blades. The use of this advanced fabrication technique allows titanium to compete successfully against composite materials in these applications.

Airframes, Missiles, and Space Vehicles. The use of titanium in the construction of early airframes parallels its use in gas turbine engines (Ref 15.1–15.3). From its beginning on the Douglas Aircraft DC-7 to the stealth aircraft used by the U.S. military, the quantities of titanium used to build military and commercial aircraft have been large.

In airframes, titanium alloys are used as forgings, extrusions, plate, sheet, tubing, castings, and fasteners. In addition to Ti-6Al-4V, which is the predominant alloy used, other alloys used in aircraft applications are Ti-13V-11Cr-3Al and Ti-3Al-8V-6Cr-4Zr-4Mo in fasteners and springs;

Ti-10V-2Fe-3Al and Ti-6Al-6V-2Sn in forgings; commercially pure, Ti-15V-3Al-3Cr-3Sn, and Ti-3Al-2.5V as hydraulic tubing; and Ti-6Al-4V extralow interstitial, Ti-10V-2Fe-3Al (e.g., on the B777), and Ti-5Al-5V-5Mo-3Cr (used on the B787) in fracture-critical applications. As aircraft became faster, larger, and more sophisticated, more and more titanium is used.

On the B777 (Fig. 15.10–15.12), Boeing used Ti-10V-2Fe-3Al for its good strength-toughness combination on the main landing gear and flap tracks; Ti-15V-3Cr-3Al-3Sn, which is cold formable, for sheet applications as well as ducts, fittings, and nut clips; and the Beta-21S alloys (Ti-15Mo-2.7Nb-3Al-0.2Si) in hot locations. Total titanium use on the B777 is approximately 15% of the gross empty weight. Recently, Boeing used the fracture-critical alloy Ti-5Al-5V-5Mo-3Cr on the 787 Dreamliner (Fig. 15.13). The 787 still predominately uses Ti-6Al-4V alloys, but high-strength Ti-5Al-5V-5Mo-3Cr is used in the landing gear, wing structure, and nacelle supports. The alloy Ti-3Al-2.5V is used for hydraulic tubing, Ti-6Al-2Sn-4Zr-2Mo has applications in higher-temperature regions of the nacelles and auxiliary power unit, and Beta C (Ti-3Al-8V-6Cr-4Mo-4Zr) is used for springs. Overall, the 787 has increased the use of titanium alloys to slightly above 20% of its gross empty weight.

Military aircraft have shown a similar increase in titanium use as the commercial aircraft discussed previously. The F-22 Advanced Tactical Fighter, with a build of 183 aircraft (powered by the P&WA F119-PW-100 engine), is shown in Fig. 15.14. It combines up to 45% Ti by weight with 35% polymer-matrix composites. The Joint Strike Fighter, with a build of 2443 aircraft (powered by the P&WA F-135 engine), is shown in Fig. 15.15. This aircraft also has more than 40% of its structural weight made of titanium. The demand for titanium should increase as the build program proceeds (Fig. 15.16).

The airframe structure of aircraft and aerospace vehicles consists primarily of sheet metal construction supported and tied together at critical points by more massive sections. Before titanium alloys were used in these massive sections, high-strength steel and aluminum alloys were used as heat treated forgings with strength-to-weight ratios over 1 million at room temperature. Titanium alloys can compete with these materials at strength-to-weight ratios of 800,000 or higher

Fig. 15.11 Titanium (Ti-15-3) nut clips for the Boeing 777. More than 9,000 (approximately half of the 18,000 used per plane) of these clips are used on the floor structure alone.

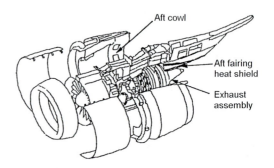

Fig. 15.12 Beta-21S applications for the nacelle on the Boeing 777, with Pratt & Whitney engines

Fig. 15.10 Titanium use on the Boeing 777 accounts for approximately 15% of its gross empty weight.

Fig. 15.13 Titanium alloys compose up to 20% of the gross empty weight of the Boeing 787.

because of weight reduction in a fixed design or in optimized new designs. In addition, titanium alloys are available with ratios of over 1 million.

There has been significant use of titanium alloys with 895 to 1240 MPa (130 to 180 ksi) tensile strengths in forged components. The alloys Ti-6Al-4V, Ti-6Al-6V-2Sn, Ti-10V-2Fe-3Al, and Ti-5Al-5V-5Mo-3Cr are used extensively in airframe forgings in either the annealed or heat treated condition, as discussed previously. The ability to be heat treated to higher strength increases the allowable design strength in heavy and light sections.

Another major application of heat treated titanium is for spherical liquid propellant tanks used in outerspace vehicles (missiles and satellites). The welding and heat treating technologies used in this application were first established for Ti-6Al-4V. The technologies are applied to Ti-15V-3Cr-3Al-3Sn to achieve strength levels of 1103 MPa (160 ksi) or higher, which are required to store liquids or gases at 21 to 28 MPa (3 to 4 ksi) internal pressure.

In advanced aircraft that encounter surface temperatures above 120 °C (250 °F), such as the B-1B bomber and the F-15, annealed titanium forgings are used extensively for heavy-section structural members. Here, titanium replaces aluminum, particularly when maximum fracture toughness is critical. Such components are machined from closed-die, open-die, and precision-forged material. Higher structural efficiency can be obtained by strengthening heat treatments. The use of high-strength titanium forgings in the

Fig. 15.14 The F-22 Raptor Advanced Tactical Fighter consists of up to 45% Ti alloy by weight.

Fig. 15.15 F-35 Joint Strike Fighter with greater than 40% of its structural weight made of titanium

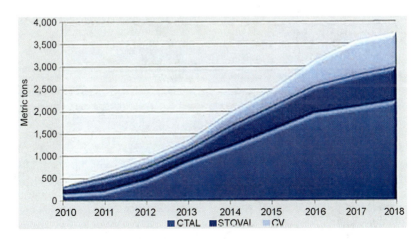

Fig. 15.16 Expected demand for titanium components as the F-35 build-out program progresses. CTAL, conventional takeoff and landing; STOVAL, short takeoff/vertical landing; CV, carrier variant. Courtesy of TIMET

Fig. 15.17 Forged Ti-6Al-4V landing gear beam

solution-treated and aged condition is common. Structural forgings, such as the landing gear beams used in commercial aircraft (Fig. 15.17), use the Ti-6Al-4V alloy at the 1170 MPa (170 ksi) tensile strength level.

Certain titanium alloys have advantages over others in heavy-section components. The depth to which an alloy can achieve high strength through heat treatment (i.e., deep hardenability) varies. The Ti-10V-2Fe-3Al alloy, being more heavily beta-stabilized than Ti-6Al-4V, achieves high strength through a section several times thicker than could be achieved with Ti-6Al-4V.

Titanium alloys extruded into L-, T-, and Z-shaped sections are commonly used as stiffening elements and wing panel members. Because titanium extrusions lack the dimensional accuracy and surface quality found in aluminum extrusions, they are generally machined on all surfaces prior to use.

Fasteners and Springs. High-strength titanium alloy fasteners are manufactured in numerous sizes and shapes. The shear strength of fasteners varies from 620 to 830 MPa (90 to 120 ksi) and tensile strength varies from 965 to more than 1380 MPa (140 to more than 200 ksi), depending on alloy and heat treatment selection. Because galling of titanium is a major concern when reassembly of threaded fasteners is required, surface treatment, lubrication, and thread design become critical to the efficient use of the fastener. An example of a titanium fastener and its installation sequence is shown in Fig. 15.18.

High-strength titanium alloys such as Ti-13V-11Cr-3Al and Ti-3Al-8V-6Cr-4Zr-4Mo are used successfully in springs, such as for closing doors, in commercial and military aircraft. High strength combined with low elastic modulus enhances the spring characteristic of these alloys.

Coatings are available that decrease the severe galling of titanium. Cold-driven titanium rivets are used with titanium, aluminum, and steel collars, depending on compatibility requirements. Beta titanium alloys are cold headed and used extensively as rivet material.

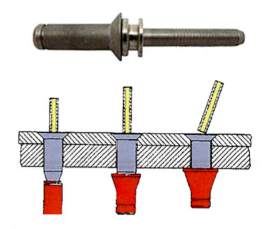

Fig. 15.18 Blind titanium fastener and installation sequence. Courtesy of Aerospace Products Division, SPS Technologies

Sheet Metal Applications

Commercially pure titanium and titanium alloys are used extensively in airframe sheet metal fabrications. The first extensive use of titanium sheet metal fabrications was in the higher-temperature portion of the airframe near the engine. At elevated temperatures, titanium alloys maintain their strength better than aluminum and can save critical weight compared with steel. Aircraft engine nacelles and firewalls are examples of titanium sheet metal fabrications that exploit the strength and elevated-temperature properties of titanium.

Titanium sheet and strip alloys are fabricated into net-shape parts to aircraft tolerances and quality by hot and cold forming processes. The alloy Ti-6Al-4V is capable of being formed to near-net shape by hot creep forming and superplastic forming, often in combination with diffusion bonding (see Chapter 11, "Forming of Titanium Plate, Sheet, Strip, and Tubing," in this book). Alloy Ti-15V-3Cr-3Al-3Sn was cold formed by Rockwell International Company into over 200 detailed parts on the B-1B bomber. Numerous titanium sheet metal applications are commonly incorporated into both commercial and military airframes. These include ducting, firewalls, shear webs, upper and lower wing panels, and floor paneling. In addition to weight savings and fire resistance, titanium provides corrosion resistance, which reduces maintenance, and provides galvanic compatibility with composites. Titanium can be used as connective material between composite sheets, or it can be used as skeletal members for composite fabrications.

Titanium and its alloys offer significant advantages above 95 °C (200 °F) for airframe construction. As aircraft speed pushed past Mach 2.5, titanium displaced aluminum sheet metal in structural members. The SR71 aircraft (Mach 3 plus) used titanium in over 90% of the structure. The National Aerospace Plane (hypersonic transport), which was to operate at Mach 3 to 25, would have used 70 to 80% Ti in its airframe as the use of gamma titanium aluminide (TiAl) is expanded. Titanium aluminides are intermetallic compounds that possess good creep and mechanical strength and improved oxidation resistance at temperatures above 595 °C (1100 °F). However, their lack of room-temperature ductility retards the acceptance of titanium aluminides in airframe components (see Chapter 6, "Mechanical Properties and Testing of Titanium Alloys," in this book).

Titanium sheet metal components are generally riveted to supporting structures. However, brazing, diffusion bonding, and welding by resistance, fusion, electron beam, and laser have also been used to fabricate titanium sheet metal structures. When properly formed, titanium sheet metal components fit better and have better appearance than aluminum counterparts and are free of distortion between rivets and of sagging between supporting members.

Titanium is used in airframe components in numerous alloy forms, product forms, and fabrications. Its unique combination of engineering properties, corrosion resistance, high strength-to-weight ratio, and fabricability made titanium a mainstay in airframe and aerospace construction.

Industrial Applications

The use of titanium and its alloys in industrial applications is primarily based on its superior corrosion resistance (Ref 15.15, 15.16). However, the same engineering properties (such as strength-to-weight ratio, high-temperature mechanical properties, and fatigue endurance) that are useful in aerospace applications are also used to advantage in nonaerospace applications. In addition, the physical properties of titanium, such as low modulus of elasticity, low thermal expansion, low paramagnetic susceptibility, short radioactive half-life, and good ballistic properties, are also used to advantage in the nonaerospace field.

The use of titanium in this market began in 1951 when Alcoa used titanium to build anodizing racks for aluminum. During the decade that followed, titanium was tried in every imaginable industrial application, but its real value was linked to its corrosion resistance. Early applications using this characteristic include pulp and paper chlorine dioxide bleach equipment, wet chlorine gas coolers in chloralkali plants, and pressure acid-leach internals for urea processing. Following the 1950s, applications were developed using the engineering-related properties of titanium, such as strength. Ordnance and ballistic armor were used in the 1960s; deep submersible submarine hulls were built in the 1970s; sour gas well pipe came along in the 1980s; and automotive valves for passenger car engines were used in the 1990s.

Engineering Properties

The nonaerospace use of titanium can be broken down into four general categories. Although each is based on the general property of titanium that represents the main incentive for its use, the following four categories are interrelated:

- Corrosion resistance to chemicals
- Corrosion resistance to seawater and other chlorides
- Mechanical properties
- Physical properties

Many factors must be considered when evaluating competing materials. After all the engineering design and properties of a nonaerospace application are analyzed and evaluated, the price of titanium is compared and evaluated against other metals of construction (Ref 15.17–15.22). Often titanium is more costly on a per weight basis. However, it is important to consider total fabrication cost and life-cycle cost. Even though titanium is priced higher per weight, properly engineered titanium systems provide maximum reliability, minimum downtime, and reduced operation costs. A life-cycle cost analysis using density-corrected cost and design is the best tool to use in the decision to purchase titanium.

Corrosion Resistance

Over the years, titanium and its alloys have proven to be technically superior, highly reliable, and cost-effective in a wide variety of chemical, industrial, marine, and other nonaerospace applications (see Chapter 14, "Corrosion," in this book).

The environmental resistance of titanium depends primarily on a very thin, tenacious, highly protective surface oxide film. Titanium and its alloys develop very stable surface oxides with high integrity and good adherence. The surface oxide of titanium, if scratched or damaged, immediately "heals" and restores itself in the presence of air, oxygen, or water, even in trace amounts.

The protective passive oxide film on titanium (mainly TiO_2) is very stable over a wide range of pH, redox potential, and temperature and is especially favored as the oxidizing character of the environment increases (see Chapter 14, "Corrosion"). For this reason, titanium generally resists environments ranging from mildly reducing to highly oxidizing, up to reasonably high temperatures. It is only under highly reducing conditions (especially strong reducing acids) where oxide film breakdown and resultant corrosion can occur.

The presence of common naturally occurring and intentionally added oxidizing species often maintains or extends the useful performance limits of titanium in many highly reducing-acid environments (e.g., HCl, H_2SO_4, and H_3PO_4). These inhibitive species include air, oxygen, ferric corrosion products, other specific metallic ions, and/or other dissolved oxidizing compounds. The range of applications for titanium can be expanded by alloying with certain noble elements (particularly in the platinum group metals) or by improved anodic potentials (anodic protection).

In addition, titanium generally exhibits superior resistance to chlorides and various forms of localized corrosion. Titanium alloys are considered to be essentially immune to chloride pitting and intergranular attack and are highly resistant to crevice and stress corrosion.

Another major benefit to the designer is the fact that weldments, heat-affected zones, and castings of many of the industrial titanium alloys exhibit corrosion resistance equal to their base metal counterparts. This is attributable to the metallurgical stability of the leaner titanium alloys (low-alloy content) and the similar protective oxide that forms on titanium surfaces despite microstructural differences.

Hostile Environments

The following sections review the use of titanium in the more common hostile environments (see Chapter 14, "Corrosion") encountered in industry.

Chlorine Chemicals and Solutions. Titanium is fully resistant to solutions of chlorites, hypochlorites, chlorates, perchlorates, and chlorine dioxide. It has been used to handle these chemicals in the pulp and paper industry for many years with no evidence of corrosion.

Titanium is used in chloride salt solutions and other brines over the full concentration range, especially as temperatures increase. Near-nil corrosion rates can be expected in brine media in the pH range of 3 to 11. Oxidizing metallic chlorides, such as $FeCl_3$, $NiCl_2$, or $CuCl_2$, extend titanium passivity to much lower pH levels.

Crevice corrosion is a possible limiting factor of titanium alloy applications in hot aqueous chlorides in metal-to-metal joints, gasket-to-metal interfaces, and under process stream deposits. Given these potential crevices in hot chloride-containing media, localized corrosion of unalloyed titanium and other alloys may occur, depending on pH and temperature. Titanium alloys with improved crevice-corrosion resistance, such as ASTM grades 7, 11, and 12, should be considered in these situations.

Chlorine Gas. Titanium is widely used to handle moist and wet chlorine gas, it and has earned a reputation in the chloralkali industry for outstanding performance in this service. The strongly oxidizing nature of moist or wet chlorine passivates titanium, resulting in low corrosion rates. Proper alloy selection offers a solution to the possibility of crevice corrosion when wet chlorine service temperatures exceed 170 °C (340 °F).

Dry chlorine can cause rapid attack of titanium and may even cause ignition, if moisture content is sufficiently low. However, as little as 0.5% water is generally sufficient for passivation or repassivation after mechanical damage to titanium in chlorine gas under static conditions at room temperature.

Halogen and halide compounds tend to react with titanium in a manner similar to chlorine or chloride compounds. However, special care and consideration should be given to gaseous fluorine and aqueous fluoride environments, which are generally highly corrosive to titanium and its alloys.

Other Salt Solutions. Titanium alloys exhibit excellent resistance to practically all salt solutions over a wide range of pH and temperatures. Good performance can be expected in sulfates, sulfites, borates, phosphates, cyanides, carbonates, and bicarbonates; similar results can be expected with oxidizing anionic salts such as nitrates, molybdates, chromates, permanganates, and vanadates, and also with oxidizing cationic salts, including ferric, cupric, nickelous, and noble metal compounds.

Fresh Water/Steam. Titanium alloys are highly resistant to water, including natural waters and steam, to temperatures in excess of 300 °C (570 °F). Excellent performance can be expected in high-purity water, fresh water, and body fluids. Typical contaminants found in natural water streams, such as iron and manganese oxides, sulfides, sulfates, carbonates, and chlorides, do not compromise titanium performance. Titanium remains totally unaffected by chlorination treatments used to control biofouling.

Seawater. Titanium is fully resistant to natural seawater regardless of chemistry variations and pollution effects (such as sulfides). Twenty-year corrosion rates well below 0.0003 mm/year (0.01 mil/year) have been measured on titanium exposed beneath the sea, in marine atmospheres, in splash or tidal zones, and buried in soils. In the sea, titanium alloys are immune to all forms of localized corrosion and withstand seawater impingement and flow velocities in excess of 31 m/s (100 ft/s), as shown in Table 15.6.

Abrasion and cavitation resistance of these alloys is outstanding, explaining why titanium provides total reliability in many marine and naval applications. In addition, the fatigue strength and toughness of many titanium alloys are unaffected in seawater, and lean titanium alloys with low alloy content are immune to seawater stress corrosion.

Titanium tubing has been used with great success since the 1960s in seawater-cooled heat exchangers in the power, chemical, oil refining, and desalination industries.

When in contact with other metals, titanium alloys are not subject to galvanic corrosion in seawater. However, titanium may accelerate attack on active metals such as steel, aluminum, and copper alloys. The extent of galvanic corrosion depends on many factors, such as anode-to-cathode ratio, seawater velocity, and chemistry. The most successful design strategies eliminate the galvanic couple by using compatible passive metals with titanium, all-titanium construction, or dielectric (insulating) joints. Other approaches for mitigating galvanic corrosion that are effective include coating, lining, and cathodic protection of the active metal and/or coating the titanium surface.

Nitric Acid. Titanium is used extensively for handling nitric acid in commercial applications. Titanium exhibits low corrosion rates compared with stainless steel in nitric acid over a wide range of conditions, as shown in Table 15.7. At boiling temperatures and above, the corrosion resistance of titanium is very sensitive to nitric acid purity. Generally, the higher the contamination and the higher the metallic ion content of the acid, the better titanium performs. This is in contrast to stainless steels, which are often adversely affected by acid contaminants. Because the corrosion product of titanium (Ti^{4+} ion) is highly inhibitive, titanium often exhibits superb performance in recycled nitric acid streams, such as reboiler loops, stripper bottoms, and storage tanks.

In one instance, a titanium heat exchanger handling 60% HNO_3 at 195 °C (380 °F) and 2 MPa

Table 15.6 Comparison of alloy erosion-corrosion resistance at various seawater locations

Location	Flow rate		Duration, months	Corrosion rate					
				Grade 2 titanium		70-30 Cu-Ni		Aluminum	
	m/s	ft/s		mm/yr	in./yr	mm/yr	in./yr	mm/yr	in./yr
Brixham Sea	9.8	32	12	<0.0025	<0.0001	0.3	0.01	1.0	0.04
	1.0	3.3	54	7.5×10^{-7}	3×10^{-8}	...	...	...	...
Kure Beach	8.5	28	2	1.2×10^{-4}	4.7×10^{-6}	0.05	0.002	...	...
	9.0	30	2	2.8×10^{-4}	1.1×10^{-5}	2.1	0.08	...	...
	7.2	24	1	5.0×10^{-4}	2.0×10^{-5}	0.12	0.005	...	...
	0.6–1.3	2.0–4.3	6	1×10^{-4}	4×10^{-6}	0.02	8×10^{-4}	...	...
Wrightsville Beach	9.0	30	2	1.8×10^{-4}	7×10^{-6}	...	...	...	...

Table 15.7 Comparison of corrosion rates in boiling 90% HNO_3

Metal temperature		Corrosion rate			
		Grade 2 titanium		304L stainless steel	
°C	°F	mm/yr	in./yr	mm/yr	in./yr
116	241	0.03–0.17	0.001–0.007	3.8–13.2	0.15–0.52
135	275	0.04–0.15	0.002–0.006	17.2–73.7	0.677–2.90
154	309	0.03–0.06	0.001–0.002	18.3–73.7	0.720–2.90

(300 psi) showed no signs of corrosion after more than 2 years of operation. Titanium reactors, reboilers, condensers, heaters, and thermowells have been used in solutions containing 10 to 70% HNO_3 at temperatures from boiling to 315 °C (600 °F).

Although titanium has excellent resistance to nitric acid over a wide range of concentrations and temperatures (including white fuming HNO_3), it should not be used in red fuming nitric acid because of the danger of pyrophoric (burning) reactions. Guidelines for minimum water content and maximum NO_2 concentration (NO_2/NO ratio) for avoiding pyrophoric reactions in this particular acid have been developed.

Reducing Acids. Titanium alloys are generally very resistant to mildly reducing acids but can display severe limitations in strongly reducing acids. Mildly reducing acids (such as sulfurous, acetic, terepthalic, adipic, lactic, and many organic acids) generally represent no problem for titanium over the full concentration range.

However, relatively pure, strong reducing acids (such as hydrochloric, hydrobromic, hydrofluoric, sulfuric, phosphoric, oxalic, and sulfuric acids) can accelerate general corrosion of titanium, depending on acid temperature, concentration, and purity. With the exception of fully metal-ion-complexed solutions, titanium experiences severe attack in hydrofluoric acid solutions. Titanium-palladium alloys (ASTM grades 7 and 11) offer dramatically improved corrosion resistance under these severe reducing-acid conditions. Titanium-palladium often compares quite favorably to nickel alloys in dilute reducing acids, as shown in Table 15.8 (see also Chapter 14, "Corrosion," in this book). Also, the presence of heavy metal ions, such as normal process-stream contaminants, can passivate titanium alloys in normally aggressive acid media, thereby extending its use. A low concentration of oxidizing metal ions, such as 20 to 100 ppm, can often provide effective inhibition.

Organic Chemicals. Titanium alloys generally exhibit excellent resistance to organic media.

Mere traces of moisture and/or air typically present in organic process streams ensure the development of a stable protective oxide film on titanium.

Titanium is highly resistant to hydrocarbons, chloro-hydrocarbons, fluorocarbons, ketones, aldehydes, ethers, esters, amines, alcohols, and most organic acids. Anhydrous methanol is unique in its ability to cause stress-corrosion cracking in titanium alloys. However, addition of more than 1% water is sufficient to eliminate this problem.

Titanium equipment has traditionally been used for production of terepthalic acid, adipic acid, and acetaldehyde. Acetic acid, tartaric acid, stearic acid, lactic acid, tannic acids, and many other organic acids also represent fairly benign environments for titanium. However, proper titanium alloy selection is necessary for the stronger organic acids, such as oxalic, formic, sulfamic, and trichloroacetic acids. Performance in these acids depends on acid concentration, temperature, degree of aeration, and possible inhibitors present. The ASTM grades 7, 11, and 12 titanium alloys are often preferred materials in these aggressive acids.

Alkaline Media. Titanium is generally highly resistant to alkaline media, including solutions of sodium, potassium, calcium, magnesium, and ammonia hydroxides. However, in the highly basic sodium or potassium hydroxide solutions (i.e., pH > 12), useful application of titanium may be limited to temperatures below 80 °C (175 °F) due to possible excessive hydrogen uptake and eventual embrittlement of titanium alloys in hot, strongly alkaline media.

Titanium often becomes the material of choice for alkaline-media-containing chlorides and/or oxidizing chloride species. Even at higher temperatures, titanium resists pitting, stress corrosion, and the conventional caustic embrittlement observed on stainless steel alloys in these situations.

Corrosion Resistance of Titanium Alloys. Most of the higher-strength titanium alloys exhibit excellent resistance to general and pitting corrosion in near-neutral environments, which are neither highly oxidizing nor highly reducing (see Chapter 14, "Corrosion," in this book). The metallurgical condition of the alloy plays a relatively minor role in corrosion performance, because oxide film stability is ensured in these situations. When severe crevices exist in hot aqueous chloride media, most high-strength titanium alloys exhibit slightly reduced crevice-corrosion resis-

Table 15.8 Comparison of corrosion rates in boiling HCl

HCl, wt%	Corrosion rate					
	Ti-Pd		6Ni-2Cr-5Mo		C-276	
	mm/yr	in./yr	mm/yr	in./yr	mm/yr	in./yr
2.0	0.05	0.002	11.5	0.453	0.96	0.04
3.0	0.07	0.003	17.5	0.689	1.65	0.065
4.0	0.12	0.005	20.7	0.815	2.21	0.087

tance compared with unalloyed titanium. However, titanium alloys rich in molybdenum exhibit excellent resistance to this form of attack. It is for this reason, along with its favorable strength and low density, that the high-molybdenum alloy Ti-3Al-8V-6Cr-4Mo-4Zr is a prime candidate for high-temperature, sour gas well, and geothermal brine well production tubulars and other downhole components.

General corrosion resistance of high-strength titanium alloys in strongly oxidizing or strongly reducing environments may diminish as aluminum and/or vanadium alloy content increases. Improvements in resistance to hot reducing acids can be achieved by increased molybdenum content. The effects of various alloying elements on titanium alloy corrosion have been studied. Suitable high-strength alloys can be selected for a wide range of aggressive environments.

Another major consideration is resistance to stress-corrosion cracking (SCC). Although the common industrial grades of titanium are immune to chloride SCC, certain high-strength alloys exhibit reduced toughness and/or accelerated crack growth rates in halide environments. Most of these alloys will not exhibit any susceptibility to SCC in smooth or notched conditions. However, above 250 °C (450 °F), resistance to hot salt SCC should be considered if chlorinated solvents, chloride salts, or other chlorine-containing compounds contact component surfaces.

High-strength titanium alloys may successfully avoid SCC-related effects by consideration of alloy chemistry and/or preferred alloy heat treatments and microstructures. Selection of extra low interstitial (ELI) grades or an alloy with transformed-beta microstructure offers significantly improved alloy toughness and resistance to SCC.

In summary, although high-strength titanium alloys have corrosion resistance generally superior to that of most common engineering alloys, consideration should be given to selecting a titanium alloy with full compatibility to a given environment. With the wide family of titanium alloys commercially available, optimum high-strength titanium alloy selection is almost always possible for a given environment. Technical consultation is available to assist the designer/user in achieving this end.

Mechanical and Physical Properties

The family of titanium alloys offers a full range of strength properties spanning from the highly formable, lower-strength to the very high-strength alloys (Ref 15.1–15.3). Most of the alpha-beta and beta alloys provide myriad strength/ductility property combinations through adjustments in alloy heat treatment (microstructure control) and/or composition. Thus, a suitable titanium alloy can always be selected for specific application needs.

One of the most attractive characteristics of the titanium alloys is their outstanding strength-to-weight ratio compared with many other engineering metals. This means increased structural efficiency in design, leading to weight, space, or cost reduction, depending on the application. High strength-to-density ratio has been the traditional driving force for the extensive use of titanium alloys in a multitude of aerospace applications. Many innovative uses outside the aerospace realm have developed over the years based on strength-to-weight requirements, often combined with the excellent corrosion resistance that titanium offers. Examples include deep sea submersible hulls, ballast tanks, deep well (downhole) tubular (Fig. 15.19), stress joints, riser pipes, and logging tools.

Many titanium alloys are approved for pressurized equipment falling under the American Society of Mechanical Engineer's Boiler and Pressure Vessel Code and the American National Standards Institute code. An example of a titanium pressure vessel appears in Fig. 15.20.

Physical Properties

Titanium and its alloys possess many unique physical properties, making them ideal for equipment design, even when strength or corrosion resistance may not be critical. These unique properties (Ref 15.1, 15.2) include:

- Low density
- Low modulus of elasticity
- Low coefficient of expansion

Fig. 15.19 Pin-end view of titanium drill pipe joints used for short-radius drilling operations

Fig. 15.20 Titanium personnel sphere for the human-occupied submersible Alvin is prepared for pressure testing. Courtesy of Woods Hole Oceanographic Institution

- High melting point
- Nonmagnetic (paramagnetic)
- Extremely short radioactive half-life
- High generic shock resistance

The low density of titanium, roughly 56% that of steel, means twice as much metal per kilogram (pound). Particularly when combined with alloy strength, this often means smaller and/or lighter components. Although strength-to-weight ratio is the basis for aerospace applications, these design advantages are also apparent for many types of nonaerospace rotating or reciprocating components, such as centrifuges, pumps, and automotive valves.

Reduced component weight is also of interest in certain aggressive environments. For example, downhole oil and geothermal well production tubules and logging tools are made of titanium alloys, based on these property combinations. In marine service, pleasure boat components, naval surface ships, and submarine cooling-water systems are growing markets for titanium, driven by its immunity to seawater corrosion and its light weight.

Titanium alloys exhibit modulus of elasticity values that are approximately 50% of those of steel. The low modulus means excellent flexibility, which is the basis for its use in stress joints for petroleum production riser pipes running between the sea floor and a floating production platform. Other industrial applications using the low modulus and high strength of titanium include springs and bellows.

Titanium possesses a coefficient of expansion significantly less than that of ferrous alloys. This property also allows titanium to be much more compatible with ceramic and glass materials than most metals, particularly when metal/ceramic/glass seals are involved.

The relatively high melting point of titanium and its resistance to ballistic penetration led to the consideration of titanium for armor applications. In addition to the favorable resistance of titanium to ballistic penetration, its higher melting point reduces susceptibility to melting and ignition (burning) during ballistic impact. Good toughness, weldability, and light weight are additional factors for considering titanium alloys in this application.

Titanium is virtually nonmagnetic, making it ideal for applications where electromagnetic interference must be minimized. These applications include use in electronic equipment housings, downhole well/logging tools, and ladle shells used in the electromagnetic stirring of molten metals.

Titanium and its isotopes have an extremely short radioactive half-life, thereby permitting its use in nuclear systems. In contrast to some ferrous alloys, many titanium alloys do not contain significant amounts of alloying elements that may become radioactive. This property, combined with resistance to corrosion, has made titanium the material of choice for canisters used to transport and store nuclear waste in subterranean repositories.

Erosion Resistance. The oxide film of titanium generally provides excellent resistance to abrasion, erosion, and erosion-corrosion in high-velocity process streams. The performance of titanium in flowing seawater is superior to that of copper and aluminum alloys. Even in heavy sand-laden seawater, insignificant metal loss is noted for titanium alloys in seawater flowing at a rate of up to 6 m/s (18 ft/s).

Therefore, in contrast to copper alloys, titanium piping, equipment, and heat exchangers are designed for high flow velocities with little or no detrimental effects from turbulence, impingement, and cavitation. This has positive implications relative to optimizing heat-transfer efficiency, minimizing equipment wall thickness and tube and piping size, improving equipment reliability, and reducing life-cycle costs. It is for these reasons that titanium alloys, in either wrought or cast form, have become prime materials for various coastal chemical and power plants and marine/naval applications.

Heat Transfer. Titanium is the material of choice for its heat-transfer characteristics in shell/tube and plate/frame exchangers and many types of heaters and coolers. In addition to excellent corrosion resistance, several other beneficial attributes of titanium interact to optimize exchanger heat-transfer efficiency. Among these attributes are:

- Good strength
- Resistance to erosion and erosion-corrosion
- Very thin, conductive oxide surface film
- Hard, smooth, difficult-to-adhere-to surface
- A surface that promotes drop-wise condensation

Based on its excellent resistance to corrosion and erosion-corrosion, titanium is designed with a zero-corrosion allowance. This feature, combined with its good strength, permits use of very thin heat-transfer walls. An example of a titanium tube bundle for a heat exchanger is shown in Fig. 15.21.

Furthermore, very high fluid flow rates can be designed for titanium exchangers, reflecting its outstanding resistance to fluid erosion, cavitation, and impingement. These higher flow rates directly increase overall heat-transfer efficiency while improving surface cleanliness. These attributes are used in titanium condensers installed in power plants. This reduces exchanger size and material

Fig. 15.21 Titanium tube bundle for heat exchanger

requirements and minimizes the total cost of titanium exchangers. Titanium is comparable in initial cost to certain copper and stainless steel alloys when full advantage is taken of the unique properties of titanium.

The use of thinner titanium heat-transfer wall thickness offsets the somewhat lower thermal conductivity of titanium. However, it is the surface characteristics of titanium that offer the greatest benefit to heat transfer. The limiting factor in heat transfer is the insulating metal-fluid interface films that often form in service. The high resistance of titanium to corrosion prevents buildup of corrosion products that rob metals of heat-transfer efficiency.

The hard, smooth surface of titanium also minimizes adherence and buildup of external fouling films, making cleaning and maintenance easier. It is not unusual to observe a 95 to 100% cleanliness factor for titanium in many services. Although titanium is not biotoxic, successful control of biofouling is achievable by periodic chlorination and/or by mechanical means (such as tube brushes and sponge balls). A fouling factor of less than 0.0005 is usually achieved for titanium in seawater using these strategies. Typically, much higher flow velocities are specified for titanium compared with copper alloys to promote cleanliness. Although the copper alloys have higher thermal conductivity and higher overall heat-transfer coefficients when new and clean, titanium exhibits the higher long-term operating coefficients in actual seawater service when properly designed.

Titanium is also unique in its ability to promote drop-wise condensation on its surface. When condensing water vapor from evaporative processes, most metals form continuous surface films of condensate. However, this mechanism of condensation is not nearly as efficient as the drop-wise mechanism of titanium, which is used effectively for brine and nitric acid distillation applications.

Taking full advantage of all the beneficial properties of titanium to optimize heat transfer leads to extremely reliable, highly efficient, cost-competitive heat exchangers. Not only is the initial cost of a titanium exchanger highly attractive relative to the more common engineering alloys, but improvements in life-cycle cost in aggressive service resulting from reduced maintenance/cleaning requirements is also evident. Titanium is readily available in welded and seamless tubing in many alloy grades for tubular exchangers and in sheet form for plate and frame exchangers.

Fabrication. The fabrication of titanium into usable systems is accomplished using all the same basic techniques used for the other common metals of construction (see Chapter 10, "Secondary Working of Bar and Billet," and Chapter 11, "Forming of Titanium Plate, Sheet, Strip, and Tubing," in this book). However, the technology required for each fabrication technique must be modified and adapted for titanium properties, namely, low density, low modulus, low ductility, galling tendency, high melting temperature, low thermal conductivity, tendency for surface contamination, and its reactive nature.

Cutting. Titanium and its alloys can be cut by shearing, sawing, oxyacetylene flame, plasma, waterjet, electric discharge machining, and abrasive cutoff wheels as long as care is taken to avoid excessive heat buildup (see Chapter 13, "Machining and Chemical Shaping of Titanium," in this book). Titanium is readily cold and warm formed on equipment used for stainless steel as long as titanium springback is taken into consideration.

Machining of titanium using carbide and high-speed steel tools is done on the same equipment as for stainless steels. However, titanium requires slow speeds, heavy feeds, rigid setups, copious quantities of coolant, and good chip removal (see Chapter 13, "Machining and Chemical Shaping of Titanium").

Welding of titanium can be accomplished by gas tungsten arc, gas metal arc, resistance, inertia (friction), laser, plasma, friction stir welding, explosive, and electron beam techniques (see Chapter 12, "Joining Titanium and Its Alloys"). All titanium welding must be accomplished such that the molten pool and solidifying weld are under inert gas (argon or helium) or vacuum. A titanium bike frame weld is illustrated in Fig. 15.22. Postweld heat treatment of commercially pure titanium grades is generally unnecessary.

Fig. 15.22 Welded titanium bike frame. Source: Ref 15.23

Casting. Standard tolerance and precision titanium castings are available as rammed-graphite and investment castings, respectively (see Chapter 8, "Melting, Casting, and Powder Metallurgy," in this book). The fabrication of titanium and its alloys is limited only by the capability of the fabricator and the imagination of the designer.

Medical Applications

The use of titanium in medical applications began in the 1930s and is based on its biocompatibility, density, mechanical properties, and corrosion resistance (Ref 15.1, 15.2). Titanium has outstanding resistance to corrosion and fatigue corrosion in the human body (body fluids are 0.9% saline solution). It also has good tissue tolerance and is compatible with human bone. As a surgical implant material, tissue compatibility and sterilization capability are key attributes, which have led to its growing acceptability. Many medical instruments are also made of titanium because of its special combination of physical, chemical, and mechanical properties. In the dental field, the high strength and low modulus of titanium have been central to its successful use.

The use of titanium as a surgical implant material began in the 1930s, well before the commercial industry was founded in 1950. The first experimental implants were laboratory-produced unalloyed titanium pegs that were implanted in a cat's femur. This trial, along with many subsequent evaluations, demonstrated titanium biocompatibility. This property, along with its favorable density and mechanical properties (high strength and modulus nearly that of bone), led commercially pure titanium and its alloy Ti-6Al-4V ELI to be included in the ASTM F 67 and F 136 specifications, respectively, for surgical implant material. These alloys are now used in applications such as total hip replacements, shoulder prostheses, finger joints, knee replacements (Fig. 15.23), heart valves, pacemaker cases, artificial ribs, nails, screws, and pins. A titanium valve assembly is a vital component of the Jarvik 7 artificial heart. In addition, titanium wire mesh and metal wool pads are used in facial reconstructive surgery.

Titanium is also used in external prostheses based on its low density and high strength. The weight reduction over stainless steel at no sacrifice in strength is a major benefit to patients who wear external prostheses. Titanium tubules are used to make lightweight crutches and wheelchairs. Based on strength and weight, titanium has replaced both aluminum and stainless steel in this application.

In dental applications, titanium is used in several ways, including orthodontic wire, medical and dental instruments, crown material, and implants placed in the jaw as anchors for artificial teeth (Fig. 15.24).

Braces used for orthodontic repositioning of teeth are made of several types of titanium alloy wire. Beta-titanium alloy Ti-3Al-8V-6Cr-4Zr-4Mo wire is used in orthodontic appliances for its low modulus and high strength. Nitinol (Ti-55Ni) shape-memory alloy is used for its high strength, low modulus, and its ability to return to a preformed shape induced by the warmth of the mouth. Both types of orthodontic wire reduce the number of adjustments required for a set amount of dental movements compared with traditional orthodontic wire.

The use of titanium dental implants is also gaining prominence. The procedure involves screwing a titanium anchor into the jawbone, forming a root (Fig. 15.24). A smaller titanium screw is then tightened into the anchor to serve as an abutment to which the artificial tooth or teeth are attached. The procedure is used to replace one or more teeth using one or more anchors. In addition to providing nearly natural teeth that do not wobble or irritate, the implants help minimize jawbone erosion, common to denture wearers.

Titanium is used increasingly in medical and dental instruments (Fig. 15.25). Because of its low density, surgical instruments are lighter when made of titanium. This lighter weight is an important factor in reducing surgeon fatigue during lengthy surgical procedures. The ability of titanium to be sterilized repeatedly and its "springy" quality are also important considerations in its selection.

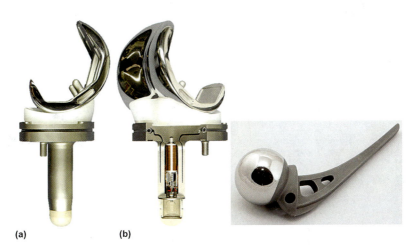

(a)   (b)

Fig. 15.23 (a) Titanium artificial knee with cut-away view. (b) Titanium artificial hip. The rough implant surface promotes bone adhesion to the implant.

Consumer Applications

Titanium is used in numerous consumer applications ranging from recreational to personal care (Ref 15.4–15.11). Each application takes advantage of one or more of the mechanical, chemical, and physical properties of titanium. Titanium is used in knife handles and carpenter's hammer heads, providing lighter weight compared with steel.

In the recreational equipment field (Ref 15.25), titanium is used in golf club shafts. The alloy Ti-3Al-2.5V is produced as a tapered, hollow, seamless tube. Its length is adjusted by trimming the ends, thereby producing the correct length, taper, and cross section for all the wood and iron clubs. Titanium was selected because of its light weight, low elastic modulus, and strength. A shaft of this type produces a controlled torque that promotes longer, straighter golf shots. Titanium is used extensively in the hollow head of golf drivers (Fig. 15.26) and in irons and putters (in all cases, both cast and forged products have been fabricated), where the weight can be distributed at the periphery of the hitting face to increase the "sweet spot" and make it easier to hit a good shot.

Frames for racing and mountain bicycles are also made using Ti-3Al-2.5V seamless and welded tubing. As in other applications, the advantage titanium bicycle frames offer over conventional thin steel or aluminum (in addition to weight savings) is low modulus of elasticity. In the case of the bicycle frame, it tends to deflect and dissipate road shock, thereby allowing the rider to suffer less fatigue.

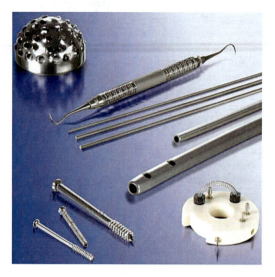

Fig. 15.25 Titanium dental tools, surgical screws, and implants. Source: Ref 15.24

IMPLANT

ABUTMENT

CROWN

Fig. 15.24 Titanium plug implanted into a human cheek bone into which is screwed an artificial tooth. The rough surface promotes bone adhesion to the titanium screw.

Fig. 15.26 Forged titanium golf club metal/wood head. Titanium club heads will spark if they strike stones when hitting a fairway shot.

Tennis racquets made of commercially pure titanium seamless and welded tubing offer advantages through light weight and low modulus. Again, these characteristics permit a truer, harder smash with improved control. Other sporting equipment includes softball bats and hockey-skate blades.

Titanium horseshoes are successfully used on harness-racing horses. The lightweight shoes are more abrasion resistant than aluminum and therefore require less reshoeing due to shoe wear.

Titanium eyeglass frames, watches, razors, and jewelry all take advantage of the light weight of titanium. Also, the ability of titanium to be anodized or heat colored to impart decorative tints, from golds to deep purples, has been used to advantage in jewelry (Fig. 15.27) and artwork.

In the electronics field, sputtered pure titanium thin films are used in diodes and switching devices on circuit boards installed in computers, microwave ovens, and alarm systems. Even though the electrical conductivity of titanium is significantly lower than that of aluminum, copper, and silver, it is used in combination with silicon to produce the desired electronic responses.

Sputtered pure titanium is also used on architectural glass as a sunlight and radiant heat reflector. The color of the titanium coating can be controlled to create interesting and complimentary effects. Titanium is also used in combination with other metallic coatings to protect them from oxidation.

The use of titanium in fireworks is based on its burning as a brilliant white flame. Finely divided titanium powder is packed in fireworks rockets and burned as a brilliant white shower during fireworks displays.

The use of titanium in water sports is a natural extension of its resistance to corrosion and erosion in fresh and salt waters. Water skis, boating propellers, hardware for sailboats and power boats, fishing reels, and underwater knives and watches are common uses for titanium and its alloys.

Armor Applications

Lightweight personal ballistic armor uses high-strength titanium sheet. Titanium alloys Ti-6Al-4V and Ti-15V-3Al-3Sn-3Cr in sheet form are used in vests, clipboards, helmets, and briefcases that are designed to withstand ballistic penetration from handguns. Titanium is also an attractive material as an armor material for military vehicles (Fig. 15.28). A number of military

Fig. 15.27 Titanium ring with anodized bands coloring its surface. Courtesy of Metallium

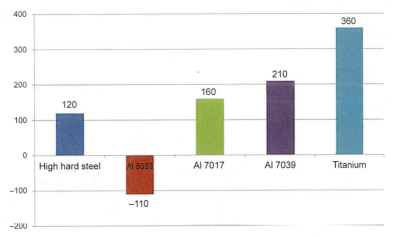

Fig. 15.28 Number of pounds reduced in weight compared to each 455 kg (1000 lb) of rolled homogeneous alloy steel at 14.5 mm (0.57 in.) armor-piercing protection. Data courtesy of Adam Gardels, ATI Wah Chang

vehicles that can benefit from titanium armor are shown in Fig. 15.29 and 15.30. The low density, moderate-to-high strength, and high toughness of titanium meet the requirements of this application.

Automotive Applications

The weight of an automobile has a big influence on fuel consumption (Fig. 15.31). Titanium is attractive for use in the automotive industry

Fig. 15.29 Active "Armor Upgrade" programs. Expanding requirements will drive upgrades and original equipment manufacturer design to titanium. Courtesy of the International Titanium Association

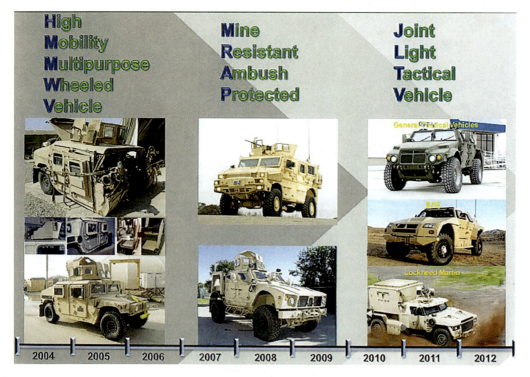

Fig. 15.30 Light utility vehicle evolution. Courtesy of the International Titanium Association

(Ref 15.26). In addition to light weight, the corrosion and strength characteristics of titanium are attractive. However, its introduction has been slowed due to initial cost and possible increased manufacturing cost, which may not offset its savings in fuel and repair cost over the lifetime of a vehicle. However, the government and market forces will continue to make auto manufacturers invest in lighter-weight materials, where titanium is a good candidate (Fig. 15.32) for engine and body applications.

Titanium valves (Fig. 15.33), connecting rods, valve springs, suspension springs (Fig. 15.34), and wheels have been used in high-performance and racing engines since the 1960s and have moved into commercial and passenger engines.

The high strength, good fatigue, good creep life, and low density of Ti-6Al-4V and Ti-6Al-2Sn-4Zr-2Mo-0.1Si make them the alloys of choice for the more fuel-efficient engines required to meet the Corporate Auto Fuel Efficiency standards set by the U.S. government.

Building Applications

Titanium is seeing increased use in architectural applications because of its excellent corrosion behavior and its attractive appearance (Fig. 15.35, 15.36).

Power Utility Applications

The use of titanium in electric-power-generating plants began in 1959 when welded titanium tubing was tested in the main turbine steam condenser at two East Coast power plants. This application was based on the resistance of titanium to corrosion in seawater and its heat-transfer efficiency. Because most coastal power plants use seawater as their major coolant, it was logical to use titanium in this service to improve reliability and reduce maintenance. Even though titanium has a lower heat-transfer coefficient than competing materials, it has proven to be a good heat-transfer material in this application for several reasons.

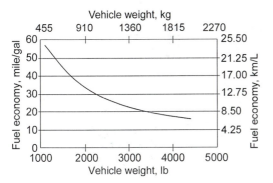

Fig. 15.31 Influence of automobile weight on fuel economy for a passenger vehicle in Japan

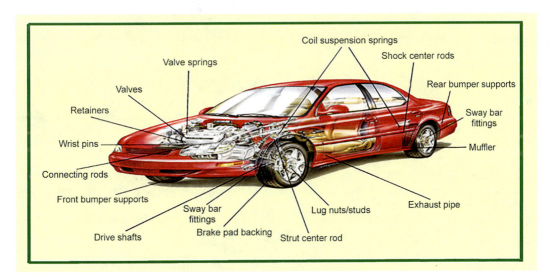

Fig. 15.32 Possible applications for titanium in a passenger automobile. Courtesy of TIMET

Fig. 15.33 The Toyota Altezza, 1998 Japanese car of the year, was the first family automobile in the world to feature titanium valves. Inset: Ti-6Al-4V intake valve (left) and TiB/Ti-Al-Zr-Sn-Nb-Mo-Si exhaust valve (right). Courtesy of Toyota Central R&D Labs, Inc.

Fig. 15.34 Titanium spring on the Volkswagen Lupo FSI

When made of titanium, thinner tubing walls can be used for several reasons:

- No corrosion by seawater eliminates the need for a corrosion factor to be added to the tubing wall thickness to allow for planned degradation.
- Because titanium does not erode in seawater, flow rates can be increased; thus, more heat is dissipated.
- Increased velocity leads to increased turbulent flow, which allows for more efficient heat transfer.

Since 1959, ASTM B338 grade 2 titanium has become the material of choice for use in heat-transfer applications, and through engineering and manufacturing developments, the wall thickness commonly used in this application has been reduced from 1.24 mm (0.049 in.) to 0.5 and 0.7 mm (0.020 and 0.028 in.). Currently, both titanium tubes and tube sheets are being used in this application, with not one failure due to corrosion of the titanium by sea, brackish, and polluted waters.

In nuclear power plants where seawater, brackish, and polluted waters are used as the main source of cooling, the service water piping that delivers the cooling water from its source to the plant has proven to be another ideal use for titanium. ASTM B337 grade 2 titanium pipe is immune to corrosion by the action of these waters.

Fig. 15.35 Seam-welded titanium roof on Maten Primary School. In coastal locations, titanium offers good resistance to seawater spray. Courtesy of Nippon Steel

Fig. 15.36 The Editor, F.H. Froes, standing in front of the Guggenheim Museum, Bilbao, Spain, which is sheathed with titanium sheet

This is commonly a retrofit application, with the steel and lined steel being replaced with titanium because of their deterioration in service. Titanium is a life-cycle cost solution in this application. An example of a pressure vessel lined with titanium is shown in Fig. 15.37.

Titanium is also used as tubes and tube sheets in auxiliary coolers, feed-water heaters, and waste steam evaporators in power plants.

The sulfur dioxide emissions created by burning high-sulfur coal and oil in plants fired by fossil fuels are scrubbed from the effluent gas in flue gas desulfurization scrubber systems, which use ASTM grades 2 and 7 titanium as corrosion-resistant linings. In North America, so-called "acid rain" legislation requires utilities to clean the sulfur dioxide from their gaseous effluents. This is a very corrosive environment, particularly in closed-loop (recycled stream) systems, where chlorides concentrate and hot wet/dry gas/liquid interfaces form. Titanium liners are successfully used in the chimney stacks and in the inlet quench and outlet ducts of these systems.

Titanium steam turbine blades withstand the rotating stresses of the turbine and the corrosion/erosion damage of the impinging steam droplets.

Fig. 15.37 Titanium grade 2 pressure vessel with titanium grade 2 half-pipe jacket. Courtesy of Titan Metal Fabricators

The Ti-6Al-4V blades are the technically correct material in this application, just as titanium was in similar blade applications in jet engines.

Marine Applications

The resistance of titanium to corrosion/erosion in seawater makes it an ideal choice for use on offshore petroleum production platforms and commercial and military marine applications (Ref 15.15, 15.16). As discussed, titanium has been successfully used in seawater-cooled utility condensers and stress joints and riser pipes for off-shore petroleum production wells. In addition to these uses, numerous other uses are common in marine environments.

Heat exchangers, declinators, and cooler condensers are common on commercial marine vessels and on offshore production platforms. This type of heat-transfer application of titanium is based on its resistance to seawater corrosion/erosion and heat-transfer efficiency, which is enhanced by increasing flow rates and turbulence. Shell, tube, plate, and frame heat exchangers are used. High reliability and low maintenance are essential in these applications, because downtime is very costly. An example of an extremely large distillation tower is illustrated in Fig. 15.38.

Titanium piping is used in offshore and commercial marine vessel applications. In addition to the high reliability resulting from the corrosion/erosion resistance of titanium, its low density is advantageous because lower structure weight translates into higher payload weight.

Titanium pumps, valves, and fittings are used along with titanium piping on offshore and commercial marine vessels. The U.S. Navy also uses a 4 m³/min (1056 gal/min) pump as their standard pump on surface ships. The main sea valves in U.S. Navy submarines are Ti-6Al-4V cast balls. Forged and machined ball valves are used on all submarine seawater ball valve applications to maximize reliability.

Trends indicate that the U.S. Navy will use increasing amounts of titanium in new design ships. Wherever reduced weight and increased reliability (i.e., no seawater corrosion) are required to keep fighting ships and support ships battle-ready, titanium is being evaluated. Exhaust stacks, jet-blast deflectors, missile canisters and launch structures, seawater piping and coolers, electrical fittings and connectors, bulkheads, hangar doors, and sonar equipment are being evaluated in various titanium alloys for U.S. Navy surface ships. The navy of the former Soviet Union built full-sized Alpha Class submarine (*The Hunt for Red October*) hulls out of titanium. The non-magnetic nature of titanium along with its weldability, toughness, and high strength-to-weight ratio make it an ideal submarine construction material. The high strength-to-weight ratio and low density of titanium allow a submarine to travel deeper than steel-hulled subs, and the non-magnetic nature may allow it to avoid detection by typical submarine-detection procedures.

Miscellaneous Applications

Titanium is used in many applications that cannot be categorized by any specific market area (Ref 15.15, 15.16). These applications show the numerous ancillary roles that titanium plays.

Fig. 15.38 Solid grades 2 and 7 titanium distillation tower used in the production of biodiesel. Courtesy of Titan Metal Fabricators

As mentioned previously, the shape memory alloy Nitinol (titanium-nickel binary) is used as wire in orthodontic fixtures. This shape-memory alloy is also used in the locking mechanism for self-cleaning oven doors and in self-sealing connectors for pipes and tubes.

A second titanium binary alloy of interest is the titanium-niobium binary, which becomes superconductive at 4 Kelvin (−269 °C, or −452 °F). These superconductors are used in applications such as electrical generators, magnetic resonance imaging, and levitation travel for mass transit. The titanium-niobium superconductors are being challenged by engineered ceramic superconductors that work at temperatures as warm as −25 °C (−13 °F).

Safe hydrogen storage can be done in a titanium-iron binary alloy. This alloy, when crushed, provides great surface-to-volume ratios and an excellent storage medium for hydrogen. Hydrogen is absorbed readily into the titanium-iron binary in the presence of a small amount of heat and pressure. When cooled to room temperature, the hydrogen is safely stored and can only be removed by reheating the mass. This idea was developed for use with hydrogen-powered vehicles that would emit pollution-free water when operated.

Other uses for titanium that are beneficial to other metals are the use of titanium as a grain refiner, oxygen getter, and carbide former in steel and aluminum. Titanium is added to liquid steel and aluminum to provide nucleation sights for grain recrystallization during heat treatment. Titanium-refined and -strengthened steels are used for catalytic convertors and automobile bodies. Fine-grained aluminum is used in building siding, screen doors, and lawn furniture.

Summary

Titanium and its alloys offer performance advantages in countless aerospace, nonaerospace, and industrial applications. These advantages are based on the low density, wide range of strengths, strength at elevated temperatures, corrosion resistance, and fabricability of titanium. In aerospace applications, the high strength-to-weight ratio and high-temperature properties of titanium are used in airframes, gas turbine engines, and missiles to make this the largest single market for titanium products.

Titanium corrosion/erosion resistance to numerous aggressive environments makes it the material of choice for industrial applications in pulp and paper manufacture, chlorine production, power utility condensers, cooling-water piping systems, oil production tubules in offshore riser pipes and stress joints, and in chemical and petrochemical plant heat exchangers, piping vessels, and other components.

The medical and dental industry successfully uses titanium in implants, external prostheses, and instruments, based on high strength, low density, and biocompatibility. Titanium bicycle frames, golf club shafts and club heads, ski poles, tennis rackets, and eyeglass frames take advantage of low density, high strength, and low elastic modulus.

In marine applications, reliability and low maintenance driven by the resistance of titanium to corrosion in seawater, its superior mechanical properties, and its low density makes it the material of choice in next-generation naval vessels. Based on its mechanical, physical, and chemical properties, the list of titanium applications will keep growing as long as designers and engineers continue to seek the technically correct solutions to materials problems.

Titanium also has numerous consumer applications, from jewelry to sports equipment and automotive applications. The corrosion resistance, strength, and light weight of titanium contribute to its acceptance in a growing number of consumer applications.

REFERENCES

15.1 F.H. Froes, *Kirk-Othmer Encyclopedia of Chemical Technology,* 5th ed., Nov 2006, p 24, 838

15.2 M.A. Imam, F.H. Froes, and K.L. Housley, Titanium and Titanium Alloys, *Kirk-Othmer Encyclopedia,* March 2010, p 1–41, online only

15.3 F.H. Froes, D. Eylon, and H. Bomberger, Ed., *Titanium Technology: Present Status and Future Trends,* Titanium Development Association, Dayton, OH, 1985

15.4 G. Lutjering, U. Zwicker, and W. Bunk, *Proceedings of the Fifth International Conference on Titanium* (Pittsburgh, PA), Metallurgical Society of AIME, Warrendale, PA, 1984

15.5 P. Lacombe, R. Tricot, and G. Beranger, Ed., *Proceedings of the Sixth International Conference on Titanium* (Nice, France), Metallurgical Society of AIME, Warrendale, PA, 1988

15.6 F.H. Froes and I.L. Caplan, Ed., *Proceedings of the Seventh International Conference on Titanium* (San Diego, CA), Metallurgical Society of AIME, Warrendale, PA, 1992

15.7 *Proceedings of the Eighth International Conference on Titanium* (San Birmingham, U.K.), Metallurgical Society of AIME, Warrendale, PA, 1995

15.8 *Proceedings of the Ninth International Conference on Titanium,* Metallurgical Society of AIME, Warrendale, PA, 2000

15.9 *Proceedings of the Tenth International Conference on Titanium,* Metallurgical Society of AIME, Warrendale, PA, 2004

15.10 *Proceedings of the Eleventh International Conference on Titanium,* Metallurgical Society of AIME, Warrendale, PA, 2008

15.11 *Proceedings of the Twelfth International Conference on Titanium,* Metallurgical Society of AIME, Warrendale, PA, 2012

15.12 G. Lutjering and J.C. Williams, *Titanium,* Springer, 2003

15.13 "General Electric GEnx," *Wikipedia,* from Paris Air Show, 2009, http://en.wikipedia.org/wiki/General_Electric_GEnx

15.14 "Aerospace Materials," *Wikipedia,* http://en.wikipedia.org/wiki/Aerospace_materials

15.15 F.H. Froes and K. Yu, Developing Applications for Titanium, *Proc. Titanium Congress 2003,* July 2003 (Hamburg, Germany)

15.16 F.H. Froes and T. Nishimura, The Titanium Industry in Japan and the USA—A Comparison, *Cost Affordable Titanium,* F.H. Froes, M.A. Imam, and D. Fray, Ed., TMS, Warrendale, PA, 2004

15.17 F.H. Froes, M.A. Imam, and D. Fray, Ed., *Cost Affordable Titanium,* TMS, Warrendale, PA, 2004

15.18 M.N. Gungor, M.A. Imam, and F.H. Froes, Ed., *Innovations in Titanium Technology,* TMS, Warrendale, PA, 2007

15.19 M.A. Imam, F.H. Froes, and K.F. Dring, Ed., *Cost-Affordable Titanium III,* Trans Tech Publications, Zurich-Durnten, Switzerland, 2010

15.20 M.A. Imam, F.H. Froes, and R.G. Reddy, Ed., *Cost Affordable Titanium IV,* Trans Tech Publications, Zurich-Durnten, Switzerland, 2013

15.21 F.H. Froes, Powder Metallurgy of Titanium Alloys, *Advances in Powder Metallurgy,* I. Chang and Y. Zhao, Ed., Woodhead Publishing, 2013

15.22 M. Qian, Ed., *Powder Metallurgy of Titanium,* Trans Tech Publications, Zurich-Durnten, Switzerland, 2012

15.23 "Titanium Welds, *Wikipedia,* http://en.wikipedia.org/wiki/File:Titanium_welds.jpg

15.24 *Adv. Mater. Process.,* Vol 163 (No. 4), April 2005

15.25. F.H. Froes and S.J. Haake, *Materials and Science in Sports,* TMS, Warrendale, PA, 2001

15.26 F.H. Froes, H. Friedrich, J. Kiese, and D. Bergoint, Titanium in the Family Automobile: The Cost Challenge, *JOM,* Feb 2004

Index

A

abundance of, 1, 353
accelerated crack growth, 16, 367
accelerated crack propagation, 129, 129(T)
acetaldehyde, 349, 366
acetone, 245, 269
acicular alpha
    defined, 155
    extruded beta titanium, 233
    microstructural changes, 153(F), 154(F), 155(F)
    overview, 145–146(F)
    SCC, 129
acid pickling, 215, 269, 287
acids
    chromic acid, 336, 336(T)
    hydrochloric acid (HCl), 336, 336(F)
    hydrofluoric acid (HF), 336–337
    nitric acid, 335–336(F), 365–366(T)
    other reducing acids, 337, 338(T)
    overview, 335
    phosphoric acid, 336
    red fuming nitric acid (RFNA), 336
    sulfuric acid, 336, 337(F)
acronyms, glossary of, 203–204
additive manufacturing (AM), 192–193(F), 194(F,T), 195(F)
adhesive bonding
    bonding procedures, 286
    joint design, 286
    mechanical properties, 287
    nondestructive tests, 287
    overview, 286
    quality control, 287
    sonic bond-testing, 287
    surface preparation, 286
ADMA. *See* Advance Materials (ADMA) Products Inc.
Advance Materials (ADMA) Products Inc.
    hydrogenated titanium process, 182–184(F,T)
    non-Kroll process, 26, 26(F)
    PM products, 176
advanced alpha alloys, 68–69(F)
AEF. *See* aluminum equivalency factor (AEF)
aerospace
    early industry/recent developments, 6
    electrochemical machining (ECM), 328
    PIM, 196–197

aerospace applications
    airframes, 359–362(F)
    competitive comparison, 358–359
    fasteners, 362, 362(F)
    gas turbine engines, 131(T), 356–358(F,T), 359(F)
    missiles, 361
    overview, 8, 354–355, 356(T), 357(T)
    space vehicles, 361
    springs, 362, 362(F)
Aerospace material Specification, 184
age hardening, 76(F), 87–89, 91(F)
aged, defined, 75
aging, 90–91(F), 92(T), 154–156(F)
Airbus
    early industry/recent developments, 5
    raw titanium mill products in airframes, 7(T)
aircraft
    Airbus A380, 6
    B1 bomber, 5
    B-1B bomber, 361, 362
    B-52, 353
    Boeing 737, 358
    Boeing 747-8
        GEnx engine, 110, 357
        titanium aluminides, 134
    Boeing 777
        Beta-21S applications, 360(F)
        nut clips titanium (Ti-15-3), 360(F)
        percent titanium, 5, 6, 360, 360(F)
        raw titanium mill products in commercial airframes, 7(T)
        Ti-10V-2Fe-3Al, 5, 360, 360(F)
        titanium use in, 360, 360(F)
    Boeing 787
        GEnx engine, 110, 357
        percent titanium, 6
        side-of-body chord manufacturing cost breakdown, 16(T)
        titanium alloy use in, 360(F)
        titanium aluminides, 134
        titanium chips generated during the manufacture, 320
        titanium use in, 360
    DC-7, 353, 360
    F-14, 5
    F-15, 5, 361

aircraft (continued)
F-22 Advanced Tactical Fighter, 5, 360, 361(F)
F-35 Joint Strike Fighter (JSF), 6, 192, 194(F), 321, 361(F)
National Aerospace Plane, 363
SR-71 Blackbird, 356, 363
U.S. SST project, 4–5
X-15 high-flying supersonic aircraft, 4
aircraft drills, 314, 316(F)
airframes
alpha-beta alloys, 131
Boeing 747-8, 131
Boeing 787, 131
casting, 169
commercial airframe demand, 354, 355(F), 355(T)
economics, 15
forging, 226
FSW, 278
increased demand for titanium, 6, 8
joining, 265
military, 5
overview, 359–362(F)
raw titanium mill products in commercial airframes, 7(T)
titanium buy weights, 356, 356(T), 357(F)
Albany Titanium Company, 3
Alcoa, 353, 363
alkaline environments, 338–339, 341(T)
Allegheny Ludlum Steel Corporation, 2, 17
Allegheny Technologies Incorporated, 132
allotropic, defined, 51
alloy solidification, 33, 33(F)
alloy stability, 121
alloying
alloying factors, 35–36(F,T)
for corrosion resistance, 345–346(T), 347(F,T), 348(F,T)
intermetallic phases, 36–37(T)
overview, 35
solid solubility, types of, 35, 35(F)
solid solutions, 35
alloying behavior, 8–9(T), 10–13(T)
alloying elements
alpha stabilizers, 16(F), 57–60(F,T)
aluminum, 4, 73
aluminum content, influence of, 57–58(F)
beta stabilizers, 57(T), 58(T), 60–65(F,T)
beta transus temperature, effect on, 66, 66(T)
classes of, 56
classification of, 58(T)
embrittlement, as cause of, 57–58(F)
eutectoid transformation, 62, 62(T)
groups, 56
hydrogen, 62–63
hydrogen embrittlement, 62
iron, 73
manganese, 4, 73
molybdenum, 73
neutral additions, 65–66(F)
overview, 51, 56–57(T)
strengthening effects, 57, 57(T)
tin, 73
titanium-aluminum phase diagram, 57(F)

titanium-hydrogen phase diagram, 63(F)
titanium-iron phase diagram, 61(F)
titanium-niobium phase diagram, 61(F)
titanium-zirconium phase diagram, 65(F)
vanadium, 4, 73
zirconium, 73
alloying titanium, principles of
alloying elements, 56–66(F,T)
atomic structure, 51–56(F,T)
overview, 51
terminal alloy formulation, 57(T), 71, 71(F)
Ti₃Al, 71–72(F)
TiAl, 72, 72(F)
titanium alloys, 66–70(F,T)
alloys, specific
5Al-2.5Sn, extrusion, 217
B120VCA (Ti-13V-11Cr-3Al), 4
Ti-0.15Pd
cathodic alloying, 346, 346(T), 347(T)
crevice corrosion, 342
Ti-0.2Pd
tube products, 219
welded tubing, 220
Ti-0.3Mo-0.8Ni
cathodic alloying, 346, 346(T), 347(T)
crevice corrosion, 342
Ti-10-2-3, quenching, phases present after, 83
Ti-10V-2Al-3Fe, forgeability, 227
Ti-10V-2Fe-3A, airframes, 361
Ti-10V-2Fe-3Al
aircraft applications, 360
applications, 6
binary phase diagram, 81, 82(F)
Boeing 777, 5
early industry/recent developments, 5
F-22 Advanced Tactical Fighter, 5
fracture toughness to yield strength, 133, 133(F)
heavy section components, 362
stress-strain curves, 81(F)
Ti-1100, gas turbine engines, 131(T)
Ti-11Mo, grain size, effect of carbon on, 108, 109(F)
Ti-12Mo-6Sn, cold reduction, effect on yield strength, 103
Ti-13V-11Cr-3Al
aircraft applications, 359
drilling speeds, 317
forming temperatures, 243
Goodman type diagrams, 122, 123(F), 124(F)
quenching, phases present after, 83
springs, 362
strip and foil rolling, 215
superplastic-forming potential, 105
testing temperature, effect on tensile strength, 118(F)
Ti-13V-11Cr-3Al alloy, equiaxed structure, 144(F)
Ti-14Al-21Nb, microstructures, 151–152(F)
Ti-15-3
aging, effect of, 91
quenching, phases present after, 83
Ti-15Mo-0.2Pd, corrosion resistance, 346, 347(T)
Ti-15Mo-2.7Nb-3Al-0.2Si, aircraft applications, 360
Ti-15Mo-2.7Zr-3Al-0.2Si (beta-21S), applications, 6

Ti-15Mo-3Nb-3Al-0.2Si, strip and foil rolling, 215
Ti-15V-3Al-3Cr-3Sn, aircraft applications, 360
Ti-15V-3Al-3Sn-3Cr, armor applications, 373
Ti-15V-3Cr-3Al-3S, welded tubing, 220
Ti-15V-3Cr-3Al-3Sn
   aircraft applications, 360
   applications, 6
   early industry/recent developments, 5
   sheet metal applications, 362
   space vehicles, 361
   tube products, 219
Ti-15V-3Cr-3Sn-3Al
   aging, effect of, 91, 92(T)
   beta prime, 82
   electron density, 80(T), 81
   solution treatment, 89, 90
   strip and foil rolling, 215
Ti-17 IMI829, gas turbine engines, 131(T), 356
Ti-17 IMI834, gas turbine engines, 131(T), 356
Ti-2Al-2Mn, Rolls-Royce Avon engine, 4
Ti₃Al
   alloying principles, 71–72(F), 73(F)
   early industry/recent developments, 5
   intermetallic phases, 37
   Young's modulus, 68–69(F)
Ti-3Al-2.5, welded tubing, 220
Ti-3Al-2.5Sn, seamless pipe and tubing, 220–221(F)
Ti-3Al-2.5V
   aircraft applications, 360
   recreational equipment field, 372–373(F)
   strip and foil rolling, 215
   tube products, 219
Ti-3Al-8V-6Cr-4Mo-4Zr
   aircraft applications, 360
   corrosion resistance, 367
   metastable beta, 148, 149(F)
   neutral additions, 65
Ti-3Al-8V-6Cr-4Zr-4Mo
   aircraft applications, 359
   cathodic alloying, 346, 347(F)
   drilling speeds, 317
   quenching, phases present after, 83
   springs, 362
Ti-4.5Al-5Mo-1.5Cr, 70, 114, 115(T), 116(T), 189
Ti-4Al-0.20
   biaxial yield strength, 101–102
   R-value, 102
Ti-4Al-2.5V-1.5Fe (alloy 425), early industry/recent
   developments, 6
Ti-4Al-3Mo-1V, uniaxial tensile, biaxial tensile, and
   R-values, 101(T)
Ti-4Al-4Mn, early history, 4
Ti-5Al-1Mo-1V, Young's modulus, 68–69(F)
Ti-5Al-2.5Sn
   biaxial yield strength, 101–102
   cold reduction, effect on yield strength, 103
   creep strength, 119, 119(T)
   Goodman type diagrams, 122, 123(F), 124(F)
   intermetallic phases, 150, 151(F)
   machining processes, effect of, 125, 125(F)
   mechanical properties, 283
   neutral additions, 65
   platelike alpha structure, 146, 146(F)
   spheroidal beta, 149, 149(F)
   strength-density ratio comparison, 14(T)
   stress relief, 107(F)
   strip and foil rolling, 215
   superplastic-forming potential, 105
   temperature, effect on grain size, 110(F)
   uniaxial tensile, biaxial tensile, and R-values, 101(T)
   valve springs and plates, 350
Ti-5Al-2.5Sn ELI
   serrated alpha structures, 145, 145(F)
   uniaxial yield strength, 102
Ti-5Al-2Sn, forgeability, 227
Ti-5Al-2Sn-2Zr-4Cr-4Mo, beta flecks, 166
Ti-5Al-5V-5Mo-3Cr
   aircraft applications, 360
   airframes, 361
   Boeing 787, 6
Ti-5Al-6Sn-2Zr-1Mo-0.25Si, creep strength,
   117(T), 119
Ti-5Ni, intermetallic phases, 150, 150(F)
Ti-679, intermetallic phases, 150, 150(F)
Ti-6.8Mo-4.5Fe-1.5Al, early industry/recent
   developments, 6
Ti-6Al-2Fe-0.1Si, early industry/recent developments, 6
Ti-6Al-2Nb-1Ta-1Mo, marine applications, 349
Ti-6Al-2Sn-4Zr-2Mo
   aircraft applications, 360
   creep strength, 119, 119(T)
   gas turbine engines, 131(T), 356
   neutral additions, 65
   testing temperature, effect on tensile strength, 117(T)
Ti-6Al-2Sn-4Zr-2Mo-0.1Si
   automotive applications, 375, 376(F)
   early industry/recent developments, 5
   gas turbine engines, 131(T), 356
Ti-6Al-2Sn-4Zr-6M, cathodic alloying, 346, 347(F)
Ti-6Al-2Sn-4Zr-6Mo
   athermal titanium martensite, 80
   forgeability, 227
   gas turbine engines, 131(T), 356
   metastable beta, 147, 148(F)
Ti-6Al-4, extrusion, 217
Ti-6Al-4V
   acicular alpha, 145–146(F)
   aircraft applications, 360
   airframes, 359, 361
   alpha case, 149–150(F)
   alpha prime, 144(F), 147
   alpha-beta and beta-processed Ti-6Al-4V forgings,
     128, 129(T)
   annealed, properties of, 70(T)
   annealing, 216
   annealing terms, 91–92
   armor applications, 373
   athermal titanium martensite, 80
   automotive applications, 375, 376(F)
   Bauschinger effect, 104, 104(F)
   cast, microstructure, 168(F), 172–174(F)
   casting, 169, 171, 171(T)

alloys, specific (continued)
  closed-die forging, 231(F)
  cold rolling, 216
  commercial aircraft, 362, 362(F)
  duplex alpha-beta anneal on microstructure, effect of,
    115(F)
  early history, 4
  elongations, 105
  extrusion, 217, 218
  extrusion ratio, 233(F)
  fastener materials, 219
  FCGR, 173
  flow stress, 133, 133(F)
  forgeability, 227
  forging reduction (50%), effect of, 153, 153(F)
  forging temperatures, 227
  forming temperatures, 243
  FSW, 278, 278(F)
  gas turbine engines, 131(T), 356
  GEnx engine, 357
  Goodman type diagrams, 122, 123(F), 124(F)
  heavy section components, 362
  hot isostatic pressing (HIP), 172
  intergranular beta, 147(F)
  marine applications, 349
  mechanical properties, 283
  metastable beta, 147–148(F)
  microstructures, 168(F)
  notches, effect of, 124
  phase diagram, 113(F)
  PIM characteristics, 197
  platelike alpha structure, 147, 147(F)
  primary alpha, 144, 144(F)
  quenching, phases present after, 83
  ring roll forging, 230, 231(F), 232(T)
  sheet metal applications, 362
  S-N curves, 126(F)
  solution treatment, 83(F), 89, 89(F), 90, 90(F)
  space vehicles, 361
  SPF, 253, 255(F)
  springback, 248, 249(F)
  steam turbine blades, 376(F)
  strain-rate sensitivity, 104, 104(F)
  stress relief, 106, 107(F)
  strip and foil rolling, 215
  surgical implants, 8
  textures, 99
  thermal treatment, 153–154, 156(F)
  titanium-hydrogen phase diagram, 63(F)
  tube products, 219
  unidirectional composites, characteristics, 137(T)
  U.S. Navy submarines, 378
  valence electrons per atom, 81, 81(T)
  VHN, 116
  welded tubing, 220
  Widmanstätten Structure, 146, 146(F)
Ti-6Al-4V ELI, aircraft applications, 360
Ti-6Al-5Zr-1W-0.25Si, creep strength, 117(T), 119
Ti-6Al-6V-2Sn
  aircraft applications, 360
  airframes, 361

beta flecks, 166
  early industry/recent developments, 5
  extrusion, 217, 218
  fastener materials, 219
  fracture toughness, 128, 128(F)
  neutral additions, 65
  uniaxial tensile, biaxial tensile, and R-values, 101(T)
Ti-7Al-12Zr, uniaxial tensile, biaxial tensile, and
  R-values, 101(T)
Ti-8Al-1Mo-1V
  anhydrous methanol/SCC, 345
  early industry/recent developments, 5
  forgeability, 227
  forging reduction (50%), effect of, 153, 153(F)
  forging temperatures, 227
  forming temperatures, 246
  heat treatment, 92
  intergranular beta, 149, 150(F)
  Larson-Miller parameter, 119, 120(F)
  mechanical properties, 283
  quenching, phases present after, 83
  uniaxial tensile, biaxial tensile, and R-values,
    101(T)
Ti-8Al-1V-1Mo
  forgeability, 227
  gas turbine engines, 131(T), 356
Ti-8Mn, early history, 4
TiAl
  alloying principles, 72, 72(F)
  early industry/recent developments, 5
  gas turbine engines, 356
  intermetallic phases, 37
  machining, advances in, 321
  National Aerospace Plane, 363
  overview, 110
  SPF, 253–254
TiB21S, early industry/recent developments, 5
Ti-Code 12
  tube products, 219
  welded tubing, 220
TIMETAL 54M (Ti-5Al-4V-0.75Mo-0.5Fe), 6
Timetal Ti-54M (Ti-5Al-4V-0.75Mo-0.5Fe), machining,
  advances in, 321
alloys partially soluble in solid state
  all alpha, 38–39(F)
  alpha-beta, 39–40(F)
  overview, 38, 39(F)
alpha, defined, 155
alpha + beta alloy castings, microstructure of
  alpha plate colonies, 168(F), 173
  beta grain size, 168(F), 172
  casting porosity, 168(F), 173
  grain refinement, 174
  grain-boundary alpha, 168(F), 172–173
  microstructure modifications, 173–174(F)
alpha 2, 133, 155
alpha alloys
  alloy classification, 142
  alloy selection, 267
  BTU/UTS ratios, 101
  corrosion, 332

cost-optimized titanium-ruthenium and titanium-palladium alpha alloys, 332(T)
creep and stress-rupture, 119
crystal structures, 207
etching, 327
heat treatment, 92
matrix, 151
overview, 66, 68, 332
oxygen contamination, 150
R-values, 101
spheroidal beta, 149
strain hardening, 103
superplasticity, 105
texture strengthening, 101, 102(T)
alpha case
chemical reactivity, 294–295(F)
defined, 155
Alpha Class submarine (Soviet Union), 378
alpha phase stabilizers, 225
alpha prime
athermal titanium martensite, types of, 79–80(F)
defined, 155
metastable phase boundaries, 78(F)
overview, 147, 148(F)
alpha stabilizers, 16(F), 57–60(F,T), 155
alpha transus
defined, 48, 155
ordered intermetallic compounds, 151
overview, 44
titanium aluminides, 135, 136
titanium martensites, 78
alpha-beta alloys, 131–135(T). *See also* specific alloys
characteristics of, 70, 70(T)
short-time tensile properties, 69(F)
weldability, 69–70
alpha-beta structure, 153, 153(F), 155, 155(F)
aluminum (Al)
as alloying element, 4, 53, 57–58(F), 73
alpha alloys, 66
boiling point, 32(T)
corrosion resistance, effect on, 332
creep and stress rupture, 120
creep strength, effect on, 119, 119(T)
embrittlement, as cause of, 57–58(F)
heat treatment, titanium-aluminum alloy, 92, 93(F)
melt charges, 163
melting point, 32(T)
stress-corrosion cracking (SCC), 345
tensile strength and strain, effect on, 65, 66(F)
terminal alloy formulation, 71
TTT characteristics of Ti-15%V alloy, effect on, 86, 87(F)
aluminum equivalency factor (AEF), 121
American National Standards Institute, 294(T), 296
American Society of Mechanical Engineers, 268, 282–283, 331–332, 333(T)
anhydrous methanol, 339, 344, 350, 366
annealing
batch annealing, 216
definition of, 106
function of, 91
grain growth, 108, 108(F), 109(F)

phases present in standard commercial alloys at various quenching temperatures, 83(F)
plate rolling, 215
purpose of, 86–87
recovery process, 106–107(F,T)
recrystallization, 107–108(F)
rod and bar, rolled, 214
terms (*see* annealing terms (Ti-6Al-4V))
Ti-5Al-2.5Sn, effect of temperature on grain size, 110(F)
annealing terms (Ti-6Al-4V)
beta anneal, 92
duplex anneal, 92
mill anneal, 91
recrystallize anneal, 91–92
stress-relief anneal, 92
anode baskets, 349
anodes
chlorine industry, 348, 350
corrosion reactions, 334
electrochemical applications, 349–350
galvanic corrosion, 342, 343(T)
platinized-titanium anodes, 350
TIMET electrolytic cell, 21, 21(F)
anodizing racks, 353, 363
applications, titanium
aerospace, 354–362(F,T)
armor, 373–374(F)
automotive, 374–375(F), 376(F)
building applications, 375, 377(F)
consumer, 372–373(F)
early applications, 353
engineering properties, 363–370(F,T)
history, 353
hostile environments
alkaline media, 366
chlorine chemicals and solutions, 364
chlorine gas, 364
fresh water/steam, 364–365
halogen and halide compounds, 364
nitric acid, 365–366(T)
organic chemicals, 366
other salt solutions, 364
reducing acids, 366, 366(T)
seawater, 365, 365(T)
industrial, 363
marine, 378, 379(F)
material availability, 353–354(T), 355(T), 356(F,T)
medical, 370–371(F), 372(F)
miscellaneous, 378–379
power utility applications, 375–378(F)
sheet metal, 362–363
argon
historical background, 1
hot isostatic pressing (HIP), 172
microstructure modifications, 173
porous structures, 200, 201(F)
SPF, 253
welding procedures, shielding, 270
armor applications, 373–374(F)
Armstrong/International Titanium Powder process, 23–24(F)
ASM International, *Titanium and Its Alloys*, 5

ASTM International, 282, 331–332, 333(T)
ASTM International/American Society of Mechanical
    Engineers titanium grades, 9(T)
athermal titanium martensite, 79–80(F)
ATI, 26, 131, 132, 132(T), 187(T)
atom vacancies, 35
atomic diameters, of elements, 36, 53(F)
atomic mass unit (amu), 31
atomic structure, titanium
    atom diameter, 33(F), 34(F), 52–54(F,T)
    atomic diameters, of elements, 53(F)
    electronic configuration of first 29 elements in atomic
        series, 52(T)
    electrons, 51–52(F,T)
    lattice structure, 54, 54(F), 55(F)
    overview, 51–52(F,T)
    physical properties, 53(T), 54–56(T)
atoms
    elementary particles, 31
    states of matter, 31–32(T)
    titanium, 51
automotive applications, 374–375(F), 376(T)

B

backing fixtures, 270–271(F)
BAE Systems, 320
baffles, 208, 270, 271
bar and billet, secondary working of
    extrusion, 231–234(F)
    forging, 226–228(F,T)
    forgings, classes of, 228–231(F), 232(T)
    microstructure and mechanical properties
        alpha-beta-processed alloys, 232(T), 234, 234(F),
            235(T)
        beta-processed alloys, 234–239(F,T)
    modeling, 240, 240(F), 241(F)
    overview, 225
    physical metallurgy, 225, 226(F)
    surface effects of heating
        hydrogen, 239–240
        oxygen contamination, 239, 240(F)
barium compounds, 300
Battelle Columbus, 2
Bauschinger effect, 104, 104(F)
bcc. See body-centered cubic (bcc)
bender, 241
beryllium, 44, 52(T), 54, 54(F), 108
beta, defined, 155
Beta 1 (Ti-13V-11Cr-3Al), 70, 84
beta alloys, 133, 133(F)
beta eutectoid, defined, 155
beta flecks, 166, 166(F)
beta isomorphous, defined, 156
beta prime, 82, 83(F)
beta stabilizers, 57(T), 58(T), 60–65(F,T), 156, 225
beta transformation
    equilibrium phase relationships, 75, 76(F)
    Lever rule, 76–77(F)
    metastable phases and metastable phase diagrams,
        77–84(F,T)

nose time, effect of alloy additions on, 86, 87(T)
transformation kinetics, 84–86(F), 87(F)
beta transus, 48, 66, 66(T), 156
beta transus temperature, 66, 66(T)
Beta-C (Ti-3Al-8V-6Cr-4Zr-4Mo), 84, 149(F), 228(T)
beta-isomorphous elements, 69, 92, 121
biaxial tensile ultimate strength to the standard uniaxial
    tensile strength (BTU/UTS), 101
billets and bars, forged
    finishing, 211
    grain refinement, 210–211(F)
    overview, 209(T), 210
    testing, 211
binary alloys, 37(F), 39(F), 43–44(F)
binary diagram, 38, 39(F)
binders, PIM, 194–196
bleed air leak-detect (BALD) bracket, 192, 193
blended-elemental (BE)
    alloy types studied, 178
    components, 184, 185(F)
    consolidation, 178, 178(T)
    direct processing, 178, 179(F)
    mechanical properties, 178–179, 180(F)
    microstructure, 178, 179(F)
    powder production, 177–178(F)
    preforms, 178
    shape-making properties, 180, 180(F)
blocker, defined, 241
body-centered cubic (bcc)
    binary alloys, intermediate phases in, 43
    crystal structures, 54(F)
    overview, 34
    slip, 96
Boeing
    CHIP process, 182
    early industry/recent developments, 5
    raw titanium mill products in airframes, 7(T)
boiling points, some metals, 32(T)
boring. See turning
boron
    grain growth, 108
    recrystallization, 108
box, defined, 241
brazed joints, mechanical properties, 285
brazing
    brazed joints, mechanical properties, 285
    brazing filler metals, 285
    chemical and metallurgical characteristics, 283–284
    defects, 285
    dissimilar-metal joints, 285
    furnace brazing, 284
    helium arc torch brazing, 285
    induction brazing, 284
    nondestructive testing, 285
    overview, 283
    resistance brazing, 285
    shielding, 284
    surface preparation, 284
    torch brazing, 284–285
    weldments, comparison, 283
brazing filler metals, 285

broaching, 319, 320(F), 321(T)
broken-up structure (BUS)
    casting, 173, 174
    powder metallurgy, 178, 179(T), 190, 190(F), 191, 203
    prealloy (PA) approach, 179(F), 190
BTU/UTS ratios. *See* biaxial tensile ultimate strength to the
       standard uniaxial tensile strength (BTU/UTS)
building applications, 375, 376(T)
bulging. *See* tube bulging
Bureau of Mines in Albany, Oregon, 1
Burgers vectors, 72, 73(F)

C

cadmium, 54(F), 344(T), 345, 350
capacity, 1
Carbide Industry Standardization Committee, 296
carbides (TiC), 1
carbon
    ductility, effect on, 119
    grain growth, 108
    recrystallization, 108
carbonitrides (TiCN), 1
cast structures, 33, 33(F), 174, 275
casting
    alloys used, cast parts, 171(T)
    casting technology, 168(F), 169–174(F,T), 175–176(F)
    casting technology, trends in, 174–175
    overview, 168–169(F)
    physical properties, 370
casting technology
    alpha + beta alloy castings, microstructure of, 168(F),
       172–174(F)
    casting alloys, 171–172(T)
    heat treatment, 168(F), 172
    hot isostatic pressing (HIP), 168(F), 172
    investment casting, 170(F), 171, 171(F)169(F)
    mechanical properties, 168(F), 173(F), 174, 174(T),
       175(F), 176(F)
    microstructure and mechanical properties, 172
    net-shape casting, 169, 169(F)
    rammed graphite mold, 169–171(F)
    sand casting, 171
    weld repair, 172
casting technology, trends in
    alloy design, 175
    melting technology, 175
    postcasting treatments, 175
    shape making, 174–175
cathodes, 21, 334, 342, 343(T)
cathodic polarization, 335, 345
Ceracon (ceramic consolidation) process, 185
ceramic molds, 170–171(F)
CermeTi (Dynamet Technology, Inc.), 182
Charpy impact energy, 125–126, 238(T)
Charpy impact strength, 7(T), 11(T), 12(T), 13(T)
Charpy V-notch energy, 126, 126(F)
chemical blanking. *See* chemical machining
chemical machining
    equipment, 327
    etching, 326–327(T)

masking, 325–326
overview, 323–325(F,T)
precleaning, 325
scribing, 326, 326(F)
chemical milling. *See* chemical machining
chemical properties/physical properties, special
    corrosion resistance (oxidizing environments), 9
    electrical conductivity, 15
    magnetic susceptibility, 15
    thermal conductivity and expansion, 14
Chinuka technique, 26, 26(F)
CHIP. *See* cold and hot isostatic pressing (CHIP)
chloride industry, 348
chlorinated oils, 299, 299(T)
chlorine chemicals and solutions, 364
chlorine compounds, 338, 340(T)
chlorine dioxide, 348, 363, 364
chlorine gas, 363, 364
chlorine industries, 348
chromium (Cr)
    corrosion resistance, 332, 346
    melt charges, 163
    recrystallization, 108
    segregation, defects, 166
    terminal alloy formulation, 71
closed-die, defined, 241
cluster mills, 212, 213(F)
cobalt high-speed steels, 296, 315
coin, defined, 241
cold and hot isostatic pressing (CHIP), 178(T), 180–182(F,T)
cold isostatic pressing, 181. *See also* cold and hot isostatic
       pressing (CHIP)
cold worked, definition of, 105–106(F)
commercially pure (CP) titanium
    alpha (or near-alpha alloys), 66
    annealing, 106–107(F), 111
    Armstrong/International Titanium Powder process, 23(F)
    casting, 169, 171, 171(T)
    chloride residues, 299
    cold rolling, 216
    corrosion behavior, 332(F)
    CSIRO technique, 24, 24(F)
    drilling speeds, 317
    drills, 317(T)
    extraction processes under development, 24
    extrusion, 217
    formability, 68
    forming, 243
    hardness, 116
    hot salt corrosion, 345
    iron, 108
    joining CP titanium (*see* joining)
    melt charges, 163
    minimum and average mechanical properties of wrought
       titanium alloys at room temperature, 11(T)
    modeled strain distribution, 240(F)
    modeling, 240, 241(F)
    pitting, 341
    plate, microstructures of, 109(F)
    postweld heat treatment, 370
    properties, 294(T)

commercially pure (CP) titanium (continued)
  rare earth additions, 200(F)
  SCC, 331(T)
  seamless pipe and tubing, 220–221(F)
  sheet metal applications, 362–363
  slab milling, 305, 310(T)
  straightening, 218
  stress-relief behavior, 111
  surgical implants, 371
  tennis racquets, 373
  threaded fasteners, 290
  tie bars, 163
  tube bending, 256
  tube products, 219
  tubing processing, 220
  typical physical properties of wrought titanium alloys, 10(T)
  welded tubing, 220
  wire, 219
  wire products, 219
compound, defined, 48
constitution diagram, 37, 48, 142, 142(F), 143(F)
constitution diagram, defined, 48. See also phase diagrams
consumable electrode vacuum arc remelt (VAR) electric furnaces, 161. See also vacuum arc remelting
consumable electrodes, 161–162(F)
consumer applications, 372–373(T)
  architectural glass, 373
  electronics field, 373, 373(F)
  eyeglass frames, 373
  fireworks, 373
  horseshoes, titanium, 373
  recreational equipment field, 372–373(F)
  water sports, 373
contamination
  heat treatment, during, 92–93
  nitrogen, 92
  oxygen, 92
continuous silicon carbide (SiC) fibers, 136–137(F)
control flatten, defined, 241
copper
  melt charges, 163
  segregation, defects, 166
  solid solutions, 35
copper coatings, 217, 219
cored structures, 38
coring, 38
Corona 5, 12(T), 70, 114, 115(T), 116(F), 189
Corporate Auto Fuel Efficiency standards, 375
corrosion
  alkaline environments, 338–339, 341(T)
  behavior, 331, 331(T), 332(F,T)
  chemical and related applications, 347–350
  environmental cracking, 342–344(T)
  erosion-corrosion, 343
  galvanic corrosion, 342–343(F,T)
  gaseous environments, 339–341
  halide solutions/metal salts, 337–338, 340(T)
  hot salt, 345
  hot salt corrosion, 345
  hydrogen embrittlement, 345

liquid/solid metal cracking, 345
  localized attack, 341–342(T)
  methanol, 344
  mixed-potential theory, 334–335(F)
  natural waters, 339
  organic compounds, 339, 341(T)
  other alcohols, 344–345
  overview, 331
  oxidizers, 344
  polarization, 333(F), 335
  potential-pH diagram, 332–333(F)
  reactions, 333–334
  seawater, 345
  titanium grades, 331–332, 333(T)
  types, 331(T)
corrosion, forms of
  acids
    chromic acid, 336, 336(T)
    hydrochloric acid (HCl), 336, 336(F)
    hydrofluoric acid (HF), 336–337
    nitric acid, 335–336(F)
    other reducing acids, 337, 338(T)
    overview, 335
    phosphoric acid, 336
    sulfuric acid, 336, 337(F)
  uniform corrosion, 335
corrosion inhibitors, 333(F), 337, 338(T), 339(F,T)
corrosion prevention, alloying for, 345–346(T), 347(F,T), 348(F,T)
corrosion rate
  alkaline environments, 339, 341(T)
  alloy erosion-corrosion resistance at various seawater locations, 365(T)
  anodic protection, 337, 339(T)
  ATI 425 corrosion test results, 132(T)
  boiling 90% HNO_3, 365(T)
  boiling HCl, 366(T)
  chlorine chemicals and solutions, 364
  chlorine compounds, 338, 340(T)
  chlorine gas, 364
  corrosion resistance of titanium alloys in boiling nitric acid solutions, effect of alloying elements on, 348(T)
  halide solutions/metal salts, 340(T)
  hydrofluoric acid (HF), 336–337
  mixed-potential theory, 334–335(F)
  nitric acid, 365
  organic acids, titanium alloys resistance to, 337, 338(T)
  seawater, 365
  titanium alloys in boiling acid solutions, 346(T)
  unalloyed titanium and certain titanium alloys, 347(T)
  unalloyed titanium in alkaline solutions, 341(T)
  unalloyed titanium in chromic acid, 336(T)
  unalloyed titanium in inorganic environments, 340(T)
  unalloyed titanium in organic compounds, 341(T)
corrosion resistance
  chemical and related applications, 347–350
  metal titanium, 8
  oxidizing environments, 9
  titanium alloys, 366–367
corrosion-resistant racks, 327

cost
    titanium, various stages of mill product fabrication, 16(T)
    titanium precursors, 15(T)
    titanium relative to other metals, 15(T)
    worldwide consumption, structural materials, 15(T)
cracking
    environmental cracking, 342–344(T)
    liquid/solid metal cracking, 345
    SCC (see stress-corrosion cracking (SCC))
    susceptibility to, 128
Cramet, Inc., 3
Crane Company, 3, 17
creep and stress rupture, 117(T), 119–121(F,T)
crevice corrosion
    alloyed versus unalloyed titanium, 366–367
    alloying for corrosion protection, 345, 346, 346(T)
    chlorine chemicals and solutions, 364
    chlorine gas, 364
    localized attack, 341–342(F)
    potential-pH diagram, 332
    resistance to, 331, 346(T)
    water purification, 349
Crucible Steel, 2, 4
crystal growth, 32, 32(F), 33(F)
crystal structure
    defects, 35
    deformation, alloy influence on, 207–208
    overview, 207
    solidification, 33–35(F,T)
crystallographic texture, 95, 98, 99, 107. See also texture
        strengthening
CSIR process, 25–26(F)
CSIRO technique, 24–25(F)
CSIRO TiRO process, 24(F)
current density, 157, 247, 285, 334–335(F)
cutting, physical properties, 370

D

deBoer, J.H., 1
defects
    brazing, 285
    chromium (Cr), 166
    copper, 166
    crystal structures, 35
    iron, 166
    manganese, 166
    porosity, 281
    riveting, 289
    weld defects, 281
defect-tolerant approach, 129. See also fracture mechanics
        approach
deformation
    annealing, 106–108(F,T), 109(F)
    internal changes, 105–106(F)
    neocrystallization, 108–110(F)
    overview, 95–96
    slip, 96–97(F,T), 98(F)
    strain effects, 104, 104(F)
    strain hardening, 102–104(F)
    superplasticity, 104–105(F,T)

    texture, development of, 98–99(F), 100(F,T)
    texture strengthening, 99–102(F,T)
    twinning, 96(F), 98, 98(F), 99(F)
degreasing, 245, 246, 269, 286
Degussa Company, 2
dendrite, defined, 48
dendrite arms, 32. See also diffusion bonding (DB)
dendrites, 32, 33
dendritic grain growth, 32, 33
diffusion bonding (DB). See also superplastic forming/
        diffusion bonding (SPF/DB)
    early industry/recent developments, 5
    sheet metal components, 363
    solid-state welding, 275–277
    SPF/DB, 277
direct-energy deposition (DED), 192
dislocations
    crystal structures, 35
    slip, 96–97
Douglas Aircraft Company, 2, 353
Douglas Mach 2 X -3 Stiletto, 2
Dow Chemical Company, 3
Dow-Howmet, 3
down (climb) milling, 300, 302
draw, defined, 241
drilling titanium
    cutting fluids, 317
    drill materials, 315, 317(T)
    drilling feeds, 315, 317, 317(F)
    drilling speeds, 317
    drills and drill design, 314–315(F), 316(F,T)
    inspection, 318
    machine tools, 313–314
    overview, 313
    techniques, 317–318, 319(T)
drop-wise condensation, 369–370
ductility, 117(T), 118–119
duplex (DM) microstructure, 136, 152
DuPont, 2, 344(T)

E

early industry/recent developments, 2–6(F,T), 7(T)
ECM process, 327–328(F)
economics, 15, 15(T), 16(F)
edger, defined, 241
E.I. du Pont de Nemours Co, 17
electrical conductivity, 15
electrochemical applications, 349–350
electrochemical machining (ECM), 327–328(F)
electrochemical potential, 332–333, 335
electrolytes, 327–328
electrolytic winning, 20–21(F)
Electrometallurgical Company of Union Carbide, 3
electron beam (EB) cold hearth melting, 167–168(F)
electron beam welding (EBW), 273–274
electron diffraction, 141
electron microscope, 141
electronegativity, 36
electron-probe microanalyzer, 141
electrons, 31, 333, 334

electrowinning processes, 17, 21, 22
elementary particles, 31
elements, 31
ELI. *See* extra low interstitial (ELI)
elongated alpha, 144(F), 156
elongated structures, 143, 144(F)
elongation, definition of, 118
elongations, 104–105(T)
embrittlement. *See also* hydrogen embrittlement
    alloying elements, 57–58(F)
    solid-solution hardening, welding, 268
    welding, 266
endurance limit, 121, 122, 125, 179, 180(F)
engineering properties
    corrosion resistance, 363–364
    hostile environments
        alkaline media, 366
        chlorine chemicals and solutions, 364
        chlorine gas, 364
        fresh water/steam, 364–365
        halogen and halide compounds, 364
        nitric acid, 365–366(T)
        organic chemicals, 366
        other salt solutions, 364
        reducing acids, 366, 366(T)
        seawater, 365, 365(T)
    overview, 363
    physical properties
        casting, 370
        cutting, 370
        erosion resistance, 369
        fabrication, 370
        heat transfer, 369–370(F)
        machining, 370
        overview, 367–369
        welding, 370, 370(F)
    titanium alloys, corrosion resistance of, 366–367
Environmental Protection Agency (EPA), 202
equiaxed structure, defined, 156
equiaxed structures, 143, 144(F)
equilibrium
    defined, 48
    use of term, 37
equilibrium diagram, 37, 48. *See also* constitution diagram
equilibrium phase diagrams, 76(F), 77–78
erosion resistance, 353, 359, 369, 378
erosion-corrosion, 343, 365(T), 369
etchants, 158(T)
etching
    chemical machining, 326–327(T)
    metallographic preparation, 158
eutectic, defined, 48
eutectic phase diagram, 38
eutectic reactions, 40–41(F)
eutectoid, defined, 48
eutectoid alloys, 44, 45(F)
explosion tear test, 126, 127(F)
extra low interstitial (ELI)
    SCC resistance, 367
    Ti-5Al-2.5Sn ELI, 102
extraction processes, 23(T)

extraction processes, under development
    ADMA Products non-Kroll process, 26, 26(F)
    Armstrong/International Titanium Powder method,
        23–24(F)
    Chinuka technique, 26, 26(F)
    CSIR method, 25–26(F)
    CSIRO technique, 24–25(F)
    FFC Cambridge process, 24, 24(F)
    MER approach, 25, 25(F)
    overview, 22–23
extractive metallurgy
    basic steps, 16–17
    costs, production, 21–22(F,T)
    electrolytic winning, 20–21(F)
    magnesium-reduction process (Kroll), 18–20(F)
    mill products shipments, trends, 27, 28(F,T)
    overview, 16–17(F)
    processes under development, 22–26(F,T)
    sodium-reduction process (Hunter), 20, 20(F)
    sponge production, trends, 26–27(F), 28(F)
    titanium sponge production, 17–18(F)
    titanium tetrachloride ($TiCl_4$) production, 17–18(F)
extrusion
    bar and billet, 231–234(F)
    extruded pipe and hollows, 218–219
    extruded shapes, 218
    lubrication, 217
    overview, 216–217
    Sejournet glass process, 217–218(F)
extrusion ratio, 233, 233(F)

F

fabrication, 152–154(F), 155(F), 156(F), 370
face-centered cubic (fcc), 34, 54(F)
facing. *See* turning
fatigue characteristics (good), 122, 124(T)
fatigue crack growth rate (FCGR)
    alpha plate colonies, 173
    BE processing, 179
    cast Ti-6Al-4V, 173
    HIP, 172
    overview, 129–130(F), 131(F)
fatigue cracks, 128–129(F)
fatigue strength, 121–125(F,T), 126(F)
fcc. *See* face-centered cubic (fcc)
FCGR. *See* fatigue crack growth rate (FCGR)
FFC Cambridge process, 24, 24(F)
filler wire, 219, 271–272(T), 281, 283, 285
fingerprints, 245, 246, 268, 281, 325
finish, defined, 241
finite-element measurement, 240
fire, 300, 319
fire hazard, 202, 300, 313, 318, 319
flame cutting, 322–323, 324(T)
flash welding, 275, 275(F), 276(F), 277(T)
flatten, defined, 241
flattening procedures, 215
flaw size, 128
Flexon (Marchon Eyewear Inc.), 138–139
fluid die compaction (FDC), 185–187, 190

fluidized-bed chlorination, 17
fluxes
    brazing, 284–285
    inert gas metal arc welds, 267
    soldering, 286
Food and Drug Administration (FDA), 200
forging, 208
    characteristics, titanium alloys, 228(T)
    forgeability, 226–227(F), 228(T)
    forging temperatures, 227–228(T)
    overview, 226, 226(F)
forgings, classes of
    blacksmith forging, 228, 229(F)
    closed-die forging, 228–229(F)
    flat-die forging, 228, 229(F)
    hand forging, 228, 229(F)
    hot-die forging, 229
    isothermal forging, 229–230
    open-die forgings, 228, 229(F)
    overview, 228
    precision forging, 229(F), 230
    ring roll forging, 230–231(F), 232(T)
forming
    cold forming versus hot forming, 244(T)
    failures in sheet forming processes, 244(T)
    forming considerations
        cleaning, 245
        forming temperatures, 243–245(T)
        grease removal, 245
        handling, 245
        scale removal, 245
    heating methods
        furnace heating, 247
        hot die heating, 247
        overview, 246(T)
        radiant heating, 247
        resistance heating, 247
    hot sizing, 248(T)
    material parameters controlling deformation, 244(T)
    overview, 243
    preparation for
        blank preparation, 246
        overview, 245–246
        surface preparation, 246
    relative formability of annealed titanium alloys for sheet
        forming operations, 244(T)
    temperatures, forming operations, 246(T)
    tooling materials, 247, 248(T)
forming (titanium plate, sheet, strip, tubing). *See* forming
forming blanks, 246, 247, 249
forming lubricants, 247, 248(T)
forming processes
    bend test data, 243, 245, 245(T)
    brake forming, 245(T), 247, 248, 249(F,T)
    compression bending, 255
    deep drawing, 250–252(F)
    dimpling, 261, 261(F)
    draw bending, 255–256(F)
    drop-hammer forming, 244(T), 256–257, 258(F),
        259(F)
    heated forming dies, 247

hot sizing, 248(T), 261–262(F,T)
    joggling, 261, 261(F)
    magnetic forming, 255
    overview, 247–248
    press bending, 255
    ram bending, 255
    roll bending, 255, 257–258
    roll bending machine, 257, 259(F)
    roll forming, 257, 259(F)
    shear forming, 258–261(F)
    SPF/DB, 254, 255(F)
    spinning, 258–261(F)
    stretch forming, 244(T), 249–250(F), 251(F)
    superplastic forming (SPF), 252–254(F,T), 255(F)
    trapped-rubber forming, 252, 252(F)
    tube bending, 255–256(F), 257(T)
    tube bulging, 254–255(F)
fracture mechanics approach
    beta-processed alloys, 235
    FCGR, 129
    flaw size, 128
    intermetallic compounds, 72
    lifing, approach to, 130, 131(F)
    titanium alloys, 113
    titanium aluminides, 135
fracture toughness, calculating, 128
fracture toughness index diagram, 126–127(F)
freckles, 166
friction stir welding (FSW)
    advantages, 280
    disadvantages, 280–281
    other alloys and thicknesses, 279–281(F,T)
    overview, 277–278(F,T)
    Ti-6Al-4V, 278, 278(F)
    titanium alloys, 278, 278(F)
fully lamellar (FL) microstructure, 135–136, 152

G

galling
    chemical reactivity, 294
    control and inspection, 312
    cutting fluids, 302
    drilling, 313
    fasteners and springs, 362
    lubricants, forming, 247
    lubrication, 217
    milling titanium, 300
    protective coatings, 208
    tube bending, 256
    turning, 305
    wire products, 219
galvanic corrosion
    aluminum, 6
    Monel fasteners, 290
    overview, 342–343(F,T)
    seawater, 365
    titanium alloys, 365
gamma, defined, 157
Gamma titanium aluminide (TiAl), 110, 363. *See also* alloys,
    specific: TiAl

gas metal arc welding (GMAW), 269, 272–273, 281, 290
gas tungsten arc and gas metal arc spot welding, 272–273
gas tungsten arc welding and gas metal arc welding, 269–273(F,T)
gas tungsten arc welding (GTAW)
  overview, 269
  tubing processing, 220
  weld repair, 172
gas turbine engines, 131(T), 356–358(F,T), 359(F)
gaseous environments, 339–341
General Electric Company
  first pure titanium metal, 1
  GEnx (General Electric Next-Generation) engine, 110
  TiAl, first commercial use, 110
  titanium aluminides, 134
GEnx engine, 110, 113, 134, 136, 357, 358(F)
German Fabrication Machines' (GFM's), 209(T), 212, 214(F)
GMAW. See gas metal arc welding (GMAW)
Goodman diagram or modified Goodman diagram, 122, 123(F), 124(F)
grain boundaries
  acicular α and prior-beta grain boundaries, 144(F)
  acicular alpha, 145, 146(F)
  acicular alpha and prior-beta grain boundaries, 155(F)
  acicular alpha (transformed beta) and prior-beta grain boundaries, 154(F)
  alloy solidification, 33
  cored structures, 38
  defects, 35
  dislocations, 96
  intermetallic phases, 150
  prior beta grain boundaries, 156(F)
  prior-beta, 153, 153(F)
  serrated alpha structures, 145(F)
  SPF, 252
grain growth, 108, 108(F), 109(F)
green manufacturing technology, 181
Gregor, Reverend William, 1
grinding, 157, 210, 318–319, 320(T)
GTAW. See gas tungsten arc welding (GTAW)

H

hafnium (Hf)
  as alloying element, 73
  binary alloys, intermediate phases in, 43, 43(F)
  tensile strength and strain, effect on, 65, 66(F)
halide solutions/metal salts, 337–338, 340(T)
halogen and halide compounds, 287, 332, 364
hardening, use of term, 99
hardness, 116
hazards, 298, 300, 349
hcp. See hexagonal close-packed (hcp)
heat transfer
  applications, titanium, 369–370(F)
  chlorine coolers, 348
  erosion resistance, 369
  heat exchangers, 350

  machinability, 293
  marine applications, 378
  power utility applications, 375
  tree rings, 166
  tubing walls, 376
heat treatment
  age hardening, 76(F), 86–89, 91(F)
  aging, 90–91(F), 92(T)
  annealing, 83(F), 86–87, 91–92
  contamination during, 92–93
  overview, 75, 86–87, 88(F,T)
  oxidation, retarding, 93
  purpose of, 86
  solution treatment, 83(F), 89–90(F)
  strength trends, 87, 88(F)
  stress relieving, 87
  temperature ranges, 87
  titanium-aluminum alloy, 92, 93(F)
helium
  electronic configuration, 52, 52(T)
  inert-gas-filled welding chambers, 271
  melting titanium, 161
  plasma arc welding (PAW), 273
  porosity, minimizing, 281
  shielding gas, 270
  welding procedures, shielding, 270
helium arc torch brazing, 285
hexagonal close-packed (hcp)
  crystal structures, 34, 54(F)
  slip, 96
  texture, development of, 98
high-density inclusions (HDIs), 165
high-temperature hydrogenation (HTH), 174
high-temperature near-alpha alloys, 130, 131(F,T)
HIP. See hot isostatic pressing (HIP)
historical background, 1–2(F), 353
horsepower, 314
hostile environments
  alkaline media, 366
  chlorine chemicals and solutions, 364
  chlorine gas, 364
  fresh water/steam, 364–365
  halogen and halide compounds, 364
  nitric acid, 365–366(T)
  organic chemicals, 366
  other salt solutions, 364
  reducing acids, 366, 366(T)
  seawater, 365, 365(T)
hot isostatic pressing (HIP)
  casting, 168
  casting porosity, 173
  casting technology, 168(F), 172
  early industry/recent developments, 5
  fatigue strength, increasing, 174
  porous structures, 200, 201(F)
  postcasting treatments, 175
  prealloy (PA) approach, 189–190
hot sizing, 244, 244(T), 248(T), 250, 261–262(F,T)
Hume-Rothery rules, 36
Hunter, M.A., 1, 16

Hunter process
    ductile sponge, 17
    Kroll process, difference between, 16
    leaching, 19
    sodium-reduction process, 20, 20(F)
hydride phase, 63, 157, 201(F)
hydrogen
    atoms, 31
    bar and billet, 240
    ductility, effect on, 119
    etching, 326–327
    hydrocarbon fuels, 93
    microstructure modifications, 173
    storage, titanium-iron binary alloy, 379
    surface effects of heating, 239–240
    as temporary alloying element, 62–63
    thermal stability, effect on, 121
    titanium-hydrogen phase diagram, 63(F)
    welding procedures, shielding, 270
hydrogen embrittlement
    early 1950s, 62
    environmental cracking, 343–344, 345
    galvanic corrosion, 343
    overview, 16
    types, 63–65(F)
hydrogen pickup
    bar and billet, 240
    etching, 327
    heat treatment, during, 93
    heating, during, 240
    hydrogen embrittlement, 345
hydrogen-powered vehicles, 379
hypoeutectoid alloys, 76, 77, 77(F), 85–87(F)

I

ICI Metals Division (subsequently IMI), 4, 17
ilmenite (FeTiO₃), 1, 7, 7(T), 27
impact energy sources, 255
Imperial Chemical Industries Ltd. (ICI), 3–4
induction heating, 215, 220, 277, 284
induction melting, 168
industrial applications, 363
inert gases
    brazing, 284
    flash welding, 275
    helium arc torch brazing, 285
    melting, 161
    PAW, 273
    pressure welding, 277
    quality control, weldments, 282
    solid-state welding, 277
    welding, 370
    welding chambers, 271, 271(F)
ingot breakdown
    intermediate conditioning, 210, 210(F)
    overview, 208, 209(F)
    precautions, 210
    protective coatings, 208–209
    titanium processing, 209, 209(T), 210(F)

ingot metallurgy (IM) products, 168, 174, 190–191(F,T)
insulating blankets, 247
intergranular beta, 147(F), 149, 150(F), 155(F), 157
intermetallic phases, 36–37(T)
internal changes, 105–106(F)
International Titanium Association, 5, 202
International Titanium Incorporated, 3
interstitial elements
    alpha stabilizers, 59, 60(F)
    alpha system, 142
    commercially pure titanium weld-metal notch toughness,
        267(F)
    defined, 35, 157
    hydrogen, 62
    resolved shear stress for slip in titanium as a function of
        purity, 97, 97(T)
    slip, 97
    solid solubility, 35
    strength and ductility of titanium, effect on, 266(F)
    strength levels, controlling in unalloyed titanium, 9
    tensile strength, 118
    welded joints, 266
    welding, 266, 266(F), 267(F)
    weld-metal notch toughness, effect on, 266(F)
iron
    as alloying element, 53, 73
    boiling point, 32(T)
    grain growth, effect on, 108, 109(F)
    melt charges, 163
    melting point, 32(T)
    segregation, defects, 166
iron titanate. See ilmenite (FeTiO₃)
iron-titanium oxide (FeTiO₃), 7
isomorphous alloys, 37(F), 43(F), 44, 88
isothermal section, 46

J

Japanese sponge production, 4(T)
jet engines
    fan-jet engine, 357
    GEnx engine (see GEnx engine)
    J-57, 353, 354
    P&WA F-135 engine, 360
    PWA J57, 4
    Rolls-Royce, Avon engine, 4
    Rolls-Royce Rb 199, 359(F)
joining
    adhesive bonding, 286–287
    brazing, 283–285
    dissimilar-metal joints, 281, 282
    mechanical fastening, 287–290(T)
    overview, 265
    soldering, 285–286
    welding, 265–283(F,T)

K

Klaproth, M.H., 1
Kroll, W.J., 1, 16, 161

Kroll process
    historical background, 2
    Hunter process, difference between, 16
    hydrogenated titanium process, 183
    leaching, 2, 2(F), 17, 19
    magnesium-reduction process, 18–20(F)
    schematic, 19(F)

L

Larson-Miller parameter, 106, 107(F), 119, 120(F)
lasers, 323
lathe tools, 296, 309
lathes, 305, 309, 311
lattice structure, 54, 54(F), 55(F)
leaching, 2, 2(F), 17, 19
lenticular alpha structure, 168, 168(F)
leucoxene, 7
Lever rule, 39, 76–77(F)
life-cycle cost analysis, 363
lifing, 113, 129, 130, 131(F)
limited solid solutions, 38
liquidus, defined, 48
localized attack
    crevice corrosion, 341–342(F)
    overview, 342
    pitting, 341, 342(T)
low-cycle fatigue, 125, 126(F)
lubricants
    deep drawing, 252
    extruded shapes, 218
    extruding titanium, 218
    extrusion, titanium, 232, 233(F)
    extrusion, tubes, 219
    extrusion process, 219
    forming lubricants, 247, 248(T)
    hot extruding titanium, 217
    mechanical fastening, 288
    molten glass, 232, 233(F)
    roll forming, 257
    stretch forming, 250
    wire processing, 219–220

M

machinability
    Boeing, 321
    chemical reactivity, 294–295(T)
    cutting fluids, 296
    heat treatment, 86
    machining behavior, 293
    modulus of elasticity, 295
    overview, 294(T)
    oxygen, 239
    stainless steel, 293
    thermal problems, 293–294(T)
    TIMETAL 54M, 6
    titanium (see titanium, machinability)
machining
    equipment (see also machining requirements;
        machining titanium)
    overview, 370

machining requirements
    cutting fluid considerations, 296, 298–299(T)
    cutting tools, 294(T), 296, 297(T)
    machine tools, 295–296(F)
machining titanium
    advances, recent, 320–322, 323(F)
    broaching, 319, 320(F), 321(T)
    chemical (see chemical machining)
    drilling (see drilling titanium)
    electrochemical machining (ECM), 327–328(F)
    flame cutting, 322–323, 324(T)
    hazards, 300
    machinability, 293–295(F,T)
    milling (see milling titanium)
    requirements (see machining requirements)
    safety considerations, 300
    scrap prevention, 299, 299(T)
    surface grinding, 318–319, 320(T)
    tapping, 319–320, 322(F,T)
macroscopic stress-strain behavior, 100–101
magnesium-reduction process (Kroll), 18–20(F)
magnetic susceptibility, 15
Mallory-Sharon Titanium Corporation, 2
mandrel bar drawing, 221
mandrels
    bar drawing, 221, 221(F)
    composition of, 219
    conical shear-forming, 259
    extruded pipe and hollows, 218, 219
    multiball mandrel, 255–256(F)
    seamless tube processing, 220
    tube bulging, 254
manganese, 4, 53, 73, 163, 166
marine applications, 349, 378, 379(F)
martensite
    athermal titanium martensite, 79–80(F)
    origin of term, 78
martensitic reaction, 78
maskants, 325–326
matrix, 136–138(F,T), 150–151(F), 157
McDonnell (later McDonnell Douglas)
    early industry/recent developments, 4
    F3H airframe, 4
mechanical fastening
    forming, 287
    machining, 287–288
    metallurgical and chemical properties, 287
    overview, 287
    riveting
        overview, 288, 288(T)
        quality control, 289
        rivets, upsetting, 288–289(T)
    stress-corrosion susceptibility, 288
    surface preparation, 287
    threaded fastening
        installation, 289–290(T)
        overview, 289
        properties, 290
        quality control, 290
        repair, 290
mechanical properties, 7(T), 9, 10–13(T), 14(F),
    15(T)

medical applications
    dental applications, 371, 372(F)
    external prostheses, 371, 371(F)
    implants, 349
    medical instruments, 370–371
    overview, 370
    surgical implants, 194, 349, 371, 371(F)
    surgical instruments, 371, 372(F)
medical field
    iodine-l25 interstitial implants, 8
    overview, 8
    SMA, 139
melting
    conventional vacuum arc remelting, 161–167(F,T)
    overview, 161
    trends, 167–168(F)
melting points, some metals, 32(T)
MER approach, 25, 25(F)
mercury, 344(T), 345, 350
metal titanium
    applications, 8
    overview, 7–8(T)
metallographic preparation
    etching, 158
    grinding, 157
    mounting, 157
    polishing, 157–158
    sectioning, 157
metallographic specimen preparation, 154
metallography, titanium and its alloys
    history, 141
    metastable phases, 142(F), 144(F), 146–149(F)
    overview, 141
    physical metallurgy (see physical metallurgy)
    related terms (see metallography, titanium and its
        alloys, related terms)
    terminology (see terminology, titanium alloy structures)
metallography, titanium and its alloys, related terms
    alpha case, 149–150(F)
    intergranular beta, 149, 150(F)
    intermetallic phases (compounds), 150, 150(F),
        151(F)
    matrix, 150–151(F)
    metallographic specimen preparation, 154
    microstructure, effect of fabrication and thermal
        treatment on, 152–154(F), 155(F), 156(F)
    ordered intermetallic compounds, 151–152(F)
    spheroidal beta, 14(F), 149, 149(F)
metal-matrix composites, 136–138(F,T)
metal-removal rates (MRRs), 320–321
metastable beta alloys. See also beta alloys
    drilling speeds, 317
    heat treatment, 89
    omega phase, 149
    overview, 70, 148
    terminal alloy formulation, 71
    welding, 268
metastable beta, defined, 157
metastable beta phase, 76(F), 78(F), 80–81(F,T), 82(F)
metastable phases
    alpha prime, 144(F), 147
    metastable beta, 91(F), 142(F), 147–148(F), 149(F)

omega phase, 148–149
    overview, 146–147
metastable phases and metastable phase diagrams
    beta prime, 82, 83(F)
    metastable beta phase, 76(F), 78(F), 80–81(F,T), 82(F)
    omega phase, 82
    overview, 76(F), 77–78
    quenching, phases present after, 82–84(F)
    titanium martensites, 78–80(F)
methanol, 269, 344, 344(T), 350, 366
methyl ethyl ketone (MEK), 245, 246
M_f, defined, 157
microstructure, effect of fabrication and thermal treatment
    on, 152–154(F), 155(F), 156(F)
microstructures. See also alloying elements; heat treatment
    annealing, 83(F), 92
    annealing, effect on, 92
    casting, 168, 174, 175
    commercially pure titanium plate, 109(F)
    ductility, 119
    metallography, 141–158(F,T)
    PA approach, 185, 189–190(F)
    powder metallurgy, 178, 179(F)
    prealloyed Ti-6Al-4V compacts, 189(F)
    preforms, 178
    SCC, 367
    terms used in interpreting, 141, 155–157
    Ti-14Al-21Nb, 151–152(F)
    Ti-6Al-4V, 168(F)
    TiAl intermetallic, 135, 135(F)
    titanium powder metallurgy, categories of, 189–190(F)
mill products
    applications, titanium, 353, 354
    Chinese titanium market in 2011, 355(F)
    commercial airframes, 7(T)
    forming, 245
    history, 2, 3, 4, 6
    hydrogen embrittlement, 16
    PA approach, 187
    primary working, 207
    secondary working, use of term, 243
    shipments, trends, 27, 28(F,T)
    texture, development of, 99
    world demand forecast, 354(T)
    world estimated division of consumption in 2012, 28(T)
    world forecast (2018), 28(T)
milling titanium
    carbide milling, 301
    cutter design, 301
    cutting fluids, 302
    cutting speed, 302
    depth of cut, 302
    end milling, 303, 306–309(T)
    face or skin milling, 302–303, 304(T), 305(F)
    face-milling operations, 302, 303(T)
    feeds, 301–302
    milling behavior, 300–301
    milling cutters, 301
    milling machines, 301
    milling techniques, 302
    overview, 300
    peripheral-milling, 303

milling titanium (continued)
    pocket milling, 303
    profile milling, 303
    slab milling, 303, 305, 310(F)
    spar milling, 303, 305
    tool materials, 301
    turning, facing and boring (see turning)
miscibility gap, 48–49
missiles, 361
mist system, 298
mixed-potential theory, 334–335(F)
modulus of elasticity, 15, 293, 295, 363, 368, 372
Moeller Manufacturing Co. Inc., 321
molecules, 32
molten salt baths, 245, 287
molybdenum (Mo)
    as alloying element, 53, 73
    binary alloys, intermediate phases in, 43, 43(F)
    corrosion resistance, 346
    corrosion resistance, effect on, 332
    creep and stress rupture, 120
    jet-engine compressor components, 68
    melt charges, 163
Monel fasteners, 290
monotectic reactions
    defined, 49
    phase diagrams, 41, 41(F)
monotectoid reaction, defined, 49
morphology, defined, 157
M_s, defined, 157

N

National Distillers and Chemical Corporation, 2–3
National Lead Company, 2, 17
natural waters, 339
near-alpha alloys, 66–68(F)
near-gamma (NG) microstructure, 136, 152
near-lamellar (NL) microstructure, 136, 152
neocrystallization, 108–110(F)
neutral additions, 51, 65–66(F)
neutrons, 31
nickel (Ni)
    boiling point, 32(T)
    corrosion resistance, effect on, 332
    melting point, 32(T)
    solid solutions, 35
Nilson, L.F., 1
niobium (Nb)
    binary alloys, intermediate phases in, 43, 43(F)
    creep and stress rupture, 120
    jet-engine compressor components, 68
    melt charges, 163
    terminal alloy formulation, 71
Nitinol, 138, 379
nitrides (TiN), 1, 163
nitrogen
    contamination, during heat treatment, 92
    ductility, effect on, 118–119
    welding procedures, shielding, 270
nitrogen tetroxide, 16, 340–341, 344, 344(T), 350

nonaerospace
    applications, 355(F)
    early applications, 353
    early industry/recent developments, 4
    engineering properties, 363–370(F,T)
    historical background, 4
    industrial applications, 363
    PIM, 196
    sponge production, trends, 26(F), 27
nondestructive testing
    adhesive bonding, 287
    brazing, 285
    welding, 282
nose time, 86, 87(T)
notches, 64, 124–125(F)
nucleation, 32

O

Oak Ridge National Laboratory, 192
Occupational Safety and Health Administration
    (OSHA), 202
omega phase, 82, 157
open die, defined, 241
ordered structure, 144, 151, 157
Oregon Metallurgical Corporation, 3
Oremet, 4, 170(F)
organic compounds, 338(T), 339, 341(T), 344(T), 350
overaged, 90
overheating
    cutting speed, 295
    cutting tools, 296
    forging temperatures, 228
    grinding, 157
    heating methods, 246
    Sejournet glass process, 217
    soldering, 286
overlays, 284
oxidation
    acetaldehyde, 349
    advanced alpha alloys, 68
    applications, titanium, 373
    brazing filler metals, 285
    contamination during heat treatment, 93
    corrosion reactions, 333
    fluxes, 284
    forging, 208
    gamma alloys, 135
    gamma titanium aluminide, 110
    heat treatment, during, 93
    hot rolling, 215
    hydrogen embrittlement, 345
    intermetallics, 135
    jet engines, 359
    low-density hard inclusions, 165
    retarding, during heat treatment, 93
    shielding, 284
    stress-relief heat treatments, 172
    titanium aluminides, 363
    vacuum arc remelting, 164
    welded tubing, 220

oxide scale, 208, 217–218, 239, 262, 287
oxidizers, 335, 344, 344(T), 350
oxidizing agents, 331(T), 337, 339(T), 342, 350
oxyacetylene flame heating, 267(T), 269, 277, 284–285, 370
oxygen
    bar and billet, contamination, 239, 240(F)
    casting, 169
    contamination, 209
    contamination, detecting, 150
    contamination, during heat treatment, 92–93
    ductility, effect on, 118–119
    fracture toughness, effect on, 129, 130(F)
    melt charges, 163
    net-shape casting, 169
    PIM, 194, 197
    stress-corrosion cracking (SCC), 345
    welding procedures, shielding, 270

P

palladium (Pd), 331, 332(F,T)
paper interleaving, 246
partitioning coefficient, 165–166(F,T)
passivation, 331, 364
PAW. *See* plasma arc welding (PAW)
PCC Airfoils, 136
peak aged condition, 90
peak strength, 90–91(F)
peritectic reactions
    defined, 49
    phase diagrams, 41–43(F)
    Type 1, 41–42(F)
    Type 2, 42, 42(F)
    Type 3, 42–43(F)
peritectoid alloys, 44–46(F)
peritectoid reaction, 44, 48, 49
perovskite (CaTiO$_3$), 7, 7(T)
Peterson, O., 1
pH, defined, 333
phase, defined, 157
phase diagrams
    alloys completely soluble in solid state, 37–38(F)
    alloys partially soluble in solid state, 38–40(F)
    binary alloys, intermediate phases in, 37(F), 39(F), 43–44(F)
    eutectic reactions, 40–41(F)
    eutectoid alloys, 44, 45(F)
    isomorphous alloys, 37(F), 43(F), 44
    monotectic reactions, 41, 41(F)
    overview, 37
    peritectic reactions, 41–43(F)
    peritectoid alloys, 44–46(F)
    solid-state transformations, 39(F), 40(F), 44
    ternary phase diagrams, 46–47(F)
    titanium-zirconium phase diagram, 65(F)
photofabrication. *See* chemical machining
physical metallurgy
    alloy classification, 141–142
    alpha system, 142, 142(F)
    beta system, 142–143(F)
    crystal structure, 141

physical properties
    casting, 370
    cutting, 370
    erosion resistance, 369
    fabrication, 370
    heat transfer, 369–370(F)
    machining, 370
    overview, 367–369
    pure titanium, 7(T)
    welding, 370, 370(F)
pickling solutions, 210, 269
Pilger mill, 220, 221(F)
pitting, 341, 342(T)
planar growth, 32
plane-strain fracture toughness, 128, 128(F), 132
plasma arc hearth cold melting, 167–168
plasma arc welding (PAW), 269, 273, 273(T)
plasma rotating electrode processing (PREP), 184, 187(T), 190, 191
plasma torches, 323
plastic deformation. *See* deformation
plastic flow, 98, 104, 127, 128
platelike alpha
    acicular alpha, 145
    defined, 157
    microstructure, effect of fabrication and thermal treatment on, 153, 153(F)
    overview, 146, 146(F), 147(F)
    Ti-6Al-4V bar, 154(F)
PM. *See* powder metallurgy (PM)
polarization, 333(F), 335
polishing
    electropolishing, 157–158
    mechanical polishing, 158
    vibratory polishing, 158
polyvinyl chloride (PVC), 327
porosity
    casting porosity, 168(F), 173
    helium, 281
    ingot porosity, 164, 165
    weld defects, 281
pot, defined, 241
potential-pH diagram, 332–333(F)
Pourbaix diagram, 332
powder injection molding (PIM), 193–197(F,T), 198(F)
powder metallurgy (PM), 176. *See also* titanium powder metallurgy; titanium powder metallurgy, categories of
powder-bed fusion (PBF), 192
power utility applications, 375–378(F)
P.R. Mallory Company, 2
Pratt & Whitney Aircraft (P&WA), 353, 357, 360
prealloy (PA) approach
    alloy types studied, 187, 189
    consolidation, 185–187
    direct processing, 187
    mechanical properties, 188(F), 190–191(F,T)
    metal cans, parts produced using, 179(F), 190(F,T), 191–192
    microstructures, 189–190(F)
    overview, 184, 186(F), 187(F,T), 188(F), 189(F)
    powder processing, 184–185, 189(F)
    preforms, 187

preblock, defined, 241
prefinish, defined, 241
preform, defined, 241
pretension, 289, 289(T)
primary alpha, 143–144(F), 157
primary working
    billets and bars, forged, 209(T), 210–211(F)
    coil rolling, 215
    cold rolling, 216, 216(F)
    crystal structure, 207–208
    extrusion, 216–219(F)
    forging, 208
    hot rolling, 215–216(F)
    ingot breakdown, 208–210(F,T)
    overview, 207
    plate rolling, 214, 215
    radial precision forging machines, 209(T), 212, 214(F)
    rod and bar, rolled, 212–214(T)
    rolling, 211–212(F), 213(T)
    sheet rolling, 213(T), 214–215(F)
    strip and foil rolling, 215
    wire and tube processing, 219–221(F)
prior-beta grain size
    acicular alpha, 145–146(F)
    defined, 157
process challenges, 15–16
properties, 8–9(T), 10–13(T)
protons, 31
punch
    brake forming, 248, 249(T)
    deep drawing, 250, 251(F)
    drilling techniques, 318
    drop-hammer forming, 256
    extrusion, 231–232
    mechanical fastening, 287, 290
    punch straightening, 218
pure titanium. *See also* commercially pure (CP) titanium
    alloying (*see* alloying elements)
    bar, 145(F)
    crystal structures, 141, 207
    density as a function of temperature, 55(F)
    early history, 1
    hydrogen embrittlement, 16
    lattice structure, 54
    magnesium-reduction process (Kroll), 20
    physical and mechanical properties, 7(T)
    physical metallurgy, 225
    rapid solidification (RS), 199
    recrystallization, 108, 108(F)
    sheet, 144(F)
    slip, 96
    sputtered, 373
    tubing, 151(F)
pyrophanite ($MnTiO_3$), 7

R

radial precision forging machines, 209(T), 212, 214(F)
rapid omnidirectional compaction (ROC), 185, 190

rapid solidification (RS)
    alloy types, 198–199(F,T), 200(F)
    powder production, 197
    processing and consolidation, 197–198
Reactive Metals, Inc., 3
reactivity with interstitial elements, 1
reaming, 287, 288, 318
recrystallization
    alloy additions, effect of, 108
    boron, 108
    carbon, 108
    chromium, 108
    overview, 95, 106, 106(F), 107–108(F)
    titanium processing, 209
red fuming nitric acid (RFNA), 344
redraw, 250, 251(F)
reduction, use of term, 334
reduction in area
    alpha-beta and beta-processed Ti-6Al-4V forgings,
        129(T)
    annealed Ti-6Al-4V, 70(T)
    Boeing 787, 132
    ductility, 118
    mechanical properties of Ti-6Al-4V castings, 174(T)
    minimum and average mechanical properties of wrought
        titanium alloys at room temperature, 11–13(T)
    tensile properties of a hydrogenated titanium compact
        (after dehydrogenation), 183(T)
    tensile properties of a Ti-Mo-Al, effect of cooling rate on,
        199(T)
    tensile properties of blended-elemental Ti-6Al-4V
        compacts compared to mill-annealed wrought
        products, 178(T)
    Ti-6Al-4V prealloyed powder compacts, properties of,
        190(T)
reduction reaction, 22, 334, 345
Rem-Cru Titanium Inc., 2
Remington Arms, 2
repair welds, 282
repassivation, 364
Republic Steel, 2, 3
residual stress
    chemical machining, 324–325
    chemical milling, 324
    cracks, 282
    cutting, 312
    cutting fluid considerations, 296
    downfeeds, 318
    heat treatment, 86, 172
    milling behavior, 300
    relief of, 95, 106, 107(F)
    repair welding, 282
    resistance heating, 218
resistance spot and seam welding, 274, 274(T)
Resistance Welder Manufacturers Association, 274
revert
    machining titanium, 299, 299(T)
    net-shape casting, 169
    vacuum arc remelting, 163

riveting
  defects, 289
  overview, 289, 289(T)
  quality control, 289
  upsetting rivets, 288–289(T)
rivets
  cold-driven, 362
  holes for, 288
  mechanical fastening, 288–289(T)
  mechanically fastened joints, 290
  repair, 290
  sheet metal components, 363
  threaded fasteners, 290
  upsetting rivets, 288–289(T)
  weldments, 283
RMI Company, 3, 4
robotics, 174–175
rod and bar, rolled, 212–214(T)
roller, defined, 241
rolling
  coil rolling, 215
  cold rolling, 216, 216(F)
  cross-rolling, 99
  hot rolling, 215–216(F)
  overview, 211–212(F), 213(T)
  plate rolling, 214, 215
  sheet rolling, 213(T), 214–215(F)
  straight-away rolling, 99
  strip and foil rolling, 215, 216
Rolls-Royce, 136, 357, 359(F)
ruthenium (Ru), 8, 332
rutile (impure TiO$_2$)
  chlorination of, processes, 17
  historical background, 1
  occurrences, 7, 7(T)
  relative cost of titanium plate production from, 22(T)
R-value, 101, 102

S

safety considerations/precautions
  flame cutting, 322–323
  machining titanium, 300
  overview, 200, 202
  surface grinding, 319
Safran-Snecma, 136
salt baths
  caustic salt bath, 239
  hot, 281
  molten salt baths, 245, 287
salts
  electrolytic winning, 21
  environmental cracking, 350
  furnace designs, 161, 162
  halide solutions/metal salts, 337–338, 340(T)
  hostile environments, 364
  ingot porosity, 165
  ingot surfaces, 167
  magnesium chloride (MgCl$_2$) salts, 161

magnesium chloride salts, 19
melting, 164
pitting, 341
purify titanium, 17
SCC, 344(T), 367
sodium-reduction process, 20
vacuum arc remelting, 162
satellites, 361
scale removal, 245
Schloemann mill, 216
scribing, 324, 326, 326(F), 327
scribing knives, 327
scribing templates, 327
seawater
  beta fabrication, effect on strength and fracture toughness, 129, 130(F)
  engineering properties, 365, 365(T)
  environmental cracking, 345
  SCC, 129(T)
  stress rupture curves, 129(F)
secondary working, bar and billet. See bar and billet, secondary working of
secondary working, forming titanium plate, sheet, strip, tubing. See forming
section size
  defined, 90
  effect of, 90
  rolling, heating times, 212
  tensile strength, 116–118(F)
  U.S. Navy's dynamic tear test, 126
Sejournet glass process, 217–218(F)
Sendzimir and Schloemann, 212
Sendzimir cluster mill, 216
serrated grain, 157
shape, defined, 241
shape memory alloys (SMA), 138–139, 379
Sharon Steel Corporation, 2
sheet metal applications, 362–363
shells, 38, 51, 171, 218, 219, 368
shielding
  brazing, 284
  flash welding, 275
  gas tungsten arc welding and gas metal arc welding, 269–272(F,T)
  GTAW, GMAW, PAW, 269–272(F,T)
  plasma arc welding (PAW), 273
  solid-state welding, 277
  welding, 282
shot blasting, 218, 219, 239
silicon
  creep strength, effect on, 117(T), 119
  grain growth, 108
  jet-engine compressor components, 68
  melt charges, 163
silver
  alloying elements in titanium, classification of, 58(T)
  anodes, 349
  binary alloys, intermediate phases in, 43
  brazing filler metals, 285

silver (continued)
  corrosion resistance, titanium, 350
  dissimilar-metal joints, 281
  eutectoid temperature and composition, 62(T)
  galvanic series in flowing water, 343(T)
  lattice structure, 54, 54(F)
  liquid/solid metal cracking, 345
  SCC, environments that promote, 344(T)
  soldering, use in, 285–286
  weld contamination, nondestructive evaluation, 282
skull, 16, 161
slip, 96–97(F,T)
slip planes, 97, 97(F)
SMA. See shape memory alloys (SMA)
smearing, 210
S-N curves, 121–122(F), 125, 125(F), 126(F), 188(F)
sodium-reduction process (Hunter), 20, 20(F)
soldering, 285–286
solid solubility, types of, 35, 35(F)
solid solutions, 35, 38, 49
solidification, metals
  alloy solidification, 33, 33(F)
  cast structures, 33, 33(F)
  crystal growth, 32, 32(F), 33(F)
  crystal structures, 33–35(F,T)
  nucleation, 32
solid-state transformations, 37, 37(F), 39(F), 40(F), 44
solid-state welding, 275–277
solidus, defined, 49
solubility, 1
solution heat treating and aging (STA), 131
solution treatment, 83(F), 89–90(F)
solvus, defined, 49
Soviet, 4, 5, 27, 378
space vehicles, 361
spars, 303, 305
SPF. See superplastic forming (SPF)
SPF/DB. See superplastic forming/diffusion bonding
    (SPF/DB)
sphene (CaTiSiO₅), 7
spheroidal structure, defined, 157
sponge producers and capacity (1947-1987), 3(T)
sponge production (excluding U.S.), 1
sports applications, 6, 226, 373
spray mist, 302, 317
spraying, 197, 198(F), 199(F)
springback
  adhesive bonding, 287
  brake forming, 248
  hot sizing, 248(T), 261–262(F,T)
  SPF, 253
  stretch forming, 249
STA. See solution heat treating and aging (STA)
stainless steel
  brake forming, 248
  brazing, dissimilar-metal joints, 285
  competitive comparison, 358
  corrosion rates in boiling 90% HNO₃ comparison, 365(T)
  crystal structures, 54
  cutting, equipment, 370
  data points and computer-generated S-N curves, 188(F)

drop-hammer forming, 256
electrical resistivity, 56
external prostheses, 371
fan-jet engine, 357
flame cutting, 322
forming considerations, 243
forming equipment, 243
galvanic series in flowing water, 343(T)
grease removal, 245
halide solutions/metal salts, 337
machinability, 293, 294(T), 321
machining, equipment, 370
nitric acid, 365
primary working equipment, 207
rolling equipment, 211
strength comparison, 9, 14(F)
thermal conductivity, 56
titanium alloys in boiling acid solutions, comparison,
    346(T)
tooling material, forming, 252
tooling materials for forming operations, 248(T)
welding conditions, 273
wire brushes, for scale removal, 245
states of matter, 31–32(T)
static chlorination, 17
strain effects, 104, 104(F)
strain hardening, 102–104(F)
strain-energizing process (SEP), 185, 190
strain-rate sensitivity, defined, 104
strengthening, 91, 91(F). See also strengthening effects;
    texture strengthening
strengthening effects
  alloying elements, 57, 57(T), 59–60
  intermetallic phases, 37
  stress-strain relationship, 103(F)
  texture strengthening, 102
  titanium martensites, 79
strength-to-weight ratio
  aerospace applications, 356
  airframes, 360–361
  beta alloys, 70
  calculated, 356
  gas turbine engines, 356–357
  industrial applications, 363
  submarine construction, 378
  titanium alloys, 367
  titanium alloys, nonaerospace applications, 368
  titanium and other aerospace metals, 9, 14(F)
stress corrosion
  process challenges, 16
  welding, 268–269
stress relieving
  milling titanium, 301
  purpose of, 87
  turning, 313
stress-corrosion cracking (SCC)
  aluminum (Al), 345
  anhydrous methanol, 344, 366
  environmental cracking, 343, 344, 344(T)
  molybdenum (Mo), 345
  niobium (Nb), 345

oxygen, 345
 process challenges, 16
 seawater, 129, 129(T)
 tantalum (Ta), 345
 titanium alloys, 367
 toughness, 128–129(F)
 vanadium (V), 345
stress-intensity factor, 127–128
stress-relief heat treatments, 168(F), 172
stress-strain curves, 81, 81(F), 103, 103(F)
submarine construction, 363, 368, 378
substitutional alloying, 33(F), 34(F), 52–53
substitutional elements, defined, 157
sulfur, 52(T), 108, 339
sulfur dioxide, 349, 350, 377
superplastic forming (SPF), 5, 251, 252–254(F,T), 255(F)
superplastic forming/diffusion bonding (SPF/DB), 5, 254,
    255(F)
superplasticity, 104–105(F,T), 252, 253. *See also* superplastic
    forming (SPF)
surface grinding, 318–319, 320(T)
swaging, 220, 261
Swarf, 202
swedge, defined, 242

T

tantalum (Ta)
 binary alloys, intermediate phases in, 43, 43(F)
 corrosion resistance, 346
 creep and stress rupture, 120
 melt charges, 163
tapping, 319–320, 322(F,T)
tapping geometry, 322(F,T)
tensile strength, 116–118(F,T)
tensile stress-strain relationship, 103, 103(F)
tensile test
 ductility, 118
 evaluating sheet material, 101
 forged billets and bars, 211
 porous structures, 200
 rolled rod and bar, 214
 stress-stain relationship, 103(F)
 superplastic tensile tests, 253–254
 tensile properties of selected weld, 280(T)
 tensile strength, 116
 texture strengthening, 99, 100, 101
 true stress, calculating, 103
 true-stress/true-strain diagram, 103(F)
 wire processing, 220
terminal alloy formulation, 57(T), 71, 71(F)
terminal solid solutions, 38
terminology, titanium alloy structures
 elongated structures, 143, 144(F)
 equiaxed structures, 143, 144(F)
 overview, 143
 primary alpha, 143–144(F)
 transformed beta, 144–146(F), 147(F)
ternary alloys, 46. *See also* ternary phase diagrams
ternary equilibrium diagram, 46(F)
ternary phase diagrams, 46–47(F)

texture, development of, 98–99(F), 100(F,T)
texture strengthening, 99–102(F,T)
 alpha alloys, 101, 102(T)
 deformation, 99–102(F,T)
 titanium, 99–102(F,T)
 use of term, 99
 yield strength, 100–101
thermal conductivity and expansion, 14
thermal stability
 definition of, 120
 titanium alloys, 121(T)
thermohydrogen processing (THP)
 BE processing, 178
 casting, 173, 174
 prealloy (PA) approach, 185
 titanium PM, 200, 201(F)
threaded fasteners
 A-286 alloy, 290
 commercially pure (CP) titanium, 290
 installation, 289–290(T)
 Monel fasteners, 290
 overview, 289
 properties, 290
 quality control, 290
 repair, 290
 rivets, 290
 titanium alloys, 290
threaded fastening
 installation, 289–290(T)
 overview, 289
 properties, 290
 quality control, 290
 repair, 290
TiCl$_4$. *See* titanium tetrachloride (TiCl$_4$)
TIMET
 early industry/recent developments, 2, 4, 6
 molten salt electrowinning process, 21
 sponge production, trends, 26, 27(F)
TIMET electrolytic cell, 21, 21(F)
time-temperature-transformation (TTT) diagram, 84–86(F),
    87(F)
tin
 as alloying element, 53, 57–58(F), 73
 creep strength, effect on, 119, 119(T)
 embrittlement, as cause of, 57–58(F)
 jet-engine compressor components, 68
 melt charges, 163
TITANflex (Eschenbach Optik), 138–139
titanium
 athermal titanium martensite, 79–80(F)
 bending limits, 257(T)
 boiling point, 32(T)
 chemical shaping (*see* chemical machining;
     electrochemical machining (ECM))
 coefficient of expansion, 368
 density of, 54–55, 368
 electrical resistivity, 56
 and isotopes, radioactive half-life, 368–369
 joining titanium (*see* joining)
 linear coefficient of thermal expansion, 56
 machining (*see* machining titanium)

titanium (continued)
melting point, 32(T), 56, 368
nonmagnetic, 368
physical properties, 56(T)
recrystallization, 110
specific heat, 56
texture, development of, 98–99(F), 100(F,T)
texture strengthening, 99–102(F,T)
thermal conductivity, 55–56
titanium, chemical and related applications
acid solutions, 349
chlorides, 348
chlorine, 348
drugs, 349
electrochemical applications, 349–350
foods, 349
marine applications, 349
medical implants, 349
miscellaneous, 350
organic petroleum and petrochemicals, 349
overview, 347
water purification, 348–349
titanium, machinability
chemical reactivity, 294–295(T)
machining behavior, 293
modulus of elasticity, 295
overview, 294(T)
thermal problems, 293–294(T)
titanium alloys
advanced alpha alloys, 68–69(F)
alpha (or near-alpha alloys), 66–68(F)
alpha-beta alloys, 69–70(F,T)
beta alloys, 70
Charpy impact energy, 125–126(F)
corrosion resistance of, 366–367
creep-resistant alloys, 119(T)
drilling with high-speed drills, 319(T)
elevated-temperature properties, 117(T)
FSW, 278, 278(F)
joining titanium alloys (see joining)
metastable beta alloys, 70
modulus of elasticity, 368
overview, 66, 66(F)
superplastic characteristics, 105(T)
thermal stability, 121(T)
threaded fasteners, 290
titanium alloys, mechanical properties and testing
alpha morphology, effect on alloy behavior, 113–114(F),
115(F,T), 116(F)
alpha-beta alloys, 131–133(T) (see also specific alloys)
beta alloys, 133, 133(F) (see also specific alloys)
creep and stress rupture, 117(T), 119–121(F,T)
ductility, 117(T), 118–119
elevated-temperature properties, 117(T)
fatigue crack growth rate (FCGR), 129–130(F), 131(F)
fatigue strength, 121–125(F,T), 126(F)
hardness, 116
high-temperature near-alpha alloys, 130, 131(F,T) (see
also specific alloys)
metal matrix composites, 136–138(F,T)
overview, 113

shape memory alloys, 138–139
tensile strength, 116–118(F,T)
titanium aluminides, 133–136(F,T) (see also specific
alloys)
toughness, 125–129(F,T), 130(F)
titanium aluminides (TiAl, Ti₃Al, TiAl₃), 133–136(F,T)
Titanium and Its Alloys (ASM International), 5
titanium applications. See applications, titanium
titanium consumption, 8(T)
Titanium Development Association. See International
Titanium Association
titanium dioxide (TiO₂), 1, 7, 8(T), 202, 334
titanium dioxide (TiO₂) film, 331, 336
titanium grades, 331–332, 333(T)
titanium martensites, 78–80(F)
titanium melting
melting facilities, 167–168
product quality, 167, 167(F)
Titanium Metals Corporation of America (TMCA), 2, 17. See
also TIMET
titanium plate, forming. See forming
titanium powder metallurgy
categories, 172–200(F,T)
future developments, 202–203
overview, 176–177(F,T)
safety, 200, 202
titanium powder metallurgy, categories of
additive manufacturing (AM), 192–193(F), 194(F,T),
195(F)
blended-elemental (BE), 172–180(F,T)
cold and hot isostatic pressing (CHIP), 178(T),
180–182(F,T)
hydrogenated titanium process (ADMA Products),
182–184(F,T)
porous structures, 200, 201(F), 202(F)
powder injection molding (PIM), 193–197(F,T), 198(F)
prealloy (PA) approach, 179(F), 184–192(F,T)
rapid solidification (RS), 197–199(F,T), 200(F)
spraying, 197, 198(F), 199(F)
thermohydrogen processing (THP), 200, 201(F)
titanium production processes, comparison, 22(T)
titanium sheet, forming. See forming
titanium sponge
cost of, 17
cost to produce 1kg (1982 to 1984), 22(T)
defined, 17
magnesium-reduced titanium sponge, 17(F)
production of, early, 17
titanium sponge production
current and predicted global sponge production, 27(F)
magnesium-reduction process (Kroll), 18–20(F)
pie chart, worldwide major titanium sponge
manufacturers, 27(F)
processes, cost differences between, 22(T)
sodium-reduction process (Hunter), 20, 20(F)
titanium tetrachloride (TiCl₄), 17–18(F)
trends, 26–27(F), 28(F)
titanium strip, forming. See forming
titanium tetrachloride (TiCl₄)
Armstrong/International Titanium Powder method, 23–24(F)
CSIRO technique, 24, 24(F)

extractive metallurgy, 16
historical background, 1, 2
ilmenite (FeTiO₃), 7
occurrences, 7
production, 17–18(F)
purification of, 18
titanium tubing. *See also* tube bulging; wire and tube
  processing
microstructures,, 151(F)
power utility applications, 375–378(F)
seawater, 365
tube bending, 255–256(F), 257(T)
water purification, 349
welded, 8
titanium-aluminum alloy, 92, 93(F)
titanium-niobium binary, 379
tool spalling, 294
toughness, 125–129(F,T), 130(F)
toughness tests
Charpy impact energy, 125–126(F)
U.S. Navy's dynamic tear test, 126
transformation kinetics, 84–86(F), 87(F)
transformed beta
acicular alpha, 145–146(F)
defined, 157
overview, 144–145
platelike alpha structure, 146, 146(F), 147(F)
serrated alpha structures, 145, 145(F)
Widmanstätten Structure, 145, 146, 146(F)
transition elements, 51, 52, 157
transition metals, 52, 69
transition zone, 324, 325(F)
tree rings, 166–167(F)
Trent engine, 136
true strain, 103–104(F)
true stress, calculating, 103
TTT curves, 84–86(F)
TTT diagram. *See* time-temperature-transformation (TTT)
  diagram
tube bulging, 254–255(F)
tungsten arc torch, 286
tungsten inert welding. *See* gas tungsten arc welding (GTAW)
turning
control and inspection, 312–313(T), 314(T)
cutting fluids, 311–312
cutting speed, 311
cutting tools, 309–310(F), 311(T)
depth of cut, 311
feeds, 311
machines, 305, 309
overview, 305
tool materials, 310–311(T), 312(T)
turnings, 163
twinning, 96(F), 98, 98(F), 99(F)

U

undercooling, 32, 33
undercutting, 252, 274, 323
uniform corrosion, 335
unit cells, 99

upset, defined, 242
upsetting rivets, 288–289(T)
U.S. Bureau of Mines
Kroll process, 2
magnesium-reduction unit, 2
sand casting, 171
sponge plant, 2
titanium sponge, first production, 17
U.S. Department of Defense, 184
U.S. Department of Energy, 184, 192
U.S. Naval Research laboratory, 126, 126(F)
U.S. Navy submarines, 378
U.S. Navy, titanium, early industry, 2
U.S. Navy's dynamic tear test, 126
U.S. production and data, current and historic, 1
U.S. titanium consumption, 8(T)
USSR, 8, 80, 80(T). *See also* Soviet

V

vacuum arc remelting
furnace designs, 161–162(F)
ingot breakdown, 208
melt charges, 162–163
melt shop practices, 162
melting, 162(F), 163–164
melts, 162
product quality
beta flecks, 166
freckles, 166
high-density inclusions (HDIs), 165
ingot porosity, 165
ingot surfaces, 167
low-density hard inclusions, 165, 165(F)
macrosegregation, 166
overview, 164
segregation, alloying elements, 165–167(F,T)
soft alpha segregates, 165, 165(F)
tree rings, 166–167(F)
revert, use of, 163
vacuum hot pressing (VHP), 187, 190
Van Arkel, A.C., 1
vanadium (V)
as alloying element, 4, 53, 73
binary alloys, intermediate phases in, 43, 43(F)
boiling point, 32(T)
corrosion resistance, effect on, 332
creep and stress rupture, 120
jet-engine compressor components, 68
melt charges, 163
melting point, 32(T)
vanadium oxytrichloride (VOCl₃), 18, 18(F)
VHN. *See* Vickers hardness number (VHN)
Vickers hardness number (VHN), 116, 270(T)
von Bolton, Werner, 161

W

warpage, 302, 303, 325
water purification, 348–349
weld cracks, 281

weld defects, 281
weld repair, 169, 172, 173, 175
weldability
    Advance Materials (ADMA) Products Inc., 26
    alpha (or near-alpha alloys), 68
    alpha-beta alloys, 69–70, 267–268(F)
    armor applications, 368
    BE processing, 180
    beta alloys, 70
    brazing filler metals, 284
    melting, 161
    submarine construction, 378
welding
    alloy selection
        alpha alloys, 267, 267(F)
        alpha-beta alloys, 267–268(F)
        dissimilar metals, 268
        metastable beta alloys, 268
    chemical and metallurgical characteristics, 265
    commercially pure titanium weld-metal notch toughness,
        267(F)
    contamination control, 265–267(F,T)
    dissimilar-metal joints, 281
    embrittlement, 266
    mechanical properties, 282–283
    metastable beta alloys, 268
    nondestructive testing, 282
    overview, 265, 370
    physical properties, 370, 370(F)
    quality control, 282
    repair welding, 282
    stress corrosion, susceptibility to, 268–269
    stress-relieving, 282, 283(T)
    surface preparation, 269
welding procedures
    electron beam welding (EBW), 273–274
    filler wire, 271–272
    flash welding, 275, 275(F), 276(F), 277(T)
    friction stir welding (FSW), 277–281(F,T)
    gas tungsten arc and gas metal arc spot welding, 272–273
    gas tungsten arc welding and gas metal arc welding,
        269–273(F,T)
    overview, 269
    plasma arc welding (PAW), 273, 273(T)
    resistance spot and seam welding, 274, 274(T)
    solid-state welding, 275–277
weld-metal notch toughness, 266, 266(F), 267(F), 268, 270(T)

Western Zirconium Company, 3
white glove handling, 246
Widmanstätten structure, 145, 146, 146(F), 157
wire and tube processing
    seamless pipe and tubing, 220–221(F)
    tube processing, 220
    tube products, 219
    wire processing, 219–220
    wire products, 219
World Titanium Conferences, 5
wrought titanium alloys
    mechanical properties, room temperature,
        11–13(T)
    physical properties, 10(T)

X

x-ray diffraction, 141, 150

Y

yield strength
    testing speed, effect on, 117(T)
    texture strengthening, 100–101
Young's modulus, 68–69

Z

zinc
    crystal structure, 54(F)
    die-casting industry, 350
    galvanic series in flowing water, 343(T)
    liquid/solid metal cracking, 345
    punch, brake forming, 248
    silver-cadmium-zinc alloys, 285
zinc chloride, 337, 340(T), 341, 348
zirconium (Zr)
    as alloying element, 53, 57–58(F), 73
    binary alloys, intermediate phases in, 43, 43(F)
    boiling point, 32(T)
    corrosion resistance, effect on, 332
    creep strength, effect on, 119, 119(T)
    embrittlement, as cause of, 57–58(F)
    jet-engine compressor components, 68
    melt charges, 163
    melting point, 32(T)
    tensile strength and strain, effect on, 65, 66(F)